Basic Electrical Engineering

Basic Electrical Engineering
Second Edition

K.N. Srinivas
Professor & Head
Vice Principal (Academic)
SRM Institute of Science and Technology
Ramapuram Campus, Chennai

I.K. International Pvt. Ltd.
NEW DELHI

Published by
I.K. International Pvt. Ltd.
4435-36/7, Ansari Road, Daryaganj
New Delhi–110 002 (India)
E-mail: info@ikinternational.com
Website: www.ikbooks.com

ISBN 978-93-90620-39-5

© 2021 I.K. International Pvt. Ltd.

All rights reserved. No part of this publication may be reproduced, stored in a retrieval system, or transmitted in any form or any means: electronic, mechanical, photocopying, recording, or otherwise, without the prior written permission from the publisher.

Published by Krishan Makhijani for I.K. International Pvt. Ltd., 4435-36/7, Ansari Road, Daryaganj, New Delhi–110 002 and Printed by Rekha Printers Pvt. Ltd., Okhla Industrial Area, Phase II, New Delhi–110 020.

Preface to the Second Edition

First edition of this book was received with encouraging positive feedback from the faculty members as well as the student community. The author and the publisher acknowledge the same.

This new second edition has been thoroughly revised with two focuses: firstly to make the textbook suitable for first year engineering students of all the branches; secondly to include four chapters which are required by the revisions that are happening in the first year basic electrical engineering course.

First edition of this book contained five chapters and in this second edition four new chapters have been included to cope up with the revisions that took place in the syllabus.

Sixth chapter deals with the introductory treatment on power electronics. History of the power semiconductor devices, converter topology, inverter topology and modulation fundamentals are given in this chapter.

Seventh chapter deals with a complete overview of power system. Components of electric power generation, transmission and distribution have been given. The utilization of electric power depends solely on the types of load and an introduction to the types of load is covered in this chapter. Other important accessories of power system, namely, the substation, switchgear, fuses, circuit breakers and batteries are also introduced.

An elaborate introduction on the measuring instruments is presented in chapter 8. The necessary forces that make a measuring instrument to work, the types of measuring instruments, their construction and principle of operation and their applications are documented in this chapter.

Chapter 9 provides practical wiring knowledge to the beginner. Types of house/industrial wiring and the tools used to undertake house wiring are taught by this chapter.

The author and the publisher wish that the encouragement geven by the engineering community for the first edition will be provided for this second edition too.

The author would like to thank and acknowledge the sovereignty provided by the Management of SRM IST Ramapuram Campus for enabling me to complete this second edition.

Dr. K.N. Srinivas
HOD-EEE and VP-Academic
SRM Institute of Science and Technology
Ramapuram Campus, Chennai 89

Preface to the First Edition

The aim of the book is to provide a consolidated text for the first year B.E. Computer Science and Engineering students and B.Tech Information Technology students of Anna University. The syllabus has been currently thoroughly revised for the non-semester yearly pattern by the University. Various aspects of fundamentals of electrical engineering, electronic devices and their applications and fundamentals of control system are the three focuses of the new syllabus. The student will have to refer to various books to complete the syllabus, which is greatly reduced by this single textbook; compilation of several branches of electrical, electronics and control engineering is precisely carried out in this textbook as prescribed by the new syllabus.

The textbook, made up of five chapters, systematically covers the five units of the syllabus. The *first chapter* provides a detailed discussion on the fundamentals of electric circuits, DC circuits, AC circuits, 3-phase circuits, resonance and the network theorems. Lecture-type presentation of the rudiments of the fundamentals in conjunction with hundreds of solved examples is the strength of this chapter. Magnetic circuits and various magnetic elements and their properties, with a large number of illustrations are presented in the chapter 2. Two major electric machines viz., DC machines and transformers are dealt with in chapter 3. Equivalent circuits of machines supported with respective photographs will ease the beginner to understand the concepts of machines much better. Two more major electric machines, viz., synchronous machines and asynchronous machines are presented in the chapter 4. Fundamentals of control systems with various practical examples and relevant worked illustrations form the chapter 5.

Hundreds of numerical illustrations and diagrammatic representations should always be the uniqueness of a textbook intended for the beginners; in this context, this textbook was planned and completed.

Care was taken with utmost keenness to avoid any type of errors, by the author as well as the publishers; in spite, if errors are noted, they may please be e-mailed to us so as to make the further editions error-free.

K.N. Srinivas
knsrinivas1967@gmail.com

Contents

Preface to the Second Edition   v
Preface to the First Edition   vii

1. **Basic Electricity**   2
   Part I: DC Circuits   2
   1.1   Definition of Current   2
   1.2   Ampere   3
   1.3   Electric Potential   3
   1.4   Electric Potential Difference (P.D.)   3
   1.5   Volt   4
   1.6   Resistance   4
   1.7   Ohm   4
   1.8   Resistance in Terms of Physical Quantities   5
   1.9   Power   5
   1.10   Watt   5
   1.11   Energy   5
   1.11-A   An Unit   6
   1.12   Joule   6
   1.13   Fundamental Laws of Electric Circuits   6
   1.14   Ohm's Law   6
   1.15   Alternative Expressions for Power & Energy   6
   1.16   Alternative Expressions for Current and Voltage   7
   1.17   Worked Illustration I   7
   1.18   Tutorial Illustrations I   11
   1.19.   Kirchhoff's Laws   11
   1.19-A   Kirchhoff's Current Law or KCL   11

1.19-B	Kirchhoff's Voltage Law or KVL	12
1.20	Electric Circuit	14
1.20-A	Series Circuit	14
1.20-B	Parallel Circuit	15
1.20-C	Series-Parallel Circuit	16
1.21	Worked Illustrations II	16
1.22	Tutorial Illustrations II	33

Alternating Current — 35

Part II: AC Fundamentals — 35

1.23	Introduction	35
1.24	A Sinusoidal Wave	35
1.25	Different Terms Associated with a Sinusoidal Wave	38
1.26	Measurement of Alternating Quantities	39
1.27	Average Value	39
1.28	Root Mean Square (RMS Value)	40
1.29	Measuring Factors	40
1.30	Representation of Alternating Quantities	41
1.31	Vector Representation	42
1.32	Polar and Rectangular Representations	42
1.33	The Time-Domain Equation Representation	44
1.34	The Waveform Representation	44
1.35	Worked Illustrations III	45
1.36	Tutorial Illustration III	54

Part III: Series and Parallel AC Circuit — 55

1.37	The Power Factor	55
1.38	Power in a Pure Resistor	56
1.39	Power in a Pure Inductor	57
1.40	Power in a Pure Capacitor	57
1.41	Power in a R-L Series Circuit	58
1.42	Power in a R-C Series Circuit	59
1.43	Power in R-L-C AC Series Circuit	59
1.44	The Reactance and the Impedance (AC Series Circuit)	60
1.45	The Impedance Triangle	61
1.46	The Ohm's Law—w.r.t. AC Elements	62
1.47	Table of Representations	63
1.48	Series and Parallel AC Circuits	63

1.49	Worked Illustrations IV	63
1.50	Tutorial Illustrations IV	92

Part IV: Resonance — 93

1.51	Resonance	93
1.52	Series Resonance	94
1.53	The Conductance, the Susceptance and the Admittance	96
1.54	Parallel Resonance	98
1.55	Worked Illustration V	99

Part V: Network Analysis and Theorems — 107

1.56	Introduction	107
1.57	Mesh Current Method	108
1.58	Nodal Analysis	109
1.59	Thevenin's Theorem (or Helmholtz's Theorem)	111
1.60	Norton's Theorem	113
1.61	Reciprocity Theorem	114
1.62	Compensation Theorem	115
1.63	Maximum Power Transfer Theorem	116
1.64	Worked Illustration VI	117

Part VI: Concept of 3-Phase EMF Generation — 134

1.65	Introduction	134
1.66	3-Phase System	134
1.67	Worked Illustrations VII	138
1.68	Tutorial Problems V	142
	Conclusion	144
	Questions	144

2 Magnetism and Electromagnetism — 148

Part I Basic Terms in Electromagnetism — 148

2.1	Introduction	148
2.2	Magnetic Circuit	148
2.3	Definitions of Terms Concerning Magnetic Circuits	149
2.4	Series Magnetic Circuits	153
2.5	Series Parallel Magnetic Circuits	153
2.6	Leakage Flux	154
2.7	Magnetic Fringing	155
2.8	Worked Illustrations VIII	155
2.9	Tutorial Illustrations VI	166

	2.10	Electromagnetic Induction	166
	2.11	Faraday's Postulations	166
	2.12	Flaming's Right Hand Rule (FRH)	167
	2.13	Demonstration of Flaming's Right Hand Rule (FRH)	167
	2.14	Electromagnetic Force	168
	2.15	Types of Magnetic Inductions	169
	2.16	Dynamically Induced EMF	169
	2.17	Statically Induced EMF	170
	2.18	Self and Mutual Inductances	171
	2.19	Energy Stored in a Magnetic Field	173
	2.20	Worked Illustration IX	173
	Part II: Magnetic Characteristics of Materials		**176**
	2.21	Magnetization Curves	176
	2.22	Diamagnetic and Paramagnetic Materials	177
	2.23	Ferromagnetic Materials	178
	2.24	Ferrimagnetic Materials	179
	2.25	Saturation	179
	2.26	The Hysteresis Loop and Magnetic Properties	180
	2.27	Hysteresis in Magnetic Recording	182
	2.28	Variations in Hysteresis Curves	183
	2.29	The Hysteresis Curve of Alloys	183
	2.30	Magnetization of Ferromagnetic Materials	184
	Conclusion		186
	Questions		186
3.	**Electric Machines**		**188**
	Part I: DC Machines		**188**
	3.1	Introduction	188
	3.2	DC Machines	189
	3.3	DC Generator	191
	3.4	Types of DC Generators	195
	3.5	EMF Equations of DC Generator	196
	3.5-A	Worked Illustrations X	196
	3.6	Current and Voltage Equations of a DC Machine	197
	3.7	Characterization of DC Generator	208
	3.7-A	No-Load Voltage Characteristics of a DC Generator	208
	3.7-B	Buildup of a Self-Excited Shunt Generator	210

	3.7-C	Critical Field Resistance	211
	3-7-D	Reasons for Failure (of Self-Excited Shunt Generator) to Build Up Voltage	212
	3.8	Load Versus Voltage Characteristic of a Shunt Generator	213
	3.9	DC Motor	214
	3.10	The Back EMF	215
	3.11	Torque Developed in a DC Motor	216
	3.12	Types of DC Motor	218
	3.13	Worked Illustrations XI	218
	3.14	Speed of a DC Motor	219
	3.15	Relations between Torque and Speed of a Motor	220
	3.16	Applications of DC Motors	223
	3.17	Characteristics of DC Motors	223
	3.18	Starters for DC Motors	235
	3.19	Speed Control of DC Motors	236
	Part II: Transformers		**239**
	3.20	Transformer	239
	3.21	Types and Construction	239
	3.22	EMF Equation of a Transformer	242
	3.23	Worked Illustrations XII	243
	3.24	Equivalent Resistance	247
	3.25	Magnetic Leakage	249
	3.26	Transformer with Resistance and Leakage Reactance	249
	3.27	Characteristics of a Transformer	250
	3.28	Voltage Regulation of a Transformer	253
	3.29	Losses in a Transformer	253
	3.30	Efficiency of a Transformer	253
	3.31	Worked Illustrations XIII	254
	3.32	Application of Different Types of Transformers	258
	3.33	Open Circuit and Short Circuit Tests	260
	3.34	Three-Phase Transformer Construction	262
	Conclusion		265
	Questions		265
4	**AC Machines**		**270**
	Part I: Synchronous Machines		**270**
	4.1	The Alternator Construction and Principle of Operation	270

4.2.	Advantages of Stationary Armature and Rotating Field Construction	274
4.3	Work Illustration XIV	275
4.4	Alternator on Load	277
4.5	Phasor Diagram of a Loaded Alternator	279
4.6.	Equivalent Circuit for a Single-Phase and/or Polyphase Synchronous Generator	280
4.7	Voltage Regulation	281
4.8	Parallel Operation of Alternators	286

Part II : Three-Phase Induction Motor — **289**

4.9	Introduction	289
4.10	Construction of a Three-Phase AC Induction Motor	290
4.11	Principle of Operation	292
4.12	Concept of Slip	292
4.13	Torque in a 3-Phase Induction Motor	294
4.14	Equivalent Circuit	296
4.15	Starting Arrangements for Motors	297
4.16	Methods of Speed Control	304

Part III: Single-Phase Induction Motor — **306**

4.17	Introduction	306
4.18	Single-Phase Induction Motors	307
4.19	How the Rotating Field is Obtained	310
4.20	The Different Types of Single-Phase Motors	312
	Conclusion	315
	Questions	315

5 Control Systems — 318

5.1.	Introduction	318
5.2	Mathematical Modeling	321
5.3.	Transfer Function	327
5.4.	Control System Components	331
5.5.	Time Response Analysis of Control System	342
5.6.	Desirable Pole Locations of Transfer Functions and System Stability	348
	Problems	353
	Conclusion	355
	Questions	355

6. Introduction to Power Electronics — **358**
- 6.1 Power Electronics: A Small Idea — 358
- 6.2 Power Semiconductor Device: History — 358
- 6.3 Power Converter Topologies — 360
- 6.4 Protection of Power Devices and Converters — 361
- 6.5 Types of Basic DC-DC Converters — 362
- 6.5-A Buck Converters (DC-DC) — 363
- 6.5-B Boost Converters (DC-DC) — 364
- 6.5-C Buck-Boost Converters (DC-DC) — 364
- 6.6 Inverter — 367
- 6.7 Modulation — 368
- Conclusion — 371
- Questions — 371

7. Power System: An Overview — **374**
- 7.1 Power System — 374
- 7.2 Basic Structure of Power System — 374
- 7.3 Substations — 379
- 7.4 Electric Power Distribution — 381
- 7.5 The Load — 384
- 7.6 Some Important Accessories in the Power System — 388
- 7.6-A Swichgear — 388
- 7.6-B Fuses and Switch Fuse Unit (SFU) — 389
- 7.6-C SFU — 390
- 7.6-D MCB (Miniature Circuit Breaker) — 392
- 7.6-E MCCB (Molded Case Circuit Breaker) — 393
- 7.6-F ELCB (Earth Leakage Circuit Breaker) — 393
- 7.6-G RCCB (Residual Current Circuit Breaker) — 394
- 7.7 Notes on Electric Batteries — 394
- Conclusion — 398
- Questions — 398

8. Measuring Instruments — **400**
- 8.1 Definition and Classification — 400
- 8.2 Calibration of Instruments — 401
- 8.3 The Major Forces (Torques) Which Operate a Measuring Instrument — 402

xvi *Contents*

8.4	Permanent Magnet Moving Coil (PMMC) Instrument	407
8.5	Moving Iron (MI) Instrument	409
8.6	Dynamometer Type Wattmeter	413
8.7	Single- and Three-Phase Wattmeter and Energy Meter	414
8.8	Instrument to Measure Resistance	415
Conclusion		417
Questions		417

9. Domestic Wiring — **420**

9.1	Domestic Wiring	420
9.2	Wiring Materials and Accessories	420
9.3	Types of Wiring	423
9.4	Simple Domestic Wiring Layouts	425
9.5	Fluorescent Tube Lighting	426
9.6	Methods of Earthing	427
9.7	Fuse	427
9.8	Earthing	428
9.9	Types of Earthing	428
Conclusion		429
Questions		429

Appendix — **431**

Index — **433**

INTRODUCTION

This chapter, in its initial portions introduces the fundamentals of electric circuit. Various terms associated with an electric circuit to undertake analysis are presented. The two fundamental laws of an electric circuit, viz., Ohm's law and Kirchhoff's laws are presented followed by a detailed presentation on the application of the same. Series and parallel electric circuits are discussed. Worked examples are presented.

In the second part of the chapter, a detailed presentation on the fundamentals of AC circuits is presented. The concept of the operator 'j', various methods of representing an alternating quantity and related introductory topics are presented with illustrations.

Part III presents a detailed theory of AC series/parallel circuits and solutions to AC circuits.

Part IV of the chapter introduces the concept of resonance in AC circuits and a discussion on the same with illustrations is given.

In part V of this chapter network theorems are presented. General procedures for solving an electric circuit using the network theorems is presented first, which is then followed by a number of worked examples.

In part VI, a brief introduction to 3-phase AC system is presented with analytical treatment on the fundamentals of the same.

CHAPTER OBJECTIVE

At the completion of this chapter, the reader will be able to:

* solve problems in AC and DC circuits of any complexity, including series and parallel connections, using Ohm's law and Kirchhoff's laws;
* handle problems of AC resonance circuits;
* solve electric circuits using widely used network theorems; and
* understand the fundamental concepts of 3-phase AC circuits.

KEYWORDS

V, I and R, Ohm's law, KV laws, Series-parallel circuits, Sinusoidal wave, Instantaneous value, Maximum value, Average value, RMS value, Form and peak factors, Phasor and other representations, R-L-C circuits, Inductive reactance, Capacitive reactance, Impedance, Series resonance, Q-factor, Current magnification, Voltage magnification, Parallel resonance, Network theorems, Phase rotation, Phase sequence, and Star & delta.

1
Basic Electricity

PART I: DC CIRCUITS

1.1 DEFINITION OF CURRENT

In conducting materials like copper, aluminium etc., enormous amount of 'free electrons' are available and are distributed in a scattered fashion. They move from one atom to the other atom—of the same material—at random. Fig. 1.1, depicts such random movements of free electrons. The arrowhead shows the moving direction of the respective free electrons. The conductor of Fig. 1.1 is still 'unexcited', (that is, it is not given any electrical supply).

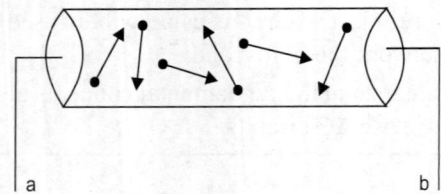

Fig. 1.1. Movement of free electrons in a conducting material. The circles at the tips are the free electrons.

However, when an electric potential is impressed across the terminals a-b of the conducting material, it is readily observed that the randomly moving—loosely packed—free electrons arrange themselves in a neatly—aligned path. It is depicted in Fig. 1.2. Now, the free electrons move from negative terminal of the source towards the positive terminal in a very systematic manner.

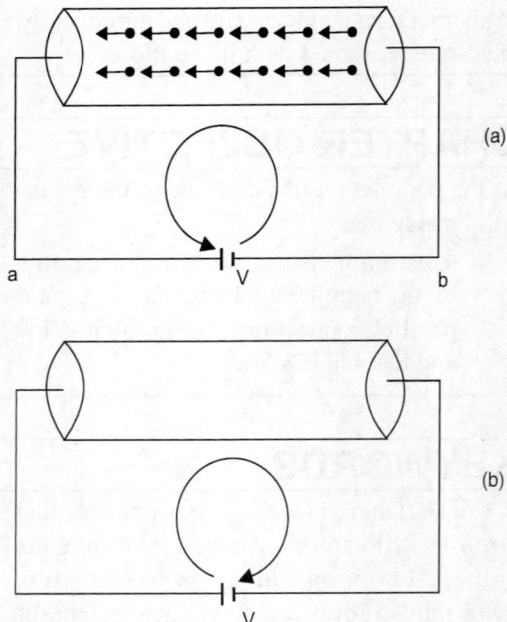

Fig. 1.2. Movement of free electrons in a conductor after a potential of 'V' volt is given.

Such a continuous flow of electrons in an electric circuit is called the ELECTRIC CURRENT.

Note: The scientists of 1800's concluded that the flow of direct current is from anode to cathode—that is, from the positive terminal to the negative terminal—of the source. Such a current flow is known to be the "Conventional current flow". It is shown in Fig. 1.2b. Whether it is electric current or conventional current, their magnitude will be the same; only the direction changes.

1.2 AMPERE

The electric current is measured by the unit 'AMPERE'. The magnitude (or the total amount) of electric current discussed in the above sec 1.1 at any part of the conducting material is directly proportional to the rate of flow of electrons. As the electrons are charged particles, this can also be said as proportional to the rate of flow of charges.

Mathematically, letting Q as the charges, it can be expressed as

$$\text{Current } I = \frac{d}{dt}(Q) \Rightarrow \text{(rate of change of charge)} \quad (1.1)$$

where t is in seconds.

The unit of current, Ampere, can now be conveniently defined, using equation 1.1, as : "WHEN A CHARGE OF ONE COULOMB PASSES THROUGH ANY PART OF WIRE IN ONE SECOND THE WIRE IS SAID TO CARRY ONE AMPERE".

1.3 ELECTRIC POTENTIAL

When a body is charged, either electrons are supplied to it or they are removed from it; but in both the

Fig. 1.3. Illustration for defining electric potential.

cases, work is done. Such ability of a charged body to do work is termed to be the 'Potential'.

To understand, let two conducting bodies A and B be assumed to be available in space as in Fig. 1.3, and let them be charged.

The Potential of the body A with respect to the body B is defined as the work done in moving a positive unit charge Q from B to A.

Note carefully that the body B is charged in such a way to remove electrons from it and the body A is charged to receive these electrons. Of course, in both the operations an amount of work is done by the bodies.

Body A attracts the charge, whereas in the body B, the charge is repelled. Thus these bodies have an ability to do work by exerting a force of attraction or repulsion on their charged particles. THIS CAPACITY OF A CHARGED PARTICLE TO DO WORK IS CALLED THE ELECTRIC POTENTIAL. Based on this discussion, it can be said for a circuit that, if an electric charge of quantity Q is passed through the circuit against a voltage of V, then the work done is equal to VQ.

$$\text{Work done} = VQ \quad (1.2)$$

1.4 ELECTRIC POTENTIAL DIFFERENCE (P.D.)

After the charge exchanging process is over (as discussed in sec. 1.3), the potentials of the bodies A and B would have changed to some other values.

THIS DIFFERENCE IN THE ELECTRIC POTENTIAL OF THE TWO CHARGED BODIES

IS CALLED THE ELECTRIC POTENTIAL DIFFERENCE (P.D.)

The unit of P.D. is. 'volt'.

1.5 VOLT

The volt is the unit for potential or P.D. or voltage or E.M.F. (Electromotive Force). It is defined as given below, following the discussions made in sec. 1.3 and 1.4.

THE POTENTIAL DIFFERENCE BETWEEN TWO CONDUCTING BODIES IS ONE VOLT IF ONE JOULE OF WORK IS DONE IN TRANSFERRING ONE COULOMB OF POSITIVE CHANGE FROM ONE CONDUCTING BODY TO ANOTHER. (Fig. 1.3 may be referred to).

1.6 RESISTANCE

It is seen that when an electric potential difference of 'V' volts is applied across the two terminals of a conducting material, the free electrons of the material align in a row and reach the positive terminal of the supply (cell) promptly.

However, in doing so, the free electrons are met by other atoms of the conductor and both these collide with each other. The collision between the free electrons and the atoms is reflected as an 'opposition' to the flow of free electrons.

That is, this collision process 'resists' the flow of free electrons; hence it is referred to as the RESISTANCE of the material. (Refer to Fig. 1.4)

The collision between the free electrons and the atoms contributes to power loss and is dissipated as heat by the conductor.

It is obvious that if the resistance of the conductor is high, the flow of free electrons will be 'resisted' much and hence the free electrons flow—that is, the current flow—will be less. Thus the resistance of a conductor and the current flow through it are of inverse proportion.

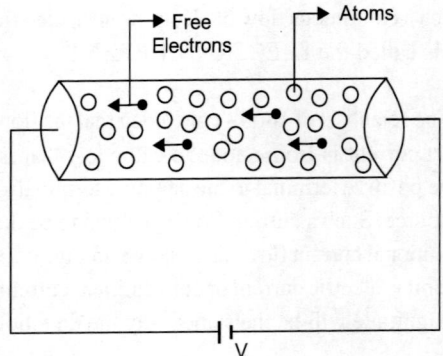

Fig. 1.4a. Illustration for discussing resistance.

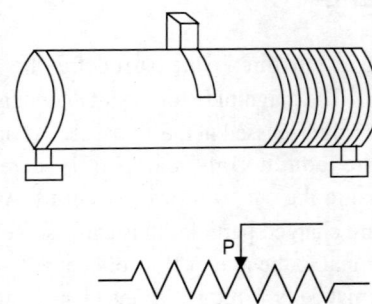

Fig. 1.4b. A rheostat. Location of pointer P in the rheostat will give different resistance values.

$$I \alpha \frac{1}{R} \qquad (1.3)$$

where R is the resistance of the conductor and
I is the current through it.

Note: An equipment which changes continuously the resistance offered to the flow of current in a circuit, is known as a rheostat. It is shown in Fig. 1.4b.

1.7 OHM

The resistance of a material has unit, ohm, and is denoted by the symbol Ω. Based on discussions made in sec. 1.6, ohm can be defined as follows:

A CONDUCTOR IS SAID TO HAVE A RESISTANCE OF ONE OHM, IF ONE AMPERE OF CURRENT PASSING THROUGH IT PRODUCES A HEAT OF 0.24 CALORIES.

1.8 RESISTANCE IN TERMS OF PHYSICAL QUANTITIES

The resistance of a material having a length of l and an uniform cross-section area of 'a' is related as:

$$R \alpha \frac{l}{a} \qquad (1.3A)$$

Introducing a proportionality constant ρ, the above expression becomes

$$R = \frac{\rho l}{a} \qquad (1.3B)$$

This constant ρ is called the resistivity or specific resistance of the material. With l in cm and 'a' in m^2 the unit of ρ will become Ω cm/m^2.

Equation (1.3B) is the value of resistance of a resistor in terms of its physical quantities and is valid only at a constant temperature.

1.9 POWER

The basic definition of power is, "the rate of doing work". In electric literature, the work done by a circuit which is supplied with a PD of V volts and carrying a charge of Q coulombs is VQ. (Refer to equation. 1.2) On this basis the power can be written as,

$$P = \frac{d}{dt} \text{ (work done)}$$
$$= \frac{d}{dt} (VQ)$$
$$= V \frac{d}{dt} (Q)$$

and by equation (1.1) d/dt (Q) is the flow of current in the circuit. Therefore,

$$P = VI \qquad (1.4)$$

Its unit is watt.

Note: The other ways of representing the electrical power are discussed in sec. 1.15.

1.10 WATT

By equation (1.4), the power P can become 1 watt, when V and I, both are 1. Watt is defined on this basis, as given below:

"THE POWER CONSUMED IN A CIRCUIT IS ONE WATT IF A POTENTIAL DIFFERENCE OF ONE VOLT CAUSES ONE AMPERE OF CURRENT TO FLOW THROUGH THE CIRCUIT."

1.11 ENERGY

Let the circuit of Fig. 1.5 be considered which is supplied with a voltage of V.

Let the resistance of the conductor be R ohm. Let a current of I amperes be available in the circuit.

Taking such circuit, as an example, the 'energy' may be defined, in general as, "the amount of work needed to maintain the current of I amperes flowing through a resistance R ohms—for a complete duration of time, t seconds".

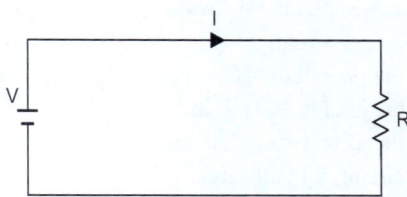

Fig. 1.5. Illustrative circuit to define energy.

6 Basic Electrical Engineering

By equation (1.4), the work done is VI; hence the energy will be the amount of work done for t seconds. Therefore,

$$\text{Energy} = (VI)\, t \tag{1.5}$$

Its unit is joles or watt second.

Note: The other ways of representing the electrical energy are discussed in sec. 1.15.

1.11-A AN UNIT

The electricity authorities charge the consumers for every 'unit' they consume. An 'unit' is nothing but one kW-Hr.

Rate (in paise) per kW-Hr is fixed by the authorities. Hence, if a consumer consumes n amount of kW-Hr, then he will be charged an amount, n times (rate per kW-Hr).

1.12 JOULE

With VI as 1 watt and t as 1 second, equation (1.5) gives the energy as 1 joule. Based on this, joule is defined as given below:

"IT IS THE ENERGY IN THE CIRCUIT WHEN ONE WATT HAS BEEN ABSORBED FOR ONE SECOND."

1.13 FUNDAMENTAL LAWS OF ELECTRIC CIRCUITS

In a closed electric circuit always the following three quantities are present.
   (i) the potential difference impressed 'across' the circuit, V, in volts,
   (ii) the flow of electric current in the closed circuit, I, in amperes.
   (iii) the resistance offered to the flow of electric current, R, in ohms.

Relating these three quantities, powerful formulae, namely, Ohm's law and Kirchhoff's laws are formed. They are discussed in the following sections.

1.14 OHM'S LAW

By Ohm's law the V, the I and the R are related as follows:

"THE RATIO OF POTENTIAL DIFFERENCE (V) BETWEEN ANY TWO POINTS OF A CONDUCTOR, TO THE CURRENT (I) FLOWING THROUGH THE CONDUCTOR IS A CONSTANT, AS LONG AS THE TEMPERATURE OF THE CONDUCTOR DOES NOT CHANGE."

It can be placed mathematically that,

$$V/I = \text{Constant}$$

and the constant is the resistance (R) of the material. Hence,

$$V/I = R \tag{1.6}$$

From equation (1.6) it can be derived that

$$V = IR \tag{1.7}$$

It is the voltage of voltage drop across the resistor, R.

$$I = V/R \tag{1.8}$$

It is the current through the resistor, R.

1.15 ALTERNATIVE EXPRESSIONS FOR POWER & ENERGY

The equation (1.4) and (1.5) can be modified with the help of ohm's law equation.

By equation (1.4), the electric power is given as, $P = VI$. Substituting for V from equation (1.7), the electric power will become,

$$P = (IR)\, I = I^2 R.$$

$$P = I^2 R \text{ watts} \tag{1.9}$$

Similar substitution can be made by inserting equation (1.8) in (1.4). The electric power will then be

$$P = VI = V(V/R) = V^2/R.$$

$$P = V^2/R, \text{ W} \tag{1.10}$$

These equations (1.9) and (1.10) are the alternative expressions for the electric power. Usually equation (1.9) and (1.10) are used for finding power loss in a resistor. [Refer to illustrative example 10]. The electrical energy, by equation (1.5), is power × time and new alternative expressions for power is now available by equations (1.9) and (1.10). Using these two equations, the alternative electrical energy equations can be readily written as,

$$\text{Energy} = (I^2 R) t \tag{1.11}$$

$$= (V^2/R) t \tag{1.12}$$

The equation (1.11) and (1.12) are the alternative expressions for electrical energy.

1.16 ALTERNATIVE EXPRESSIONS FOR CURRENT AND VOLTAGE

By Ohm's law it is known that $I = V/R$,

That is $I = V \cdot \dfrac{1}{R}$

By the same law, equation (1.6) gives, $R = V/I$. Substituting this R in the expression of I,

$$I = V \times \frac{1}{R} = V \times \frac{1}{(V/I)} = \frac{VI}{V}$$

and by equation (1.4), VI is the electrical power. Hence,

$$I = \frac{P}{V} \tag{1.13}$$

Similar substitution, if made for the voltage V, it can be seen that,

$$V = \frac{P}{I} \tag{1.14}$$

Of course, substitution can be made like this and other different expressions can be obtained.

1.17 WORKED ILLUSTRATION I

Illustration 1.1

A current of 5.559 A passes through a resistance of 100 Ω for a duration of 5 seconds. How much coulombs pass through any section of the resistor for the given duration.

Solution

By equation (1.1), $I = \dfrac{dQ}{dt}$

or simple, $I = \dfrac{Q}{t}$

$$\therefore Q = It$$
$$= (5.559)(5)$$
$$= \mathbf{27.795 \text{ C (Ans)}}.$$

Illustration 1.2

In a circuit, 100 C of charge circulates at a constant rate for every 5 seconds. Find the current.

Solution

Yet another simple problem, which can be solved with the help of equation (1.1).

$$I = \frac{dQ}{dt}$$

$$= \frac{100}{5} = 20 \text{ A (Ans)}.$$

Illustration 1.3

Calculate the resistance of 100 m length wire having a uniform cross-sectional area of 0.1 mm² if the wire is made up of a material having a resistivity of 50×10^{-8} Ω m.

Solution

Given data: $l = 100$ m, $a = 0.1$ mm²

$$= 0.1 \times 10^{-6} \text{ m}^2$$

resistivity $\rho = 50 \times 10^{-8}$ Ω m

Needed: R

When Physical quantities of a wire are known, the resistance is given by,

$$R = \frac{\rho l}{a} \text{ (by equation 1.3B)}$$

$$= 50 \times 10^{-8} \times \frac{100}{0.1 \times 10^{-6}} = 500 \text{ W (Ans.)}$$

Note: The given resistivity of 50×10^{-8} Ω m is of manganin. Instead of giving the value of ρ, the material could also have been given. In such case, from log book the ρ of manganin can be collected and substituted.

Illustration 1.4

A piece of silver wire has a resistance of 1 ohm. What will be the resistance of manganin wire of one-third length and one-third diameter if the specific resistance of manganin is 3 times that of silver.

Solution

Let suffix 's' denote silver value and suffix 'm' denote manganin values.

By equation (1.3B)

$$R_s = \frac{\rho_s l_s}{a_s}; \quad R_m = \frac{\rho_m l_m}{a_m}$$

$$\therefore \frac{R_m}{R_s} = \frac{\rho_m l_m}{a_m} \cdot \frac{a_s}{\rho_s l_s}$$

$$= \frac{\rho_m}{\rho_s} \cdot \frac{l_m}{l_s} \cdot \frac{a_s}{a_m}$$

hence $R_m = \left(\frac{\rho_m}{\rho_s} \cdot \frac{l_m}{l_s} \cdot \frac{a_s}{a_m} \right) R_s$

Assuming wire to have a circular cross-section,

$$\frac{a_s}{a_m} = \frac{\left(\frac{\pi d_s^2}{4} \right)}{\left(\frac{\pi d_m^2}{4} \right)} = \frac{d_s^2}{d_m^2} = \left(\frac{d_s}{d_m} \right)^2$$

Given that,

$$l_m = \frac{1}{3} l_s \text{ or } \frac{l_m}{l_s} = \frac{1}{3}$$

$d_m = \frac{1}{3} d_s$ or $\frac{d_s}{d_m} = 3$

$\frac{\rho_m}{\rho_s} = 30$.

$R_s = 1\Omega$.

using the above equation for R_m, and substituting the respective values,

$R_m = 30 \times \frac{1}{3} \times (3)^3 = \mathbf{90\,W\ (Ans.)}$

Illustration 1.5

A potential difference of 10 volts is applied across a 2.5 Ω resistor. Calculate the current, power dissipated, and the energy transformed into heat in 5 min.

Solution

The problem is reproduced in Fig. 1.5A.

By ohm's law, current $I = V/R = \frac{10}{2.5} = \mathbf{4\,A}$ (Ans.)

By equation (1.9), $P = I^2 R$

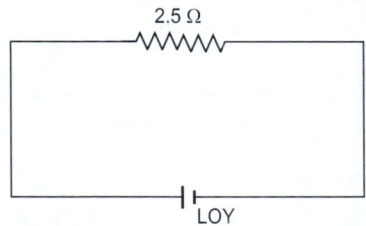

Fig. 1.5A. Circuit for illustration 1.5

$= 4^2 \times 2.5 = \mathbf{40\,W\ (Ans.)}$

(or by equation (1.10),

$P = \frac{V^2}{R} = \frac{10^2}{2.5} = 40\,W$

By equation (1.11), Energy = $I^2 R t$

$= (40)(5)(60) = 24000\,W\text{-Sec or Joules}.$

Illustration 1.6

Find the voltage and current ratings of a 5 K Ω, 0.5 watts resistor.

Solution

Using the equation (1.10), $P = \frac{V^2}{R}$. Thus

$V = \sqrt{PR} = \sqrt{0.5 \times 5 \times 10^3} = \mathbf{50\,V\ (Ans.)}$

Note that, voltage must be substituted in volts, resistance in ohms, current in amps, power in watts. It is given that the resistance is 5 K Ω (kilo ohms) and it is 5000 Ω. This Ω value is substituted above.

$I = \frac{V}{R}.$

∴ Current rating $I = \frac{50}{5000} = 0.01\,A$ or

$= \mathbf{10\,milli\,A\ (mA)}$ (Ans.)

Illustration 1.7

An electric heater draws 12 A at 230 V for a period of 3 hrs. If electrical energy costs 40 paise per unit, find the cost of operating the heater.

Solution

Refer to Sec. 1.11-A.

It this problem total kW-Hr is $(230 \times 12 \times 3)$; it is 8280 W-Hr, which is $\dfrac{8280}{1000} = 8.28$ kW-Hr.

Because one unit (or one kW-Hr) is charged at 40 paise, the charge for 8.28 kW-Hr will be, $8.280 \times 40 =$ **paise 31.2 or Rs. 3.312** **(Ans.)**.

Illustration 1.8

A building is supplied at 200 V. The load being as follows:

100 W lamps 60 Nos.
2 kW radiators 10 Nos.
50 A motors 2 Nos.

Find

(i) Total current
(ii) Cost of energy consumed in a week of 48 hours at 10 paise per unit.

Solution

Load means different electrical equipment which when connected to a supply will consume current.

(i) The wattage of lamps is 100 W and are connected to supply. By using equation (1.4) the current in each lamps is P/V, $100/200 = 0.5$ A. Hence, the total current observed by 50 lamps is, $50 \times 0.5 = 25$ A.

Similarly, current consumed by each radiator is $2000/200$, 10 A. Total current consumed by the ten radiators is $10 \times 10 = 100$ A.

Each motor consumes 50 A; the total current consumed by the two motors is thus, 50×2, which is 100 A.

Therefore, the total current consumed by the building is the summation of all the individual load currents—i.e., $25 + 100 + 100 = 225$ A **(Ans)**.

(ii) Total energy consumed in the week of 48 hrs by each load is given below: (by using equation 1.5)

Lamps load : No. of lamps × wattage of each lamps × time

$= 50 \times 100 \times 48 = 2,40,000$ W-Hr

$= 2,40,000/100 = 240$ kW-Hr

Similarly for, radiators $= 10 \times 2 \times 48 = 960$ kW-Hr
For motors, $2 \times (50 \times 200) \times 48 = 960000$ W-Hr

$= 960$ kW-Hr

Hence, total energy consumed by all loads is, $240 + 960 + 960$ equal to 2160 kW-Hr. (or units)

It is given per unit the cost is 10 paise. Hence for 2160 units, the cost will be $2160 \times 10 = 21600$ paise or **Rs. 216** **(Ans)**.

Illustration 1.9

A battery is charged at 15 amps for 10 hrs. What is the quantity of electricity supplied

(i) in coulombs
(ii) in amp-hr
(iii) If the charging voltage is 120 V what is the cost at 10 paise per unit.

Solution

(i) By equation (1.1) the charge Q coulomb can be written as, $Q = It$ where t is in seconds.

$= 15\ (10 \times 60 \times 60,\ \text{sec})$

$= 5,40,000$ coulombs **(Ans)**.

(ii) In terms of ampere-hours, the quantity of electricity is ampere × hours = 15 × 10 = **150 amp-hr (Ans).**

(iii) By equation (1.5), the energy consumed by any system is (VI) t. Here it will be (120 × 15 × 10), and is 18000 W-Hr.

In terms of kW-hr, energy is 18. Per unit (or kW-Hr), the cost is 10 paise. Hence for 18 units the cost is 18 × 10; that is, 180 paise or **Rs. 1.8 (Ans).**

1.18 TUTORIAL ILLUSTRATIONS I

Illustration T1-1

The following are the details of the load on a circuit:
(i) Six lights of 60 watts each working for 1 hour per day.
(ii) Four fluorescent tubes of 40 watts each working for 4 hours per day.
(iii) Two heaters of 1000 watts each working for 1 hour per day.
(iv) One electric iron of 750 watts working for 2 hours per day.
(v) Six fans of 60 watts each working for 18 hours per day.

If each unit of energy costs Rs. 5, what will be the total bill in Rs. for a month of 30 days?

Illustration T1-2

An electric bulb is rated 230 volts, 60 watts. Determine (i) the current through the bulb and (ii) the cost of energy for 30 days if the bulb is switched 'ON' for 6 hours per day. Assume that the cost of electrical energy is Rs. 5 per unit.

1.19. KIRCHHOFF'S LAWS

The two Kirchhoff's laws are, (i) the Kirchhoff's current law and (ii) the Kirchhoff's voltage law, they are elaborately discussed below.

1.19-A KIRCHHOFF'S CURRENT LAW OR KCL

Kirchhoff's current law (KCL) states that THE ALGEBRAIC SUM OF ALL THE CURRENTS MEETING AT A JUNCTION IS ZERO.

$$\Sigma i = 0 \quad \text{at a junction} \tag{1.15}$$

To analyse this law, let the circuit of Fig. 1.6 be considered.

It is assumed that the current entering any junction or node is positive and the current leaving any junction or node is negative.

The nodes in Fig. 1.6 are circled. The current in each element is marked arbitrarily.

Let the Kirchhoff's current law be applied to the junction 1. As far as junction 1 is concerned, the currents 'meeting' this junction are i_1, i_2 and i_3. Out of these currents, according to the above mentioned conventions, the current i_1 will be positive as it is 'entering' the junction 1, whereas the currents i_2 and i_3 will be negative as they are 'leaving' the junction 1.

Hence, the KCL equation, (equation. 1.15) at junction 1 is:

$$i_1 + (-i_2) + (-i_3) = 0$$

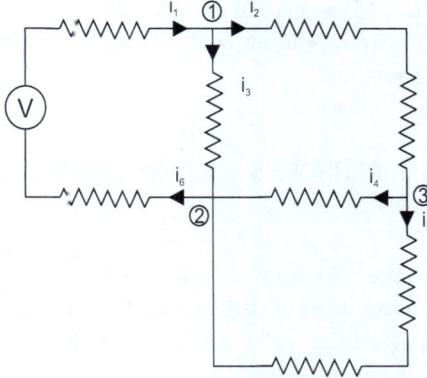

Fig. 1.6. Illustrative example for Kirchhoff's current law.

i.e., $i_1 - i_2 - i_3 = 0$ (1.16)

Similarly at junction 2, the KCL equation is:

$$i_3 + i_4 + i_5 + (-i_6) = 0$$

i.e., $i_3 + i_4 + i_5 - i_6 = 0$ (1.17)

And at junction 3, the KCL equation is:

$$i_2 + (-i_4) + (-i_5) = 0 \quad (1.18)$$

i.e., $i_2 - i_4 - i_5 = 0$ (1.19)

The KCL can also be alternatively stated as:

"THE SUM OF THE CURRENTS ENTERING A NODE WILL BE EQUAL TO THE SUM OF THE CURRENTS LEAVING THE SAME NODE."

This can be readily proved with any of the above three equations. For instance, with equation (1.16) with reference to the junction 1,

$$i_1 = i_2 + i_3$$

i.e.

Sum of the currents entering a node =
Sum of the currents leaving the same node   (1.20)

Equations (1.15) and (1.20) are the KCL equations.

KCL is a powerful law used to solve complicated electric circuits.

1.19-B   KIRCHHOFF'S VOLTAGE LAW OR KVL

This law states that in any closed circuit the algebraic sum of voltage drops of all the elements in that circuit plus the algebraic sum of all the source EMF's in that closed circuit is zero.

That is,

$$\sum_{k=1}^{\text{no. of element}} I_k R_k + \sum_{k=1}^{\text{all source}} V_K = 0 \quad (1.21)$$

in a closed circuit or loop

In applying KVL, the selection of 'Signs' for the voltage sources and for the voltage drops of elements plays a vital role, as they enter the summation.

The sign convention is given below. Note that, initially, in the concerned closed circuit, the loop current directions will be chosen arbitrarily; the following conventions are then applied.

(a) Sign convention for voltage sources

It is obvious that by following the loop current direction, when we go from negative terminal (of the battery) to the positive terminal (of the same battery), it is a potential raise. Hence, in such case, the potential of the battery is considered with plus sign (i.e., positive). On the other hand if we go from positive terminal (of the battery) to the negative terminal (of the same battery), there is obviously a drop in potential. In such case the potential is treated to be a negative, and such potential is considered a negative sign. (Refer to Fig. 1.7)

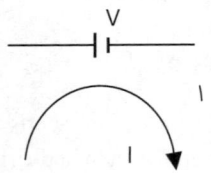 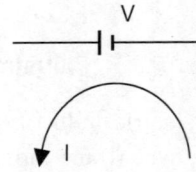

'V' will be taken as –V as the loop current flows from + to – of the battery i.e. a drop in potential. | Here, V will be taken as + V as the loop current passes from – to + of the battery; i.e. a raise in potential.

Fig. 1.7.   Illustration for battery sign convention (Loop current chosen by us

(b) Sign convention for the voltage drops, (IR), of the elements

(i) To begin with, in each element of the concerned circuit, the current entering side is marked with + sign (positive) and current leaving side is marked with a – sign (negative). It is obvious because the current always flows from a higher potential to a lower potential.

(ii) Then, in this concerned closed circuit, the loop current path is traced. In doing so, in any element, if we go from + to – then the voltage drop of this Kth element will be $-I_k R_k$ because it is a drop in potential; whereas, if we go from – to +, then the voltage drop will be $+I_k R_k$, because it is a raise in potential.

Throughout this book, only this convention is used.

Both the above conventions are clearly explained with the example below, (Fig. 1.8). There are two loops (closed paths) and the loop currents I_1 and I_2 are chosen arbitrarily.

By considering these two loop currents, the current through each element is marked. For instance, it can be seen that, in the resistor R_2, the currents I_1 and I_2 together flows and are in the "Same" direction. Hence, the total current in R_2 is $I_1 + I_2$. The other element currents can also be identified by a similar analysis.

[*Note:* The above current I_2 is in anti clock-wise direction. Let us suppose that, it is assumed to be clock-wise. If so, the current I_1 and I_2 together flow in R_2 and are in "opposite" direction in it. Hence the total current in R_2 would be $I_1 - I_2$.)

Consider the resistor R_1. I_1 is entering its left hand side and leaves its right hand side; hence the respective sides are marked with a + and a –. Similarly, the + – are marked in all the other elements.

Let now the loop ABEFA be considered. In this loop the total voltage source is V and there are two elements whose voltage drops are the product of their respective element current and element resistance. For the resistor R_1 the voltage drop will be $I_1 R_1$ and for the resistor R_2 will be $(I_1 + I_2) R_2$.

In travelling through the loop ABEFA, we go from + to – in R_1 and also in R_2. Therefore, the respective drops given above will be considered with a – sign. In tracing the same loop ABEFA, we go from – to + in the source V. So, obviously due to the raise in potential, it will be treated as +V.

With this arrangement when KVL of equation (1.21) is applied to the loop ABEFA, the equation will be,

$$\{-I_1 R_1 + [-(I_1 + I_2) R_2]\} + V = 0$$

$$-I_1 R_1 - (I_1 + I_2) R_2 + V = 0$$

$$-I_1 R_1 - I_1 R_2 - I_2 R_2 + V = 0$$

$$-I_1 [R_1 + R_2] - I_2 R_2 + V = 0$$

or $\quad I_1 [R_1 + R_2] + I_2 R_2 = V \quad$ (1.22)

When a similar analysis is carried in the closed loop BCDEB, the KVL equation will be

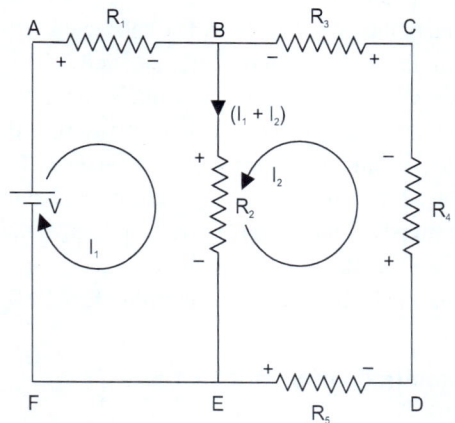

Fig. 1.8. Illustrative example for sign conventions and KVL.

$$R_3 I_2 + R_4 I_2 + R_5 I_2 + (I_1 + I_2) R_2 = 0$$

RHS is zero because in this concerned loop there is no voltage source, and is (LHS) all positive because in this concerned BCDEB loop we travel in all elements from – to +. (a raise in potential).

$$I_2 [R_3 + R_4 + R_5 + R_2] + I_1 R_2 = 0. \qquad (1.23)$$

Solving the equation (1.22) and (1.23) the loop currents I_1 and I_2 will be got and hence the individual element currents.

The worked illustrations 23, 24 etc., may be referred now to understanding further in detail, the sign conventions and KVL.

1.20 ELECTRIC CIRCUIT

Having discussed different terms connected with an electric circuit (like voltage, power, energy etc.,) and the powerful laws, namely, Ohm's law and Kirchhoff's laws which find utility in solving any complicated circuits, now at the final stage, an actual-practical 'electric circuit' is introduced.

1.20-A SERIES CIRCUIT

To define in simple words, an electric circuit is a circuit in which elements (like resistor, inductor, capicitor, etc.) are connected across an energising voltage source and allows a current to flow if it is closed (known as 'closed circuit') and do not allow current to flow if the circuit is open (known as 'open circuit').

This chapter concentrates on DC, that is, Direct Current. A DC is a one which maintains same magnitude of current indefinitely. Refer to Fig. (1.9A); all the above discussions are valid only for a DC. The other types of currents are discussed in chapter 3.

Depending on the fashion of connecting the elements, an electric circuit can be classified into,

(i) Series circuit

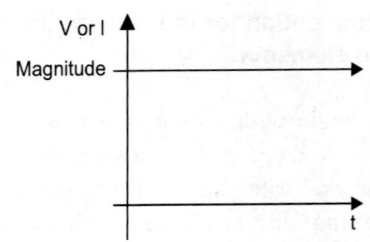

Fig. 1.9A. A DC illustration.

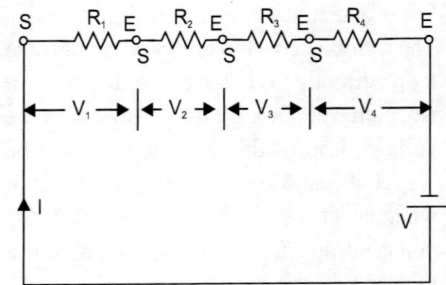

Fig. 1.9B. A Series Circuit
S: Start of an element
E: End of an element
V: The energising voltage source
I: The current in the single closed ciruit.

(ii) Parallel circuit
(iii) Series-Parallel circuit

An electric circuit in which all the elements are connected serially, that is, continuously, is known as a series circuit.

End of the first element will be connected to the start of the second; end of the second will be connected to the start of the third; and so on.

In such arrangement, of course, the start of the first element and the end of the last element will be left free which will respectively be connected to the two terminals of the energising voltage source, thus forming a "single" closed circuit.

The above discussion is depicted in Fig. 1.9B.

Characteristic of a series circuit

It is readily seen from Fig. 1.9B that, the current I forced by the voltage source V, will pass through all the elements $R_1, R_2, R_3, R_4 \ldots R_N$. Hence in a series circuit, "**current in all the elements will be the same**". Note carefully also that, depending on the value of individual resistances R_1, R_2 etc., the voltage drop (i.e., IR) in each element will change. Hence, in a series circuit, "**voltage drop across each element could be different**".

Let V_1 be the voltage drop in the element R_1. Similarly V_2, in R_2 etc.,

Apply KVL of equation (1.21) to the series circuit of Fig. 1.9B the following equation can be obtained.

$$(I_1 R_1) + (I_2 R_2) + (I_3 R_3) + (I_4 R_4) = V$$

As in a series circuit, the current in each element is same, letting $I_1 = I_2 = I_3 = I$ the above equation can now be rewritten as,

$$I R_1 + I R_2 + I R_3 + I R_4 = V$$

or $I [R_1 + R_2 + R_3 + R_4] = V$

or $V/I = (R_1 + R_2 + R_3 + R_4)$ (1.24)

i.e., $V/I = R_{net}$ (1.25)

By comparing equations (1.24) and (1.25), it is clear that, "**in a series circuit the net resistance is the sum of all the individual resistances of that circuit.**"

That is,

$$R_{net} = R_1 + R_2 R_3 + R_4 + \ldots + R_N \quad (1.26)$$

1.20-B PARALLEL CIRCUIT

Instead of joining the start and end of successive elements, when all the starts and all the ends are respectively joined together and connected across an energising voltage source, it will be now a parallel electric circuit. The parallel electric circuits are otherwise known as shunt circuits.

In other words one end of all resistors will meet at a common node (say P) and the other ends will meet at another common junctions (say Q), to form a parallel electric circuit and voltage source will be applied across P and Q.

Characteristics of parallel electric circuits

(i) Voltage across each element will be the same and will be equivalent to the applied source voltage V. That is,

$$V_1 = V_2 = V_3 = V_4 = V_6 = \ldots = V_N = V \quad (1.27)$$

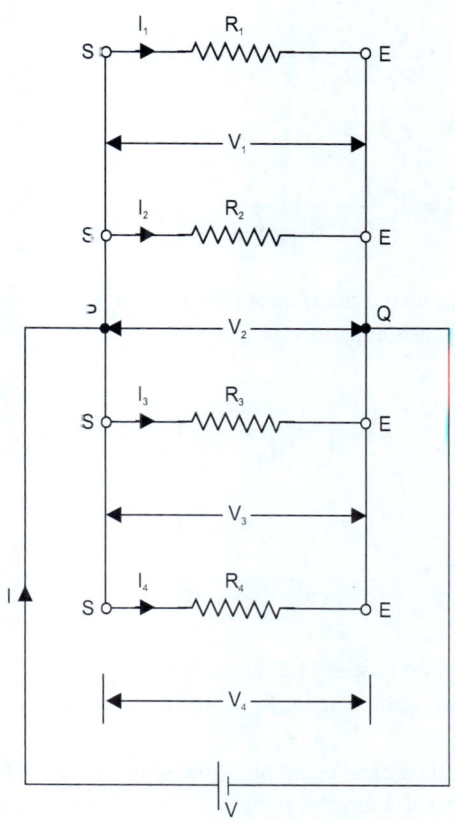

Fig. 1.10 A parallel circuit

(ii) Currents in each element will differ according to the respective resistance value, and will be given by the Ohm's law of equation (1.8). Applying KCL of equation (1.20) to the Fig. 1.10 at the junction P, the equation for this circuit will be,

$$I = I_1 + I_2 + I_3 + I_4$$

By ohm's law of equation (1.8)

$$I = \frac{V_1}{R_1} + \frac{V_2}{R_2} + \frac{V_3}{R_3} + \frac{V_4}{R_4}$$

As $V_1 = V_2 = V_3 = V_4 = V$,

$$I = \frac{V}{R_1} + \frac{V}{R_2} + \frac{V}{R_3} + \frac{V}{R_4}$$

i.e., $I = V \left[\dfrac{1}{R_1} + \dfrac{1}{R_2} + \dfrac{1}{R_3} + \dfrac{1}{R_4} \right]$

I is the total source current and is nothing but V/R_{net}
Substituting this for I,

$$\frac{V}{R_{net}} = V \left[\frac{1}{R_1} + \frac{1}{R_2} + \frac{1}{R_3} + \frac{1}{R_4} \right]; \text{ therefore}$$

$$\frac{1}{R_{net}} = \frac{1}{R_1} + \frac{1}{R_2} + \frac{1}{R_3} + \frac{1}{R_4} \quad (1.28)$$

using the equation (1.28), the complete net resistance of the parallel circuit, R_{net}, can be found out.

Note: In a parallel circuit of the shape as Fig. 1.11, if the circuit current I is known, the branch currents are found out by the formulae given below.

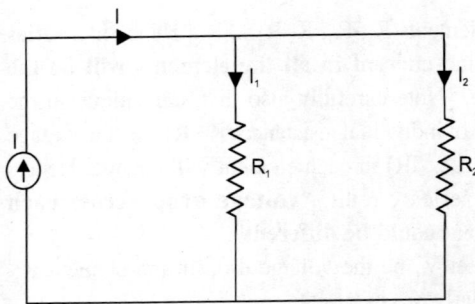

Fig. 1.11. Circuit for current division illustration.

$$I_1 = I R_2/(R_1 + R_2)$$
$$I_2 = I R_1/(R_1 + R_2) \quad (1.29)$$

This method is the current division method.

1.20-C SERIES-PARALLEL CIRCUIT

Combination of series circuit and parallel circuit will form a series-parallel circuit. In such circuits the series analysis must be applied to the series section and parallel analysis must be applied to the parallel section.

> **Reflection Section**
> 1. Differentiate between a resistor, its resistance and resistivity.
> 2. State the basic laws of electric circuits: Ohm's and Kirchhoff's laws.
> 3. Conceptually differentiate these three quantities. Voltage, EMF and potential difference.

1.21 WORKED ILLUSTRATIONS II

Illustration 1.10

A 250-volts supply in series with a resistor of 0.5 Ω is used to charge 100 cells each of EMF 2.2 volts and internal resistance of 0.01 Ω. Calculate the charging current, and the power dissipated in heating the cells and charging the cells.

Solution

The total voltage of all the cells will be 100 × 2.2 = 220 volts as all are in series.

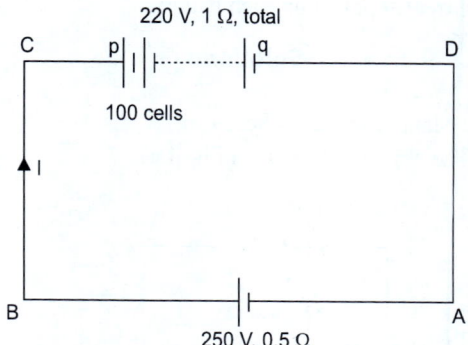

Fig. 1.12. Circuit for illustration 1.10.

The total resistance of all the cells will be 0.01 × 100 = 1 Ω. (by equation 1.26)

The situation of the circuit is in Fig. 1.12.

Tracing the loop ABCDA and applying the sign convention rule (a) discussed in KVL (Sec. 1.19B), the total source voltage is + 250 + (– 220) = 30 volts; otherwise the 30 volts is the P.D. of the full circuit; the net resistance is obviously 0.5 + 1 = 1.5 Ω.

Hence the charging current I will be, (by Ohm's law of equation (1.8))

$$I = \frac{V}{R} = \frac{30}{1.5} = 20 \text{ A} \ldots \text{(Ans)}.$$

When current passes through a conductor, the resistance of the conductor opposes the flow of current (refer to Sec. 1.6). This causes electric power loss and this power loss is dissipated as heat. Indirectly the problem needs the power loss in the cells **alone**.

By equation (1.9)

$$P = I^2 R$$

As we need to find power loss in the cells only, R and I of cell must be supplied to the above equation

$$P = (20)^2 \, 1$$

$$= 400 \text{ W} \quad \text{(Ans)}.$$

The voltage of all the cells is 220 volt (that is between p – q of Fig. 1.12) and the current through all the cells is 20 amps, whose product VI (by equation 1.4) would have been the power used or dissipated to charge the cells.

$$= 4400 \text{ W} \quad \text{(Ans)}.$$

Note: It is a classic example to represent the fact that, $I^2 R$ will always give the power wasted in a resistor as loss or heat, whereas, VI will give the source power utilised to do appreciable work—in this case, to charge the cells.

Illustration 1.11

What is the difference of potential between the points X and Y, in the network of Fig. 1.13.

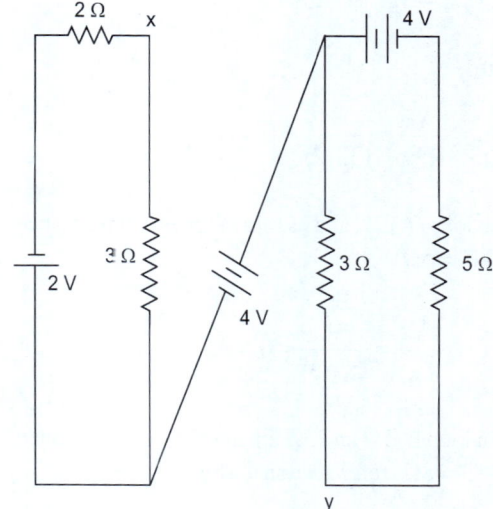

Fig. 1.13. Circuit for illustration 1.11.

Solution

Hint: Reference sections: 1.14, 1.19B.

The current i_1 and i_2 are assumed arbitrarily, as indicated in Fig. 1.14. The polarity of voltage drop

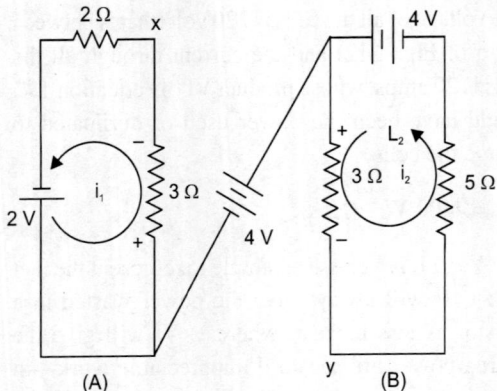

Fig. 1.14. Solution diagram for illustration 1.11.

in the respective resistors are marked by taking the sign conventions rule (b) of sec. 1.19B. Applying this together with rule (a), the following equation is obtained.

$$V_{x-y} = 3i_1 + 4 - 3i_2$$

where

$$i_1 = \frac{V}{R_{net}} = \frac{2}{(3+2)} = 0.4 \text{ A}$$

(As in loop (A) 2Ω and 3Ω are in series, net resistance is $2 + 3 = 5\Omega$)

$$i_2 = \frac{V}{R_{net}} = \frac{9}{(5+3)} = 0.5 \text{ A}$$

(As in loop B, 5Ω and 3Ω are in series, net resistance is $5 + 3 = 8\Omega$; refer section 1.18b)

$$\therefore V_{x-y} = 3(0.4) + 4 - 3(0.5) = 3.7 \text{ V}$$

Illustration 1.12

Two 12-volt batteries with internal resistors 0.2Ω and 0.25Ω respectively are joined in parallel and a resistor of 1Ω is placed across the terminals. Find the current supplied by each battery.

Solution

Hint: Main reference is Sec. 1.8B.
The circuit will be as in Fig. 1.15.

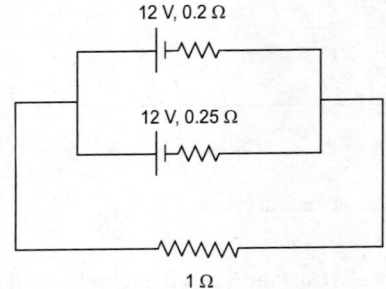

Fig. 1.15. Circuit of illustration 1.12.

Illustration 1.13

A heater rated at 2 kW, 230 V is connected to a 230 volts supply through a cable having a total resistance of 0.3Ω. Calculate the power dissipated by the heater and the power wasted in the cable. Also calculate the total energy consumed in 3 hrs.

Solution

A heater is connected to a supply through a cable. It is shown in Fig. 1.16 and the given data are marked in it.

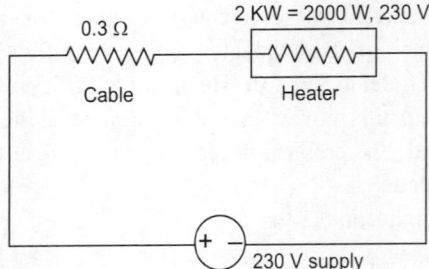

Fig. 1.16. Circuit for illustration 1.13.

It is clear from Fig. 1.16 that the cable and the heater are in 'series' across the supply of 230 V.

It is seen in Sec. 1.20-A that in a series circuit current in all the elements is the 'same'; but the voltage across each element will be different.

By knowing the power and voltage of the heater element, its resistance is found out, from the equation (1.10), as given below:

$$P = \frac{V^2}{R}$$

$$\therefore R = \frac{V^2}{P} \text{ (all refers to cable)}$$

$$= \frac{230^2}{2000} = 26.45 \, \Omega$$

The net resistance in this series circuit is, $R_{cable} + R_{heater}$ (Refer to equation 1.26)

$$\therefore R_{net} = 0.3 + 26.45 = 26.75 \, \Omega$$

and the current in the circuit is

$$I = \frac{V}{R} \text{ (all are net values)}$$

$$= \frac{230}{26.75} = 8.598 = 8.6 \text{ A, aprox.,}$$

This current will be same in all the elements, because it is a series circuit.

(i) Power dissipated by the cable is:

$P = I^2 R$ (using equation 1.9)

$= (8.6)^2 \, 0.3 = 22.178$

$= 22.2$, W approx., **(Ans)**

Note that the resistance of the cable alone is considered because the power in the cable alone is needed. Similarly,

(ii) Power dissipated in the heater will be, $(8.6)^2 (26.45) = 1956.242 = $ **1956, W (Ans)**

(iii) Energy = $(I^2 R)$ t. As the energy is needed for the whole circuit, the total current and net resistance are considered.

Energy $= (8.6)^2 (26.75) (3)$

$= 5935.29$

$= 5935.3$, W.Hr

or $= $ **5.9353, kW-Hr (Ans)**

Illustration 1.14

A bulb rated 110 volts, 60 watts is connected with another bulb rated 110 volts, 100 watts across a 220 volts mains. Calculate the resistance which should be joined in parallel with the first bulb so that both the bulbs may take their rated power.

Solution

The situation of the problem is depicted in Fig. 1.17.

Power, VI, of bulb 1 is 60 watts and its voltage rating, V, is 110 volts.

Hence the current in bulb 1 (I_1) is

$$\frac{VI}{V} = \frac{60}{110} = 0.545$$

$$= 0.545 \text{ A.}$$

Similarly, current in bulb 2 is the ratio between its power and its voltage ratings. That is,

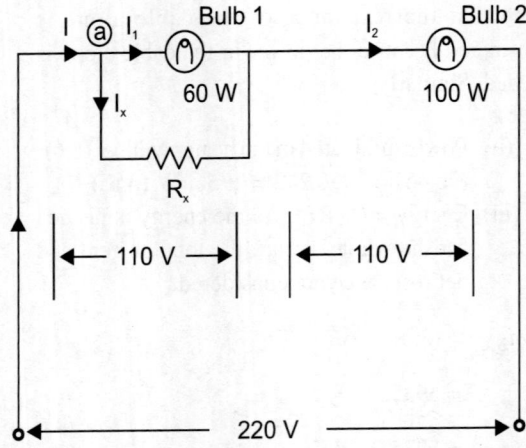

Fig. 1.17. Circuit for illustration 1.14.

$$I_2 = \frac{VI}{V} = \frac{100}{110} = 0.909 \text{ A}$$

This current I_2 will flow through the main source 220 V and complete the circuit. Therefore, $I = I_2 = 0.909$ A. (Refer to Fig. 1.17).

It can be seen that, from Fig. 1.17, the bulb 1 and the resistor R_x (whose value is to be found out) are in parallel, and this combination is in series with the bulb 2.

Because of the parallel combination of bulb 1 and R_x the current I gets divided into I_1 and I_x. Applying KCL to the junction 'a', by equation (1.15)

$$I - I_x - I_1 = 0$$

or $I_x = I - I_1$

$$= 0.909 - 0.545$$

$$= 0.3636 \text{ A}.$$

This is the current through the resistor R_x. And, as R_x is in parallel with bulb 1, its voltage will also be that of bulb 1; i.e., 110 volts (because in parallel circuit, voltage across all the elements is the same. Refer to sec. 1.20-B).

By applying Ohm's law of equation (1.6)

$$R_x = \frac{V}{I}; \text{ values of element x are considered.}$$

$$= 110/0.3636 = 302 \text{ } \Omega \text{ (Ans)}$$

Illustration 1.15

A shop is lighted by 30 lamps in parallel rated 110 volts, 100 watts. The distance from the shop to the supply mainboard is 55 m. If the cable resistance is, 1.2 ohm/km find the voltage necessary at the mainboard.

Solution

The condition is depicted in Fig. 1.18.
Current in each lamp is, by the equation (1.4),

$$I = P/V = 100/100 = 0.909 \text{ A}.$$

All the 30 lamps are in parallel, and in a parallel branch the total current is the sum of individual currents. Of course, voltage across each lamps will be the same at 110 votls.

Hence the total current consumed by the shop is, $30 \times 0.909 = 27.27$ amps.

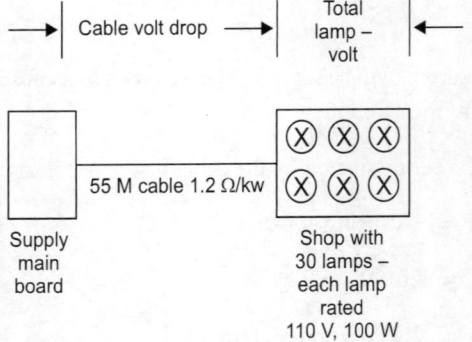

Fig. 1.18. Circuit representation for illustration 1.15.

As cable supplies the whole shop, this will be the current in the cable also.

The resistance of the cable is given to be 1.2 ohm/km and its length is 55m. Hence the total resistance of the cable for 55 m is

$1.2 \times 55/1000 = 0.066\ \Omega$.

By applying ohm's law of equation (1.7) the voltage drop in the cable is the product of its resistance and current; i.e., $0.066 \times 27.27 = 1.8$ V.

The voltage at the supply main board is, the sum of the voltage across the lamps of the shop and the cable voltage drop, which is,

$110 + 1.8 = 111.8$ V **(Ans)**.

Note: All the above problems are of different typical types. Each problem is dealt with numerous discussions in between, to teach clearly, a sense of approach, to the reader. The forthcoming problems will directly be dealt without much theoretical discussions in between, expecting that the reader must have undergone all the theory sections of this chapter discussed earlier.

Illustration 1.16

The resistance of two wires is 25 Ω when connected in series and 6 Ω when connected in parallel. Calculate the resistance of each wire.

Solution

Refer sections 1.20A and 1.20B.

Let the resistance of the first wire with one be R_1 and that of the other be R_2 in ohms. It is given that.

$R_1 + R_2 = 25\ \Omega$ (series connection)   (I)

and

$R_1 R_2/(R_1 + R_2)$ is 6 Ω (II)

From equation (II),

$R_1 R_2 = 6 (R_1 + R_2)$

using equation (I), we wet $R_1 R_2 = 6 \times 25$
Hence

$$R_1 = \frac{150}{R_2} \qquad (III)$$

Substituting R_1 value from equation (III) in (I),

$$\frac{150}{R_2} + R_2 = 25$$

i.e., $R_2^2 + 150 - 25\ R_2 = 0$

Solving this quadratic, $R_2 = 15\ \Omega$ or $10\ \Omega$ and substituting this in equation (1), $R_1 = 10$ or 15 ohms.

Illustration 1.17

Find the effective resistance of the circuit shown in Fig. 1.19; All resistances are in ohms.

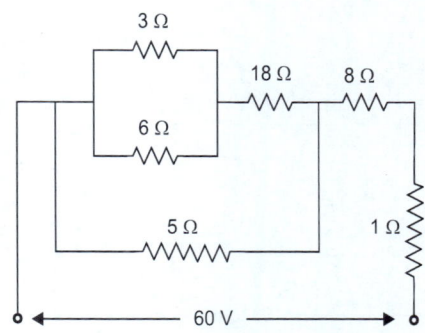

Fig. 1.19. Circuit for the illustration 1.17.

Solution

It is seen that, the resistor 3 and 6 are in parallel. Hence the combination gives a net resistance of

$$\frac{3\times 6}{3+6} = 2\,\Omega.$$

The circuit thus becomes as in Fig. 1.20.

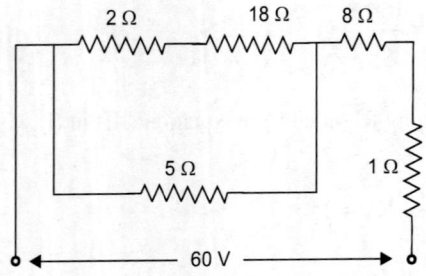

Fig. 1.20.

It is seen from Fig. 1.20 that 2 and 18 are in series and this combination is in parallel with 5 Ω. Moreover the 8 and 1 Ω resistors are in series.

$2 + 18 = 20\,\Omega$. This is in parallel with 5 Ω. Hence this net resistance is $(20 \times 5)/(20 + 5) = 4\,\Omega$, and

$8 + 1 = 9\,\Omega.$

The reduced circuit is in Fig. 1.21.

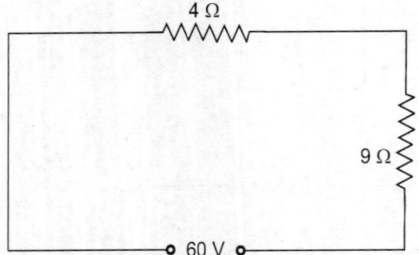

Fig. 1.21.

It is clear that the above two resistors are in series, whose addition will give the net resistance of the circuit.

$R_{net} = 13\,\Omega$ **(Ans)**

Illustration 1.18

Obtain the total power supplied by the 60 V source and the power absorbed in each resistor shown in Fig. 1.22; Resistors are in Ω.

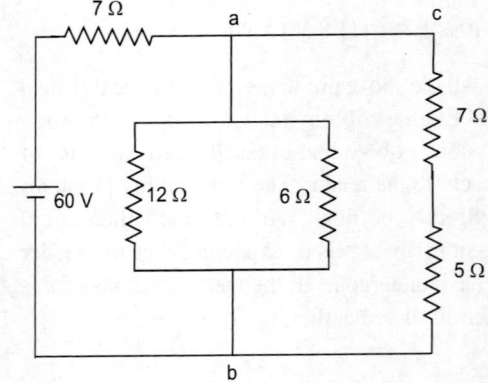

Fig. 1.22. Circuit for illustration 1.18.

Solution

If the current delivered by the source (which will be the total current) is known, the product of this current and the total voltage (60 V), will give the power supplied by the source. By knowing the individual branch currents then, the power absorbed by each resistor can be found out, by $I^2 R$ formula.

Net resistance across 'cd' is $7 + 5 = 12\,\Omega$. Net resistance across 'ab' is $(12 \times 6)(12 + 6) = 4\,\Omega$.

These two are in parallel to give a resistance of, $(12 \times 4)/(12 + 4) = 3\,\Omega$. The reduced circuit is in Fig. 1.23.

The total resistance of the circuit is, $7 + 3 = 10\,\Omega$

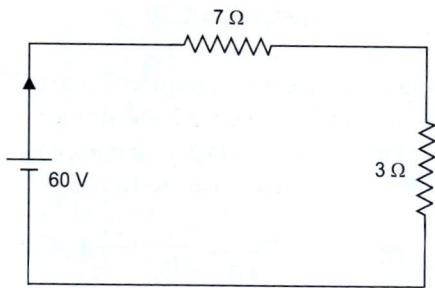

Fig. 1.23.

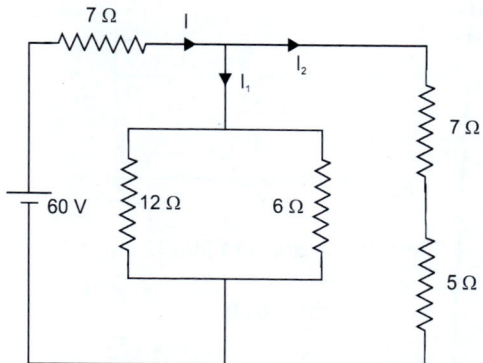

Fig. 1.23A.

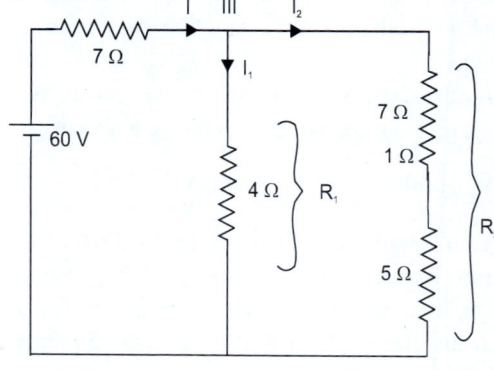

Fig. 1.23B. Circuit for illustration 1.18 with current directions marked, for evaluation of I_2

The current I hence is, $60/10 = 6$ A

The power delivered by the source is the product of its V and I. That is $60 \times 6 = 360$ W (**Ans**).

Basic Electricity 23

The current I will flow through 7 ohms resistor also. The power absorbed by it will be

$(6^2)(7) = 252$ W (**Ans**).

Referring to Fig. 1.23-B and applying the current division technique of equation (1.29).

$$I_2 = \frac{I(R_1)}{R_1 + R_2} = \frac{6 \times 4}{4 + 12} = 1.5 \text{ A}$$

Current in 7 ohm resistor = current in 5 Ω resistor
= I_2, as they are in series

∴ Power absorbed by 7 Ω resistor = $(1.5^2) 7$

= **15.75 W** (**Ans**).

Power absorbed by 5 Ω resistor = $(1.5)^2 \times 5$

= **11.25 W** (**Ans**).

Illustration 1.19

Three resistors of 2, 5 and 10 Ω are joined in parallel and a total current of 24 A is passed through them. Find the current through each resistor.

Solution

The problem is dipicted in Fig. 1.24.

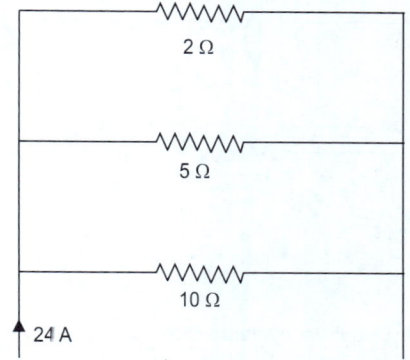

Fig. 1.24. Circuit for illustration 1.19.

24 Basic Electrical Engineering

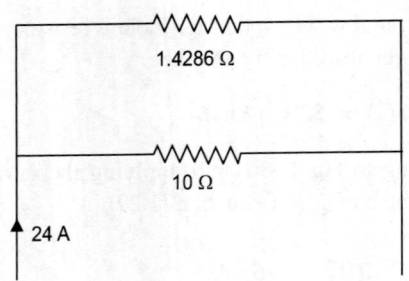

Fig. 1.25. Circuit to calculate current through a 10-ohm resistor.

The note given in section 1.20-B and equation (1.29) are used here.

Current through a 10 Ω resistor: Retaining 20-Ω resistor, the circuit can be reduced as in Fig. 1.25 (2 and 5 Ω are in parallel reducing to 1.4286 Ω)

Using equation (1.29), current through 10 Ω resistor is, $(24 \times 1.4286)/(1.4286 + 10) = 3$ A

Current through a 5 Ω resistor: The 5 ohms resistor is retained and the other two parallel resistors reduce to, $(2 \times 10)/(2 + 10) = 1.66$ Ω. This is depicted in Fig. 1.26.

∴ Current through 5 ohm resistor is, $(24 \times 1.66)/(1.66 \times 5)$ and it is, 6 A.

Similar procedure will yield current through the 2 Ω resistor as, 15 A (Readers may try this).

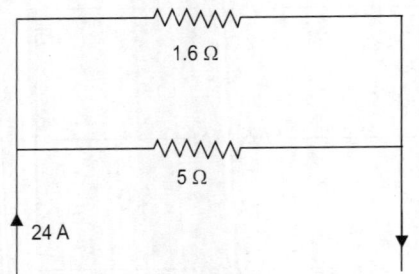

Fig. 1.26. Circuit to evaluate current through 5 ohm resistor.

Illustration 1.20

In the parallel arrangement of resistors shown in Fig. 1.27, The current flowing in the 8 ohm resistor is 2.5 amperes. Find (i) Currents in the other resistors (ii) the resistor x (iii) the equivalent resistance.

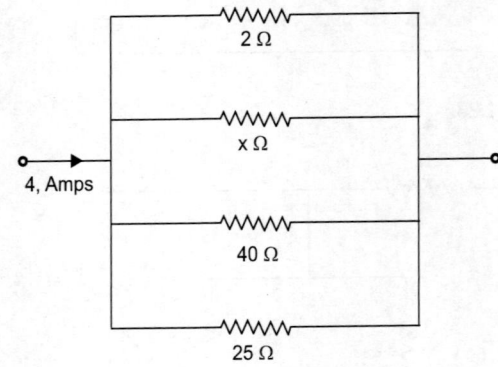

Fig. 1.27. Circuit for illustration 1.20.

Solution

It is a classic problem for beginners, which involves the Ohm's law, the current division technique, Kirchhoff's current laws and parallel branch technique.

As the current in 8 Ω resistor in known, the voltage across it can be found out by Ohm's law.

$$V_{8\Omega} = IR_{of\ 8\ \Omega} = 2.5 \times 8 = 20 \text{ V}$$

It is seen from Fig. 1.27 that all the resistors are in parallel. Hence all the resistors will have same voltage 20 V across them. Therefore using Ohm's law again, current through 40 ohms and 25 Ohms can be found out. They are respectively,

$$I_{40\ \Omega} = \frac{V}{R} \text{ (of 40 Ω)}$$

$$= \frac{20}{40} = \mathbf{0.5\ A\ (Ans)}$$

and $I_{25\,\Omega} = \dfrac{V}{R}$ (of 25 Ω)

$$= \dfrac{20}{25} = 0.8 \text{ A (Ans)}$$

As already stated, the four resistors are in parallel, the total current of 4 A will get divided across them. It is shown in Fig. 1.28. Applying KCL to the junction 'a' of Fig. 1.28 (in which current directions are marked),

$$I = I_{8\,\Omega} + I_{x\,\Omega} + I_{40\,\Omega} + I_{25\,\Omega}$$

i.e., $4 = 2.5 + I_{x\,\Omega} + 0.5 + 0.8$

∴ $I_{x\,\Omega} = 4 - (2.5 + 0.5 + 0.8) = 0.2$ A **(Ans)**

The voltage across the resistor × Ω (= 20, V) and the current through it (= 0.2 A) are known; its resistance value can be found out, by applying ohm's law, as:

$$X = \dfrac{V}{I} \text{ (of } x\,\Omega)$$

$$= \dfrac{20}{0.12} = 100 \text{ W (Ans)}.$$

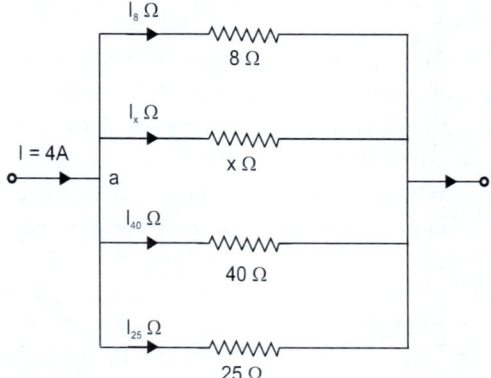

Fig. 1.28

Let the equivalent resistance of the parallel circuit be R. By using the equation (1.28)

$$\dfrac{1}{R} = \dfrac{1}{8} + \dfrac{1}{100} + \dfrac{1}{40} + \dfrac{1}{25}$$

$R = 5$ W **(Ans)**.

Illustration 1.21

Three resistors 8 Ω, 6 Ω and A Ω are in parallel drawing a total current of 15 A from supply. If the current in the A Ω resistor is 1 A, determine:
 (i) the current in each of the other resistor
 (ii) the value of A

Solution

The circuit condition is shown in Fig. 1.29.

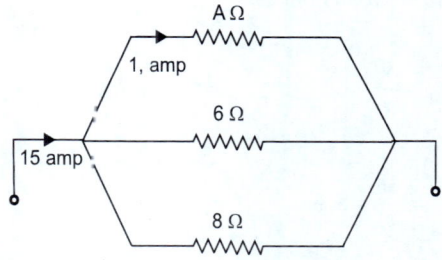

Fig. 1.29. Circuit for the illustration 1.21.

(The reader may develop technical ideas to solve this problem, before the below given solution is traced. It will increase the technical reasoning power of the reader in solving such typical problems.)

The current division method of equation (1.29) is used here.

As the resistors 6 Ω and 8 Ω are in parallel the circuit is reduced to the shape shown in Fig. 1.30.

Current through A Ω resistor, using equation (I.29), can be written as,

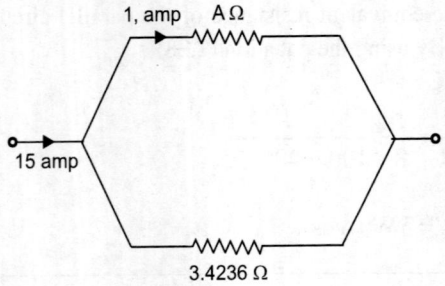

Fig. 1.30.

$$I = \frac{15 \times 3.4286}{3.4286 + A}$$

which gives the value of the resistor A as: 48 Ω (**Ans to part (ii).**)

Thus, the voltage across it is 1 × 48 = 48 V. As all the resistors are in parallel, all will have the same voltage of 48 V across them.

By using Ohm's law, current in each of the other resistors are found out as:

$$I_{6\Omega} = \frac{V}{R} \qquad \text{(of 6 Ω)}$$

$$= \frac{48}{6} = 8 \text{ A}$$

and $I_{8\Omega} = \dfrac{V}{R}$ (of 8 Ω)

$$= \frac{48}{8} = 6 \text{ A}$$

Illustration 1.22

The resistance of a galvanometer is 720 ohms. Find resistance of the parallel resistor kept across the meter (known as shunt) so that only one-tenth of the total current will pass through the meter.

Solution

The situation of the problem is depicted in Fig. 1.31.

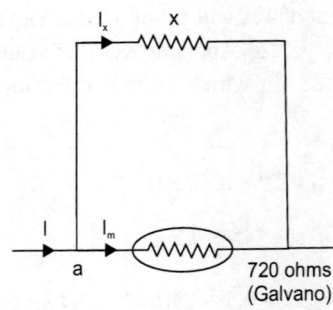

Fig. 1.31. Circuit for the illustration 1.22.
 X: parallel resistor, whose value is needed
 I: The total current
 I_m: the meter current, given to be 1/10 = 0.1 I

It is a classic problem which makes the reader to use Ohm's law in fashions.

Applying KCL to the junction 'a',

$$I = I_m + I_x$$

Thus $I_x = I - I_m = I - 0.1 \text{ I} = 0.9 \text{ I}$

By Ohm's law, voltage across the meter is the product of I and R of the meter. Hence it is 0.1 I × 720 = 72 I, V. This will be the voltage across the shunt x also, as they are in parallel.

Again by using Ohm's law, the resistance value of the shunt X is the ratio between V and I of that resistor X.

thus, $X = \dfrac{72 I}{0.9 I} = 80 \Omega$ **(Ans)**.

Illustration 1.23

In the circuit shown in Fig. 1.32, find the magnitude and direction of the current in each resistor. Internal resistance of the cell is negligible.

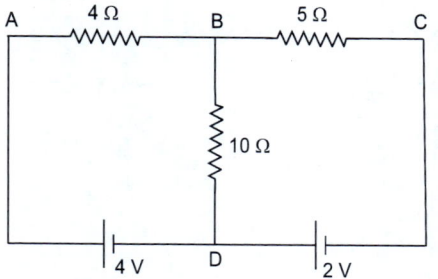

Fig. 1.32. Circuit for illustration 1.23.

Solution

Refer to the section 1.19-B. sign convention rules (a) and (b) discussed in this sections are used here.

The loop current directions are assumed and the element potentials are marked as discussed in KVL. It is shown in Fig. 1.33.

Applying KVL of equation (1.21) to the loop ABDA:

$$-4 I_1 - 10(I_1 - I_2) + 4 = 0$$

or $4 I_1 + 10 I_1 - 10 I_2 = 4$

or $14 I_1 - 10 I_2 = 4$ \hfill (I)

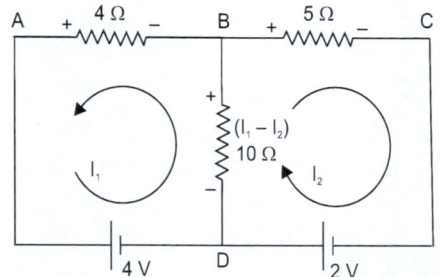

Fig. 1.33. Solution circuit for illustration 1.23.

Applying KVL of equation (1.21) to the loop BCDB:

$$-5 I_2 + 2 + 10 (I_1 - I_2) = 0$$

[*Note:* Voltage drop in 5 Ω resistor is 5 I_2 (refer to equation 1.7); as we travel from + to −, it is a drop in potential; so $-5 I_2$.

Similarly in 10 Ω resistor, we travel from − to +; it is a raise in potential. The current through 10 Ω resistor is $I_1 - I_2$. Thus the voltage drop is

$$+ 10 (I_1 - I_2).]$$

i.e., $-5 I_2 + 2 + 10 I_1 - 10 I_2 = 0$

or $10 I_1 - 15 I_2 = -2$ \hfill (II)

Solving equations, (I) and (II), $I_1 = 0.727$ A and $I_2 = 0.612$ A.

Using the above loop currents the branch currents are marked as in the below Fig. 1.34.

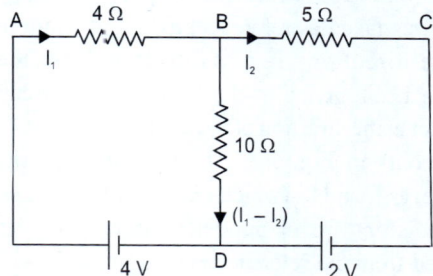

Fig. 1.34

Thus,

Current in 4 Ω resistor = I_1 = **0.727 A from A to B** (Ans).

Current in 3 Ω resistor = I_2 = **0.618 A, from B to C** (Ans).

Current in 10 Ω resistor = $I_1 - I_2$ = **0.1088 A from B to D** (Ans).

Important Note: From the same loop current directions shown in Fig. 1.33, the current through

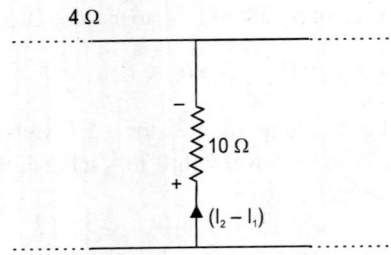

Fig. 1.35.

10 Ω resistor is 'also' $(I_2 - I_1)$, from D to B, in which case + will be at the bottom of 10 Ω resistor. This portion alone is in Fig. 1.35.

It is to be carefully noted that, when more than one current is flowing in an element then one of the currents met by the element must be taken as the reference, with respect to which, the net current must be marked in the element; the direction of the net current will be in the direction of the reference.

The 10 Ω resistor meets with I_1 and I_2. In Fig. 1.33, I_1 was taken as reference; as I_1 and I_2 are in opposite directions I_2 is subtracted from the reference, i.e., I_1 giving $I_1 - I_2$; this is the net current; this is kept in the direction of the reference current I_1.

Whereas in Fig. 1.35, I_2 was taken as the reference; as I_1 and I_2, (which are the current met by 10 Ω resistor) are in opposite directions I_1 is subtracted from the reference i.e., I_2 giving $I_2 - I_1$; this is the net current in the 10 Ω resistor; this net current is kept in the direction of the reference current I_2. The + and − are accordingly marked.

The reader may solve the same problem now, by adopting the notation of Fig. 1.35. The equations will change, but the answer will be the same as the above, both in magnitude and direction.

Illustration 1.24

In the circuit shown in Fig. 1.36, find the current in each branch and in the battery.

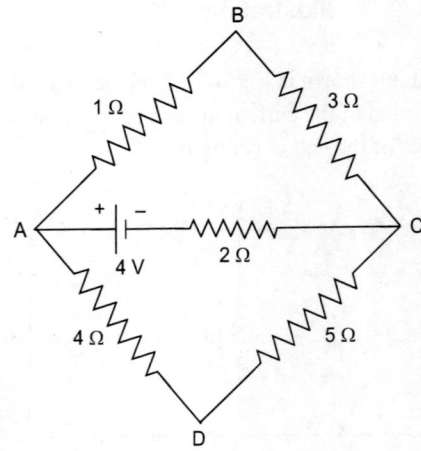

Fig. 1.36. Circuit for illustration 1.24.

Solution

Loop currents are assumed as shown in Fig. 1.37. There are two closed loops viz. ABCA and ACDA.

The element potentials are marked by tracing the loop currents. In the 2 Ω resistor. I_1 and I_2 are flowing in the "same" direction and I_1 is taken to be the reference. Thus the net current $I_1 + I_2$ in the 2 Ω

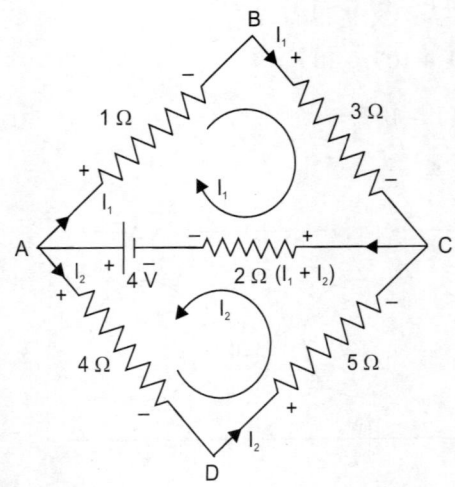

Fig. 1.37.

resistor will be in the direction of the reference current I_1.

KVL equations (by equation 1.21) for the loop ABCA:

$$-1 I_1 - 3 I_1 - 2(I_1 + I_2) + 4 = 0$$

or $6 I_1 + 2 I_2 = 4$

or $3 I_1 + I_2 = 2$ \hfill (I)

KVL equations for the loop ACDA:

$$-4 + 2(I_1 + I_2) + 5 I_2 + 4 I_2 = 0$$

or $-4 + 2 I_1 + 11 I_2 = 0$

or $2 I_1 + 11 I_2 = 4$ \hfill (II)

(*Note:* If the path ADCA is traced, the KVL equation will be, $-4 I_2 - 5I_2 - 2(I_1 + I_2) = -4$, which is, of course, the equation (II).]

Solving equation (I) and (II), $I_1 = 0.571$ A and $I_2 = 0.286$ A.

Answers:

Current through 1 Ω and 2 Ω resistors is $I_1 = 0.571$

Current through 4 Ω and 5 Ω resistors is $I_2 = 0.286$

Current through 2 Ω resistor is $I_1 + I_2 = 0.857$

The units are in amps.

Illustration 1.25

A 20 V battery and an internal resistance of 1 Ω is connected in parallel with another battery of 30 V and internal resistance of 2 Ω. An external resistance of 10 Ω is connected across the above parallel circuit. Applying Kirchhoff's laws find the current in each branch and the power consumption in the external circuit.

Solution

The circuit is depicted in Fig. 1.38.

Let the loop current be chosen as indicated; let them be I_1 and I_2 in amps.

Applying KVL to the loop abcd:

$$-20 - 1(I_1) - 2(I_1 + I_2) + 30 = 0$$

or $3 I_1 + 2 I_2 = 10$ \hfill (I)

Applying KVL to the loop d c p e f q d:

$$-30 + 2(I_1 + I_2) + 10 I_2 = 0$$

or $2 I_1 + 12 I_2 = 30$ \hfill (II)

Solving equations (I) and (II) the value of currents,

$I_1 = 1.875$ A

$I_2 = $ **2.1875 A (Ans)**

The power consumed in the external circuit is
$= I_2^2 (10) = (2.1875)^2 (10) = 47.85$ W **(Ans).**

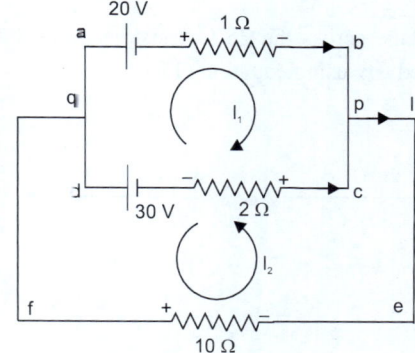

Fig. 1.38. Circuit for illustration 1.25.

Illustration 1.26

A 36 Ω resistor is connected in parallel with a series combination of 10 Ω and 26 Ω. If the voltage across the 10 Ω is 50 V, find
(a) the total voltage applied and
(b) the current in each resistor

Solution

The problem can be handled easily if it is reproduced as a diagram. Refer to Fig. 39.

As the voltage across the 10 Ω resistor is known, the current following in it can be found out by using equation (1.8)

$$I_{10\Omega} = \frac{V}{R} \quad \text{of } 10\ \Omega$$

$$= \frac{50}{10} = 5\ A \quad \textbf{(Ans to Part b)}$$

This current will also flow through the 20 Ω resistor as both are in series. Thus,

$$I_{26}\ \Omega = 5\ A \quad \textbf{(Ans to part b)}$$

The voltage across the 26 Ω resistor can now be calculated. By using equation (1.7),

$$V_{26}\ \Omega = IR \quad \text{(of 26 Ω)}$$

$$= 5 \times 26 = 130\ V$$

In a series circuit the total voltage is the sum of individual element voltages. (Refer to sec. 1.20-A).

Hence,

$$V_{ab} = V_{10\ \Omega} + V_{26\ \Omega} = 50 + 130 = 180\ V.$$

As the branch ab and 36 Ω resistor are in parallel, the voltage across both of them will be the same. (Refer to sec. 1.20-B). Therefore,

$$V_{36\ \Omega} = V_{ab} = 180\ V \quad \textbf{(Ans to part a)}$$

This will be the total voltage applied.
The current in the 36 Ω resistor is,

$$I_{36\ \Omega} = \frac{V}{R} \quad \text{(of 36 Ω)}$$

$$= \frac{180}{36} = 5\ A \quad \textbf{(Ans to part b)}$$

Illustration 1.27

Using Kirchhoff's laws find the current in the 3 Ω resistor of the circuit shown in Fig. 40.

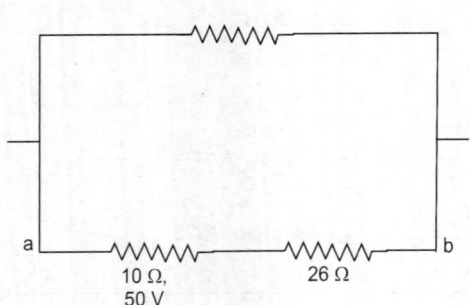

Fig. 1.39. Circuit for illustration 1.29.

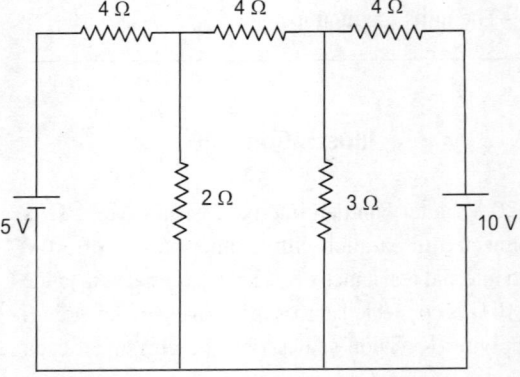

Fig. 1.40. Circuit for illustration 1.27.

Solution

See section 1.19-B and the note given in illustration 23 may be referred to.

The loop currents are chosen as in Fig. 1.41.

The branch currents with their potential polarities are shown in Fig. 1.42.

To mark current in 2 Ω resistor I_1 is taken as reference; the net current in it is $I_1 - I_2$; the direction of this net current in the 2 Ω resistor is placed in the direction of the referred current I_1.

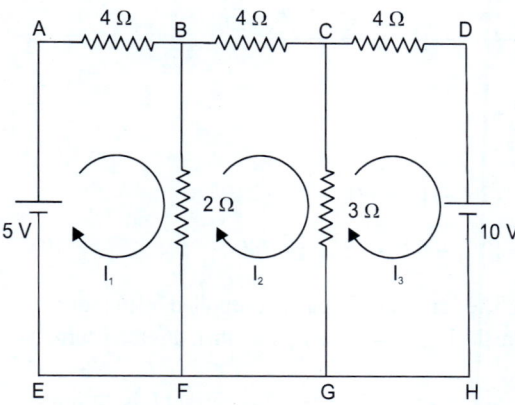

Fig. 1.41. The loop currents for illustration 1.27.

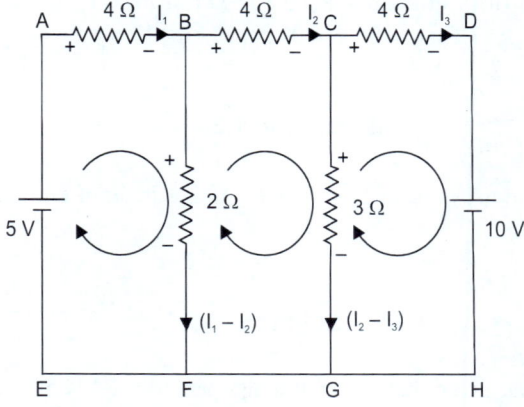

Fig. 1.42. The circuit to write the KVL equations

Similarly, the reference current for 3 Ω resistor is I_2; the net current in it is $I_2 - I_3$, and is placed in the direction of the reference current, I_2.

The KVL equations are written using equation (1.21):

For loop ABFEA:

$$-I_1 - 2(I_1 - I_2) + 5 = 0$$

or $6I_1 - 2I_2 = 5$ \hfill (I)

For loop BCGFB:

$$-4I_2 - 3(I_2 - I_3) + 2(I_1 - I_2) = 0$$

or $2I_1 - 9I_2 + 3I_3 = 0$ \hfill (II)

For loop CDHGC:

$$-4I_3 - 10 + 3(I_2 - I_3) = 0$$

or $3I_2 - 7I_3 = 10$ \hfill (III)

As the current in the 3 Ω resistor is only needed (which is $I_2 - I_3$ from C to G), it is enough if the above equations I, II and III are solved for I_2 and I_3.

Solving the equations the loop currents

$I_2 = -0.3716$ A, and $I_3 = -1.588$ A.

Note: The – sign in the currents says nothing but, the assumed directions of I_2 and I_3 are opposite to the actual flow of these currents. This I_2 and I_3 are flowing in the anticlockwise directions and 'not' in the clockwise directions as we assumed.

Thus the current through, the 3 Ω resistor is,

$I_2 - I_3$.

$I_{3\,\Omega} = (-0.3716) - (-1.588) = 1.2164$ **(Ans)**.

Note: This $I_{3\,\Omega}$ is plus. That means, the branch current $I_{3\,\Omega}$ is properly assigned by us. Therefore, it can be found out from the answer itself whether, the assumed current direction is correct or not!

Illustration 1.28

In the circuit shown in Fig. 1.43 determine
(i) the total current
(ii) the current through the 18 Ω resistor.

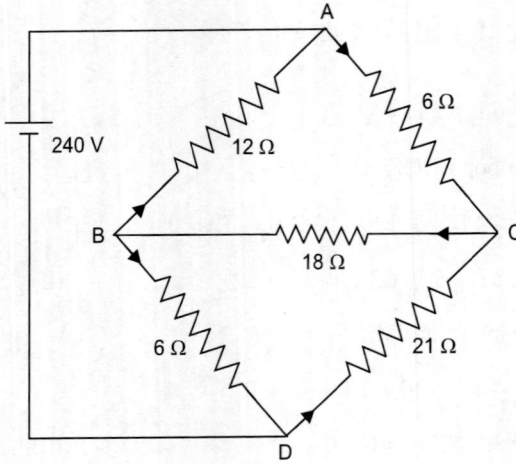

Fig. 1.43. Circuit for illustration 1.28.

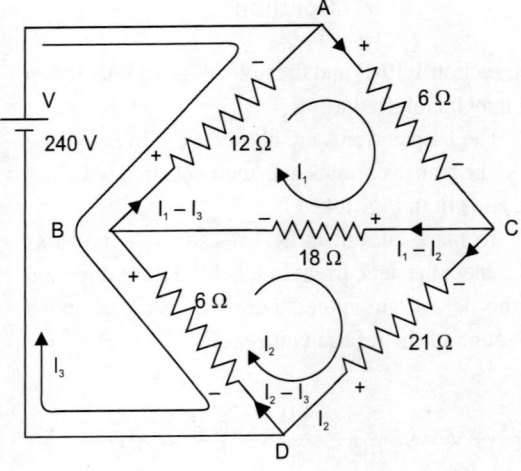

Fig. 1.44.

Solution

There are three loops, namely, ACBA, BCDB and ABDVA. The three loop currents thus formed are I_1, I_2 and I_3 respectively. These loop currents along with the branch currents thus formed and the branch potential polarities are shown in Fig. 1.44.

KVL equation for loop ACBA

$$-6I_1 - 18(I_1 - I_2) - 12(I_1 - I_3) = 0$$

or $-36 I_1 + 18 I_2 + 12 I_3 = 0$ \hfill (I)

KVL equation for loop BCDB:

$$18(I_1 - I_2) - 21 I_2 - 6(I_2 - I_3) = 0$$

or $18 I_1 - 45 I_2 + 6 I_3 = 0$ \hfill (II)

KVL equation for loop ABDVA:

$$12(I_1 - I_3) + 6(I_2 - I_3) + 240 = 0$$

or $12 I_1 + 6 I_2 - 18 I_3 = -240$ \hfill (III)

The current is the current supplied by the source. From the Fig. 1.44 it can be seen that, this total current is I_3.

Solving the I, II and III, the current I_3 is 24 amps.
(i) Thus, the total current is 24, amps **(Ans.)**.

I_1 and I_2 are respectively, 8 amps, each

(ii) Current through the 18 Ω resistor is $I_1 - I_2$. That is, $8 - 8 = 0$, amps **(Ans)**.

Illustration 1.29

Find the value of E watts in R_3 of the circuit of Fig. 1.45.

Solution

Idea: given that the power dissipation in R_3 is 50 watts. Out of the power expressions of equations (1.4), (1.9) and (1.10), if the equation (1.9) is used

Basic Electricity 33

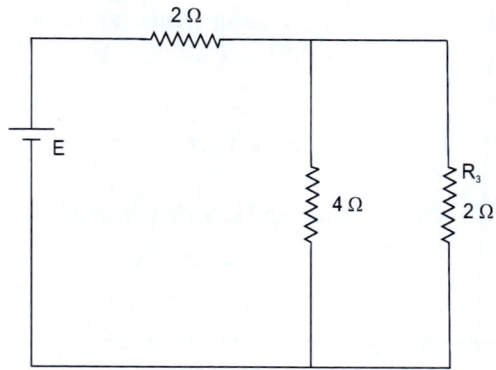

Fig. 1.45. Circuit for illustration 1.29.

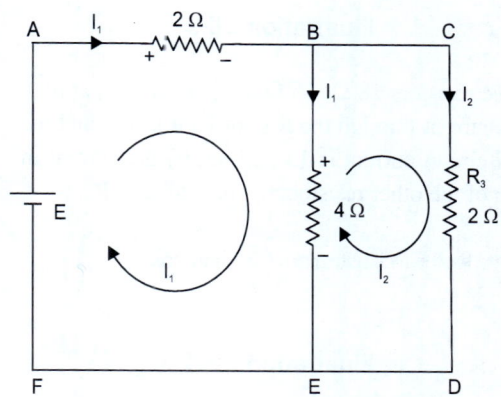

Fig. 1.46. Solution circuit of illustration 1.29.

the current through R_3 can be found out. Then, by writing KVL equation for this circuit, E can be found out.

$$I_{R3}^2 \, R_3 = 50$$

$$\therefore \; I_{R3} = \sqrt{\frac{50}{R_3}} = \sqrt{\frac{50}{2}} = 5 \text{ A}$$

Totally there are two loops. Let the loop current be chosen as in the Fig. 1.46. The branch currents and their potentials are also marked.

KVL equation for loop ABEFA:

$$-2 \, I_1 - 4 \, (I_1 - I_2) + E = 0$$

or $\quad 6 \, I_1 - 4 \, I_2 = E \quad\quad$ (I)

KVL equation of loop BCDEB:

$$-2 \, I_2 + 4 \, (I_1 - I_2) = 0$$

or $\quad 3 \, I_1 - 6 \, I_2 = 0 \quad\quad$ (II)

It can be seen from Fig. 1.46 that, the loop current I_2 is the current through the resistor R_3. Thus, $I_2 = I_{R3}$ = 5 A. Substituting this value of I_2 in the equation II, the I_1 is found out as, 7.5 A.

Substituting the values of I_1 and I_2 in equation (I), the value of E can be found out. Thus,

$$E = 6 \, (7.5) - 4 \, (5) = \mathbf{25 \text{ V} \, (Ans)}$$

1.22 TUTORIAL ILLUSTRATIONS II

Illustration TII-1

Evaluate the net resistance across A-B of the circuit of Fig. 1.47. All resistors in ohms.

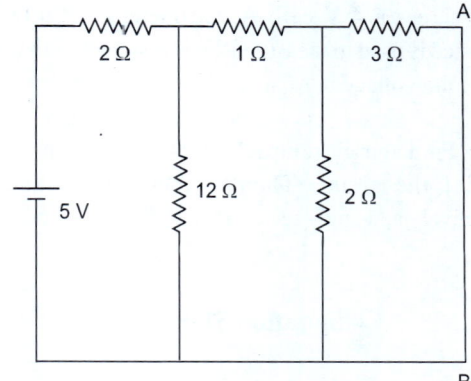

Fig. 1.47. Circuit for illustration TII-1

Illustration TII-2

Three resistors of 4.2, 6.3 and 16.8 Ω are connected in parallel. If the total current taken is 4.6 A, find out current through each resistor.

(Hint: Refer to illustration 1.19).

Illustration TII-3

Three resistors 18 Ω, 16 Ω are R Ω are in parallel. The current through the R Ω resistor is 1 A and the total circuit current is 15 A. Find (i) the current in each of the other resistors (ii) the value of R.

(Hint: Refer to illustration 1.21 and solve)

Illustration TII-4

The resistance of certain meter is 720 Ω. Find the resistance of the short, so that one-hundredth of the total current will pass through the meter

(Ans. 7.27 Ω).

Illustration TII-5

The resistors of 100 Ω and 200 Ω are connected in series across a 4 V cell. A volt meter of 200 Ω resistance is used to measure P.D. across each. What will be the voltage in each case?

(Hint: Find the current delivered by the source without the shunt. Then connect the shunt respectively and apply the current division method.)

Illustration TII-6

The resistors of value 4 and 5 Ω are connected in parallel and this combination is in series with a 2 Ω resistor. The circuit is connected to a 15 V battery. Determine (a) the current supplied by the battery (b) the current in each resistor [**Ans:** (a) 3.6 A, (b) $I_{2\,\Omega} = 3.6$ A, $I_{4\,\Omega} = 1.97$ A, $I_{5\,\Omega} = 1.56$ A].

Illustration TII-7

Find the current through the 3 Ω resistor in the given circuit of Fig. 1.48.

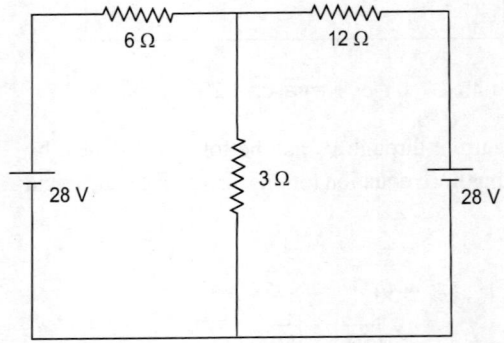

Fig. 1.48. Circuit for illustration TII-7

(Reader may proceed taking assistant from the illustration 1.23)

Illustration TII-8

Two batteries are connected in parallel and resistance of 1.5 Ω is connected across the terminals. Battery No. 1 has an emf of 120 V and an internal resistance of 0.1 Ω and battery No. 2 has emf of 110 V and internal resistance of 0.2 Ω. Calculate the current supplied by each battery and the power dissipated in 1.5 Ω resistor.

ALTERNATING CURRENT

PART II: AC FUNDAMENTALS

1.23 INTRODUCTION

It is brought out in (sec. 1.20) that a current which, without depending on time, maintaining same magnitude of current, indefinitely, is a DC.

Another type of current used in electrical history is the Alternating Current, abbreviated simply as an AC. Contrary to DC, an AC will have different magnitude of current (or voltage) at different time instants. Thus it becomes a **time-dependent**.

Few types of AC waves are presented in Fig. 1.49.

In a circuit, the alternating voltage source is denoted by the symbol, (~).

The x-axis is always considered to be the time-axis and the y-axis is considered to be the magnitude axis. The magnitude axis may contain current values, if the wave is a current wave; may contain voltage values, if the wave is a voltage wave; and so on.

Note: It has become an international practice to consider the sine wave for analysing the characteristics of an alternating circuit. Hence, this book will consider such wave to understand the essence of an alternating source fed circuit.

1.24 A SINUSOIDAL WAVE

Movement of electric conductors in a magnetic field induces an emf in it.

Refer to Fig. 1.50. Let one of the N conductors, say 'a', be considered for analysis and its movement in the magnetic field be traced. Let the rotation be clock-wise. In doing so, it can be seen that:

- When the conductor **a** is in position (i), the flux linked with it is maximum; but, here as the side **a'** and **a"** are parallel to the flux lines, the rate of change of flux linkage will be minimum (thus, a zero induced emf)
- When the conductor moves towards position (ii) the flux linkage will gradually decrease; but the rate of change of flux linkage will increase. Hence the induced emf in the conductor will gradually increase.
- At position (iii), the flux linkage with the conductor **a** reaches a minimum; but, here, as **a'** and **a"** move perpendicular to the flux lines, rate of change of flux linkage will be maximum (thus, the maximum induced emf).

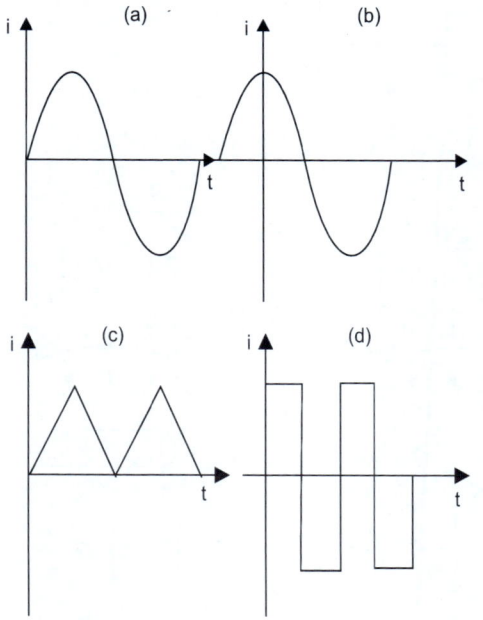

Fig. 1.49. Different AC waves
(a) A sine wave, (b) A cosine wave (c) A triangular wave (d) a rectangular wave

36 Basic Electrical Engineering

- To say precisely, when the flux linkage (Nφ) is maximum rate of change of flux linkage [d/dt (Nφ)] will obviously be minimum and vice-versa.

This clearly brings out the fact that, the conductor 'a' is experiencing a change in flux linkage. Thus an emf will be induced in it. At position (i) minimum flux linkage hence a zero emf, which will increase gradually reaching a maximum at position (iii). The above analysis covers 0 to 90°. From greater than 90° to 180°, the flux linkage with the conductor will gradually decrease, thus reducing the emf in it. At 180°, the flux linkage will become the least hence the induced emf will become zero. Observe Figs. (iv), (v).

Thus

$0 \rightarrow 90°$ = emf increase from 0 to positive maximum

$90° \rightarrow 180°$ = emf falls again to zero, from this maximum

The reader may analyse further in a similar way to observe the following facts.

$180° \rightarrow 270°$ = emf in the reverse path approaches a maximum

$270° \rightarrow 360°$ = emf reaches a zero from the negative path.

All these happen during a certain time.

Note: What happened between 0 to 180° is very similarly, happening in 180 to 360°, in reverse manner; it needs a keen observation now. From Figs. (i) to (v) side **a'** was in the influence of the North pole, which gives out flux (and side **a"** was in the influence of the South pole, which takes in flux). This covers 0 to 180°. It can readily also been seen that in the other 180° (i.e. from 180° to 360°) side **a'** in the influence of South pole; hence the entire action discussed above would be reversed. Thus the induced emf, would also be reversed.

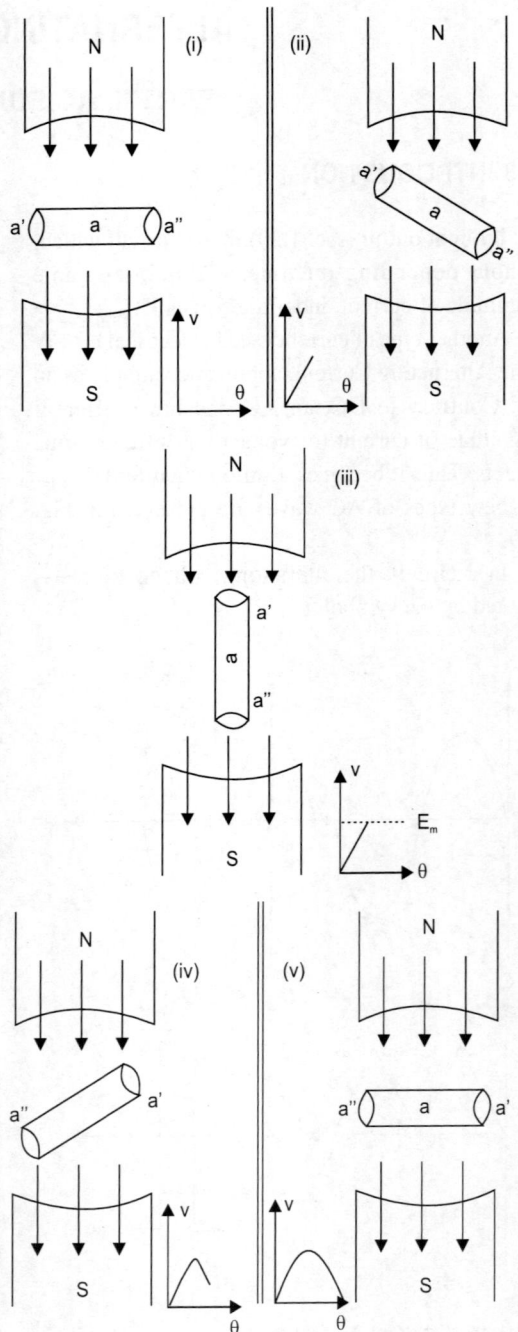

Fig. 1.50. Illustrative diagrams to discuss generation of sinusoidal voltage

Basic Electricity 37

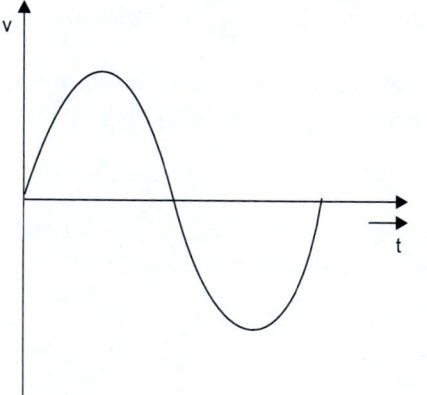

Fig. 1.51 EMF induced in the conductor 'a' in one complete rotation (0 to 360°) of it in the magnetic field.

The path of emf between 0 → 180° is shown in Fig. 1.50. The reverse action exactly takes place from 180° to 360°. As a whole the emf induced in the conductor 'a' for one complete rotation is in Fig. 1.51.

Such an emf is known as the **sinusoidal** or **sine** or **simple harmonic EMF**.

It is derived in equation (2.24) that the emf induced in a coil (or conductor) is Bl v sin θ. The flux density B, the length of the conductor l, the velocity at which it rotates v are constants (provided they are not distorted.) Thus the voltage equation (2.24) can be written as,

$$e = Bl\, v\, \sin\theta$$

$$= Bl\, \omega\, \sin\omega t, \text{ in terms of angular velocity, } \omega$$

Therefore,

$$e = E_M \sin \omega t \qquad (1.30)$$

E_M is the maximum value of emf experienced by the conductor during a complete 360° rotation or this is the maximum emf that could be induced in the conductor.

e is the instantaneous value of the emf at a time t.

Similarly the current equation can be written as,

$$i = I_m \sin \omega t \qquad (1.31)$$

The equations (1.30) and (1.31) will satisfy the Fig. 1.51. For instance, taking equation (1.30),

at θ = 0   e = 0

at α = 90°   e = E_m (i.e, maximum)

at θ = 180°   e = 0

at θ = 270°   e = – E_m (i.e., maximum in the negative side)

at θ = 360°   e = 0

Incorporating these values, the Fig. 1.52 is arrived at.

Of course, if the conductor is allowed to rotate continuously in the magnetic field, sinusoidal voltage would also induce continuously. It is appearing in Fig. 1.53.

It can be seen from Fig. 1.51 that the induced emf of the conductor 'a' for one revolution, is getting repeated in the other revolutions also **without any change** [p – c, voltage in one revolution q – r, voltage in the other revolution; r – s voltage in the next

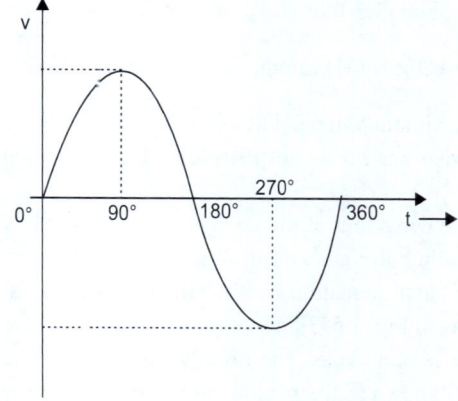

Fig. 1.52. A Sinusoidal wave with different emf's marked at the respective times instants.

38 Basic Electrical Engineering

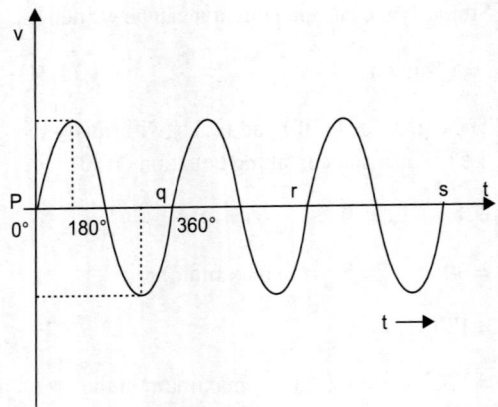

Fig. 1.53. EMF induced in a conductor, set into motion continuously.

successive revolution. Note that there is no change in the shape in these periods.)

As the shape of the sinusoidal wave (and hence the emf induced) are repeated again and again maintaining every time a constant period (like p – q q– r, r – s etc.) such wave is otherwise known as a **periodic wave**.

1.25 DIFFERENT TERMS ASSOCIATED WITH A SINUSOIDAL WAVE

By equation (1.30) or (1.31) it is known that,

Instant value = (Maximum value) × sin θ (1.32)

The maximum value is known in Fig. 1.52. This is otherwise known as **amplitude** of the sinusoidal wave.

The behaviour of the voltage wave obtained by substituting all values from 0 to 360° (i.e., 0 to 2π radians) in the equation (1.30), will give a sinusoidal wave as in Fig. 1.52; such one complete alternative is known as a cycle. The time taken to cover one cycle is known as the **period**, and is denoted by the symbol T.

If there are f such cycles covered in one second, then the f is said to be the **frequency** of the sinusoidal wave. Its unit is Hertz (Hz), or cycles per second C.P.S.

The relation between the angular velocity ω and the frequency f is given by the equation (1.33)

$$\omega = 2\pi f, \text{ radians} \qquad (1.33)$$

and the relation between the time period T and the frequency f is given by the equation (1.34)

$$T = \frac{1}{f} \text{ second} \qquad (1.34)$$

Substituting equation (1.34) in equation (1.33)

$$\omega = \frac{2\pi}{T} \qquad (1.35)$$

The above sinusoidal terms are placed in the Fig. 1.54.

Note: In the equation (1.30) and (1.31) the arguments of the 'sine' function—i.e., ωt or θ—can be in degrees or radians. But usually, as far as ω is concerned, it is given in radians. Thus care must be taken to solve a problem either in degree fully or radians.

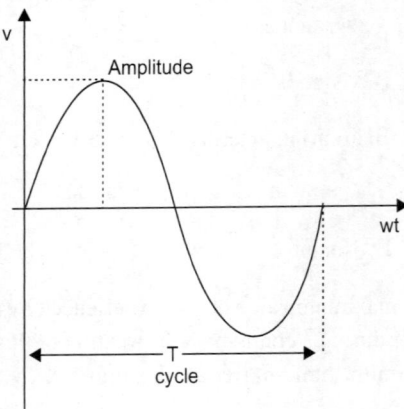

Fig. 1.54.

1.26 MEASUREMENT OF ALTERNATING QUANTITIES

There are notations by which an alternating quantity can be designated. They are
   (i) the maximum value
   (ii) the average value and
   (iii) the root mean square value

The treatment of the maximum value was explained in the previous section. The definitions of other quantities and their relation with the maximum value are brought out in this section.

1.27 AVERAGE VALUE

It is the average of mean of all the values determined over a half cycle. It gives the mean amplitude for a sinusoidal wave.

(It is of no meaning in finding out the average value for the whole cycle (0 to 360°) because what ever happened in positive half cycle, would have happened without any change, in the negative half cycle, whose average will obviously be a zero.)

Consider the current waveform of Fig. 1.55. Let a small strip of height i and width $d\theta$ be considered.

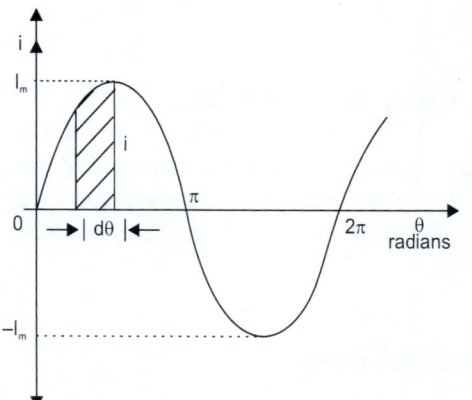

Fig. 1.55. Waveform illustration for average value discussion.

As stated, the average value is the mean of all values taken over half-a-cycle.

Suppose 5 different values of i_1, i_2, i_3, i_4 and i_5 are taken in the positive half cycle, its average will be,

$(i_1 + i_2 + i_3 + i_4 + i_5)/5$.

But, for a sine wave, by mathematics, the average value is,

$$I_{av} = \frac{\text{area of half wave}}{\text{base}} \qquad (1.36)$$

$$= \frac{1}{\pi}\int_0^\pi i\, d\theta$$

As it is a sine wave, by equation (1.32)

$i = I_m \sin \theta$. Thus,

$$I_{av} = \frac{1}{\pi}\int_0^\pi I_m \sin\theta\, d\theta$$

$$= \frac{I_m}{\pi}[-\cos\theta]_0^\pi$$

which gives,

$$I_{av} = \frac{2I_m}{\pi} \qquad (1.37)$$

Similarly if it is a sine voltage,

$$E_{av} = \frac{2E_m}{\pi} \qquad (1.38)$$

So, in general,

$$\text{average} = \frac{2}{\pi} \text{ times the maximum value} \qquad (1.39)$$

1.28 ROOT MEAN SQUARE (RMS) VALUE

RMS value is the root of mean of squared values over a full cycle. (rms value over a full cycle is meaningful as we square the values). It is otherwise known as the effective value.

As the full cycle is considered, the equation for the rms current will be, in analogy with equation (1.36),

$\dfrac{1}{\text{base}}$ rms of full wave

$$I_{rms} = \sqrt{\dfrac{1}{2\pi}\int_0^{2\pi} i^2 d\theta}$$

Squared radius

Mean

Root of mean of squared values (rms)

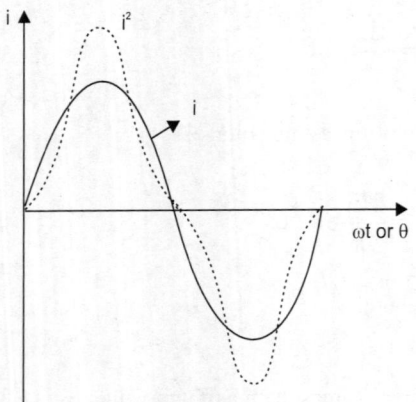

Fig. 1.56. A sine wave and its square.

Letting again, $i = I_m \sin\theta$

$$I_{rms} = \sqrt{\dfrac{1}{2\pi}\int_0^{2\pi} I_m^2 \sin^2\theta\, d\theta}$$

$$= \sqrt{\dfrac{I_m^2}{2\pi}\int_0^{2\pi} \sin^2\theta\, d\theta}$$

$$= \sqrt{\dfrac{i_m^2}{2\pi}\int_0^{2\pi} \dfrac{(1-\cos 2\theta)}{2}\, d\theta}$$

($\mathbf{Q} \cos 2\theta = 1 - 2\sin^2\theta$)

which will give, $I_{rms} = \dfrac{I_m}{\sqrt{2}}$ \hfill (1.40)

Similarly for a voltage wave

$$E_{rms} = \dfrac{E_m}{\sqrt{2}} \quad (1.41)$$

In general,

rms value = $\dfrac{1}{\sqrt{2}}$ times the maximum value (1.42)

1.29 MEASURING FACTORS

Having defined the rms and average values for a sinusoidal wave in terms of the maximum value, relating them the following factors are introduced. They are:

(a) The form factor

It is the ratio of rms value to the average value. It is denoted by f.f. Thus by using equation (1.37) and equation (1.40), form factor = $\dfrac{I_{rms}}{I_{av}} = \dfrac{I_m/\sqrt{2}}{2I_m/\pi} = \dfrac{I_m}{\sqrt{2}} \cdot \dfrac{\pi}{2I_m}$

$\therefore$ f.f. = $\dfrac{\pi}{2\sqrt{2}} = 1.11$

form factor = 1.11 (1.43)

(b) Peak factor

It is the ratio of maximum value to the rms value. It is also known as maximum factor or amplitude factor.

Peak factor = $\dfrac{I_m}{I_{rms}} = \dfrac{I_m}{I_m/\sqrt{2}} = \sqrt{2}$

Peak factor = $\sqrt{2}$ (1.44)

Note: The rms value and the average value are derived by considering a pure sinusoidal wave. Hence the f.f (= 1.11) and peak factor (= $\sqrt{2}$) are valid only for a pure sinusoidal wave. If the wave form shape changes to any other shape as in Fig. 1.49, then these values will also change.

1.30 REPRESENTATION OF ALTERNATING QUANTITIES

There are several ways by which an AC quantity can be described. To quote a few, (i) the phasor representation (ii) polar representation (ii) wave form representation, etc.

Such representations are the heart of AC analysis because the voltage and current across different elements **may** not pass through—the zero, positive maximum etc.—at the same time instant. They will be displaced by a definite time. This is one of the main characteristics of an AC circuit and is to be accounted for.

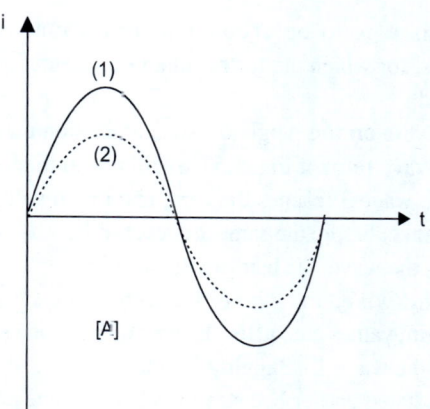

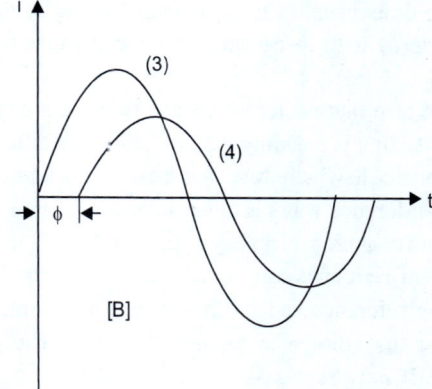

Fig. 1.57. (A) In-phase wave forms and (B) Out of phase wave forms.

Observe Fig. 1.57-A. The wave forms, 1 and 2 pass through zero, positive maximum, negative maximum etc., at the same time instant in every cycle. (of course, their magnitude can be different). Such waves are said to be **in-phase**.

Observing Fig. 1.57-B, it can readily be seen that the waves 3 and 4 **do not** pass through the zero, positive maximum etc., at the same time instant but at different time instants. Such waves are said to be **out-of-phase** by the respective angle, ϕ degrees or radians.

42 Basic Electrical Engineering

This φ is to be accounted throughout the analysis, for which the forthcoming representations are used.

A Note on the Angle φ: An examination made on the wave form of Fig. 1.57-B will reveal the fact that, the wave 3 reaches the zero, maximum values in **advance**, before the same are reached by wave 4. That is, the wave 3 is **leading** the wave 4.

Otherwise, the wave 4 reaches the zero, maximum values etc., **after** the wave 3 has done so. That is the wave 4 is **lagging** behind wave 3.

As stated earlier because the voltage wave and the current wave of an element need not pass through the same time instants, the lagging or leading angle (i.e. angle φ) with respect to a reference **must be** notified.

Note also that declaring a wave is lagging or a wave is leading is meaningless unless it is specified with reference to which wave, it is lagging or leading. Hence a reference wave is always needed. Usually, the quantity which remains common for all the elements in a circuit is considered to be the reference. Thus, the reference can be the current in a series circuit or the voltage in a parallel circuit of any element; if not the x-axis is considered to be the reference axis.

As a summary,
- a leading alternating quantity is one which reaches its maximum, zero value etc., earlier as compared to other quantity.
- A lagging alternating quantity is one which reaches its maximum, zero etc., later than the other quantity.

1.31 VECTOR REPRESENTATION

A vector representation is a line which represents the magnitude of the quantity and whose inclination to the reference axis is proportional to the time elapsed after passing its zero value.

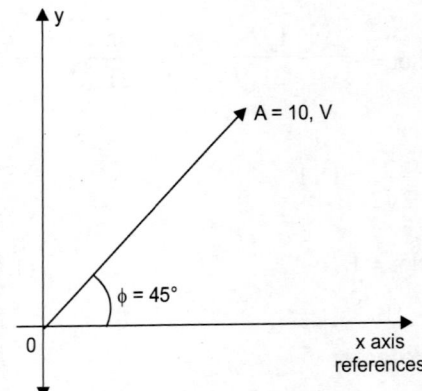

Fig. 1.58. Vectorial representation.
OA = 10 volts. Leading the reference by 45°.
OA is placed in anticlockwise rotation from the reference as it is leading.

Let there be a voltage of magnitude 10 V leading the reference vector by 45°.

Usually, the anticlockwise rotation of the vector from the reference axis will represent the leading nature and the clockwise rotation of the vector from the reference axis will represent the lagging nature.

Thus the representation of the 10 V by vector form will be as in Fig. 1.58.

Let another case be considered in which a current I_1 of 20 A leads the reference by 45° and a current of 15 A lags the reference by 10°. The vectorial representation of this is in Fig. 1.59.

This case will be considered a general case in the other types of representations also, in the forthcoming discussions.

1.32 POLAR AND RECTANGULAR REPRESENTATIONS

The general form of polar representation is a $\angle \pm \phi$, where 'a' is the magnitude of the quantity concerned and φ the phase difference between this quantity and the reference vector in which a '+' is prefixed if φ is

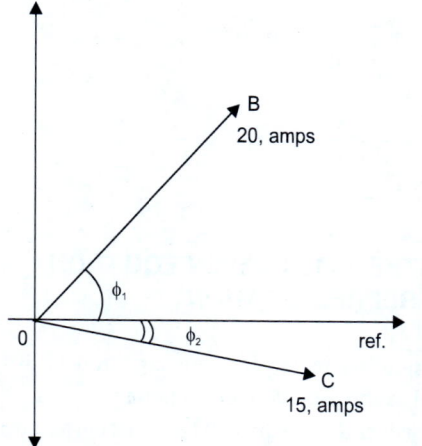

Fig. 1.59. Vectorial representation.
OB = 20 amps, leading the reference by 45° (= I1)
OC = 15 amps, lagging the reference by 10° (= I2). Note that this is placed in clock-wise direction from the reference, as it is lagging.

a leading angle and a '–' is prefixed if ϕ is a lagging angle.

Thus for Fig. 1.59, the polar representation will be

$$I_1 = 20 \angle + 45°$$

and

$$I_2 = 15 \angle -10°$$

This polar representation of a $\angle + \phi$ can be split into two components, namely, a real and an imaginary component. Real part is **a cos** ϕ and the imaginary part is **a sin** ϕ and are related by the operator 'j'.

In general, the assembled form of the real and imaginary parts will be,

a cos ϕ + j a sin ϕ.

This form is the rectangular form. The operator j denotes that, the y component is 90° ahead of the x-component.

Thus the above polar representation of I_1 and I_2 can be put in rectangular form, as given below:

For I_1:

$$I_1 = 20 \cos(45) + j\, 20 \sin 45$$
$$= 14.14 + j\, 14.14$$

for I_2:

$$I_2 = 15 \cos(-10) + j\, 15 \sin(-10)$$
$$= 14.77 - j\, 2.6$$

Note 1: As a polar form is converted into a rectangular form, a rectangular form also can be equivalently transferred into a polar form.

A rectangular form of a + jb could be transferred as C $\angle \theta$, where

$$C = \sqrt{a^2 + b^2} \text{ and } \theta = \tan^{-1} b/a.$$

We may try this conversion with the rectangular representation of I_1 and I_2 given above and check whether they are respectively 20 $\angle 40°$ and 15 $\angle -10°$.

Note 2: When alternating quantities, are added or subtracted, they could be kept in rectangular form;

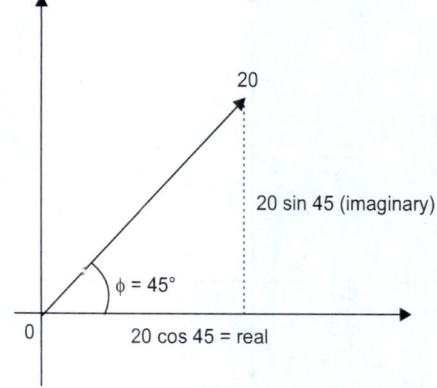

Fig. 1.60. Rectangular representation—description.

the operation becomes easy then. For example $(a + jb) + (c + jd)$ is $(a + c) + j(b + d)$. That is, the real parts are added together; and the imaginary parts are added together.

When alternating quantities are multiplied or divided, it is suggested to keep them in polar form. Let there be two alternating quantities, a $\angle \theta_1$ and b $\angle \theta_2$. If they are multiplied together then the result will be ab $\angle \theta_1 + \theta_2$; if they are divided, the result will be (a/b) $\angle \theta_1 - \theta_2$.

Note 3: Significance of operator j:
- j is an operator suggesting that the vector is rotating 90° ahead from x-axis, in anticlockwise direction.
- Its value $\sqrt{-1}$, denoting imaginary.
- If the operator j is operated once on a vector 'a', it will give **ja** as a new vector, kept ahead in anticlockwise direction by 90°. If it is operated twice, new vector will be $j^2 a$, displaced by 180° a head and so on. (Refer to Fig. 1.61).

$$j^2 = (\sqrt{-1})^2 = -1;$$

$$j^3 = (\sqrt{-1})^3 = \sqrt{-1}(\sqrt{-1})^2 = -j$$

$$j^4 = j^2 \cdot j^2 = 1$$

$$j^5 = j^4 \cdot j = j \text{ and so on.}$$

and $\dfrac{1}{j} = \dfrac{j}{j^2} = \dfrac{j}{(-1)} = -j$

1.33 THE TIME-DOMAIN EQUATION REPRESENTATION

The equation in the form of equation (1.30) are known as the time-domain equations.

In general its shape will be, (for a current wave),

$$i = I_m \sin[\omega t \pm \phi],$$

with a +, if the wave leads the reference and a –, if the wave lags the reference.

Thus the I_1 and I_2 of Fig. 1.59 can be placed in the form of time-domain equation as

$$i_1 = I_{m_1} \sin \overline{\omega t + 45°}$$

and $i_2 = I_{m_2} \sin \overline{\omega t - 15°}$

1.34 THE WAVEFORM REPRESENTATION

The lagging and leading nature of the concerned sinusoidal waveform with respect to the reference will be indicated in a common graph. Such representation is the wave form representation. In such representation, by giving a look at the graph, the nature and behaviour of the waveforms can be analysed.

Referring to the note given under sec. (1.30), and considering Fig. 1.59, the I_1 and I_2 can be placed in a wave form notation as given in Fig. 1.62. Here a wave, as shown in dotted line, is considered to be the reference wave.

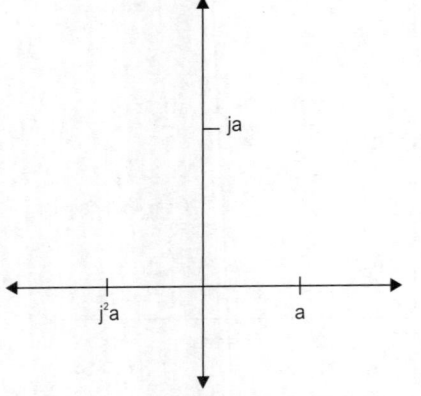

Fig. 1.61. The movement of operator j(a), j²(a) etc.

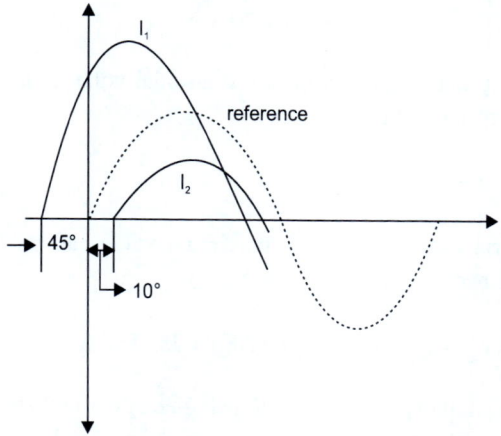

Fig. 1.62. Waveform representation.
I_1 leads the reference by 45°.
I_2 lags the reference by 10°.

1.35 WORKED ILLUSTRATIONS III

Illustration 1.30

What is the peak value of a sinusoidal alternating current of 4.78 rms A. If the periodic time of this wave in 0.03 sec, find the frequency and the angular velocity.

Solution

Given the rms = 4.78 A,

Using the peak factor relation for a sunsioidal quantity from equation (1.44).

Peak factor $= \dfrac{I_m}{I_{rms}} = \sqrt{2}.$

∴ The peak value $I_m = \sqrt{2}\ I_{rms} =$ **6.78 A (Ans.)**

By equation (1.34), the time period T and frequency f are related as, T = 1/f. Thus f = 1/T = 1/0.03 = **33.33 Hz (Ans)**.

The angular velocity ω, using equation (3.4) is $2\pi f$; that is $2\pi (37.33) =$ **209 rad/sec. (Ans.)**

Illustration 1.31

A current wave is represented by the equation i = 10 sin 250 t.
What are the
(i) maximum value
(ii) rms value of current.
(iii) frequency
(iv) the time period.

Solution

(i) The general notation for representing an alternating quantity is given is equation 1.32). Thus the maximum value is **10 A (Ans.)**

(ii) For a sine wave, peak factor is maximum value/rms value and is equal to $\sqrt{2}$. [Refer to equation (3.15).] Thus the rms value is, maximum value/$\sqrt{2}$; i.e., $I_{rms} = 10/\sqrt{2} =$ **7.7 A (Ans.)**

(iii) From equation (1.33), $\omega = 2\pi f$. Comparing the given current equation i, with the equation (3.1), it can readily be seen that, ω = 250. That is,

$$\omega = 2\pi f = 250.$$

Thus,

Frequency, $f = 250/2\pi$

$$= \textbf{39.78 Hz. (Ans.)}$$

(iv) By equation (1.34), the time period T = 1/f

$$= \dfrac{1}{39.78} = \textbf{0.025 sec (Ans.)}$$

Illustration 1.32

If a voltage waveform has a form factor of 1.15, a peak factor of 1.5 and a peak value or 5.6 kV, calculate the average and rms values of the voltage waveform. Will this wave be a sinusoidal wave? Justify.

Solution

Given that V_m = 5.6 kV and form and peak factors respectively 1.15 and 1.5. From the equations (1.44), the rms value can be written as,

$$V_{rms} = \frac{V_m}{\text{peak factor}}$$

$$= \frac{5.6}{1.5} = 3.73 \text{ kV (Ans.)}$$

Similarly, from equation (1.43),

$$V_{av} = \frac{V_{rms}}{f.f}$$

$$= \frac{3.73}{1.15} = 3.2464 \text{ kV (Ans.)}$$

It can't be a sinusoidal wave because for a sinusoidal wave the form factor and peak factors are not 1.15 and 1.5 respectively, as given.

Illustration 1.33

A sinusoidal alternating voltage has an rms value of 200, volts and a frequency of 50 Hz. It crosses the zero axis in the positive direction at time t = 0. Determine
  (i) the time when voltage first reaches the instantaneous value of 200 volts

Solution

The general equation of any sinusoidal voltage is, by equation (1.32),

$$v = V_m \sin \omega t$$

As rms value is known, maximum value can be obtained by equation (1.44); i.e.

$$V_m = \sqrt{2} \; V_{rms} = \sqrt{2} \; (200) = 282.8 \text{ V}.$$

Similarly as f is known, using equation (1.33) the angular velocity ω can be obtained; i.e., $\omega = 2\pi f$

$$= 2\pi (50) = 313.16 \text{ rad/sec}.$$

Thus equation (1.32), for this problem, is:

$$v = 282.8 \sin 314.4 \; t \qquad \text{(I)}$$

(i) When the instantaneous value v = 200 V, the equations I becomes,

$$200 = 282.8, \sin 314.16 \; t$$

Thus t is, 0.1433 sec. Thus the 200 V occurs when the time **t = 0.1422 sec. (Ans.)**

Note: In sections 1.27 and 1.28, the average and rms values for a sine wave were evaluated. The same values for other types of periodic waves, like appearing in Fig. 1.49 are evaluated in the below illustrations. Readers are thus expected to be clear with section 1.27 and 1.28, because the same fundamental procedure is still adopted here also.

Illustration 1.34

Find the average and effective values of the sawtooth wave shown in Fig. 1.63.

Basic Electricity 47

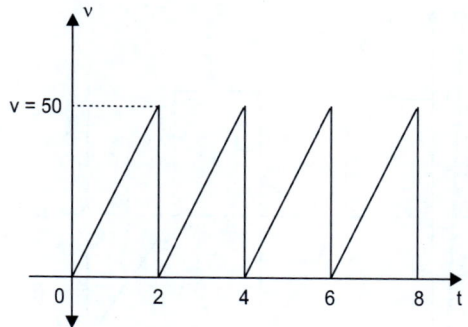

Fig. 1.63. Waveform for illustration 1.5.

Solution

As the equation of a sinusoidal wave was written and solved for average, effective values, here again the equation for this saw-tooth wave must be written.

Observe keenly that this a periodic wave, repeating **similarly** for every 2 width of x axis.

Referring to wave form of Fig. 1.64 and taking assistance from mathematics, the equation of the wave, i.e., the equation of OP is:

$$\frac{x - x_1}{x_1 - x_2} = \frac{y - y_1}{y_1 - y_2}$$; here x is t axis and y is v axis.

Thus,

$$\frac{t - 0}{0 - 2} = \frac{v - 0}{0 - 50}$$

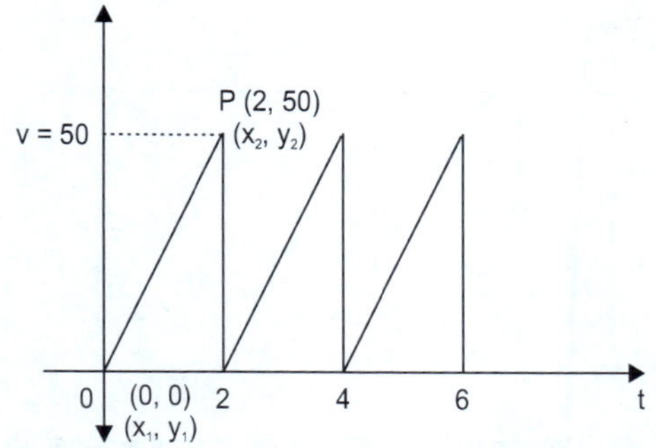

Fig. 1.64. Solution wave form for illustration 1.5.

$$\frac{t}{-2} = \frac{v}{-50}$$

or v = 25t; it is the equation.

Average value:

V_{av} = 1/base area of the curve. (By equation (3.7),)

(*Note:* here the full curve is taken, not half-curve, because there is no negative cycle).

$$V_{av} = \frac{1}{2} \int_C^c v \, dt$$

∴

$$= \frac{1}{2} \int_0^2 [25t] \, dt$$

Thus,

$$V_{av} = \frac{1}{2} \left[\frac{25}{2} (2^2 - 0^2) \right] = \textbf{25 V (Ans.)}$$

rms value: (By section (1.28),)

$$V_{rms} = \sqrt{\frac{1}{base} \int_0^{base} v^2 \, dt}$$

$$= \sqrt{\frac{1}{2} \int_0^2 v^2 \, dt}$$

$$= \sqrt{\frac{1}{2} \int_0^2 625 t^2 \, dt}$$

$$= \sqrt{\frac{1}{2} \cdot \frac{625}{3} [2^3 - 0^3]}$$

$$V_{rms} = \sqrt{833.33} = \textbf{28.86 V (Ans.)}$$

48 Basic Electrical Engineering

Note 1: Thus, only the understanding of the wave form matters; and then by taking the assistance of mathematics, equation **can** be written. Thereafter, the procedure is the same.

Note 2: The form factor, peak factor etc., can now be found out for this saw-tooth wave, As a trial,

$$\text{f.f} = \frac{\text{rms value} 25}{\text{average value}} = 28.86 = 1.15$$

Note: It was concluded in Sec. 1.29 that the f.f (and peak factor) for different waves will be of different values. In this section it is proved that for a sine wave. f.f. = 1.11. For this saw-tooth wave we see that f.f. = 1.15

Illustration 1.35

The reader may try this to evaluate the average and rms values of the saw-tooth wave if the 0 (0, 0) is moved to 0 (0, 25). Let the amplitude be 25 V.

(**Ans:** Equation is $v = 25 + 12.55\,t$ average value is, **37.5 V**)

Illustration 1.36

Calculate the rms and the average values for the voltage wave of shape shown in Fig. 1.65.

Solution

The wave has two segments; 0 to 1 and 1 to 2; and these two segments periodically repeat for base of 2. Thus there will be two equations. One for the positive part (i.e. $0 < t < 1$) and the other for the negative part (i.e., $1 < t < 2$).

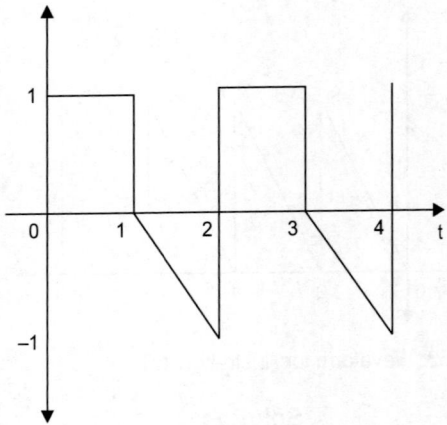

Fig. 1.65. Waveform for illustration 1.36.

Equation: In the time range 0 to 1, the voltage wave is just like a DC, maintaining a constant magnitude of 1 V, irrespective of time.

$$\therefore V = 1 \text{ is the equation} \qquad (A)$$

Using the Fig. 1.66, the equation for the negative portion can be found out.

Equation of the line **ab** is,

$$\frac{t-1}{1-2} = \frac{v-0}{0-(-1)}$$

$$\frac{t-1}{-1} = \frac{v}{1}$$

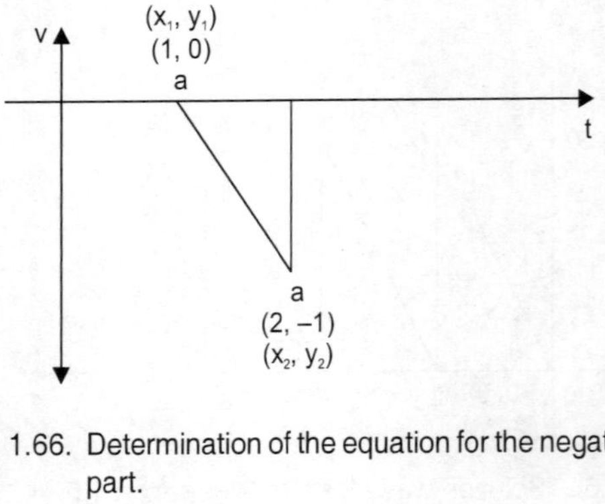

Fig. 1.66. Determination of the equation for the negative part.

$$-v = (t-1)$$

or $v = 1 - t$ \hfill (B)

The full equation of the wave of Fig. 1.65 is, from equations (A) and (B)

$v_1 = 1$ \hfill $0 < t < 1$

$v_2 = (1-t),$ \hfill $1 < t < 2$

Average value:

By equation (3.36)

$$V_{av} = \frac{1}{2}\int_0^2 v(t)\,dt$$

$$= \frac{1}{2}\left[\int_0^1 v_1\,dt + \int_1^2 v_2\,dt\right]$$

$$= \frac{1}{2}\left[\int_0^1 1\,dt + \int_1^2 (1-t)\,dt\right]$$

$$= \frac{1}{2}\left[(t)_0^1 + \left(t - \frac{t^2}{2}\right)_1^2\right]$$

$$= 0.25, \text{ volts (Ans.)}$$

rms value:

$$V_{av} = \sqrt{\frac{1}{2}\left[\int_0^2 v_1^2\,dt + \int_1^2 v_2^2\,dt\right]}$$

or

$$V_{rms}^2 = \frac{1}{2}\left[\int_0^2 1^2\,dt + \int_1^2 (1-t)^2\,dt\right]$$

$$= \frac{1}{2}\left[(t)_0^1 + \left(t - 2\frac{t^2}{2} + \frac{t^3}{3}\right)_1^2\right]$$

$$V_{rms}^2 = 0.166 \text{ V}$$

$$\therefore V_{rms} = \sqrt{0.166} = \mathbf{0.41 \text{ V (Ans.)}}$$

Illustration 1.37

Reader may try when the magnitude of the amplitude is changed to 4 in the waveform of Fig. 1.65.

And: $v_1 = 4$ \hfill $0 < t < 1$

$v_2 = -4t + 4,$ \hfill $0 < t < 2$

$V_{rms} = 3.265$ V

$V_{av} = 1$ V.

Illustration 1.38

Two currents of each 12 A and 20 A are available. The former current is in phase with the voltage and the latter is lagging it by 30°. Find the resultant sum current.

Solution

Reference sec. 1.32.

In polar form, $I_1 = 12\angle 0°$ and $I_2 = 20\angle -30°$.

As I_1 and I_2 are to be added, let them be converted into rectangular forms.

$I_1 = 12\cos 0 + j\,12\sin 0 = 12 + j\,0$

($\theta = 0°$, as I_1 is in phase with the reference voltage)

$I_2 = 20\cos(-30) + j\,20\sin(-30) = 17.32 - j\,10$.

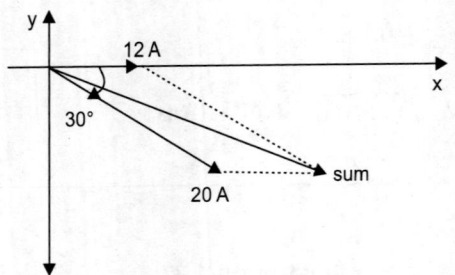

Fig. 1.67. Vectorial solution for illustration 1.9.

($\theta = -30°$, as I_2, is lagging the reference by 30°)

$\therefore I_1 + I_2 = (12 + 17.32) + j\,(0 + (-10))$

$\qquad = 29.32 - j\,10$ **A (Ans.)**

Note: This can also be solved vectorially as shown in Fig. 1.67.

Illustration 1.39

In the equation $V_m = V - IR$, $V = 100 \angle 0°$ V, $R = 10 \angle 0°\;\Omega$ and $I = 8 \angle -30°$ A. Express V_m in polar form.

Solution

R and I are to be multiplied which is to be subtracted from V. Thus two mathematic operations are to be made.

(Refer to note 2 of Sec. 1.32)

$RI = (10 \angle 0)(8 \angle -30) = 80 \angle -30$ V

$V - IR = 100 \angle 0 - 80 \angle -30$.

While adding or subtracting, it is suggestible to keep in rectangular form. So, $100 \angle 0$ in rectangular form will be, $100 \cos 0 + j\,100 \sin 0 = 100 + j0$.

$80 \angle -30$ is rectangular form will be, $80 \cos(-30) + j80 \sin(-30) = 69.28 - j40$. So,

$V - IR = (100 + j0) - (69.28 - j40)$

$\qquad = (100 - 69.28) + j\,(0 - (-40))$

$\qquad = \mathbf{30.72 + j\,40}$ **(Ans.)**

In polar form:

Magnitude = $\sqrt{30.72^2 + 40^2} = 50.43$

Angle = $\tan^{-1} 40/30.72 = 52.47$

$\therefore \mathbf{V - IR = 50.43 \angle 52.47}$ **(Ans.)**

Illustration 1.40

Find the rms value of the resultant current in a wire which carries simultaneously a direct current of 10 A and a sinusoidal alternating current of peak value 10 amps. What will be the average value.

Solution

The wave forms are in Fig. 1.68. The resultant wave is obtained by adding the DC and AC components at each time instant, and plotting the sum at the respective time instant.

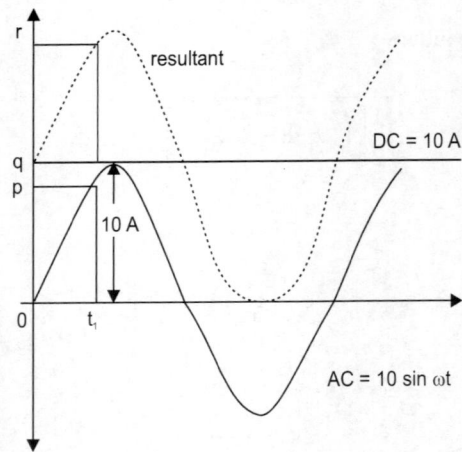

Fig. 1.68. Solution waveform for illustration 1.40.

That is at time instant t_1, DC component is 10 amps. (of course, at all the time instants it will be the same, because it is independent of time) let this be **oq**. The AC component is, **op**. The summation of these two is **or** at this same time instant t_1.

Similar summation is carried out for the whole cycle at different time instants to get the resultant wave, as shown in Fig. 1.68.

The general equation:

DC: As it is independent of time, $i_{DC} = 10$.
AC: By using the equation (1.32), $i = I_m \sin \omega t = 10 \sin \omega t$.

Thus the resultant current will be,

$$i_r = 10 + 10 \sin \omega t = 10 + 10 \sin \theta \quad (I)$$

RMS value of the resultant wave

The general form of equation for the rms value is

$$I_{rms} = \sqrt{\frac{1}{2\pi}\int_0^{2\pi} i^2 \, d\theta}$$

or

$$I_{rms}^2 = \frac{1}{2\pi}\int_0^{2\pi} i^2 \, d\theta$$

$$= \frac{1}{2\pi}\int_0^{2\pi}(10+10\sin\theta)^2 \, d\theta$$

$$= \frac{1}{2\pi}\left[\int_0^{2\pi}(100+100\sin^2\theta + 200\sin\theta)d\theta\right]$$

$$= \frac{1}{2\pi}\left[\int_0^{2\pi}100 \, d\theta + \int_0^{2\pi}100\sin^2\theta \, d\theta + \int_0^{2\pi}200\sin\theta \, d\theta\right]$$

$$= \frac{1}{2\pi}[100(2\pi - 0) + 100(\pi) + 200(0-0)]$$

$$I_{rms}^2 = 150$$

$\therefore I_{rms} = \sqrt{150} = $ **12.247 A (Ans.)**

The maximum value of the resultant wave is, by equation (3.13), $\sqrt{2} \, I_{rms} = \sqrt{2} \, (12.243) = 17.32$, amps.

$$I_{av} = \frac{2}{\pi}I_m = \frac{2}{\pi}(17.32) = \textbf{11.06 A (Ans.)}$$

Illustration 1.41

Two sine waves are represented by the equations

$$e = 141 \sin [314 t - \pi/6] \text{ V}$$

and $i = 20 \sin [314 t + \pi/6]$ A.

Find the maximum and rms value of each and the net angle of phase difference between them.

Solution

Comparing these two equations with that of the equation (3.1), it can readily be seen that,

The maximum value of the EMF is **141 V (Ans.)**

The maximum value of the current is **20 A (Ans.)**

$\omega = 314$ (for both waves)

Using the expression of peak factor of equation (1.44), the rms-value is found out.

$$E_{rms} = E_m/\sqrt{2} = 141/\sqrt{2} = \textbf{99.7 V (Ans.)}$$

$$I_{rms} = I_m/\sqrt{2} = 20/\sqrt{2} = \textbf{14.14 A (Ans.)}$$

52 *Basic Electrical Engineering*

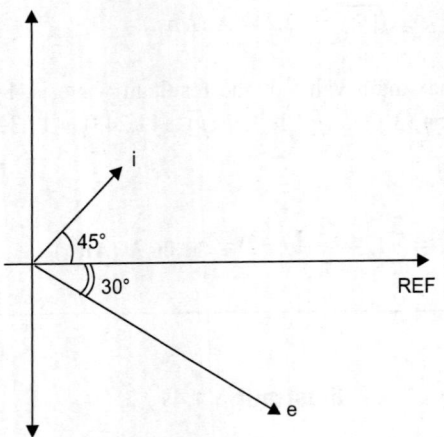

Fig. 1.69. Vectorial representation of vectors e and i or illustration 1.12.

It is clear from the expression given for i and e that, e is lagging the reference by $\pi/6$ redians (or 30°) and i is leading the reference by $\pi/4$ radians (or 45°).

It is shown vectorially in Fig. 1.69.

Thus the phase difference between v and i (or more understandably their angular displacement) is **75° (Ans.)**

Note: Mathematically, the difference in angle between the two alternating quantities will be the phase difference.

The angle of e is $314\,t - \pi/6$ and the angle of i is $314\,t + \pi/4$, and their difference is

$$314t - \frac{\pi}{6} - \left(314 + \frac{\pi}{4}\right).$$

The answer is $10\,\pi/24$ rad; which is 75°.

Illustration 1.42

An alternating current is represented by the equation $i_1 = 10 \sin 50\,t$, where t is in seconds.

A second current of the same frequency but half the amplitude lags behind the first current by 30 degrees.

(i) Find the value of the second current when the first current is at the positive peak, and also (ii) the values of both currents 0.02 sec later.

Sketch the waveform of both these currents in their correct relationships.

Solution

It is very classic example which extracts the complete aspects of lagging and leading theory.

The two current waveform are sketched in Fig. 1.70.

(i) It is clear that when i_1, is at its positive peak, its sine argument will be 90°; this is because a sinusoidal wave attains its positive peak at 90°.

For i_2 at this instant, the 'sine' argument will be, $50\,t - 30$, i.e., $90 - 30$. Thus,

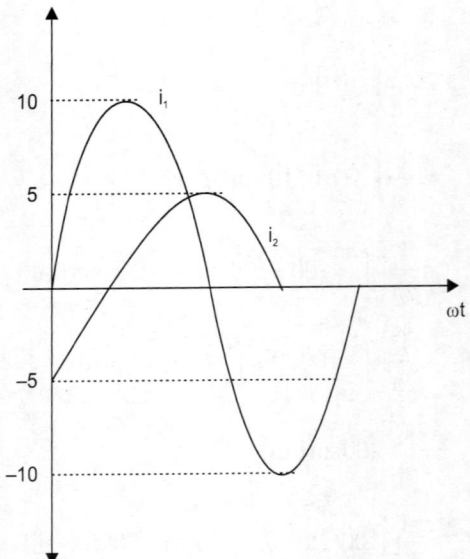

Fig. 1.70. Waveform representation of illustration 1.13
$i_1 = 10 \sin 50\,t$
$i_2 = \sin(50\,t - 30)$; this lags i_1 by 30°.

$i_2 = 5 \sin(90 - 30)$

$= 4.33$ A (Ans.)

(ii) It is given that, $\omega t = 50\,t$ radians. At the time 0.02 sec, it will be, $50 \times 0.02 = 1$ radian or $57.3°$. Thus the evaluation of i_1 and i_2 at 0.02 sec means the evaluation of the same at $57.3°$, of course, by properly considering their phase relations.

Note that, the i_1 and i_2 values are needed **after the occurrence of positive peak of** i_1. Thus

$i_1 = 10 \sin(90 + 57.3) = $ **5.4 A (Ans.)**

(*Note:* The '90 +' gives the current value after its positive peak, which is needed) and

$i_2 = 5 \sin(60 + 57.3) = $ **4.44 A (Ans.)**

(*Note:* i_2 lags i_1 by $30°$. When i_1 was at $90°$, obviously i_2 would have been at $60°$. So when i_1 is evaluated at $90 + \phi$, at the same instant, i_2 would be at $60 + \phi$.)

The reader is suggested to go through this problem in depth as the author finds this problem to be the best fit problem to understand the concept of lag and lead theory.

Illustration 1.43

Two sinusoidal alternating voltage v_1 and v_2 have peak values of 25 V and 20 V respectively and the same frequency. If v_1 leads v_2 by $40°$ plot to scale the voltage on a vector diagram on the same axis. From this evaluate (i) the sum $v_1 + v_2$, (iii) the difference $v_1 - v_2$ of the two voltages.

State also the peak values of the resultants and their phase angles relative to v_1.

Solution

v_2 is taken as the reference; v_1 leads it by $40°$. Both, in a vector diagram, are appearing in the Fig. 1.71.

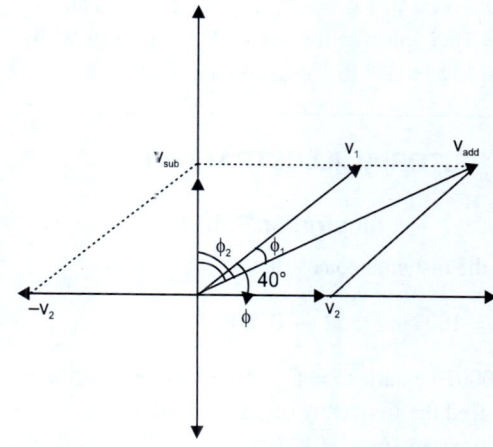

Fig. 1.71. Representation of V_1 and V_2 of illustration 1.14 in a vector diagram.

Note 1: Always two vectors are added by completing a parallel with their sides and completing the diagonal; this diagonal will give their sum.

Note 2: This could also the done analytically taking help from mathematics; consider the parallelogram, in which V_{add} is given by,

$$V_{add} = v_1 + v_2 = \sqrt{v_1^2 + v_2^2 + 2v_1 v_2 \cos\theta}$$

where θ is the angle between v_1 and v_2. So,

$V_{add} = \sqrt{25^2 + 20^2 + 2(25)(20)\cos 40}$

$= \sqrt{1790}$

$= 42.5$ V.

The reader may try this for $v_1 - v_2$.

Note: By completing the parallelogram as shown in Fig. 1.71

$v_1 + v_2 = 42.5$, volts (by measuring V_{add})

It lags v_1 by an angle $\phi_1 = $ **17.5° (by measurement) (Ans.)**

$v_1 - v_2 = v_1 + (-v_2)$; v_2 reversed is $-v_2$. Thus, $v_1 - v_2 = 16.5$ volts (by measuring V_{sub}). It leads v1 by an angle $\phi_2 = 50°$ (by measurement (Ans.)

1.36 TUTORIAL ILLUSTRATION III

Illustration T.III.-1

Find the instantaneous value of

$$i = 100 \sin(1200 t + 0.25)$$

at 0.0001 sec after t = 6 and increasing positively. Also find the frequency of this current wave.

(Ans: 36.16 A; 191 Hz).

Illustration T.III.-2

What is the value of average value over half a cycle of a sinusoidal alternating current whose maximum value in 31 amps.

(Ans: 19.75 A).

Illustration T.III.-3

An alternating current varying sinusoidally with a frequency of 60 c/s has a rms value of 20 A. Write down the equation for the instantaneous value and find this value at (a) 0.0025 sec (b) 0.0125 sec after passing through a positive maximum value. At what time, measured from a positive maximum value, will the instantaneous current be 14.14 A?

(28.28 sin 377t; (a) 22.9, A)

Illustration T.III.-4

Four emf's,

$e_1 = 100 \sin \omega t$; $e_2 = \sin(\omega t - \pi/6)$

$e_3 = 120 \sin(\omega t + \pi/4)$;

$e_4 = 100 \sin(\omega t - 2\pi/3)$

are available. Vectorially find the resultant emf and it's phase difference with (a) e_1 (b) e_2.

(Ans: 208 sin (ωt –0.202); 11°34′ leading; 18°26′ lagging.)

Illustration T.III.-5

illustration 1.14

Four Emf's are given by

$$e_1 = 100 \sin \omega t, \; e_2 = 173.2 \cos\left(\omega t - \frac{\pi}{3}\right)$$

$$e_3 = 70.7 \sin\left(\omega t + \frac{3\pi}{4}\right), \; e_4 = -50 \sin(\omega t + \pi)$$

Calculate the resultant emf expressing it in a similar form.

Reflection Section

1. Comment on the resultant when two sinusoids of the same frequency but of different amplitudes and phase angles are subtracted.
2. Form factor varies for different alternating waveforms. Give justification.
3. Peak factor is the ratio of _____.
4. What is the period of an alternating wave?
5. What is the difference between an instantaneous value and the maximum value, in an alternating waveform?
6. In a square wave can the RMS value and the mean value be the same? Verify and prove.
7. Phase angles are referred with reference to space rotation. Explain.
8. For a frequency of 200 Hz, what will be the time period?
9. In AC circuits power is dissipated is resistance only / inductance only / capacitance only / none. Pick the right answer and justify.
10. The domestic voltage is 220 V. This represents which quantity of AC voltage? Mean or RMS or average?
11. Write down the equation of 50 Hz current sine wave having RMS value of 60 A.
12. Prove that in a pure inductive coil when the frequency is halved the current in it will be doubled.
13. A sinusoidal wave has a frequency of 50 Hz and an RMS current of 30 A. The equation of this wave is_____.

Basic Electricity 55

PART III: SERIES AND PARALLEL AC CIRCUIT

1.37 THE POWER FACTOR

It is clearly enlightened that the alternating quantities will always **deviate** from each other in the sense that they may not attain specified value at a particular instant. (Sec. 1.30, Fig. 1.57). This deviation—known as the phase displacement—affects the AC power.

In DC circuits the power in any element in given by the simple product of V and I. (Refer to equation (1.4)). This is otherwise known as the **apparent power**.

But in AC circuit the power is **not** given by the mere product of voltage and current. This is because of the fact that, in AC, the voltage and current are **vectors** (but in DC they are scalars); if they do not reach their positive maximum and subsequent zeroes simultaneously then the voltage and current will be phase displaced by ϕ (refer to sec. 1.30); note carefully now that, the element in which the AC power is to be measured in experiencing a voltage and a current **not** having synchronised flow (i.e., reaching maxima's zeroes etc., at the same time) and are displaced $\phi°$; this is to be accounted for.

The $\phi°$ displacement is the **phase angle** of the element and the cosine of $\phi°$—i.e. cos ϕ—the **power factor** of the element. It is also denoted as **p.f.** Then the actual power in the AC circuit (or any element) is given by the product, VI cos ϕ; it is known as the **active power**. That is,

$$P_{ac} = VI \cos \phi \tag{1.45}$$

where, power factor is the ratio of actual power to apparent power. That is,

$$\text{p.f.} = \frac{VI\cos\phi}{VI} = \cos\phi \tag{1.46}$$

Note 1: As the numerator and denominator both are powers (or watts), p.f. has no unit.

Note 2: It will be shown later that the p.f. of an element (or circuit) can also be given as the ratio of resistance to impedance—i.e., R/Z.

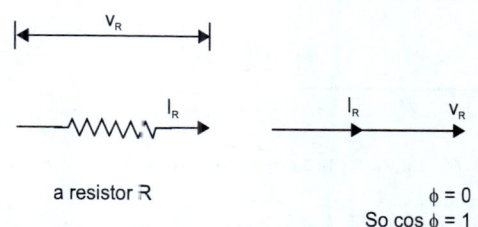

Fig. 1.72. p.f. of a pure resistive circuit.

Note 3: The three constants of an electric circuits are—the resistance, the inductance and the capacitance. The p.f. varies in these elements as explained below.

Note 4: The product VI sin ϕ is known as the reactive power, symbolized by KVAr.

(i) p.f. of a pure resistive circuit

In a pure resistive circuit the voltage across it and current through it will be in phase with each other, giving the phase angle (i.e., the angle between V and I) a zero. (refer to Fig. 1.72). Hence the p.f. is, cos 0 = 1. This is known as the **unity power factor** or **u.p.f.**

(ii) p.f. of a pure inductive circuit

In a pure inductive circuit, the current through the inductor will lag the voltage across it by 90° exactly. Thus the phase angle $\phi = 90°$ and the p.f., cos ϕ = cos 90 = 0, lag; Refer to Fig. 1.73.

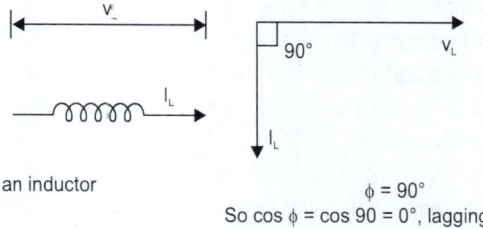

Fig. 1.73. p.f. of a pure inductive circuit

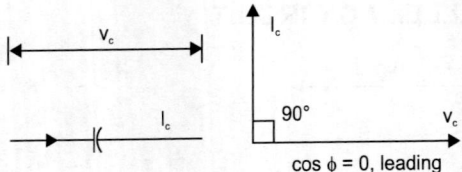

Fig. 1.74. p.f. of a pure capacitive circuit.

Note that it can also stated out, in an inductive circuit, the voltage leads the current by 90°. But the usual convention is the former. (Later in Sec. 1.48, the problems are solved by the former convention)

(iii) p.f. of a pure capacitive circuit

In a pure capacitive circuit, the current through the capacitance will lead the voltage across it by 90°. Here again the phase angle $\phi = 90°$; but the p.f. is $\cos \phi = \cos 90° = 0$, lead. Refer to Fig. 1.74.

Note 5: It is not possible to form a pure—ideal—inductor or capacitor. For example, the formation of inductor necessitates winding a few number of turns on an iron piece. This number of turns will cumulatively form an amount of resistance, whose presence will not allow the current through the inductor to lag exactly by 90° behind the voltage across it; but by an angle lesser than 90°. This is the case even with the capacitor, where the current through the capacitor leads the voltage across it by an angle lesser than 90°.

1.38 POWER IN A PURE RESISTOR

Consider the Fig. 1.75. The resistor R is supplied by AC source.

In a pure resistive circuit, as the voltage across the resistor and the current through it are in-phase with each other, when current is written as $i = I_m \sin \omega t$, then the voltage will be $v = V_m \sin \omega t$.

The instantaneous power = product of instantaneous voltage and current

i.e., $P = vi$ watts.

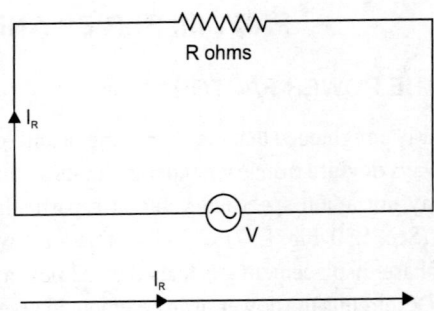

Fig. 1.75. A pure resistive circuit.

Average power drawn P_{av} = average value of $v\,i$ curve.

$$\therefore \quad P_{av} = \frac{1}{T} \int_0^T v\,i\,dt$$

where T = period = $1/f = 2\pi/\omega$

$$\therefore \quad P_{av} = \frac{\omega}{2\pi} \int_0^{2\pi/\omega} V_m \sin \omega t \, I_m \sin \omega t \, dt$$

$$= \frac{V_m I_m \omega}{2\pi} \left[\int_0^{2\pi/\omega} \sin^2 \omega t \, dt \right]$$

$$= \frac{V_m I_m \omega}{2\pi} \left[\int_0^{2\pi/\omega} \left(\frac{1 - \cos 2\omega t}{2} \right) dt \right]$$

$$= \frac{V_m I_m \omega}{4\pi} \left[\int_0^{2\pi/\omega} dt - \int_0^{2\pi/\omega} \cos 2\omega t \, dt \right]$$

which becomes,

$$P_{av} = \frac{V_m I_m}{2}$$

Writing this as $\dfrac{V_m}{\sqrt{2}} \cdot \dfrac{I_m}{\sqrt{2}}$, and using the equation (1.42),

$V_m/\sqrt{2} = V_{rms}$ and $I_m/\sqrt{2} = I_{rms}$. Thus,

$$P_{av} = V_{rms} I_{rms} = VI, \text{ watts} \qquad (1.47)$$

1.39 POWER IN A PURE INDUCTOR

In a pure inductive circuit as the voltage across the inductor leads the current through it by 90°, when current is

$i = I_m \sin \omega t$

then voltage is given by

$v = V_m (\sin \omega t + 90) = V_m \cos \omega t$.

instantaneous power $P = vi$
   Then the average power is,

$P_{av} = \dfrac{1}{T} \int_0^T vi\, dt$

$= \dfrac{\omega}{2\pi} \int_0^{2\pi/\omega} V_m \cos \omega t\, I_m \sin \omega t\, dt$

$= \dfrac{V_m I_m \omega}{2\pi} \left[\int_0^{2\pi/\omega} \sin \omega t \cos \omega t\, dt \right]$

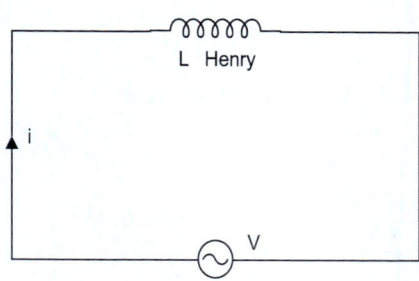

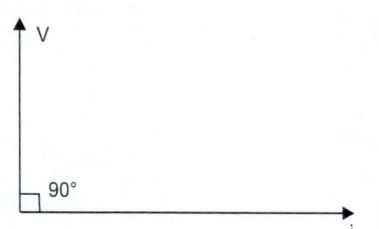

Fig. 1.76. A pure inductor circuit to analyse the power across it.

$= \dfrac{V_m I_m \omega}{2\pi} \left[\int_0^{2\pi/\omega} \dfrac{\sin(\omega t + \omega t) + \sin(\omega t - \omega t)}{2}\, dt \right]$

$= \dfrac{V_m I_m \omega}{4\pi} \left[\int_0^{2\pi/\omega} \sin 2\omega t\, dt \right]$

$= \dfrac{V_m I_m \omega}{4\pi} \left[\left(\dfrac{-\cos 2\omega t}{2\omega} \right)_0^{2\pi/\omega} \right]$

$= \dfrac{V_m I_m \omega}{4\pi} \left[\dfrac{\cos 4\pi - \cos 0}{2\omega} \right]$

$= \dfrac{V_m I_m \omega}{2\omega 4\pi} [1 - 1]$

$= \dfrac{V_m I_m \omega}{2\omega 4\pi} [0]$

$= 0$

Thus power absorbed by a pure inductive AC circuit is zero. (1.48)

1.40 POWER IN A PURE CAPACITOR

Consider the Fig. 1.77. In a pure capacitor circuit, as the current leads the voltage across it by 90°, when the voltage

$v = V_m \sin \omega t$, then the current will be,

$i = I_m (\omega t + 90) = I_m \cos \omega t$

Instantaneous power $P = vi$ and the average power is

$P_{av} = \dfrac{1}{T} \int_0^T vi\, dt$

$= \dfrac{\omega}{2\pi} \int_0^{2\pi/\omega} V_m \sin \omega t\, I_m \cos \omega t\, dt$

58 Basic Electrical Engineering

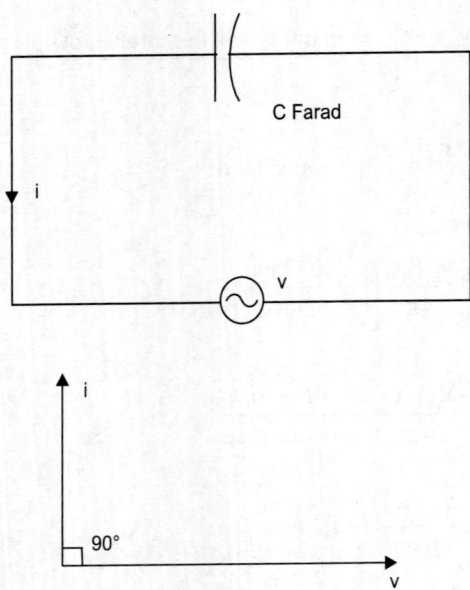

Fig. 1.77. A capacitor circuit for the analysis of power across it.

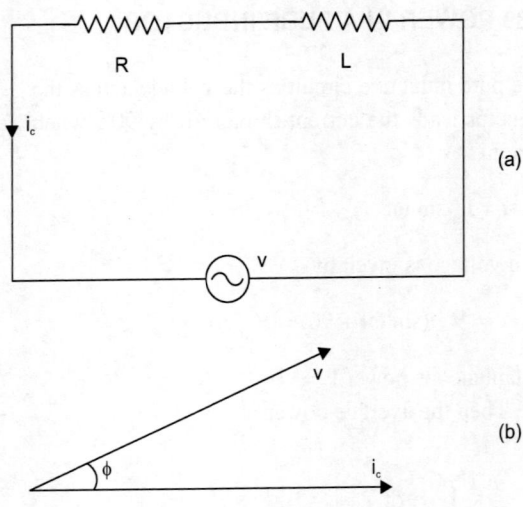

Fig. 1.78. An R-L Circuit—Power derivation.
(a) Circuit (b) Vector Diagram

So when current is written as $i = I_m \sin \omega t$ voltage v shall be written as $v = V_m \sin(\omega t + \phi)$

Then the average power is

$$= \frac{1}{T} \int_0^T v i \, dt$$

$$= \frac{\omega}{2\pi} \int_0^{2\pi/\omega} V_m I_m \sin(\omega t + \phi) \sin \omega t \, dt$$

$$= \frac{V_m I_m \omega}{2\pi} \left[\int_0^{2\pi/\omega} \sin(\omega t + \phi) \cdot \sin \omega t \, dt \right]$$

$$= \frac{V_m I_m \omega}{2\pi} \left[\int_0^{2\pi/\omega} \frac{\cos(\omega t + \phi - \omega t) - \cos(\omega t + \phi + \omega t)}{2} dt \right]$$

$$= \frac{V_m I_m \omega}{4\pi} \left[\int_0^{2\pi/\omega} \cos \phi \, dt - \int_0^{2\pi/\omega} \cos(2\omega t + \phi) \, dt \right]$$

$$= \frac{V_m I_m \omega}{4\pi} \left[\cos\phi (t)_0^{2\pi/\omega} - \left(\frac{\sin(2\omega t + \phi)}{2\omega} \right)_0^{2\pi/\omega} \right]$$

$$= \frac{V_m I_m \omega}{2\pi} \left[\int_0^{2\pi/\omega} \sin \omega t \cos \omega t \, dt \right]$$

Similar proceeding as indicated in the pure inductive case, will yield,

$P_{av} = 0$.

Thus,

The power absorbed by a pure capacitor when supplied by an alternating quantity is zero (1.49)

1.41 POWER IN A R-L SERIES CIRCUIT

As brought out in Sec. 1.35, note 4, the availability of resistor in conjunction with an inductor will not allow the voltage across it to lead the current through it exactly by 90° but by an angle ϕ lesser than it. It is shown in Fig. 1.78.

$$= \frac{V_m I_m \omega}{4\pi}\left[\cos\phi\left(\frac{2\pi}{\omega}-0\right)-\frac{1}{2\omega}\left(\sin[2\omega\left(\frac{2\pi}{\omega}\right)+\phi]\right.\right.$$

$$\left.\left. - \sin[2\omega(0)+\phi]\right)\right]$$

$$= \frac{V_m I_m \omega}{4\pi}\left[\cos\phi\left(\frac{2\pi}{\omega}\right)-(0)\right]$$

$$= \frac{V_m I_m \cos\phi}{2}$$

$$= \frac{V_m}{\sqrt{2}}\frac{I_m}{\sqrt{2}}\cos\phi$$

Using the equation (3.13),

$$= V_{rms} I_{rms} \cos\phi \text{ W. Thus,}$$

P = VI cos φ, in an R-L Series Circuit   (1.50)

1.42 POWER IN A R-C SERIES CIRCUIT

In a R-C series circuit current leads the voltage by φ°. So when the voltage is written as

$$v = V_m \sin\omega t$$

then the current has to be

$$i = I_m \sin(\omega t + \phi).$$

The Power

$$P_{av} = \frac{1}{T}\int_0^T vi\, dt$$

$$= \frac{\omega}{2\pi}\int_0^{2\pi/\omega} V_m \sin\omega t \cdot I_m \sin(\omega t + \phi) dt$$

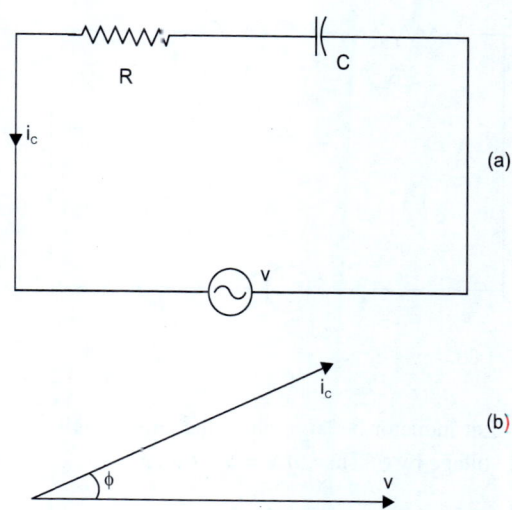

Fig. 1.79. Power in an R-C Series AC circuit.
(a) Circuit (b) Vector diagram

$$= \frac{V_m I_m \omega}{2\pi}\left[\int_0^{2\pi/\omega} \sin(\omega t + \phi)\cdot \sin\omega t\, dt\right]$$

Similar proceeding as done with as R-L circuit will yield,

$$P_{av} = \frac{V_m I_m \cos\phi}{2} = \frac{V_m}{\sqrt{2}}\cdot\frac{I_m}{\sqrt{2}}\cos\phi$$

$$= V_{rms} I_{rms} \cos\phi$$

∴ P = VI cos φ is an R-C AC series circuit   (1.51)

1.43 POWER IN R-L-C AC SERIES CIRCUIT

Power in R-L-C AC Series Circuit

Consider Fig. 1.80. As this circuit has L and C together, whichever dominates, will have the upper hand in the circuit.

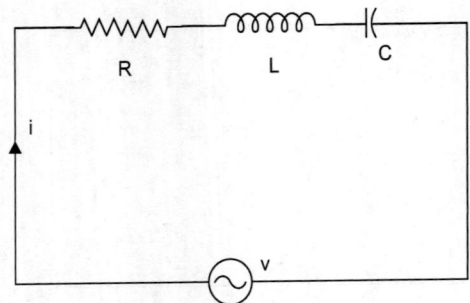

Fig. 1.80.

Let inductor be large, thus, the current will lag the voltage by ϕ: Thus, if $v = V_m \sin \omega t$,

$i = I_m \sin(\omega t - \phi)$

i can also be $I_m \sin(\omega t + \phi)$ if the resistance equivalent of the capacitor (known as capacitive reactance—discussed later) is more. So capacitor will have upper hand, thus making the current to lead the voltage by ϕ.

Let us say $i = I_m \sin(\omega t + \phi)$

(and of course, $v = V_m \sin \omega t$)
Then, the average power is,

$$P = \frac{1}{T}\int_0^T v\, i\, dt$$

$$= \frac{\omega}{2\pi}\int_0^{2\pi/\omega} V_m I_m (\sin \omega t + \phi) \sin \omega t\, dt$$

which will reduce to,

$$= \frac{V_m I_m \cos\phi}{2} = \frac{V_m}{\sqrt{2}} \frac{I_m}{\sqrt{2}} \cos\phi$$

$$= V_{rms} I_{rms} \cos\phi$$

Thus $P = VI \cos\phi$ in an RLC series circuit (1.52)

Reader may try with $i = I_m \sin(\omega t - \phi)$ and $v = V_m \sin \omega t$. The answer will be the same.

Thus a pure inductor and a capacitor absorb zero power.

A pure resistor absorbs a power of VI, watts. A resistor in conjunction with an inductor or a capacitor or both absorbs VI cos ϕ, watts.

In all the cases V and I are rms values.

1.44 THE REACTANCE AND THE IMPEDANCE (AC SERIES CIRCUIT)

It is clearly seen in Sec (1.6) that the opposition offered to the flow of current in the resistance. This opposition gets different names depending on the circuit constant used.

A circuit may contain only pure resistance; then the opposition offered to the flow of current is only the resistance, R ohms. (This is, of course, in DC circuits, as explained in chapter 1.)

If a circuit contains only inductance 'L' henrys or Capacitor C farads, then the opposition offered to the flow of current is known as the **reactance**. The unit is ohms, and the symbol is X.

If it is a pure inductive AC fed circuit, then the opposition is known as the **inductive reactance** denoted by X_L and valued as ωL. Thus,

$$X_L = \omega L = 2\pi f L\ \Omega \quad (1.53)$$

Similarly, when the circuit is a pure capacitive alternating circuit, then the opposition offered to the flow of current is the capacitive reactance X_C, valued as $1/\omega c$. Thus,

$$X_C = \frac{1}{\omega C} = \frac{1}{2\pi f C}\ \Omega \quad (1.54)$$

When the circuit has both the inductance and the capacitance in the circuit, then the net opposition to the flow of current would be $X_L \sim X_C$, Thus,

$$X = X_L \sim X_C, \text{ if both L \& C are present} \quad (1.55)$$

Moreover, if a circuit contains either inductance or capacitance or both in addition to the resistance the combined opposition offered to the flow of current is called the **impedance**. Its symbol is Z, unit is ohm (Ω) and valued as $\sqrt{R^2 + X^2}$. Thus,

$$Z = \sqrt{R^2 + X^2}$$

where X may be X_L alone in R-L AC circuit
may be X_C alone in R-C AC circuit
may be $X_L \sim X_C$ in R-L-C AC circuit (1.56)

As a conclusion, the opposition offered to the flow of a current in a DC fed circuit (i.e., a pure resistive circuit) is the resistance; the opposition offered to the flow of current in AC fed circuit is the impedance, whose value can be anything as given by the equation (1.56).

1.45 THE IMPEDANCE TRIANGLE

The resistance is a DC element; the reactance, as suggested, can be of inductive or capacitive thus displaced 90° from the resistance; the impedance Z, as seen by the equation (1.56) is the **vector** sum of these two.

Based on this justification, a triangle formed by the three sides namely—the resistance, the reactance and the impedance—is called the **impedance triangle**.

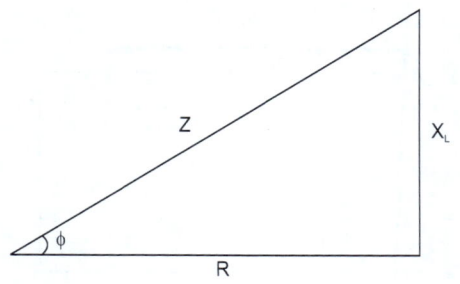

Fig. 1.81. Impedance triangle with R, X_L and Z.

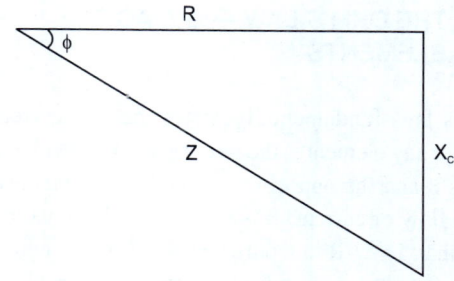

Fig. 1.82. Impedance triangle with R, X_C and Z.

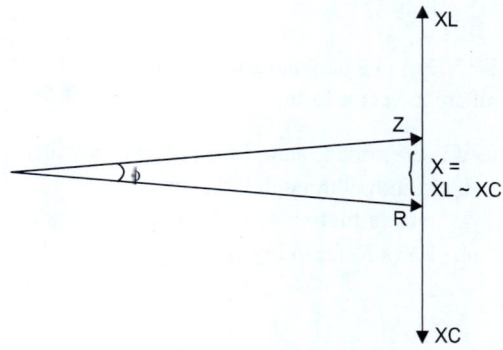

Fig. 1.83. Impedance triangle with R, X_L, X_C and Z.

The impedance triangle with R, X_L and Z is in Fig. 1.81.

The impedance triangle with R, X_C and Z is in Fig. 1.82.

The impedance triangle with X_L and X_C will have the net X on the inductive side or capacitive side, depending on which is greater. Fig. 1.83 depicts the impedance triangle with X_L and X_C where $X_L > X_C$.

Note: In all these figures that the angle ϕ is the phase angle (discussed earlier in sec. 1.37). From these three figures, the p.f., cos ϕ can be found out to be, the ratio between R and Z. that is

$$\cos \phi = R/Z \qquad (1.57)$$

(Refer to note 2 of sec. 1.30)

1.46 THE OHM'S LAW—w.r.t. AC ELEMENTS

Ohm's law, fundamentally, states that the current through any element is the ratio between the voltage across it and the opposition offered by the element to the flow of current. It is brought out by equation (1.8) that, I = V/R in a pure resistive (DC) circuit.

Using the same definition, the equation (1.8). i.e. the ohm's law, will attain a different shape when it is applied to different AC circuits, as given below.

(a)

$I = V/X_L$, in a pure inductive circuit; all are in vector form (1.58)

where V is the voltage across the inductor in volts;

I, the current through it, in amps;

X_L, its inductive reactance given by the equation (1.53). Refer to Fig. 1.84.

(b)

$I = V/X_C$, in a pure capacitive circuit; all are in vector form (1.59)

where V is the voltage across the capacitor, in volts; I, the current through it, in amps; X_C, its capacitive reactance, in ohms, given by the equation (3.25). Refer to Fig. 1.85.

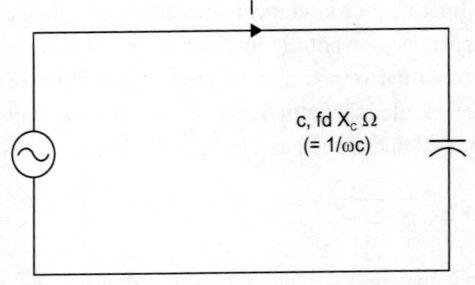

Fig. 1.85.

(c)

I = V/Z, in an RL or RC or RLC circuit; all are in vector form (1.60)

Refer to Fig. 1.86.

Note: All the above are in vector form; The problems of Sec. (1.49) will explain the vector form handling properly.

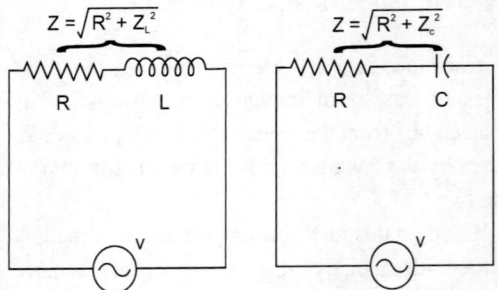

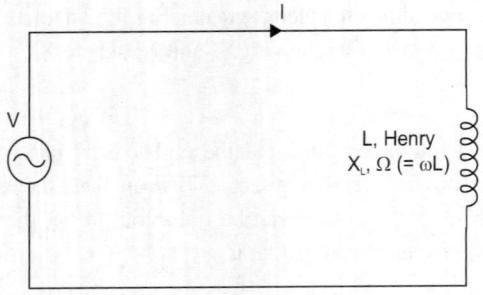

Fig. 1.84.

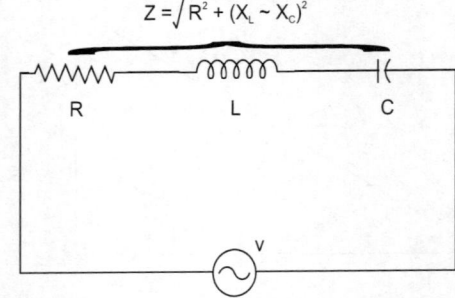

Fig. 1.86.

Basic Electricity 63

1.47 TABLE OF REPRESENTATIONS

Table 1.1. Summary of the above discussions:

Type of connection	Resistance	Reactance	Impedance (in ractangular form)	Impedance in polar form)	Phase angle ϕ	pf
Inductive circuit (R-L)	R	$X_L \, (=\omega L)$	$Z \, (= R + j X_L)$	$Z \angle \phi$ where $Z = \sqrt{R^2 + X_L^2}$	$\phi = \tan^{-1} \dfrac{X_L}{R}$	$\cos \phi$
Capacitive circuit (R-C)	R	$X_C \, (= -\dfrac{1}{\omega c})$	$Z \, (= R - j X_C)$	$Z \angle \phi$ where $Z = \sqrt{R^2 + X_C^2}$	$\phi = \tan^{-1} \dfrac{X_C}{R}$	$\cos \phi$
Inductive, capacitive series circuit (R-L-C)	R	X_L and X_C	$Z = [R + j (X_L \sim X_C)]$	$Z \angle \phi$ where $Z = \sqrt{R^2 + (X_L \sim X_C)^2}$	$\phi = \tan^{-1} \left(\dfrac{X_L \sim X_C}{R} \right)$	$\cos \phi$

Note: It is usual practice to denote the inductive reactance X_L in + Y-axis and the capacitive reactance X_C in − y-axis, as indicated in the above figures. Thus if X_L is to be put in rectangular or polar form, then it will be respectively $+j X_L$ and $X_L \angle 90°$. On the other hand, X_C in rectangular and polar forms will be, j X_C and $X_C \angle -90$, respectively.

Of course, R being in phase with X-axis, will always be denoted in rectangular form as $R + j0$ and in polar form as $R \angle 0$.

1.48 SERIES AND PARALLEL AC CIRCUITS

As far as the voltage, current, power relations are concerned, there is no difference between the DC and AC circuits. So the characteristics discussed in sec (1.20-A) and (1.20-B) will be identical in AC circuit also, except the fact that

(i) in the AC circuits the opposition to the flow of current is the impedance and thus the ohm's law of sec. (3.18) will be used

and

(ii) all will be treated in vectorial form; so the specification of current, voltage, impedance etc. will be made either in rectangular or polar form.

1.49 WORKED ILLUSTRATIONS IV

Illustration 1.44

When a coil is connected to a 200 V, 50 c/s supply it takes a current of 1.5 A and the power consumption in 70 W. Calculate the resistance and the inductance of the coil.

Solution

The circuit is depicted in Fig. 1.87. A coil inherently will have resistance, as explained in Sec. 1.37, note 4. This coil resistance is R in the Fig. 1.87.

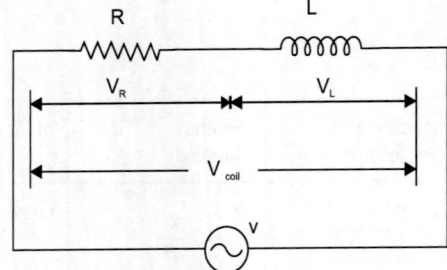

Fig. 1.87. Diagrammatic representation illustration 1.1.

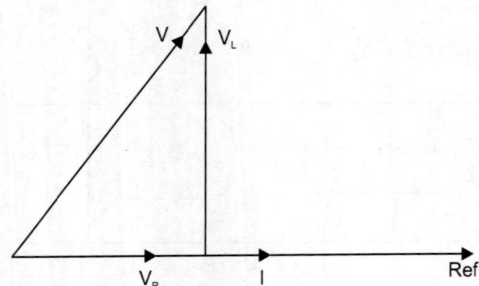

Fig. 1.88. Vector diagram for the illustration 1.1.

In drawing the vector diagram, the quantity which is common in the circuit is usually taken as the reference. Because in a series circuit, current is same in all the elements, let the current be chosen as the reference. (Fig. 1.88)

The voltage across the resistor (V_R) will be in-phase with the current through it; the inductor voltage (V_L) will lead the current through it; the vector sum of these two, by KVL, must be the supply voltage V. Thus,

$$\overline{V} = \overline{V}_R + \overline{V}_L = \overline{V}_{coil}$$

As the power in the resistor and the current through it are known, by using the equation (1.9), the resistance of the resistor could be found out. That is,

$$R = P/I^2 = \frac{70}{1.5^2} = \textbf{31.11 Ω (Ans.)}$$

Thus V_R, by ohm's law, is, $V_R = IR$ {of the resistor R.}

$$V_R = (1.5 \times 31.11) = 46.66 \angle 0 \text{ V.}$$

V = 200, given. Thus from Fig. 1.98.

$$V_L = \sqrt{V^2 - V_R^2}$$

$$= \sqrt{200^2 - 46.66^2} = 194.4 \text{ V.}$$

Note: It is a vector difference and **not** an arithmetic difference. It is, indeed, the case in AC because the phase difference enters the discussion.

By using equation (3.29), the inductive reactance X_L is,

$$X_L = V/I \text{ (of the inductor)}$$

$$= 194.4/1.5 = 129.65 \text{ Ω}$$

By using equation (3.24), the inductance is,

$$L = \frac{X_L}{2\pi f} = 129.65/2\pi \times 50 = \textbf{0.413 henry}$$

(Ans.)

Illustration 1.45

A current of 1.0 A at a frequency of 1,00,000 c/s is passed through a coil of resistance 20 Ω and inductance 400 micro henry. What is the voltage across the coil.

Solution

It is series circuit, hence the current will remain the same in both the elements.

The coil will comprise of R and L and thus the voltage across the coil would be the vector sum of V_R and V_L

$V_R = IR = 1 \times 20 = 20$ V

$V_L = 1 \times (2\pi \times 100000 \times 400 \times 10^{-6})$

(by equation 1.53)

$= 251.327$ V

$\therefore V_{coil} = \sqrt{V_R^2 + V_L^2} = \sqrt{20^2 + 251.727^2}$

$= 252.12$ V (Ans.)

Illustration 1.46

Two coils are connected in series. With 2 A DC through the circuit, the p.d.'s across the coils are 20 V, and 30 V respectively. With 2 A AC at 40 c/s the p.d.'s across the coils are 140 V and 100 V respectively. If the two coils in series are connected to a 230 V, 50 c/s supply, calculate,

(i) The current, (ii) the power, (iii) the p.f.

Solution

It is a classic example which extracts the AC fundamentals. Readers are suggested to form the complete technical reasoning to solve this problem, before the solution is traced.

The circuit is depicted in Fig. 1.89.

As the only opposition to the flow of DC is the resistance, with the DC data the resistance of the coils (R_1 and R_2 respectively) can be found out using the equation (1.6).

(Suffix refers to the coil)

$\therefore R_1 = \dfrac{V_1}{I_1} = \dfrac{20}{2} = 10\,\Omega$

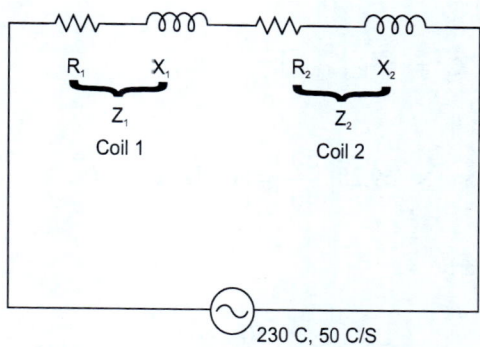

Fig. 1.89. Circuit for illustration 1.3.

$R_2 = \dfrac{V_2}{I_2} = \dfrac{30}{2} = 15\,\Omega$

($I_1 = I_2$; series circuit)

The opposition offered to the flow of current in AC is the impedance Z; therefore, with the AC data given, the impedance of the coils (Z_1 and Z_2 respectively) can be found out.

$Z_1 = \dfrac{V_1}{I_1} = \dfrac{140}{2} = 70\,\Omega$ at 40 c/s

$Z_2 = \dfrac{V_2}{I_2} = \dfrac{100}{2} = 50\,\Omega$ at 40 c/s

From the impedance diagram of Fig. (1.81), the reactance of coil 1 could be said as, $X_1 = \sqrt{Z_1^2 - R_1^2}$. Substituting the values

$X_1 = \sqrt{70^2 - 10^2} = 69.28\,\Omega$ at 40 c/s.

$X_2 = \sqrt{50^2 - 15^2} = 47.7\,\Omega$ at 40 c/s.

The current, power etc. are needed at 50 c/s and what we have is at 40 c/s; as X_1, the inductive reactance, is directly proportional to the frequency, the respective reactance at 50 c/s would be,

$$X_1 = [X_{1(\text{at }40c/s)}] \cdot \frac{50}{40}$$

$$= 69.28 \left(\frac{50}{40}\right) = 86.6\,\Omega \text{ at 50 c/s.}$$

Similarly

$$X_2 = [X_{2(\text{at }40c/s)}] \cdot \frac{50}{40}$$

$$= 47.7 \left(\frac{50}{40}\right) = 59.6\,\Omega \text{ at 50 c/s.}$$

Total resistance, $R = R_1 + R_2 = 10 + 15 = 25\,\Omega$

Total reactance, $X = X_1 + X_2 = 86.6 + 59.6$

$$= 146.2\,\Omega$$

Thus, the total impedance is,

$$Z = \sqrt{(\text{total resistance})^2 + (\text{total reactance})^2}$$

$$Z = \sqrt{R^2 + X^2}$$

$$= \sqrt{25^2 + 146.2^2} = 148.32\,\Omega$$

(*Note:* This is the way of finding out the net impedance. The impedance is a vector.)

(i) The current in the circuit, I is V/Z, of the circuit

$$= \frac{230}{148.32} = \mathbf{1.551\ A\ (Ans.)}, \text{ at 50 Hz}$$

(ii) As proved in Sec. (1.39), an inductor consumes zero power. Thus power must have been consumed only by the resistors. Using $I^2 R$ formula,

Net Power $P_{net} = I^2 R_{net} = 1.551^2 (10 + 15)$

$$= \mathbf{60.11\ W\ (Ans.)}$$

(iii) The net circuit p.f., by equation (1.57) is

$$\cos\phi = \frac{R}{Z} \text{ (of the circuit)}$$

$$= \frac{(10+15)}{148.32} = \mathbf{0.1685\ (Ans.)}$$

Note: In an R-L circuit the power was proved to be the product $V_L I_L \cos\phi$, in sec. 1.41. For this circuit, it will be,

$$P_{net} = VI \cos\phi \qquad \text{(of the circuit.)}$$

$$= 230 \times 1.551 \times 0.1685$$

$$= 60.13\ W$$

almost the same as before.

Illustration 1.47

Calculate the current in an inductive circuit of resistance 5 Ω and inductance 0.02 henry, when supplied at 200 V, 50 c/s. What is the angle of phase difference.

Solution

The circuit is depicted is Fig. 1.90.
From equation (1.53), the inductive reactance X_L is

$$X_L = 2\pi f L$$

Basic Electricity 67

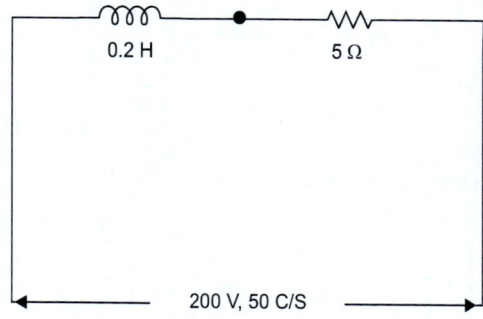

Fig. 1.90. Circuit for illustration 1.47.

$= 2 \times 31.4 \times 50 \times 0.02$

$= 6.28\ \Omega.$

$R = 5\ \Omega$, given

Using equation (1.56),

$Z = \sqrt{R^2 + X_L^2}$

$= \sqrt{5^2 + (6.28)^2}$

$= 8.029\ \Omega$

Now, by Ohm's law,

$I = \dfrac{V}{Z}$

$= \dfrac{200}{8.029} = 24.91\ A\ (Ans.)$

Using equation (1.57),

$\cos\phi = \dfrac{R}{Z} = \dfrac{5}{8.029}$

$\phi = \cos^{-1}\left(\dfrac{5}{8.029}\right)$

$= 51.5°\ (Ans.)$

Illustration 1.48

Calculate the reactance and impedance of a circuit having a resistance of 8 Ω and inductance 0.08 henry, when connected to a supply of which the frequency is 25 c/s. If the supply voltage is 100 V, find the current taken. What is the angle of phase difference.

Solution

The inductive reactance can be calculated from equation (1.53) as

$X_L = 2\pi f L$

$= 2 \times 3.14 \times 25 \times 0.08$

$= 12.57\ \Omega\ (Ans.)$

and $R = 2\ \Omega$ (given)

The impedance, Z, of the circuit is,

$Z = \sqrt{R^2 + X_L^2}$

$= \sqrt{8^2 + 12.57^2}$

$= 14.9\ \Omega\ (Ans.)$

The applied voltage is 100 V, given; Using ohm's law, the current I is,

$I = \dfrac{V}{Z} = \dfrac{100}{14.9} = 6.711\ A\ (Ans.)$

The p.f., by the equation (1.57) is,

$\cos\phi = \dfrac{R}{Z} = \dfrac{8}{14.9}$

$\phi = \cos^{-1}\left(\dfrac{8}{14.9}\right)$

= cos⁻¹ (0.5369)

= **57.52° (Ans.)**

Illustration 1.49

An inductive circuit has a resistance of 40 Ω and an inductance of 0.2 henry; it is connected to a 500 V, 50 c/s supply. Find:
(a) Its reactance
(b) Its impedance
(c) The current taken
(d) The angle of lag

Solution

The circuit is depicted (i) Fig. 1.91.

(a) The inductive reactance is,

$$X_L = 2\pi f L, \Omega$$

$$= 2 \times 3.14 \times 50 \times 0.2$$

$$= \mathbf{62.8\ \Omega\ (Ans.)}$$

(b) $R = 40\ \Omega$, given. Thus the impedance of the circuit is

$$Z = \sqrt{R^2 + X_L^2}$$

$$= \sqrt{40^2 + (62.1)^2}$$

$$= \mathbf{74.47\ \Omega\ (Ans.)}$$

(c) Using the Ohm's law, current

$$I = \frac{\text{voltage}}{\text{impedance}}$$

$$= \frac{500}{74.47}$$

$$= \mathbf{6.714\ A\ (Ans.)}$$

(d) The p.f.,

$$\cos\phi = \frac{R}{Z} = \frac{40}{74.47}$$

Thus the phase angle,

$$\phi = \cos^{-1}\left(\frac{40}{74.47}\right)$$

$$= \mathbf{57.5°\ (Ans.)}$$

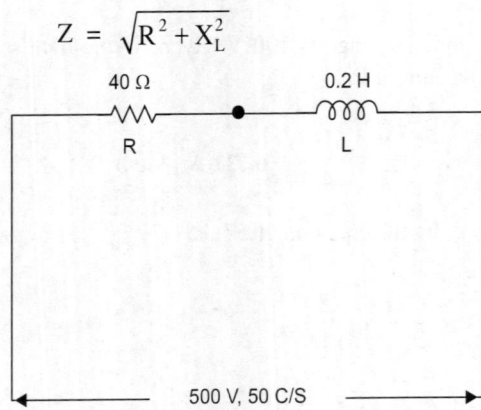

Fig. 1.91. Circuit for illustration 1.49.

Illustration 1.50

An inductive coil of resistance 4 Ω takes a current of 8 A when connected to a 100-V, 50 c/s supply.

Calculate:

(a) The impedance
(b) The reactance
(c) The inductance of the circuit.

What is the angle of phase difference between current and voltage.

Basic Electricity 69

Solution

(a) Using the Ohm's law,
The impedance,

$$Z = \frac{\text{voltage}}{\text{current}} = \frac{100}{8}$$

$$= 12.6 \, \Omega \text{ (Ans.)}$$

(b) The inductive reactance, X_L, is given by

$$X_L = \sqrt{Z^2 - R^2}$$

$$= \sqrt{(12.5)^2 - 4^2}$$

$$= 11.84 \, \Omega \text{ (Ans.)}$$

(c) As, $X_L = 2\pi fL$ and is 11.84, the inductance, L is:

$$L = \frac{X_L}{2\pi f} = \frac{11.84}{2\pi \times 50}$$

$$= 0.037696$$

$$= 0.0377 \, H \text{ (Ans.)}$$

(d) Angle can be found out from p.f. relation

$$\cos\phi = \frac{R}{Z} = \frac{4}{12.5}$$

$$\phi = \cos^{-1}\left(\frac{4}{12.5}\right)$$

$$= 71.3° \text{ (Ans.)}$$

Illustration 1.51

A choking coil of negligible resistance takes a current of 10 A when connected to a 220 V, 50 c/s supply. A non-inductive resistor under the same conditions takes 12 A. If the two are connected in series and placed across the same supply, find the current taken.

Solution

Usually a coil (or choking coil) will be represented with a resistor R and an inductor L in series.

As stated, here, the choking coil has a resistance of negligible value; it has only inductance. So the inductive reactance, X_L, using the equation (1.58) is,

$$X_L = \frac{V}{I} = \frac{220}{10} = 22 \, \Omega.$$

and by the second condition the resistance is,

$$R = \frac{V}{I}$$

$$= \frac{220}{12} = 18.34 \, \Omega \text{ (by the equation 1.4)}$$

Now when they are connected in series

$$Z = \sqrt{R^2 + X_L^2}$$

$$= \sqrt{(11.34)^2 + 22^2}$$

$$= 28.636 \, \Omega$$

In an AC circuit with an impedance Z, the current is,
I = V/Z (equation 1.60)

i.e., $$I = \frac{220}{28.636}$$

$$= 7.68 \, A \text{ (Ans.)}$$

Illustration 1.52

An inductive circuit of resistance 16.5 Ω and inductance 0.14 H takes a current of 25 A. If the frequency is 50 c/s, find the voltage applied.

Solution

The inductive reactance X_L is

$X_L = 2\pi f L = 2 \times 3.14 \times 50 \times 0.14$

$\quad\quad = 43.96\ \Omega$

$R = 16.5\ \Omega$, given

Thus the impedance

$Z = \sqrt{R^2 + X_L^2}$

$\quad = \sqrt{(16.5)^2 + (43.96)^2}$

$\quad = 46.95\ \Omega$

The current taken is 25A, given

∴ Voltage $= I.Z$ (refer to equation 3.31)

$\quad\quad\quad\quad = 25 \times 46.95$

$\quad\quad\quad\quad = 1173.75$

$\quad\quad\quad\quad = $ **1174 V approx. (Ans.)**

Illustration 1.53

A coil of inductance 0.45 H takes a current of 0.85 A, when connected to a 250 V, 40 c/s supply. What is the resistance of the coil.

Solution

The impedance,

$Z = \dfrac{\text{voltage}}{\text{current}}$

$\quad = \dfrac{250}{0.85} = 294.12\ \Omega$

Inductive reactance,

$X_L = 2\pi f L$ (by equation 1.53)

$\quad\quad = 2 \times 3.14 \times 40 \times 0.45$

$\quad\quad = 113.1\ \Omega$

Both inductance and impedance are known; so the resistance, R, is

$R = \sqrt{Z^2 - X_L^2}$

$\quad = \sqrt{294^2 - (113.1)^2}$

$\quad = $ **271.5 Ω (Ans.)**

Illustration 1.54

An inductive circuit of resistance 4.5 Ω takes a current of 1.2 A when connected to a 220 V, 50 c/s supply. What is the inductance of the coil.

Solution

The current drawn = 1.2 A

∴ Impedance = $\dfrac{V}{I} = \dfrac{220}{1.2} = 183.3 \, \Omega$

The resistance is given to be 4.5 Ω

So the inductive reactance $= X_L = \sqrt{Z^2 - R^2}$

$$= \sqrt{183.3^2 - 4.5^2}$$

$$= 183.19 \, \Omega$$

And, it is also equal to $2\pi f L$

i.e., $2\pi f L = 183.19$

Thus, the inductance L is,

$$L = \dfrac{183.19}{2\pi f} = \dfrac{183.19}{2\pi \times 50}$$

$$= 0.5388 \text{ H (Ans.)}$$

Illustration 1.55

A circuit of resistance 10 Ω and inductance 0.1 H takes a current of 5.37 A when connected to a 100 V supply. What is the frequency.

Solution

The impedance,

$$Z = \dfrac{V}{I} = \dfrac{100}{5.37}$$

$$= 18.622 \, \Omega$$

The impedance in this series circuit is also,

$$= \sqrt{R^2 + X_L^2}$$

or $X_L = \sqrt{R^2 - R^2}$

$$= \sqrt{(18.62)^2 - 10^2}$$

$$= 15.71 \, \Omega$$

and $X_L = 2\pi f L$

∴ $f = \dfrac{15.71}{2 \times 3.14 \times 0.1}$

$$= 25 \text{ c/s (Ans.)}$$

Illustration 1.56

A circuit is connected to an AC supply as shown in the Fig. 1.92, the voltage of which is kept constant at 200 Volts. If the frequency is increased from 50 Hz to 100 Hz, what change in current will take place.

Solution

Refer to the Fig. 1.92.

The inductive reactance X_L, when frequency is 50 c/s, is

$$X_L = 2\pi f L$$

$$= 2 \times 3.14 \times 50 \times 0.06$$

$$= 18.84 \, \Omega \text{ at 50 HZ.}$$

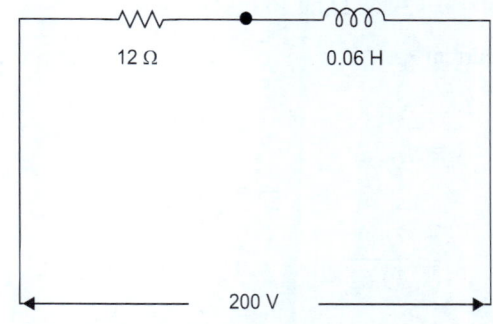

Fig. 1.92. Circuit for illustration 1.56.

The resistance as shown is 12 Ω; so the impedance, Z, can be

$$Z = \sqrt{R^2 + X_L^2}$$

$$= \sqrt{12^2 + 18.84^2}$$

$$= 22.34 \ \Omega$$

Thus the current,

$$I = \frac{V}{Z}$$

$$= \frac{200}{22.34}$$

$$= 8.952 \text{ A at 50 HZ}$$

When the frequency is doubled, because the X_L is directly proportional to the frequency, it will also be doubled.

$$X_L = 2\pi f L = 2 \times 3.14 \times 100 \times 0.06$$

$$= 37.68 \ \Omega, \text{ at 100 HZ.}$$

The impedance,

$$Z = \sqrt{R^2 + X_L^2}$$

$$= \sqrt{12^2 + (37.68)^2}$$

$$= 39.62 \ \Omega \text{ (at 100 HZ)}$$

The current,

$$I = \frac{V}{Z}$$

$$= \frac{200}{39.62}$$

$$= \textbf{5.055 A at 100 Hz (Ans.)}$$

Note: This problem is a classic example to narrate the importance of X_L's dependence on the frequency; the author suggests the reader to analyse this illustration in depth.

Illustration 1.57

An inductive circuit takes a current of 10.5 A when connected to a 500 V, 50 c/s supply. This current rises to 12.5 A when the frequency is reduced to 40 c/s, the voltage being maintained constant. Find the resistance and inductance of the circuit.

Solution

The suffix 1 denotes 50 c/s conditions and the suffix 2 denotes 40 c/s conditions.

The impedance, when the supply is 50 c/s, is

$$Z_1 = \frac{V}{I}$$

$$= \frac{500}{10.5} = 47.6 \ \Omega.$$

The impedance when the supply frequency is 40 c/s is

$$Z_2 = \frac{500}{12.5} = 40 \ \Omega$$

Inherently the inductive load contains the resistance and inductance. (refer to Fig. 1.93)

Thus the inductive reactances are,

$$X_{L1} = 2\pi f_1 L = 2 \times 3.14 \times 50 \times L$$

$$= 314 \text{ L } \Omega, \text{ at 50 c/s}$$

$$X_{L2} = 2\pi f_2 L = 2 \times 3.14 \times 40 \times L$$

$$= 251.20 \text{ L } \Omega, \text{ at 40 c/s.}$$

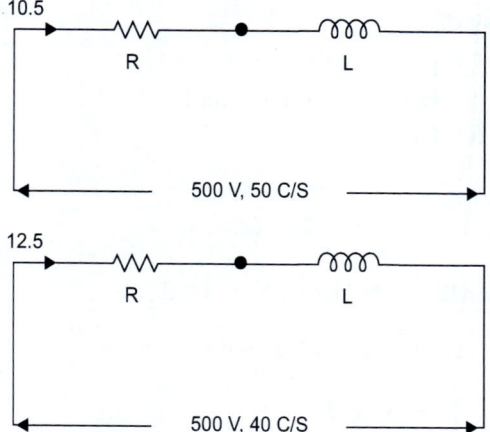

Fig. 1.93. Circuit for illustration 1.14.

Also,

$$R^2 + X_{L1}^2 = Z_1^2$$

and

$$R^2 + X_{L2}^2 = Z_2^2$$

So substituting for X_{L1} and X_{L2},

$$R^2 + (314L)^2 = 47.6^2$$

$$R^2 + (251.2L)^2 = 40^2$$

Solving both the equations the resistance and inductance are found to be,

R = 20.4 Ω

L = 0.137 H (Ans.)

Illustration 1.58

When connected to a 220-V, 50-c/s supply, an inductive circuit takes 5 A, but this current falls to 4.4 amps when 10 Ω are added in series. Find the resistance and inductance of the circuit.

Solution

Let the resistance be R Ω and the inductance L, henry

$$\text{The impedance, } Z = \frac{\text{voltage}}{\text{current}}$$

$$= \frac{220}{5} = 44\,\Omega$$

The inductive reactance,

$$X_L = 2\pi f L$$
$$= 2 \times 3.14 \times 50 \times L$$
$$= 314\, L\, \Omega$$

The impedance can also be written, using the equation (1.56) as,

$$R^2 + (314L)^2 = 44^2 \quad \text{(I)}$$

When 10 Ω is added the net resistance becomes R + 10 Ω. Then the impedance is, Z = 220/4.4 = 50 Ω; this can also be written as

$$(R + 10)^2 + (314\,L)^2 = 50 \quad \text{(II)}$$

Solving equation (I) and (II), **R = 23.2 Ω and L = 0.119 H (Ans.)**

Illustration 1.59

A circuit connected to a 200 V supply takes a current of 4.8 A. If the angle of lag is 25°, find the power absorbed.

Solution

The power absorbed = $V \times I \times \cos\phi$ for single phase supply. It is given that current is 4.8 A, voltage is 200 V and the angle is 25° lag.

$$\therefore = W = VI \cos\phi = 200 \times 4.8 \times \cos 25°$$
$$= 200 \times 4.8 \times 0.9063$$
$$= \mathbf{870\ W\ (Ans.)}$$

Illustration 1.60

The current taken by a circuit is 1.2 A when the applied p.d. is 250 V and the power taken is 135 W. What is the p.f. and the angle of lag.

Solution

Power absorbed =

$$W = V.I. \cos\phi$$

or $\quad \cos\phi = \dfrac{W}{V.I}$

$$= \dfrac{135}{250 \times 1.2}$$

$$= 0.45$$

$\cos\phi$ = Power factor = 0.45

and angle, $\phi = \cos^{-1} 0.45 = \mathbf{63.2°\ (Ans.)}$

Illustration 1.61

An inductive circuit of resistance 10 Ω and inductance 0.1 H is connected to a 100 V, 50-c/s supply.

Calculate:
(a) The current
(b) The total power absorbed
(c) The p.f.

Solution

The inductive reactance, X_L is $2\pi fL$

i.e. $X_L = 2 \times 3.14 \times 50 \times 0.1$

$$= 31.4\ \Omega$$

$\therefore$ Impedance,

$$Z = \sqrt{R^2 + X_L^2}$$

$$= \sqrt{10^2 + 31.4^2}$$

$$= 32.95\ \Omega$$

(a) The current

$$I = \dfrac{\text{voltage}}{\text{impedance}} = \dfrac{100}{32.95}$$

$$= \mathbf{3.03\ A\ (Ans.)}$$

(b) Power

$$W = I^2 R$$

$$= (3.03)^2 \times 10$$

$$= \mathbf{91.809\ W\ (Ans.)}$$

(It is to be noted that the inductor consumes zero power. Hence the power must have been consumed by the resistor only. So, the power absorbed is found out only in the resistor using the $I^2 R$ formula.)

Power factor,

$$\cos\phi = \frac{R}{Z}$$

$$= \frac{10}{32.95}$$

$$= 0.303 \text{ Ans.}$$

Note: In a 1—phase circuit power absorbed is VI cos φ; in this illustration it is 100 × 3.0 × 0.303, which is 92 W, as got in part (b).

Illustration 1.62

An inductive circuit takes a current of 27.5 A at a p.f. 0.32 lagging when connected to a 2,200 V, 25 c/s supply. Find the resistance and inductance of the circuit.

Solution

The impedance of the circuit,

$$Z = \frac{\text{voltage fed}}{\text{current drawn}}$$

$$= \frac{2{,}220}{27.5} = 80\,\Omega$$

The power consumed by a 1-phase circuit is

$$= V \times I \times \cos\phi$$

$$= 2200 \times 27.5 \times 0.32$$

$$= 19360 \text{ W.}$$

Also the power consumed = $I^2 R$; inductance is not considered for power, as it consumes zero power.

i.e., $(27.5)^2 R = 19360$

$$R = \frac{19360}{27.5 \times 27.5} = \mathbf{25.6\,\Omega \text{ (Ans.)}}$$

The inductive reactance, X_L, is given by

$$X_L = \sqrt{Z^2 - R^2}$$

$$= \sqrt{80^2 - (25.6)^2}$$

$$= 75.79\,\Omega$$

Also,

$$X_L = 2\pi f L$$

$$= 2 \times 31.4 \times 25 \times L$$

Thus,

$$L = \frac{75.79}{2 \times 3.14 \times 25}$$

$$= \mathbf{0.4825\ H \text{ (Ans.)}}$$

Illustration 1.63

A choking coil of effective resistance 2 Ω is used in series with a resistor (non-inductive) of 12 Ω, which needs a current of 5 A. If the supply voltage is 100 V at 50 HZ,

Find:

(a) The voltage across the non-inductive resistor
(b) The voltage across the choke.
(c) The p.f. of the choke.
(d) p.f of the whole circuit.

Solution

Refer to Fig. 1.94

(a) The voltage across non-inductive resistance is IR (of 12 Ω)

$= 5 \times 12 = $ **60 V (Ans.)**

(b) Now the value of voltage drop in the resistance of the choke is IR (of 2 Ω).

$= 5 \times 2 = 10$ V.

Thus voltage 60 V and 10 V are the drops in the resistance portion of the circuit only.

Supply voltage = 100 V.

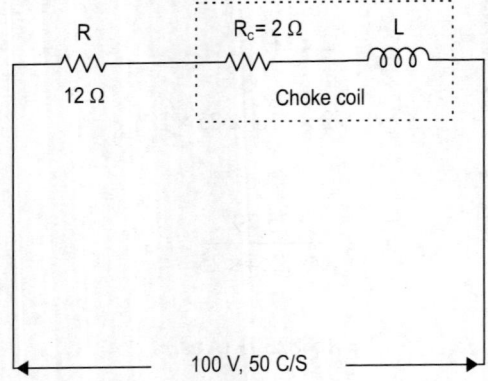

Fig. 1.94. Circuit for illustration 1.20.

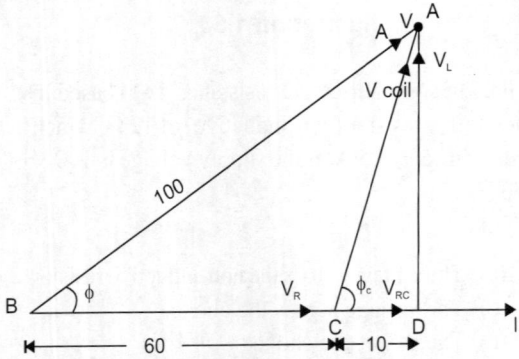

Fig. 1.95. Vector diagram for the circuit of Fig. 1.94.

Refer the vector diagram in Fig. 1.95 for this illustration. The common quantity is to be chosen as the reference in a series circuit as the current is common, it is taken as the reference. As the voltage drop in a resistor will be in phase with the current through it (refer to Fig. 1.72), 60 V (BC) and the 10 volts (CD) are kept in phase with the current.

Now voltage AD can be had from the Fig. 1.95 as,

$$\sqrt{100^2 - 70^2} = 71.4 \text{ V}$$

This voltage is the reactive voltage in quadrature with current. That is, the voltage across the inductor L alone; hence only it leads the current exactly by 90°.

Now the voltage across coke is the vector sum of it's resistance voltage and the inductor voltage i.e. AC.

$$AC = \sqrt{DC^2 + AD^2}$$

$$= \sqrt{10^2 + 71.4^2}$$

$$= 72.1 \text{ V}.$$

So the voltage across choke

$= $ **72.1 V (Ans.)**

Now, p.f. of choke $\cos \phi_C = \dfrac{DC}{AC} = \dfrac{10}{72.1} = $ **0.14**

(Ans.)

Power factor of whole circuit is $\cos \phi$. That is,

$$\cos \phi = \frac{BD}{AB}$$

$$= \frac{70}{100} = \mathbf{0.7 \text{ (Ans.)}}$$

Illustration 1.64

A non-inductive resistor of 10 ohms requires a current of 8 A and is to be fed from a 200 volts, 50 c/s supply. If a choking coil of effective resistance 1.2 ohms is used to cut down the voltage, find:

(a) Its reactance
(b) Its inductance
(c) Voltage across the choking coil
(d) Power absorbed by non-inductive resistor

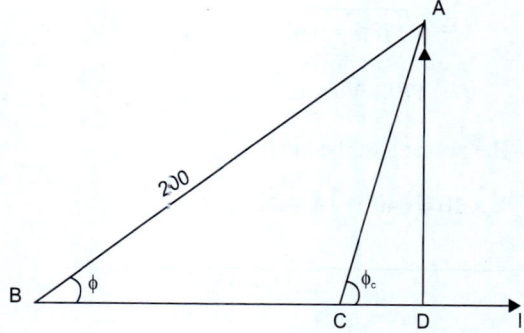

Fig. 1.97. Vector diagram for circuit of Fig. 94.

Solution

(Refer to Fig. 1.96)

The voltage drop across the non-inductive resistor is the product of current through it and its resistance. $8 \times 10 = 80$ V.

Now voltage drop in choking coil resistance is,

$= I \times R_c = 8 \times 1.2 = 9.6$ V.

∴ Total resistance drop in the circuit is,

$= 80 + 9.6 = 89.6$ V

Refer to Fig. 1.97.

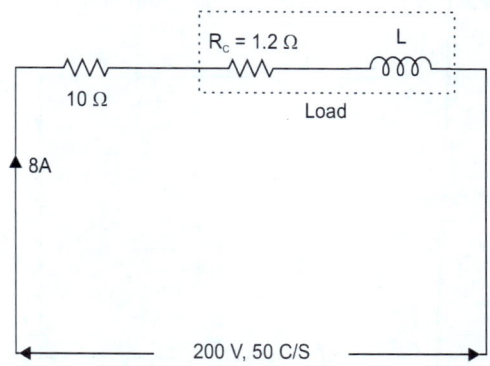

Fig. 1.96. Circuit for illustration 1.21.

AD, the voltage drop in the inductor, in quadrature to the current is

$$= \sqrt{AB^2 - BD^2}$$

$$= \sqrt{200^2 - 89.6^2}$$

$$= 178.8 \text{ V}.$$

The inductive reactance, X_L is the ratio of the voltage across it to the current through it.

i.e., $X_L = \dfrac{178.8}{8} =$ **22.35 Ω (Ans.)**

(b) The reactance

$X_L = 2\pi f L$

$= 2 \times 3.14 \times 50 \times L = 22.35$, Ω

$$L = \dfrac{22.35}{2 \times 3.14 \times 50}$$

$= $ **0.07115 H (Ans.)**

(c) Voltage across choking coil

$AC = \sqrt{AD^2 + DC^2}$

78 Basic Electrical Engineering

$$= \sqrt{178.8 + 9.6^2}$$

$$= \mathbf{179.1\ V\ (Ans.)}$$

(d) The power absorbed is $I^2 R$.

$8^2 \times 10 = \mathbf{640\ W\ (Ans.)}$

Illustration 1.65

A capacitor has a capacitance of 20 µF. Find its capacitive reactance for frequencies 25 and 50 c/s. Find in each case the current if the supply voltage is 1100 volts.

Solution

The capacitance of the capacitor is of 24 µ F. So at 25 c/s the capacitive reactance, using the equation (1.54) is,

$$X_{C1} = \frac{1}{2\pi f_1 C} = \frac{1}{2 \times 3.14 \times 25 \times 20 \times 10^{-6}}$$

$$= \mathbf{318.3\ \Omega\ (Ans.)}$$

and in case of 50 c/s supply frequency,

$$X_{C2} = \frac{1}{2\pi f_2 C} = \frac{1}{2 \times 3.14 \times 50 \times 20 \times 10^{-6}}$$

$$= \mathbf{159\ \Omega\ (Ans.)}$$

The current in the 25 c/s case is

$$I_1 = \frac{V}{X_{C1}}$$

$$= \frac{100}{318.47}$$

$$= \mathbf{3.45\ A\ (Ans.)}$$

and in 50 c/s case is

$$I_2 = \frac{V}{X_{C1}}$$

$$= \frac{1100}{159.2}$$

$$= \mathbf{6.91\ A\ (Ans.)}\ \text{(using the equation 1.59)}$$

Illustration 1.66

An inductive circuit of resistance 50 Ω and inductance 0.08 H is connected in series with a capacitor. When connected across a 200 V 50 c/s. supply, the current taken is 3.8 A leading. What is the value of the capacitor?

Solution

Refer to Fig. 1.98.

As the current is known to be 3.8 A, the impedance Z, of the circuit can be found out.

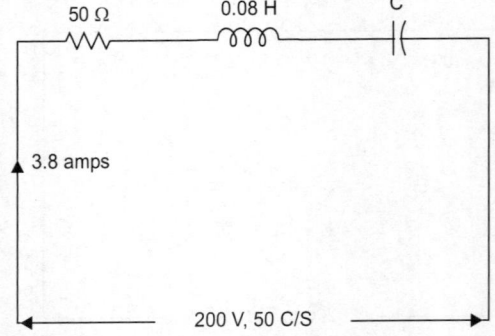

Fig. 1.98. Circuit for illustration 1.23.

$$Z = \frac{200}{3.8} = 52.64 \, \Omega$$

Also, the impedance,

$$Z = \sqrt{R^2 + X^2}$$

where R is the resistance and X the net reactance, $(= X_L \sim X_C)$

So

$$X = \sqrt{Z^2 - R^2} = X_L \sim X_C$$

$$= \sqrt{52.64^2 - 50^2}$$

$$= 16.41 \, \Omega$$

The inductive reactance

$$X_L = 2\pi f L$$

$$= 2 \times 3.14 \times 50 \times 0.08$$

$$= 25.12 \, \Omega$$

$X = X_L \sim X_C$; thus, $X_C = X + X_L$

i.e., $X_C = 16.41 + 25.12 = 41.53 \, \Omega$

and $X_C = 41.53 = \dfrac{1}{2 \times 3.14 \times 50 \times C}$

(by equation (1.54))

$$C = \frac{1}{2 \times 3.14 \times 50 \times 41.53}$$

$$= 76.6 \, \mu F \text{ (Ans.)}$$

Illustration 1.67

A circuit of resistance 10 Ω and inductive reactance 20 Ω is connected in series with an adjustable capacitor across a 220 V 50 c/s. supply. To what value must the capacitor be adjusted so that the power absorbed shall be 800 W.

Solution

Refer to Fig. 1.99.

The power dissipated is 800 W and voltage given is 220 V; let the current taken by the circuit be I A.

(Current)^2 × Resistance = Power.

The capacitor and inductor absorbs zero power.
Thus,

$$I^2 \times 10 = 800$$

$$I = \sqrt{800/10}$$

$$= 8.944 \, A.$$

The impedance of the circuit when taking 8.944 A current is

$$Z = \frac{\text{voltage}}{\text{current}} = \frac{220}{8.944}$$

$$= 24.61 \, \Omega$$

Also the impedance Z is,

$$Z = \sqrt{R^2 + X^2}$$

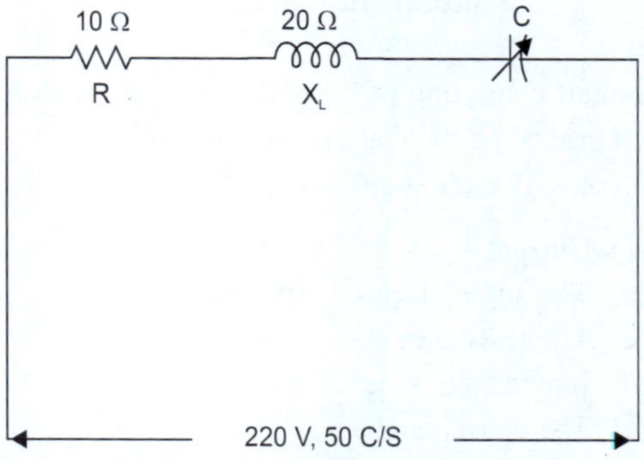

Fig. 1.99. Circuit for illustration 1.24.

or $X = \sqrt{Z^2 - R^2}$

$\quad = \sqrt{24.61^2 - 10^2}$

$X = 22.49 \ \Omega$

$X = X_L \sim X_C$

$\therefore X_C = X + X_L$

$X_C = 22.49 + 20$

$\quad = 42.49 \ \Omega$

Also $X_C = \dfrac{1}{2\pi f C}$

That is,

$X_C = 42.49 = \dfrac{1}{2 \times 3.14 \times 50 \times C}$

(by equation 1.54)

and $C = \dfrac{1}{2 \times 3.14 \times 50 \times 42.49}$

$\quad = 75 \ \mu F$ **(Ans.)**

Illustration 1.68

The circuit consisting of a resistor in series with a capacitance takes 80 W, at a power factor of 0.4 from a 100 volts 50 V c/s supply find:

(a) Current
(b) The angle of phase difference
(c) The resistance
(d) Impedance
(e) The capacitance

Solution

The power = 80 = volts × Current × Power factor

$\therefore I = \dfrac{W}{V \cos \phi} = \dfrac{80}{100 \times 0.4} = 2$ A.

So that current drawn is **2 A (Ans.)**
The power factor, $\cos \phi = 0.4$, given

$\therefore \phi = \cos^{-1}(0.4) = \mathbf{66.4°}$ **(Ans.)**

The power consumed is only by the resistor, and it is, $I^2 \times R$

i.e., $I^2 R = 80$ W.

or $\therefore R = \dfrac{80}{I^2} = \dfrac{80}{4} = 20 \ \Omega$ **(Ans.)**

Total impedance of the circuit.

$= \dfrac{\text{voltage}}{\text{current}}$

$= \dfrac{100}{2} = 50 \ \Omega$ **(Ans.)**

The capacitive reactance,

$X_c = \sqrt{Z^2 - R^2}$

$\quad = \sqrt{50^2 - 20^2}$

$\quad = 45.83 \ \Omega$

also, $X_C = \dfrac{1}{2\pi f C}$

Thus, $C = \dfrac{1}{2\pi f \times X_C}$

$= \dfrac{1}{2 \times 3.14 \times 50 \times 45.83}$

$C = 69.5\ \mu F$. (Ans.)

Illustration 1.69

A resistor of 40 ω is connected in series with a capacitor of 50 μ F across 110 V supply. At what frequency will the current taken be 2 amps.

Solution

The current drawn is 2 A; so the impedance of the circuit

$Z = \dfrac{\text{voltage}}{\text{current}}$

$= \dfrac{110}{2} = 55\ \Omega$

The capacitive reactance

$X_C = \sqrt{Z^2 - R^2}$

$= \sqrt{55^2 - 40^2}$

$= 37.75\ \Omega$

and it is also equal to $\dfrac{1}{2\pi f C}\ \Omega$

or $f = \dfrac{1}{2\pi X_C C} = \dfrac{1}{2\pi \times 39.95 \times 50 \times 10^{-6}}$

$= 84.321$ Hz or Cps. (Ans.)

Illustration 1.70

When 1 A is passed through three coils A, B and C in series, the voltage drops are respectively, 6, 3 and 8 volts on DC and 7, 5, 10 volts on AC. Determine.

(i) p.f. and power dissipated in each coil.
(ii) the p.f. of the whole circuit when AC flows.

Solution

Refer to Fig. 1.100 for DC approach.

In DC the coils offer only 'resistance' to the flow of current, where the opposition by the inductive reactance is nil. Thus, by Ohm's law of equation (1.6)

$R_1 = V_1/I_1 = 6/1 = 6\ \Omega$

$R_2 = 3/1 = 3\ \Omega$

$R_3 = 8/1 = 8\ \Omega$

(Note, in a series circuit, current is the same in all the elements.)

For AC approach consider Fig. 1.101. In this case the opposition offered will be impedance, respectively, Z_1, Z_2 and Z_3.

Thus using the ohm's law of equation (1.58)

$Z_1 = 7/1 = 7\ \Omega$

$Z_2 = 5/1 = 5\ \Omega$

$Z_3 = 10/1 = 10\ \Omega$

Fig. 1.100 Circuit for illustration 1.27.

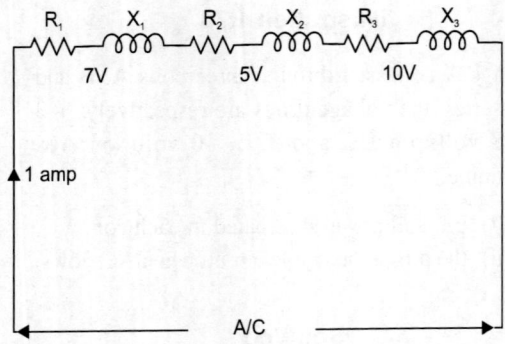

Fig. 1.101.

Thus, using the equation (1.56)

$$X_1 = \sqrt{Z_1^2 - R_1^2} = \sqrt{7^2 - 6^2} = 3.6 \ \Omega$$

$$X_2 = \sqrt{Z_2^2 - R_2^2} = \sqrt{5^2 - 3^2} = 4 \ \Omega$$

$$X_3 = \sqrt{Z_3^2 - R_3^2} = \sqrt{10^2 - 8^2} = 6 \ \Omega$$

So,

total resistance $(R_T) = R_1 \ R_2 + R_3$

$$= 6 + 3 + 8 = 17 \ \Omega$$

total reactance $(X_T) = X_1 + X_2 + X_3$

$$= 3.6 + 4 + 6 = 13.6 \ \Omega$$

Hence the total impedance, Z_T is,

$$Z_T = \sqrt{R_T^2 + X_T^2}$$

$$= \sqrt{17^2 + 13.6^2} = 21.77 \ \Omega$$

Note: Do not add impedances directly if they are not in vector form; they must be added vectorially, as stated above.

(Analyse also the illustration 30 in which the impedances were added directly because they are in vectorial form.)

The overall p.f. hence is, $\dfrac{R_T}{Z_T}$

$$\therefore \cos \phi_{over\ all} = \frac{17}{21.8} = 0.78 \ \text{lag}.$$

Illustration 1.71

Chocking coil with an effective resistance of 20 Ω and an inductance of 0.08 H is connected in series with a condenser of 100 μF. Calculate (a) the current taken by the circuit when connected to a 200 V, 50 Hz mains and (b) the cosine of the angle of phase difference.

Solution

Refer to Fig. 1.102.

The inductive reactance,

$$X_L = 2 \pi f L$$
$$= 2 \pi \times 50 \times 0.08$$
$$= 25.12 \ \Omega$$

The capacitive reactance,

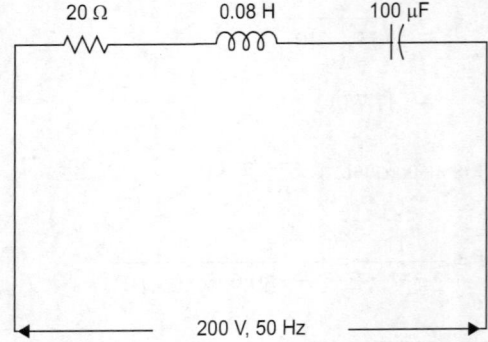

Fig. 1.102. Circuit for illustration 1.71.

$$X_C = \frac{1}{2\pi f C}$$

$$= \frac{1}{2\pi 50 \times 100 \times 10^{-6}}$$

$$= 31.85 \ \Omega$$

Thus, the net reactance

$$= X_L \sim X_C$$

$$= 25.12 \sim 31.85$$

$$= 6.73 \ \Omega$$

So, impedance, $Z = \sqrt{R^2 + X^2} = \sqrt{20^2 + 6.73^2}$

$$= 21.1 \ \Omega$$

Hence the current = V/Z = 200/21.1 = **9.48 A.**

(Ans.)

The cosine of the angle of difference, is the p.f.

$$\text{p.f.} = \cos\phi = \frac{R}{Z} = \frac{20}{21.1} = 0.95 \text{ lead.}$$

Note: p.f. is LEADING since the capacitive reactance, X_C is greater than the inductive reactance, X_L.

Illustration 1.72

A current of 5 A flows through a non-inductive resistance in series with a choking coil when supplied at 250 V, 50 Hz mains. If the voltage across the resistance is 125 V and across the coil 200 V, calculate (a) the impedance, reactance and resistance of the coil; (b) power absorbed by the coil; (c) the total power. Draw the vector diagram.

Solution

Refer to Fig. 1.103 and Fig. 1.104.

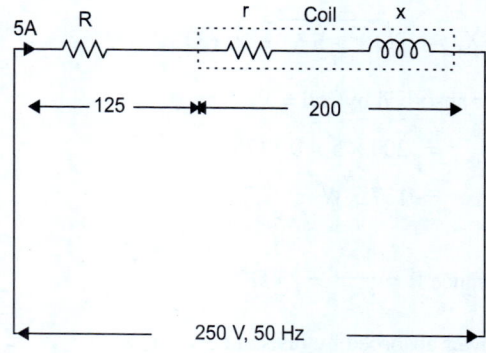

Fig. 1.103. Circuit for illustration 1.29.

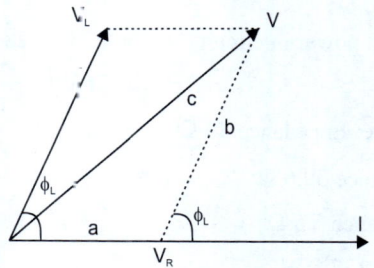

Fig. 1.104. Vector diagram.

In Fig. 1.104.
We know that $c^2 = a^2 + b^2 + 2ab\cos\theta$

Likewise $V^2 = V_R^2 + V_L^2 + 2V_R V_L \cos\phi_L$

i.e., $250^2 = 125^2 + 200^2 + 2 \times 125 \times 200 \times \cos\phi_L$

$$\therefore \cos\phi_L = \frac{250^2 - 125^2 - 200^2}{2 \times 125 \times 200} = 0.1375$$

i.e., Power factor of choke coil = 0.1375.

Impedance of choke = $\dfrac{200}{5} = 40 \ \Omega$

i.e., $\dfrac{r}{Z_L} = 0.1375$

so $r = 0.1375 \times 40 = 5.5 \ \Omega$

and $X_L = \sqrt{40^2 - 5.5^2} = 39.6\,\Omega$

Power absorbed by coil = $V_L I \cos \phi_L$

$\quad\quad\quad = 200 \times 5 \times 0.1375$

$\quad\quad\quad = 137.5$ W

Resistance R = $\dfrac{125}{5} = 25\,\Omega$

∴ Power absorbed by resistance = $5^2 \times 25$

$\quad\quad\quad\quad\quad = 25 \times 25$

$\quad\quad\quad\quad\quad = 625$ W

Hence total power in circuit $\quad = 625 + 128$

$\quad\quad\quad\quad\quad = 763$ W

Ans: Choke: Impedance 40 Ω

Reactance 39.6 Ω

Resistance 5.5 Ω

Power 137.5 W

Overall circuit power 763 W

Illustrations in Full Vector Form

Illustration 1.73

In the Fig. 1.105 below calculate, (i) the net impedance, (ii) the current, (iii) the voltage drops V_1, V_2 and V_3 (iv) Power absorbed by each impedances and (v) the total circuit power.

Solution

All are ohms. Thus the resistances and reactances are given. In vector form

$Z_1 = 4 + j3\,\Omega = 5\,\angle 36.87°\,\Omega$

$Z_2 = 6 - j8\,\Omega = 10\,\angle -53.13°\,\Omega$

and $Z_3 = 4 + j0\,\Omega = 4\,\angle 0\,\Omega$

(i) The net impedance is the **vector** sum of all the Z's Thus,

$\bar{Z} = \bar{Z}_1 + \bar{Z}_2 + \bar{Z}_3$

$= (4 + j3) + (6 - j8) + (4 + j0)$

$= 14 - j5\,\Omega = 14.87 \angle -19.65\,\Omega$

$= 14.87\,\angle -19.7\,\Omega$

(ii) By ohm's law, I = V/Z, where I, V and Z are the total circuit values.

∴ I = 100 $\angle 0$/14.87 $\angle -19.7$ = 6.73 $\angle 19.7$ A.

(iii) Again by using the ohm's law

$\bar{V}_1 = \bar{I}\,\bar{Z}_1,\ \bar{V}_2 = \bar{I}\,\bar{Z}_2$ and $\bar{V}_3 = \bar{I}\,\bar{Z}_3$

In all V's, I = 6.73 $\angle 19.7$, as it is a series circuit. Substituting the respective values,

$\bar{V}_1 = (6.73 \angle 19.7)(5 \angle 36.89)$

$\quad = 33.65 \angle 56.59 = 18.53 + j\,28$ V

$\bar{V}_2 = (6.73 \angle 19.7)(10 \angle -53.13)$

$\quad = 67.3 \angle -33.43 = 56.16 - j\,37$ V

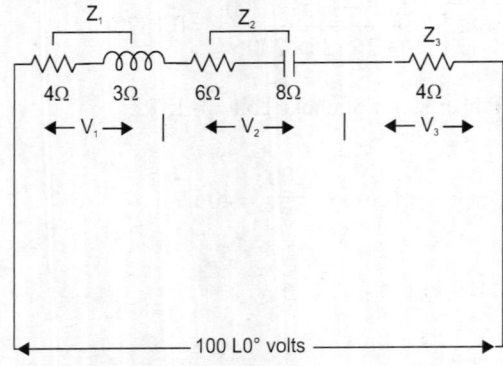

Fig. 1.105.

and $\bar{V}_3 = (6.73 \angle 19.7)(4 \angle 0)$

$= 26.92 \angle 19.7 = 25.34 + j\,9.07$ V

Check: by KVL, $\bar{V}_1 + \bar{V}_2 + \bar{V}_3 = \bar{V}_{supply}$

$= (18.53 + j\,28) + (56.16 - j\,37) + (25.34 + j\,9.07)$

$= 100 + j\,0$

$= 100 \angle 0$ volts, which is the supply voltage.

(iv) In all the Z's, only the resistor will absorb power. Thus in Z_1, the power will be $I^2R_1 = (6.73^2) \cdot 4 = 181.2$ W.

In Z_2, the power lost is $I^2 R_2$; that is, $(6.73)^2 \cdot 6 = 271.8$ W.

In Z_3, it is $(6.73)^2 \cdot 4 = 181.20$ W.

(v) The total power is the sum of all the three individual powers; it will be, $181.2 + 271.8 + 181.2 = 633.8 = 634$ W.

Illustration 1.74

Let the illustration 1.61 be considered to deal this problem vectorially. Given that,

L = 0.1 H

R = 10 Ω

V = 100 V, at 50 Hz.

Let V be taken as the reference. So, $\bar{V} = 100 \angle 0$ volts.

The inductive reactance $X_L = 2\pi f L = 31.4$ Ω. This vector form will be jX_L (refer to the note, under sec. 1.44). That is, $j\,31.4$ Ω.

Net impedance is $R + jX_L = 10 + j\,31.4$ Ω, in rectangular form. The same in polar form will be, $32.95 \angle 72.33$ Ω. (refer to the 'note 1' of sec. 1.32)

(a) The current

$$\bar{I} = \frac{\bar{V}}{\bar{Z}} = \frac{100 \angle 0}{32.95 \angle -72.33}\text{ A}$$

$= 3.03 \angle -72.33$ A

In rectangular form,

$I = 0.919 - j\,2.88$ A.

The angle -72.33 says nothing but the current is lagging the reference, that is the voltage, by 72.33°. Thus the angle between V and I is 72.33.

(b) The total power absorbed is VI cos φ, where φ is the angle between V and I. So,

P = 100 × 3.03 × cos 72.33

= 91.97

≅ 92 W

(c) The p.f. is, of course in this problem, cos 72.33; that is 0.303.

The vector diagram for the circuit is in Fig. 1.106.

Note: The readers can compare these two illustrations (18 and 31) and observe that the magnitudes are the same in both the cases; but if vectorially analysed, the phase relation will also be known. All AC circuit problems must be solved like this only, in vector form. The above problem from 1 to 29 are introduced in 'magnitude-wise' approach only, to avoid vector complexity and to make the reader to understand in basic consumptions easily.

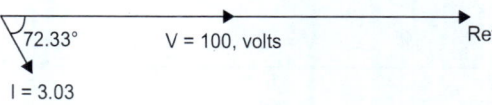

Fig. 1.106. Vector diagram for the illustration 1.31.

Illustration 1.75

Let the illustration 1.65 be considered for vectorial type approach.

Given that, C = 20 µF, and V = 1100 V

Let v be the reference; thus, V in vector form will be, V = 1100 + j0 = 1100 ∠0, V

At 25 Hz:
The capacitive reactance is,

$$X_C = \frac{1}{2\pi f C}$$

$$\therefore \quad X_C = \frac{1}{2\pi \times 25 \times 20 \times 10^{-6}} = 318.3 \, \Omega$$

In vector form,

$$X_C = -j\, 318.3 \, \Omega \text{ or } 318.3 \angle -90, \, \Omega$$

Thus the current is,

$$V/X_C = 1100 \angle 0 / 318.3 \angle -90$$
$$= 3.45 \angle 90 \, \Omega$$

From the angle of current it is clear that this current leads the reference (volts) by 90°. In a vector diagram, it is depicted in Fig. 1.107.

Readers may try other illustration dealt above in a similar vectorial sense, to get further trained in AC fundamentals.

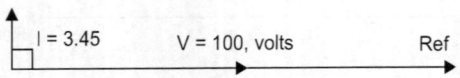

Fig. 1.107. Vector diagrams for illustration 1.32.

Illustration 1.76

A 100 V, 80 W lamp is to be operated on 230 V, 50 Hz AC supply. Calculate the inductance of the choke required to be connected in series with lamp for its operation. The lamp can be taken as equivalent to a non-inductive resistance.

Solution

Current through the lamp when connected across 100 V supply,

$$I = \frac{W}{V} = \frac{80}{100} = 0.8 \, A$$

Resistance of the lamp,

$$R = \frac{V}{I} = \frac{100}{0.8} = 125 \, \Omega$$

If a choke of inductance L henry is connected in series with the lamp to operate it on 230 V, the current through the choke will also be 0.8 A.

The impedance of the circuit when choke is connected in series with the lamp,

$$Z = \frac{V}{I} = \frac{230}{0.8} = 287.5 \, \Omega$$

Reactance of choke coil,

$$X_L = \sqrt{Z^2 - R^2} = \sqrt{287.5^2 - 125^2} = 258.5 \, \Omega$$

But $X_L = 2\pi f L$

or $\quad L = \dfrac{X_L}{2\pi f} = \dfrac{258.5}{2\pi \times 50} = 0.825 \, H$

Hence inductance of choke coil, L = **0.825 H. (Ans.)**

Illustration 1.77

An iron-cored coil has a DC resistance of 6 Ω. When it is connected to 230 V, 50 Hz mains, the current taken is 3.5 A at a power factor of 0.5. Determine:
(i) Effective resistance of the coil.
(ii) Inductance of the coil.
(iii) Resistance which represents the effect of the iron loss.

Solution

Given: D.C. resistance (True resistance), R = 6 Ω; supply voltage = 230 V, f = 50 Hz, I = 3.5 A; p.f. = 0.5.

(i) **Effective resistance of the coil, R_e:**

Total power consumed by the iron-cored choke coil,

P = Power loss in ohmic resistance + Iron loss in core = $I^2 R + P_i$

or

$\frac{P}{I^2} = R + \frac{P_i}{I^2}$, where $\frac{P}{I^2}$ is known as *effective resistance* of the coil.

∴ Effective resistance,

$R_e = \frac{P}{I^2} = \frac{VI \cos\phi}{I^2} = \frac{230 \times 3.5 \times 0.5}{(3.5)^2} = \textbf{32.86 Ω. Ans}$

(ii) **Inductance of the coil, L:**

Impedance of the coil,

$Z = \frac{V}{I} = \frac{230}{3.5} = 65.7 \Omega$

Inductive reactance of the coil,

$X_L = \sqrt{Z^2 - R_e^2} = \sqrt{(65.7)^2 - (32.86)^2} = 56.9 \Omega$

$L = \frac{X_L}{2\pi f} = \frac{56.9}{2\pi \times 50} = \textbf{0.1811 H. (Ans.)}$

(iii) **Resistance representing iron loss:**

Since $\frac{P}{I^2} = R + \frac{P_i}{I^2}$

Effective resistance, R_e = True resistance + Resistance representing iron loss 32.86 = 6 + Resistance representing iron loss

∴ Resistance representing iron loss = 32.86 − 6 = **26.86 Ω. (Ans)**

Illustration 1.78

Determine the current through the circuit of Fig. 1.108 and find V_1, V_2, V_3. Also calculate the total power dissipated in the circuit.

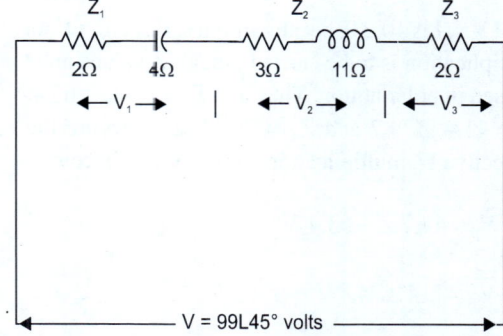

Fig. 1.108. Circuit for illustration 1.33.

Solution

Given, the applied voltage to be, 99 ∠45° V. The resistance and the reactances in vector form are, respectively 2 ∠0 Ω, −j 4 (or 4 ∠ −90) Ω, 3 ∠0 Ω, j 11 (or 11 ∠90) Ω and 2 ∠0 Ω.

The net impedance Z = Z + j X. If more resistors and reactances are in series, then net impedance will be Z = Σ R + Σ j X.

In this case,

Σ R = 2 + 3 + 2 = 7 Ω. (i.e., 7 ∠0 Ω)

and

Σ X = −j 4 + j 11 = j 7 Ω (i.e., 7 ∠90 Ω)

Thus, net impedance is,

$Z = (7 + j\,7)\,\Omega = 9.9\,\angle 45\,\Omega$

By the ohm's law, the current drawn by circuit is, $I = V/Z$. In this case, $I = 99\,\angle 45 / 9.9\,\angle 45 = 10\,\angle 0$ A.

Again by using the ohm's law, the V_1, V_2, and V_3 are evaluated.

$\bar{V}_1 = \bar{I}\bar{Z}_1$ where $Z_1 = 2 - j4\,\Omega$

$\bar{V}_2 = \bar{I}\bar{Z}_2$ where $Z_2 = 3 + j11\,\Omega$

(ref. to the Fig. 1.108)

$\bar{V}_3 = \bar{I}\bar{Z}_3$ where $Z_3 = 2 + j0\,\Omega$

In all V's, I is $10\,\angle 0$, because it is a series circuit. As multiplication is to be carried out, all Z values must be used in polar notion. They are, $Z_1 = 4.47\,\angle -63.4$, $Z_2 = 11.4\,\angle 74.7$ and $Z_3 = 2\,\angle 0\,\Omega$. Carrying the respective IZ multiplications, the V's will become,

$\bar{V}_1 = 44.7\,\angle -63.4\,V$

$\bar{V}_2 = 114\,\angle 74.7\,V$

$\bar{V}_3 = 20\,\angle 0\,V$

Check: No doubt $\bar{V}_1 + \bar{V}_2 + \bar{V}_3$ must be equal to the supply voltage, by KVL. As addition is to be done, let $\bar{V}_1$, $\bar{V}_2$ and $\bar{V}_3$ be in rectangular form; respectively they are,

$\bar{V}_1 = 19.9 - j39.7$, $\bar{V}_2 = 30 + j110$, $\bar{V}_3 = 20 + j0\,V$

$\bar{V}_1 + \bar{V}_2 + \bar{V}_3 = (19.9 - j39.7) + (30 + j110) + (20 + j0)$

$= 69.9 + j\,70.3\,V$

$= 99.1\,\angle 45.1\,V$

$\cong 99\,\angle 45\,V$

which is the supply voltage.

Illustration 1.79

Two coils A and B are connected in series across a 240 V, 50 c/s supply. The resistance of A is 5 Ω and the inductance of B is 0.015 H. If the input from the mains is 3 kW and 2 KVAr, find the inductance of A and the resistance of B.

Solution

Refer to Fig. 1.109.

It is pointed out in sec. 1.37 that the kW is $VI \cos \phi$ and the reactive power KVAr is $VI \sin \phi$; respectively these values are given to be 3 and 2.

The ratio between kVAr and kW is

$$\frac{VI \sin \phi}{VI \cos \phi} = \tan \phi = \frac{2}{3} = 0.666$$

$\therefore \phi = 33.69$. Thus the circuit p.f. is, $\cos \phi$;

It is $\cos 33.69 = 0.832$.

The circuit current I is, $VI \cos \phi / V \cos \phi$ that is,

$$I = \frac{3000}{240 \times 0.832} = 15\,A$$

Applying ohm's law, the whole circuit impedance is found out.

$Z = V/I = 240/15 = 16\,\Omega$

The pf can also be written as, R/Z. That is,

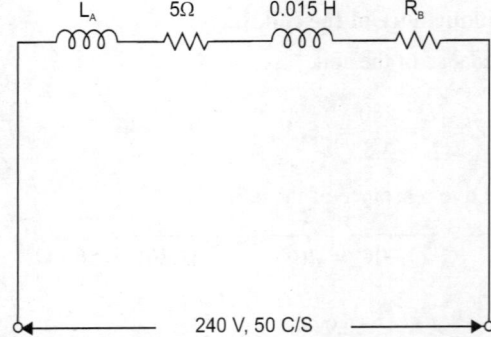

Fig. 1.109. Circuit for illustration 1.79.

$\cos \phi = R/Z$

$\therefore R = Z \cos \phi = (16)(0.832) = 13.3 \, \Omega$

Because the Z and $\cos \phi$ are for the whole circuit, the resistance R is also for the full circuit. That is,

$R = R_A + R_B.$

Thus, R_B, the resistance of coil B is,

$R_B = R - R_A = 13.3 - 5 = 8.3 \, \Omega$ **(Ans.)**

The impedance Z can also be written as

$Z = \sqrt{R^2 + X^2}$ or $X = \sqrt{Z^2 - R^2}$

using the whole circuit values,

$X = \sqrt{16^2 - 13.3^2}$
$= 8.89 \, \Omega$

This is, $X = X_A + X_B$; given that,

$X_B = 2 \pi \times 50 \times 0.015 = 4.7 \, \Omega$

Thus X_A is

$X_A = X - X_B = 8.89 - 4.7 = 4.18 \, W$ **(Ans.)**

[The reader may try for L_A, which is 0.0133 H.]

Classic Examples in AC Series Circuits

Illustration 1.80

Series RL Circuit

Consider the circuit shown in Fig. 1.110.
As 141.4 is maximum value,

$\Rightarrow$ rms value $= \dfrac{141.4}{\sqrt{2}} = 100 \angle 0 \, V$

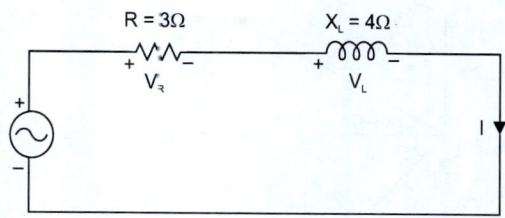

Fig. 1.110. Circuit for illustration 1.35
Let e = 141.4 sin wt ($\therefore$ maximum value = 141.4 volts) (polarities are instantaneous)

The net impedance

$Z = Z_1 + Z_2 = 3 + j4 = 5 \angle 53.13° \, \Omega$

The current in the circuit is given, by Ohm's law,

$I = \dfrac{V}{Z} = \dfrac{100 \angle 0}{5 \angle 53.13} = 20 \angle -53.13° = 12 - j16 \, A$

The voltage drop across the resistor

$V_R = IR = 20 \angle -53.13 \times 3 \angle 0$
$= 60 \angle -53.13$
$= 36 - j 48 \, V$

The voltage drop across the inductor

$V_L = IX_L = 20 \angle -53.13 \times 4 \angle 90$
$= 80 \angle 36.7$
$= 64 + j 48$

Cross checking for voltage across the circuit

$E = V_R + V_L = 36 - j 48 + 64 + j 48 = 100 \angle 0 \, V$

The power dissipated in the circuit:

Power $= VI \cos \phi$
$= 100 \times 20 \times \cos(-53.13)$
$= 1200 \, W$

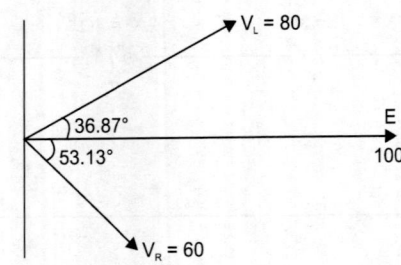

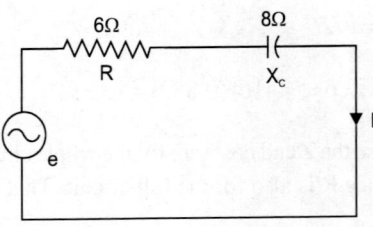

Fig. 1.112. Circuit for illustration 1.81.

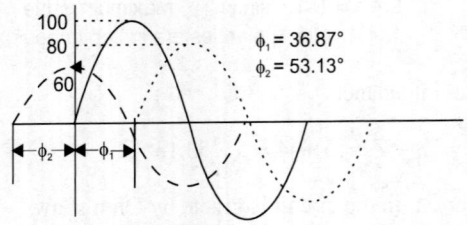

Fig. 1.111. Waveform representation for illustration 1.35.

(or $I^2 R$; that is $(20^2)(3)$, which is 1200 W)
The total power may also be given as

$$P_T = P_R + P_L$$
$$= VI \cos \phi_R + VI \cos \phi_L$$
$$= 60 \times 20 \times \cos 0 + 60 \times 20 \times \cos 90$$
$$= 1200 \text{ W}.$$

Illustration 1.81

Series RC circuit

Consider circuit of Fig. 1.112.
Given, the instantaneous value of current

$$i = 7.07 \sin(\omega t + 53.13)$$

$$I_{rms} = \frac{\text{max value of } i}{\sqrt{2}} = \frac{7.07}{\sqrt{2}}$$

$$= 5 \text{ A (magnitude)}$$

$$= 5 \angle 53.13 \text{ A}$$

The voltage drop across resistance

$$V_R = IR = 5 \angle 53.13 \times 6 \angle 0$$
$$= 30 \angle 53.13 = 18 + j\,24 \text{ V}$$

$$V_C = IX_C = 5 \angle 53.13 \times 8 \angle -90$$
$$= 40 \angle -36.87 = 32 - j24 \text{ V}$$

$$V = V_R + V_C = 18 + j\,24 + 32 - j\,24$$

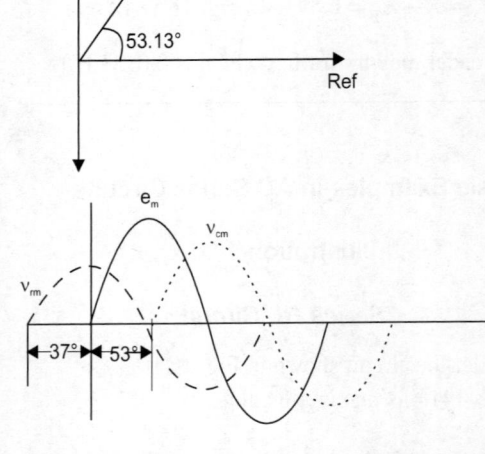

Fig. 1.113. Waveform or time domain representation for illustration 1.81.

$$= 50 = 50 \angle 0 \text{ V} = e$$

$$P = VI \cos \phi = 50 \cos (53.13)$$

$$= 150 \text{ W.}$$

$$\cos \phi = \cos (53.13) = 0.6 \text{ leading}$$

$$E_m = E_{rms} \sqrt{2} \sin \omega t$$

$$V_{cm} = V_{c\,rms} \sqrt{2} \sin (\omega t + 53)$$

$$V_{rm} = V_{r\,rms} \sqrt{2} (\omega t - 37)$$

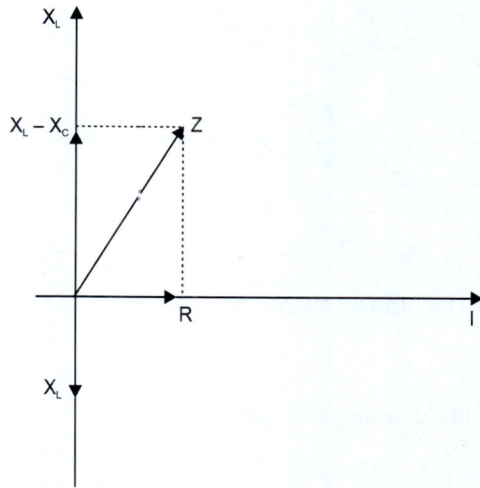

Fig. 1.115. Impedance diagram for illustration 1.82.

Illustration 1.82

RLC series circuit

Consider the circuit of Fig. 1.114

$$E_m = 70.7 \text{ volts} \Rightarrow E_{rms} = \frac{70.7}{\sqrt{2}} = 50 \text{ V}$$

The impedance of the circuit

$$Z = \sqrt{R^2 + (X_L - X_C)^2} = \sqrt{3^2 + (7-3)^2}$$

$$= \sqrt{9 + 16}$$

$$= \sqrt{25} = 5 \; \Omega$$

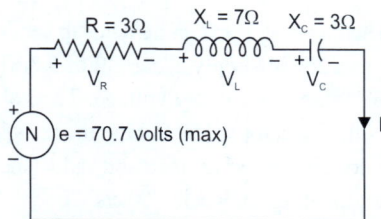

Fig. 1.114. Circuit for illustration 1.82.

or in vector form,

$$\bar{Z} = 3 + j4 = 5 \angle 53.13°$$

The current flowing in the circuit

$$I = \frac{V}{Z} = \frac{50 \angle 0}{5 \angle 53.13°} = 10 \angle -53.13°$$

$$= 6 - j8 \text{ A}$$

The voltage across the various elements are given as

$$V_R = IR = 10 \angle -53.13° \times 3 \angle 0°$$

$$= 30 \angle -53.13° \text{ V}$$

$$V_L = IX_L = 10 \angle -53.13° \times 7 \angle 90°$$

$$= 70 \angle 36.87° \text{ V}$$

$$V_C = IX_C = 10 \angle -53.13° \times 3 \angle -90°$$

$$= 30 \angle -143.13° \text{ V}$$

Using KVL, the reader may check the following equation:

$$V_R + V_L + V_C = V_{supply}$$

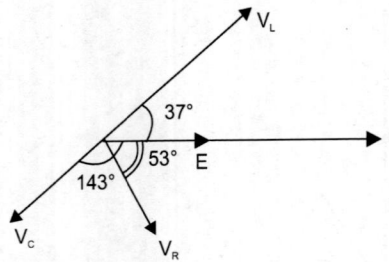

Fig. 1.116. Phasor diagram for illustration 1.82.

Time domain equations:

$v = 70.7 \sin \omega t$

$i = 10\sqrt{2} \sin(\omega t - 53.13)$

$v_R = 30\sqrt{2} \sin(\omega t - 53.13)$

$v_L = 70\sqrt{2} \sin(\omega t + 36.87)$

$v_C = 30\sqrt{2} \sin(\omega t - 143.13)$

Power dissipated in the circuit

$P = VI \cos \phi$

$= 50 \times 10 \times \cos(53.13)$

$= 300 \text{ W}$

1.50 TUTORIAL ILLUSTRATIONS IV

T.IV-1: The PD applied to a circuit is 240 volts the current taken is 2 amps. The power absorbed is 384 watts. Determine pf and the phase angle of the circuit. Is the angle lag or lead.

[Ans: 0.8, 36°, lag]

T.IV-2: An AC electromagnet takes 8 A of current and dissipates 800 W; when placed in series with a non-inductive resistor of 10 Ω across a 220 V, 50 Hz supply, the current taken is 8 A. find (i) the voltage across the electromagnet.

[184.5 V]

T.IV-3: The power taken by an inductive circuit when connected to a 100 volts, 50 c/s is 500 W. The current taken is 8 A. Determine the following quantities for the circuit at these stated conditions.

(a) The resistance (b) The impedance (c) The reactance (d) The p.f. of the circuit.

[Ans: 7.812 Ω, 12.5 Ω, 9.76 Ω, 0.625]

T.IV-4: A 100 V, 100 W lamp is supplied from a 200 V AC supply through a capacitor. Determine (a) the capacitance needed (b) angle of phase difference between the current and the voltage.

[Ans: 18.5 μF, 60°]

T.IV-5: What capacitance in series with a resistor of 40 Ω will allow a current of 15.5 A to flow in a circuit applied with 1100 V. The frequency of operation is 50 c/s.

[Ans: 54.4 μF]

T.IV-6: The capacitive reactance of 3 Ω is connected in series with a resistance of 4 Ω across a 141.4 V supply, Evaluate the current.

[Ans: 28.3 A]

T.IV-7: A non-inductive resistor in series with the capacitor takes 2.5 A, when the supply is 120 V at 50 c/s. The power dissipated in the resistor is 250 W. Find the value of the resistor and the capacitor.

[Ans: 40 Ω, 120 μf]

T.IV-8: When a resistor and an inductor in series are connected to a 250 V supply, a current of 5 A flows lagging 35° behind the supply voltage. The voltage drop across the inductor is 150 V. Find the resistance of the resistor and the resistance and inductance of the inductor, the frequency being 50 c/s.

[Ans: 31.8 Ω; 9 Ω, 0.087 H]

Reflection Section

1. What is true power and what is apparent power?
2. In a series circuit containing R, L and C, power loss occurs in which element?
3. Justify that in any AC circuit apparent power is more than the actual power.
4. The apparent power drawn by an AC circuit is 10 kVA and the active power drawn is 8 kW. Draw the power triangle to scale and find out the reactive power in the circuit. Mention its unit?
5. Discuss the relation between the value of inductance and the relative permeability of the material used on a coil.
6. The RMS value of the AC current is equivalently given in terms of steady DC current. Justify. (Hint: Heat produced in the circuit.)
7. Mention R, X_L and X_C in polar form, phasor form, waveform and rectangular form.
8. Derive expressions for active power, apparent power and true power.
9. The phase difference between current and voltage in an element is 30°. Comment on the essential condition for the frequency, with justification.
10. What is the power factor of an inductive circuit and a capacitive circuit?
11. Discuss the role of introducing power factor for AC circuits?
12. What will be the phase angle between two alternating waves of equal frequency, when one wave attains maximum value the other wave is at zero value? Comment on the same question, if the frequency of one of the waves is altered.

PART IV: RESONANCE

1.51 RESONANCE

The resonant circuit is a combination of the resistance R in ohms, the inductance L in henry and the capacitance C in farads having a frequency response characteristics as shown in Fig. 1.117. The response could be a voltage, a current, an acoustic (sound) wave etc.

The characteristic behaviour resonance can occur only in AC circuits in which R, L and C all are present and is of two types viz., **Series resonance and Parallel resonance**.

It could be noted from Fig. 1.117 that the response is maximum only for a particular frequency, f_r, known as the **resonant frequency**, and this response decreases to the right and left of this resonant frequency.

It is seen from Fig. 1.117 that the response is almost maximum for a frequency range say from f_1 to f_2. On this basis, a resonant circuit could be said as a circuit which relates a range of frequencies (f_1

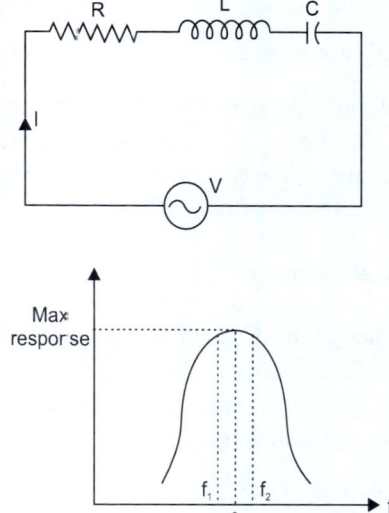

Fig. 1.117. A frequency—response curve

to f_2) for which the response will be near or equal to the **maximum**. Note particularly that the frequencies to the far left or far right are of no significance as far as the maximum response is concerned; hence for all practical purposes, such frequencies are nullified or neglected with respect to the maximum system response.

The apt example for the resonant circuit is the radio or television receiver. The radio or television receiver has a frequency—response curve as indicated in Fig. 1.117 for **each** broadcast station. The broadcast stations broadcast their programmes (which gets transformed in terms of an electrical equivalent response) at a particular frequency; when the tuner of the radio set is moved and if such tuning reaches the frequency of the broadcast station, the receiver receives the station's programmes (of course, this 'electrical equivalent' programme will be again transformed into human hearable form). Hence the resonant circuit is also known as the **tuned circuit**.

1.52 SERIES RESONANCE

The property of cancellation of reactance in a R-L-C series AC circuit is known as the resonance. Mathematically $X_L - X_C = 0$, at resonance. That is,

$$X_L = X_C \text{ at resonance} \quad (1.61)$$

In an R-L-C series AC circuit, the impedance is given by, $Z = \sqrt{R^2 + X_L - X_C)^2}$. (Refer to equation (1.50)). As at resonance. $X_L = X_C$, the impedance reduces to the resistance itself. Thus,

$$Z = R, \text{ at resonance} \quad (1.62)$$

By ohm's law in an AC circuit the current is given by, $I = V/Z$. At resonance, as $Z = R$, the current I can be written as, $I = V/R$. More precisely, with V and the reference, in vetorial form,

$$I = V \angle 0 / R \angle 0$$
$$= (V/R) \angle 0° = I \angle 0.$$

This implies that the current in the series circuit is in phase with the voltage applied.

Thus, when a series AC circuit is experiencing resonance, the voltage applied and the current in the circuit will be in-phase. (1.63)

Recollect the equation (1.52), which says that the ratio of the total resistance to the total impedance of a circuit is the power factor. As at resonance, the resistance and the impedance are the **same** (refer to the equation 1.62), the power factor will become one- i.e., **unity**. Thus,

$$\cos \phi = R/Z = R/R = 1, \text{ at resonance} \quad (1.64)$$

Yet another interesting observation, which can occur only at resonance in an AC circuit, is the 180° phase shift between the voltage across the inductance and the voltage across the capacitance.

From equation (1.53), the voltage across the inductance is, $V_L = I X_L$. At resonance current in polar form is $I \angle 0$ [by equation (1.63)]; the inductive reactance X_L in polar form is $X_L \angle 90$ (refer to the note of sec 1.47) Thus,

$$V_L = I \angle 0 \, X_L \angle 90 = I X_L \angle 90 \quad (I)$$

Similarly using the equation (1.53), the voltage across the capacitance is, $V_C = I X_C$; the capacitive reactance X_C in polar form is $X_C \angle -90$ (refer to the note of Section 1.47). Thus,

$$V_C = I \angle 0 \, X_C \angle -90 = I X_C \angle -90 \quad (II)$$

It is clear from equation (I) and (II) that the V_L, voltage across the inductance, leads the reference by 90° and V_C, the voltage across the capacitance, lags the reference by 90°. Thus they are displaced by 180°.

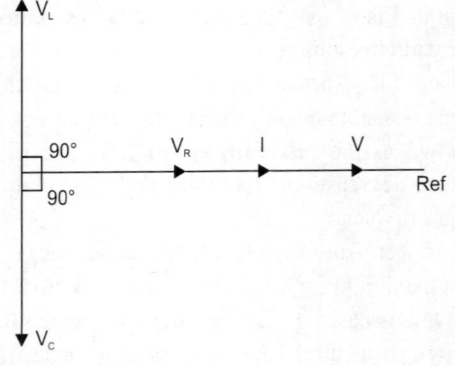

Basic Electricity 95

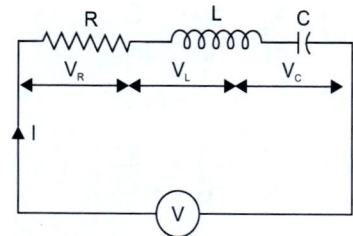

Fig. 1.118. (A) The voltage, current relation of an R-L-C series AC circuit under resonance. (B) RLC series AC circuit.

This occurs only at resonance in AC series circuit. Moreover, as the current in a series circuit is same and at resonance $X_L = X_C$, the voltage across the inductor $V_L = IX_L$ and the voltage across the capacitor $V_C = IX_C$ will be equal is magnitude. These conditions and the equation (1.63) are in Fig. 1.118.

Resonant frequency

Consider the Fig. 1.118(B) for which, at resonance, the impedance diagram is in Fig. 1.119.

Using the condition of resonance from equation (1.61) and the equations (1.53) and (1.54)

$$X_L = X_C$$

$$2\pi f L = \frac{1}{2\pi f C}$$

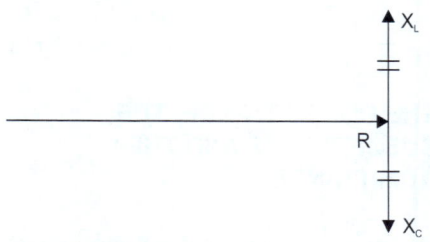

Fig. 1.119. Impedance diagram of RLC series AC circuit at resonance.

Since this balance can occur only at resonance let the frequency of operation f be changed with the resonant frequency f_r.

i.e., $2\pi f_r L = \dfrac{1}{2\pi f_r C}$

cross multiplying, $4\pi^2 f_r^2 LC = 1$

$$f_r^2 = \frac{1}{4\pi^2 LC}$$

$$\therefore f_r = \frac{1}{2\pi\sqrt{LC}}, \text{HZ} \qquad (1.65)$$

Current in a Series Resonant Circuit

In any AC circuit, the current I is V/Z, where $Z = \sqrt{R^2 + (X_L \sim X_C)^2}$. At resonance, $X_L - X_C = 0$; thus $Z = R$.

Under non-resonant condition it is known that, with X_L and X_C present, the impedance Z will be greater than the resistance R, because $Z = \sqrt{R^2 + (X_L \sim X_C)^2}$ and $X_L \sim X_C$ is not zero. But, at resonance, as $X_L \sim X_C = 0$, the impedance reaches the resistance value itself thus becoming less. This makes the current I (= V/Z = V/R) to reach a maximum. Thus, a series AC circuit will experience a maximum current only at **resonance**; at all the other times, the current will be lesser.

Voltage magnification

When a series R-L-C series AC circuit undergoes resonance, as derived in the above discussion, $V_L = V_C$. That is,

$$I X_L = I X_C$$

But, at resonance, I = V/R, as Z = R. Substituting the value of the series circuit current I in any one of the voltages V_L or V_C, it can be seen that

$$V_L = IX_L = \frac{V}{R} \cdot X_L = V\left(\frac{X_L}{R}\right)$$

$$V_C = IX_C = \frac{V}{R} \cdot X_C = V\left(\frac{X_C}{R}\right)$$

The above derivations reveal that the voltage across the inductor V_L or the voltage across the capacitor V_C is few times greater than the total supply voltage itself, by the factor X_L/R or X_C/R respectively. As these factors magnify the voltages V_L or V_C. These factors are known as the **voltage magnifications**.

The variation of impedance and current with frequency is shown in Fig. 1.120.

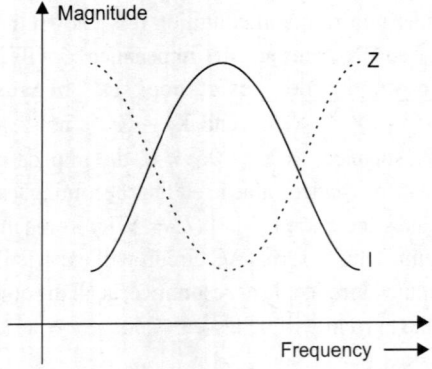

Fig. 1.120. I and Z Vs f_r.

Quality Factor

The ratio of the reactive power P_Q (of either inductor or capacitor) to the average power P of the resistor, at resonance, is the quality factor, denoted by Q_S for series resonant circuit. So,

$$Q_S = \frac{P_{q-L}}{P} = \frac{I^2 X_L}{I^2 R} = \frac{X_L}{R} = \frac{\omega L}{R}$$

$$Q_S = \frac{P_{q-C}}{P} = \frac{I^2 X_C}{I^2 R} = \frac{X_C}{R} = \frac{1}{\omega CR}$$

So,

$$Q_S = \frac{\omega L}{R} = \frac{1}{\omega CR} \qquad (1.66)$$

Other way of defining the quality factor is, at the series resonance, as $V_L = V_C$, the V_L and the V_C will be greater than the supply voltage by the factor, voltage magnification. The ratio V_L/V is referred to be this quality factor. Thus,

$$Q_S = \frac{V_L}{V} = \frac{IX_L}{IZ} = \frac{IX}{IR}$$

i.e., $$Q_S = \frac{I.2\pi f_r L}{IR} = \frac{2\pi f_r L}{R}$$

Substituting for f_r from the equation (1.65)

$$Q_S = \frac{2\pi L}{R} \cdot \frac{1}{2\pi\sqrt{LC}}$$

$$Q_S = \frac{1}{R}\sqrt{\frac{L}{C}} \qquad (1.67)$$

1.53 THE CONDUCTANCE, THE SUSCEPTANCE AND THE ADMITTANCE

Let an AC circuit be considered in which the circuit current of I amps lags the supply voltage V by an angle φ as in Fig. 1.121.

Real part of the current I is, I cos φ. By Ohm's law, I = V/Z and the pf, cos φ is R/Z. Thus,

$$I\cos\phi = \frac{V}{Z} \cdot \frac{R}{Z}$$
$$= VG$$

where $G = R/Z^2$, known as the **conductance**. This is something equal to the reciprocal of the resistance; the unit is mho. Symbol is ℧.

Imaginary part of the current is I sin φ, and sin φ, from the fig. 1.91, is X/Z.

$$I\sin\phi = \frac{V}{Z} \cdot \frac{X}{Z}$$
$$= VB$$

where $B = X/Z^2$, known as the **susceptance**, unit is mho.

Susceptance, $B = X/Z^2$;

Admittance, $Y = 1/Z$;

$$Y = \sqrt{G^2 + B^2};$$

$$X = G \pm j B; \qquad (1.68)$$

Note 1: In series circuits, the net impedance is the vector sum of all the individual impedances. Similarly, in parallel circuits the net admittance, Y, is the vector sum of all individual branch admittances.

Note 2: In series circuits, the circuit current I is V/Z_{net}. In parallel circuits, each branch will experience a current of VY_1, VY_2 etc., (as V is common for all branches). The net current will be the vector sum of all these currents. That is,

$$VY = VY_1 + VY_2 + ... + VY_N$$

$$VY = V[Y_1 + Y_2 + ... + Y_N]$$

$$\therefore Y_{net} = Y_1 + Y_2 + ... + Y_N$$

Refer to Fig. 1.122.

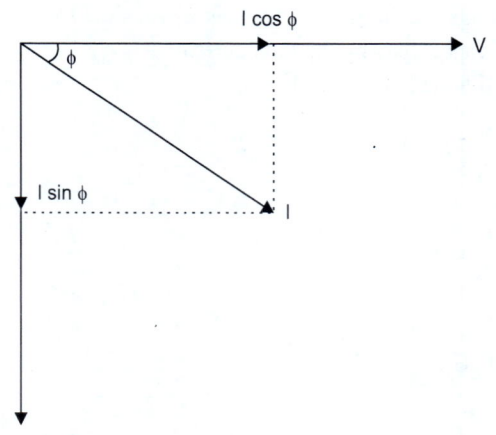

Fig. 1.121.

The admittance is the reciprocal of the impedance, Z; that is, Y = 1/Z, where Y is the admittance; unit is mho. Thus,

Conductance $\quad G = R/Z^2$;

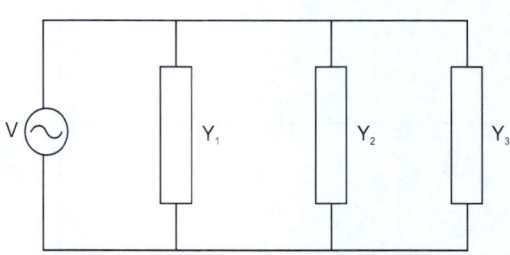

Fig. 1.122. A parallel circuit

Note 3: The p f. in a series circuit R/Z. In parallel circuit, equivalent to R is G and Z is Y. Thus p.f. is G/Y.

1.54 PARALLEL RESONANCE

The property of cancellation of susceptance of a R-L-C parallel AC circuit is known as the parallel resonance. In parallel resonance also, the p.f. is 1, i.e., unit,

Or

Resonance in a parallel circuit is said to occur when the quadrature component of the current of one branch is equal to the quadrature component of the current in the other branch.

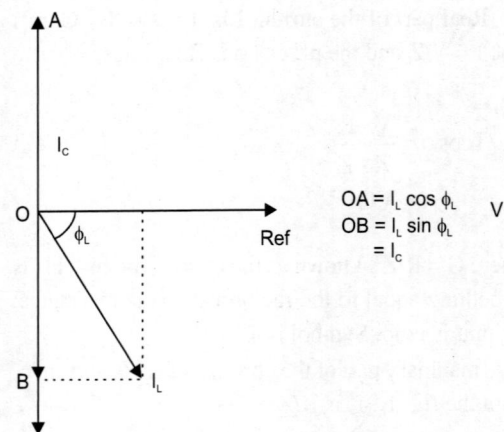

Fig. 1.124. Vector diagram for Fig. 1.123 at resonance.

Current in a Parallel Resonant Circuit

An AC parallel circuit experiences a minimum possible current only at resonance; at all the other working conditions, the current will be greater than the resonant current.

For a proof, consider the circuit of Fig. 1.123.

The total current is I, which gets divided into I_L and I_C; these currents are in vector form in the Fig. 1.124.

At resonance, the quadrature component of the currents will be the same. That is,

$$I_C = I_L \sin \phi_L \quad \text{(I)}$$

Also, as the quadrature component of the currents are the same at resonance, vectors OA and OB cancel leaving the total current,
I to be $I_L \cos \phi_L$. That is,

$$I = I_L \cos \phi_L$$

i.e. $$I = \frac{V}{Z_L} \cdot \frac{R}{Z_L} = \frac{VR}{Z_L^2} \quad \text{(II)}$$

From (I), $$\frac{V}{X_C} = \frac{V}{Z_L} \cdot \frac{X_L}{Z_L}$$

i.e. $$X_L X_C = Z_L^2$$

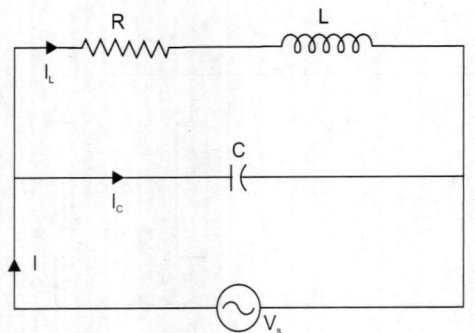

Fig. 1.123. A parallel AC circuit for parallel resonance discussion.

$$2\pi f_r L \frac{1}{2\pi f_r C} = Z_L^2$$

$$\frac{L}{C} = Z_L^2$$

This value of Z_L^2 in equation (II) will get

$$I = \frac{VR}{L/C}$$

or

$$I = \frac{V}{(L/CR)} \qquad (1.69)$$

The quantity L/CR is known as the **dynamic resistance** which is much larger than simple resistance R. Thus, a parallel resonant circuit offers maximum impedance and thus the current is minimum.

Quality Factor

For a parallel AC circuit, the quality factor, Q, can be defined as the ratio between I_C and I.
For the Fig. 1.123.

i.e., $\quad Q = \dfrac{I_C}{I} = \dfrac{V/X_C}{V/(L/CR)}$

$\qquad = \dfrac{V/(1/\omega C)}{V/(L/CR)}$

$\qquad = \dfrac{V\omega C}{\left(\dfrac{VCR}{L}\right)}$

$\qquad = \dfrac{V\omega CL}{VCR}$

$\qquad = \dfrac{\omega L}{R}$

This ω is at resonance; therefore,

$\qquad = 2\pi f_r \cdot \dfrac{L}{R}$

$\qquad = 2\pi \cdot \dfrac{1}{2\pi\sqrt{LC}} \cdot \dfrac{L}{R}$

$\qquad = \dfrac{1}{R}\sqrt{\dfrac{L}{C}}$

which is nothing but the equation (1.67).
So in both the series resonant circuit and parallel resonant circuit, the quality factor Q, is the same.

1.55 WORKED ILLUSTRATION V

Illustration 1.83

When the capacitor is set to 500 μF in a circuit, the current reaches its maximum. Find the inductance of the circuit. The frequency of operation is 1 M c/s.

Solution

The current in an AC series circuit reaches a maximum, only at resonance. So, the above given circuit is at resonance.
As per the equation (1.61), at resonance,

$$X_L = X_C$$

i.e., $\quad 2\pi f_r L = \dfrac{1}{2\pi f_r C}$

$$4\pi^2 f_r^2 LC = 1$$

Thus

$$L = \dfrac{1}{4\pi^2 f_r^2 C}$$

Substituting the values given, the inductance becomes,

$$L = \frac{1}{4\pi^2 (1\times 10^6)^2 (500\times 10^{-6} \times 10^{-6})}$$

$$= 5.06 \times 10^{-5} \text{ H (Ans.)}$$

Illustration 1.84

A coil having a resistance of 10 Ω and an inductance of 0.15 H is connected in series with a 100 μF capacitor. At what frequency will the current by a maximum? If the applied voltage is 200 V, find the current. Calculate also the p.d. across the inductor and the capacitor.

Solution

The current will be maximum only at resonance. Thus, the frequency is to be found out at the resonant conditions, which by equation (1.65) is,

$$f_r = \frac{1}{2\pi\sqrt{LC}}$$

$$= \frac{1}{2\pi\sqrt{0.15 \times 100 \times 10^{-6}}} = \textbf{41.1 Hz. (Ans.)}$$

At resonance, the current is simply V/R.
It is, 200/10 = 20 A.
The capacitive reactance X_C is

$$\frac{1}{(2\pi \times 41.4 \times 100 \times 10^{-6})}.$$

It is, 38.4 Ω, = 39 Ω.
The inductive reactance X_L is $2\pi \times 41.4 \times 0.15$.
It is, 39 Ω.
Therefore the voltage across the inductor is IX_L and the capacitor is IX_C. They can respectively be found out as 780 V each.

Note: In full vector form, letting V as the references, it will be denoted as, 200 ∠0 V.

R = 10 ∠, Ω.

Thus the current will be 200 ∠0/10 ∠0 = 20 ∠0, amps.

X_C = j 39 Ω or 39 ∠–90 Ω.

X_L = j 39 Ω or 39 ∠90 Ω.

Then the respective voltage drops will be

(20 ∠0) (39 ∠–90) = 780 ∠–90 V

and

(20 ∠0) (39 ∠90) = 780 ∠90 V

All these quantities as a vector diagram is in Fig. 1.125.

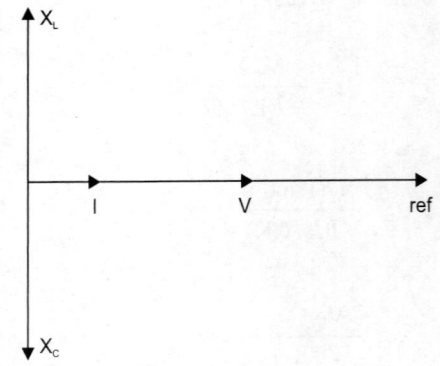

Fig. 1.125. Vector diagram to the scale, for illustration 1.84.

Illustration 1.85

Calculate the Q factor of a series circuit consisting of a resistance of 20 Ω, an inductor of 0.3 H; f_r is 50 c/s.

Solution

Using the relation (3.42),

$$Q = \frac{\omega L}{R} = \frac{2\pi f L}{R}$$

$$= (2\pi \times 50 \times 0.3)/20$$

$$= \mathbf{4.712 \text{ (Ans.)}}$$

Illustration 1.86

A resistor and a capacitor are connected in series with a variable inductor. When the circuit is connected to a 240 V, 50 c/s. supply, the maximum current given by the inductance is 0.5 A. At this current the voltage across the capacitor is 250 V. Calculate the values of (a) the resistance, (b) the capacitance and (c) the inductance.

Solution

As the circuit experiences a maximum current, it is at resonance. At resonance, it is known that $I = V/R$.

From this, $R = V/I = 240/0.5 = \mathbf{480 \; \Omega.}$ **(Ans.)**

Also given the drop in the capacitor to be 250 V. This is IX_C. From this the capacitive reactance X_C is found to be,

$$250/0.5 = 500 \; \Omega.$$

That is,

$$500 = 1/2\pi f C.$$

So, $C = 1/(2\pi \times 50 \times 500)$

$$= \mathbf{6.37 \; \mu F \text{ (Ans.)}}$$

From the equation, (1.65)

$$fr = \frac{1}{2\pi\sqrt{LC}}$$

or

$$L = \frac{1}{f_r^2 \, 4\pi^2 \, C} = \frac{1}{50^2 \, 4\pi^2 (6.37 \times 10^{-6})}$$

$$= \mathbf{1.59 \; H \text{ (Ans.)}}$$

[The reader even may get L from, X_L expression.]

Note: The 50 c/s is taken to be the resonant frequency.

Illustration 1.87

A series circuit consists of resistance of 4 Ω, an inductance of 0.5 H, and a variable capacitor across a 100 V, 50 Hz supply. Find (a) the capacitance for getting resonance, (b) p.d. across the inductance and capacitance, (c) the Q-factor of the series circuit.

Solution

The readers may technically reason the problem before the below solution is traced. It is a classic problem.

At resonance, Z reduces to R itself, thus making the current to be V/R. So,

$$I = 100/4 = 25 \text{ A.}$$

This will be the current through all the elements. Let the voltage drop across the inductor be found out.

$$V_L = IX_L = 25 \, (2\pi \times 50 \times 0.5)$$

$$= 3927 \text{ V.}$$

This also will be the voltage drop across the capacitor, as at resonance $V_L = V_C$. That is,

IX$_C$ = 3927.

Thus, X$_C$ = 3927/25 = 157 Ω.

So, C = 1/(2 π f X$_C$) = 1/(2 π × 50 × 157)

C = 20 μF (Ans.)

[or, at resonance X$_L$ = X$_C$. From this also, C can be got)

Pd across the inductance and capacitance is 3927 volts each, displaced by 180°. (Ans.)

$$Q = \frac{\omega L}{R} = \frac{2\pi f L}{R} = (2\pi \times 50 \times 0.5)/4$$

Q = 39.26 (Ans.)

Examples in Parallel Circuits

Illustration 1.88

A coil having a resistance of 5 Ω and an inductance of 0.02 H is arranged in parallel with another coil having a resistance of 1 Ω and an inductance of 0.08 H. Calculate the current through the combination and the power absorbed when a voltage of 100 V at 50 Hz is applied.

Solution

The circuit is depicted in the Fig. 1.126.

Let V be the reference. So, V = 100 ∠0 V.

The impedance of the branch 1 is R$_1$ + j X$_1$. The reactance of the first branch is, X$_1$ = 2 π f (0.02). It is, 6.28 Ω. Thus

Z$_1$ = 5 + j 6.28 = 8.027 ∠51.4 Ω

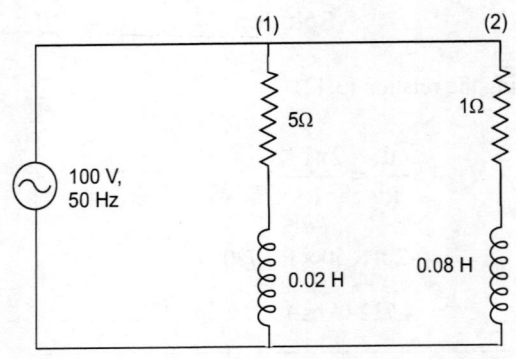

Fig. 1.126. Circuit for illustration 1.88.

and

$$Y_1 = \frac{1}{Z_1} = \frac{1}{8.027 \angle 51.4}$$

= 0.1245 ∠–51.47 = 0.0776 – j 0.097 mho

Similarly, Z$_2$ = R$_2$ + j X$_2$ = 1 + j (2 π f × 0.08)

= 1 + j 25.13 = 25.153 ∠ 87.7 Ω

and

$$Y_2 = \frac{1}{Z_2} = \frac{1}{25.153 \angle +87.77}$$

= 0.03975 ∠–87.7

= 1.588×10^{-3} – j0.0397 mho

Thus the net admittance, Y, of the whole circuit is the vector sum of individual branch admittances, Y$_1$ + Y$_2$

∴ $\bar{Y} = \bar{Y}_1 + \bar{Y}_2$ = (0.0776 – j0.097) +

(1.588×10^{-3} – j0.0397)

= 0.0792 –j 0.137

= 0.1583 ∠–60, mho

The total current is VY. That is,

$I = (100 \angle 0)(0.157 \angle -60)$

$= \mathbf{15.7 \angle -60 \text{ A (Ans.)}}$

Illustration 1.89

In a series-parallel circuit, the parallel branches A and B are in series with C. The impedances are:

$Z_A = (4 + j\,3)\ \Omega$

$Z_B = (4 - j\,5.3)\ \Omega$

$Z_C = (2 + j\,8)\ \Omega$

If the current $I_C = (25 + j\,0)$ amps draws the complete phasor diagram determining branch currents and branch voltages and the total voltage. Determine also the total admittance.

Solution

Observe the Fig. 1.127.

Using the current division technique of sec. (1.20B)

$I_A = \dfrac{I_C \cdot Z_B}{Z_A + Z_B}$, all in vector form.

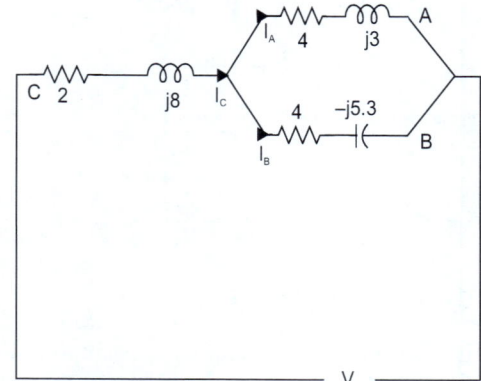

Fig. 1.127.

$= \dfrac{(25 + j0)(4 - j5.3)}{(8 - j2.3)}$

$= 19.94 \angle -36.92 = 15.94 - j\,11.97\ \text{A}.$

Similarly

$I_B = \dfrac{I_C Z_A}{Z_A + Z_B} = \dfrac{(25 + j0)(4 + j3)}{(8 - j2.3)}$

$= 15.01 \angle 52.9$

$= 9.056 + j\,11.97\ \text{A}.$

The voltage across the branch A is,

$V_A = I_A Z_A$

$= 19.94 \angle -36.92\ (4 + j\,3)$

$= 99.7 \angle -0.027$

$\cong 100 \angle 0\ \text{V}.$

As branch B is parallel to branch A, $V_B = V_A$ (reader may check this with $I_B Z_B$ formula).

Voltage across the branch C is,

$V_C = I_C Z_C$

$= (25 \angle 0)(2 + j\,8)$

$= (25 \angle 0)(8.25 \angle 76)$

$= 206.25 \angle 76.$

$= 50 + j\,200\ \text{V}.$

Thus the total voltage applied is the vector sum of V_A or V_B and V_C. That is,

$V = V_A + V_C$

$= (100 + j\,0)(50 + j\,200)$

$= 150 + j\,200$

$= \cong 250 \angle 53.2\ \text{V}.$

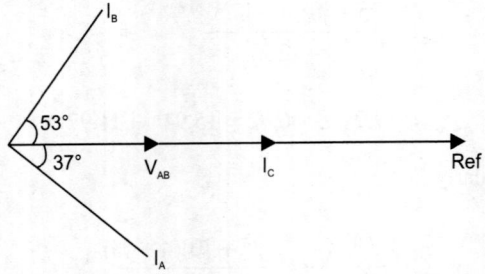

Fig. 1.128. Vector diagram for illustration 1.89.

All the voltages and current, with I_C as reference is in Fig. 1.128.

Admittances:

$$Y_A = \frac{1}{Z_A} = \frac{1}{4+j3} = \frac{1}{5\angle 36.86}$$

$$= 0.2 \angle -36.86 \text{ mho}$$

$$Y_B = \frac{1}{Z_B} = \frac{1}{4-j5.3}$$

$$= 0.15 \angle 52.96$$

$$Y_C = \frac{1}{Z_C} = \frac{1}{2+j8}$$

$$= 0.1213 \angle -75.96$$

Illustration 1.90

A 100 Ω resistor shunted by a 0.4 H inductor is in series with a variable capacitor C. A voltage of 250 V at 50 Hz is applied to the circuit. Determine (i) the value of C to give upf and (ii) the total current now.

Solution

The circuit is depicted in Fig. 1.129.

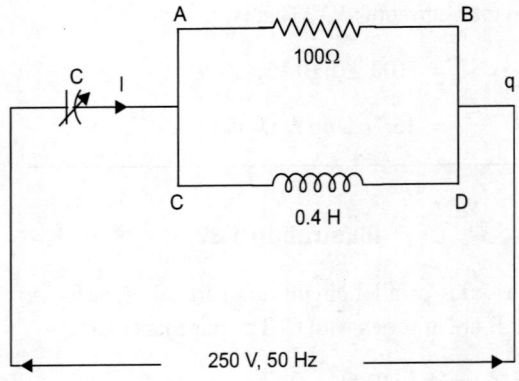

Fig. 1.129. Circuit for illustration 1.8.

Upf—the unity power factor—occurs at parallel resonance or in a series resonance.

Let the impedance between **pq**, Z_{pq}, be found out.

$$Y_{AB} = \frac{1}{100} \angle 0 \text{ mho and}$$

$$Y_{CD} = \frac{1}{2\pi \times 50 \times 0.4} = 7.96 \times 10^{-3} \angle -90 \text{ mho}$$

Thus net admittance,

$$Y_{pq} = Y_{AB} + Y_{CD}$$

$$= 0.01 - j\,7.96 \times 10^{-3} \text{ mho}$$

$$= 0.0128 \angle -38.5 \text{ mho}$$

So, the net impedance,

$$Z_{AB} = \frac{1}{Y_{AB}} = 78.23 \angle 38.5 \text{ Ω}$$

$$= 61.21 + j\,48.72 \text{ Ω}$$

The capacitive reactance of the capacitor must be equal to the reactance of the net impedance, for resonance to occur (so that the circuit will be at upf).

∴ $X_C = 48.72\ \Omega$, which gives **C = 65.33 μF. (Ans.)**

(Reader can find the easiness of Y, the admittance method; otherwise the evaluation of Z_{pq} would have been lengthier).

At his condition with C, the nett impedance would only be R.

So, = V/R = 250/61.21 = **4.08 A (Ans.)**

Illustration 1.91

A coil having an impedance of $(8 + j\ 6)\ \Omega$ is connected across a 200 V supply. Expenses the current in the coil in (i) polar and (ii) rectangular co-ordinate forms.

If a capacitor having a susceptance of 0.1 mho is placed in parallel with the coil, find (iii) the magnitude of the current taken from the supply.

Solution

Using the ohm's law, I = V/Z. Given that

$$Z = (8 + j\ 6)\ \Omega = 10\ \angle 36.86\ \Omega$$

Let V be the reference. So, $V = 200\ \angle 0$

∴ $I = 200\ \angle 0/10\ \angle 36.86 = $ **20 ∠–36.86 A (Ans.)**

= **16 – j12 A (Ans.)**

The conductance of the capacitor, B_C, is given to be 0.1 mho.

Already existing admittance,

$$Y = \frac{1}{8 + j6} = \frac{1}{10\angle 36.86}$$

$$= 0.1\ \angle{-36.86}$$

$$= 0.08 - j\ 0.06 = G - j\ B\ \text{mho}$$

$B_C = 0.1\ \angle 90 = j\ 0.1$

Thus the net admittance is $(0.08 - j\ 0.06) + (j\ 0.1)$; it is, $0.08 + j\ 0.04 = 0.089\ \angle 26.56$

So, the current taken from the supply now is VY.

i.e., $(200\ \angle 0)\ (0.089\ \angle 26.56) = 17.8\ \angle$ **26.56 A.**

Illustration 1.92

A coil A of inductance of 0.08 H and resistance 12 Ω is connected in parallel with a capacitor of 120 μF. The combination is connected to a supply at 240 V, 50 Hz. Determine the total current from the supply and its p.f. Illustrate the answers with a phasor diagram.

Solution

Refer the Fig. 1.130.

This problem is solved by using the parallel circuit theory. Let the voltage applied be the reference; so, $V = 240\ \angle 0\ V$.

$$Z_1 = R + j\ X_L$$

$$= 12 + j\ (2\pi \times 50 \times 0.08)$$

$$= 12 + j\ 25.13\ \Omega = 27.85\ \angle 64.47.\ \Omega$$

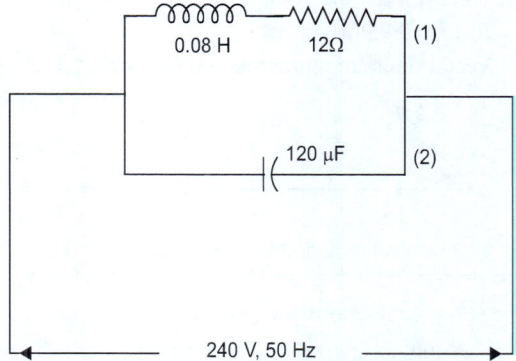

Fig. 1.130. Circuit for illustration 1.92.

Thus,

$$Y_1 = \frac{1}{Z_1} = \frac{1}{27.85\angle 64.47}$$

$$= 0.0359\ \angle{-64.47}$$

106 Basic Electrical Engineering

$$= 0.0155 - j\,0.032 \text{ mho}$$

Similarly,

$$Z_2 = -j\,X_C \quad (\because R = 0)$$

$$= -j\dfrac{1}{2\pi \times 50 \times 120 \times 10^{-6}}$$

$$= -j\,26.53\,\Omega = 26.53\,\angle{-90}\,\Omega$$

Thus, $Y_2 = \dfrac{1}{Z_2} = \dfrac{1}{26.53\angle{-90}}$

$$= 0.0377\,\angle 90 = j\,0.0377 \text{ mho.}$$

Nett Y of the circuit is,

$$Y = Y_1 + Y_2 = 0.0155 - j\,0.32 + j\,0.0377$$

$$= 0.0165\,\angle 20.17$$

Current is,

$$I = VY = (240\,\angle 0)(0.0165\,\angle 20.17)$$

$$= 3.96\,\angle 20.17 \text{ A.}$$

The current leads the voltage by 20.17°. The p.f. is cos 20.17 = 0.938 lead.

Vector diagram representation is in Fig. 1.131.

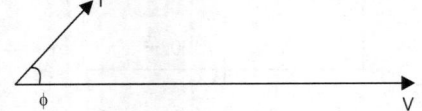

Fig. 1.131. Vector diagram to the illustration 1.92.

Reflection Section

1. When compared to a non-resonant circuit, in a series resonant circuit, the impedance of the circuit is _____.
2. Why a resonant circuit is called a tuned circuit? Will the power factor of R-L-C parallel resonant circuit lead or lag or be at unity? Justify.
3. In R-L-C series resonant circuit, the magnitude of the resonant frequency can be changed by changing the values of (i) L or C (ii) L and C. Which is correct and why?
4. Why a parallel AC resonant circuit has high impedance?
5. To tune a parallel resonant circuit to a higher frequency, what should happen to the value of the circuit's capacitance? Justify.
6. A 12 Ω resistor, a 40 µF capacitor and a 8 mH inductor in series across an AC circuit. Find out the resonant circuit. (Ans: 281 Hz)
7. Give the quality factor in terms of two powers of an AC circuit.
8. When the resistance in parallel with a resonant circuit is reduced, comment on the circuit's bandwidth.
9. Give two other famous technical names to represent a resonant circuit.
10. If the resonant frequency in a series R-L-C circuit is 50 kHz along with a bandwidth of 1 kHz, find the quality factor. (Ans: 50)
11. Why the Q-factor has no units?
12. Does the bandwidth and Q-factor vary proportionally? Justify.
13. Comment on the change in quality factor and bandwidth when the resonant frequency is reduced.
14. In a circuit consider two conditions, (i) the current leads the supply voltage and (ii) the current lags the supply voltage. Comment on the relation between the circuit's frequency and the resonant frequency.
15. Prove that the quality factor is V_L/V_s. Hence, find the source voltage when the voltage across the inductor is 2000 V and the Q-factor is 20.
16. What happens to the voltage across the inductor and the capacitor in a resonant circuit, when the Q-factor is decreased?
17. What is Q-factor in terms of reactive power and average power? If Q = 2 and the average power = 5 W, find out the value of reactive power.
18. What is the unit of bandwidth?
19. At resonance, bandwidth includes the frequency range that allows 70.7% of the maximum current to flow. Prove.
20. What is half power frequency? Relate it with bandwidth.

PART V: NETWORK ANALYSIS AND THEOREMS

1.56 INTRODUCTION

Generally, *network analysis* is any structured technique used to mathematically analyze a circuit (a "network" of interconnected components, made up of V_s, R, L and C). Quite often the technician or engineer will encounter circuits containing multiple sources of power or component configurations which defy simplification by series/parallel analysis techniques. In those cases, he or she will be forced to use other means. This part of the chapter presents a few techniques useful in analyzing such complex circuits.

To illustrate how even a simple circuit can defy analysis by breakdown into series and parallel portions, take start with the series-parallel circuit, of Fig. 1.132.

To analyze this circuit, one would first find the equivalent of R_2 and R_3 in parallel, then add R_1 in series to arrive at a total resistance. Then, taking the voltage of battery B_1 with that total circuit resistance, the total current could be calculated through the use of Ohm's law (I = E/R), then that current figure used to calculate voltage drops in the circuit. All in all, a fairly simple procedure.

However, the addition of just one more battery as shown in Fig. 1.133, could change all of that.

Resistors R_2 and R_3 are no longer in parallel with each other, because B_2 has been inserted into R_3's

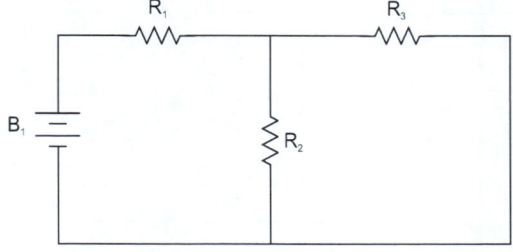

Fig. 1.132. A simple series-parallel circuit.

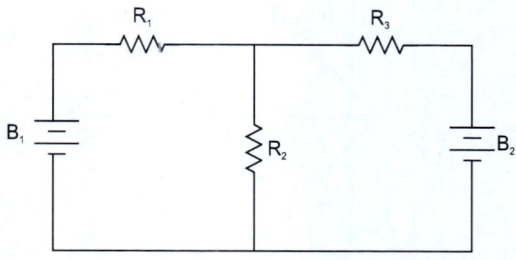

Fig. 1.133. An illustration to defy a simple series-parallel circuit.

branch of the circuit. Upon closer inspection, it appears there are *no* two resistors in this circuit directly in series or parallel with each other. This is the crux of the problem: in series-parallel analysis, we started off by identifying sets of resistors that *were* directly in series or parallel with each other, and then reduce them to single, equivalent resistances. If there are no resistors in a simple series of parallel configuration with each other, then what can we do?

It should be clear that this seemingly simple circuit, with only three resistors, is impossible to reduce as a combination of simple series and simple parallel sections: it is something different altogether. However, this is not the only type of circuit defying series/parallel analysis.

Here we have a bridge circuit, in Fig. 1.134 which also cannot be simply solved by series-parallel circuit approach.

Thus, to handle the electric circuits which are otherwise cannot be solved by the simple series-parallel approach, systematic procedures with the help of mathematics have clean developed and are called the network theorems.

Various familiar network theorems are:

(i) Mesh analysis
(ii) Nodal analysis
(iii) Thevenin's theorem

108 Basic Electrical Engineering

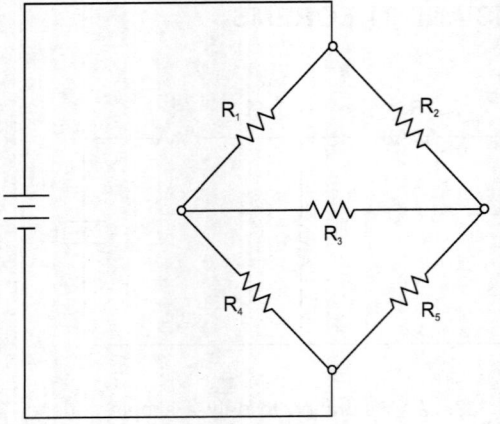

Fig. 1.134. A Bridge circuit.

(iv) Norton's theorem
(v) Superposition theorem
(vi) Reciprocity theorem
(vii) Maximum power transfer theorem.

Every theorem is first stated and explained with a numerical example. A set of illustrations are provided then.

1.57 MESH CURRENT METHOD

The *Mesh Current Method* is quite similar to the Branch Current method in that it uses simultaneous equations, Kirchhoff's Voltage Law, and Ohm's law to determine unknown currents in a network. It has been discussed in sec. 1.19 and in illustrations 24 and 25. However, for the completion of the theorem, example of Fig. 1.135 is considered here. It differs from the Branch Current method in that it does *not* use Kirchhoff's Current Law, and it is usually able to solve a circuit with less unknown variables and less simultaneous equations, which is especially nice if you're forced to solve without a calculator.

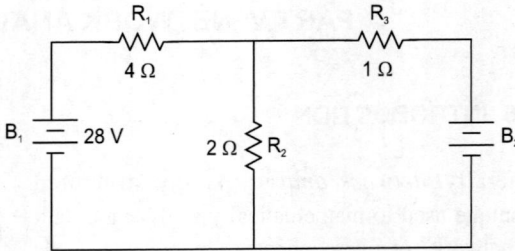

Fig. 1.135. Illustration to demonstrate mesh analysis.

Let's see how this method works on the example problem.

The first step in the Mesh Current method is to identify "loops" within the circuit encompassing all components. In our example circuit, the loop formed by B_1, R_1, and R_2 will be the first while the loop formed by B_2, R_2, and R_3 will be the second. The strangest part of the Mesh Current method is envisioning circulating currents in each of the loops. In fact, this method gets its name from the idea of these currents meshing together between loops like sets of spinning gears.

The choice of each current's direction is entirely arbitrary.

The next step is to label all voltage drop polarities across resistors according to the assumed directions of the mesh currents using the sign convention procedure of sec. 1.19-B. The potentials are marked as in Fig. 1.136.

Using Kirchhoff's Voltage Law, we can now step around each of these loops, generating equations

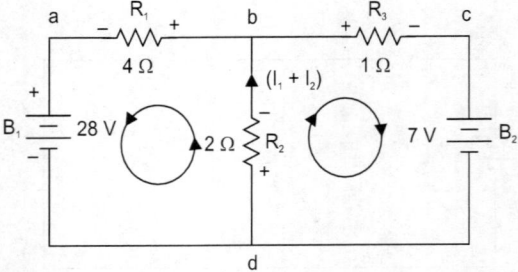

Fig. 1.136. Assumed loop currents.

representative of the component voltage drops and polarities. We will denote a resistor's voltage drop as the product of the resistance (in ohms) and its respective mesh current (that quantity being unknown at this point). Where two currents mesh together, we will write that term in the equation with resistor current being the *sum* of the two meshing currents, if the two loop circuits are additive in that branch; otherwise, we will subtract.

Selecting the path **a-d-b-a** and applying KVL:

$$-28 - 2(I_1 + I_2) - 4I_1 = 0$$

Notice that the middle term of the equation uses the sum of mesh current I_1 and I_2 as the current through resistor R_2. This is because mesh current l_1 and l_2 are going the same direction through R_2, and thus complement each other. This becomes,

$$+ 6I_1 + 2I_2 = -28 \tag{1}$$

At this time we have one equation with two unknowns. To be able to solve for two unknown mesh currents, we must have two equations. If we trace the other loop of the circuit, we can obtain another KVL equation and have enough data to solve for the two currents. Select **c-b-d-c** and apply KVL.

$$+2(I_1 + I_2) + 7 + 1I_2 = 0$$

Simplifying the equation as before, we end up with:

$$2I_1 + 3I_2 = -7 \tag{2}$$

Note that, even though the simultaneous equations (1) and (2) can be solved directly, the use of matrix algebra will ease the computational efforts. We have,

$$6I_1 + 2I_2 = -28 \text{ and}$$

$$2I_1 + 3I_2 = -7$$

In matrix form, the above two equations will look like:

$$\begin{bmatrix} 6 & 2 \\ 2 & 3 \end{bmatrix} \begin{bmatrix} I_1 \\ I_2 \end{bmatrix} = \begin{bmatrix} -28 \\ -7 \end{bmatrix}$$

i.e., $[R][I] = [V]$ \hfill (3)

which of course is Ohm's law. The matrix form of equation (3) is the mesh (or) loop equation.

Let Δ be the determinant of $[R]$.

Let Δ_1 be the determinant of $[R]$ after replacing the first column along with $[V]$.

Let Δ_2 be the determinant of $[R]$ after replacing the second column alone with $[V]$.

Now, by applying Cramer's rule I_1 and I_2 are obtained, as:

$$I_1 = \Delta_1/\Delta \text{ and } I_2 = \Delta_2/\Delta.$$

1.58 NODAL ANALYSIS

We now illustrate the application of KCL for solving the networks. Here whenever a voltage source is given it should be converted into a current source before nodal equations using KCL are written.

Consider Fig. 1.137.

First we transform the voltage source into a current source and the diagram becomes, as shown in Fig. 1.138.

Taking node O as reference, we write down KCL equations.

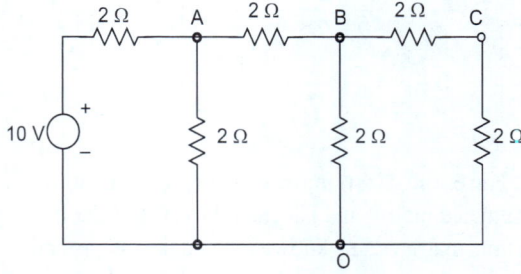

Fig. 1.137. Circuit to demonstrate nodal analysis.

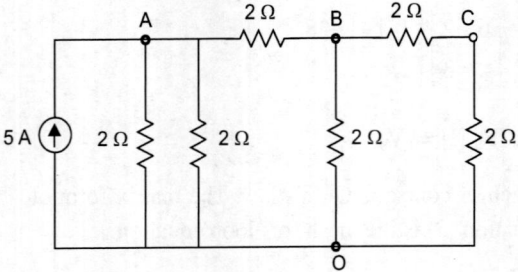

Fig. 1.138.

At node A

$$\frac{V_A}{2} + \frac{V_A}{2} + \frac{V_A - V_B}{2} = 5$$

$$V_A\left[\frac{1}{2} + \frac{1}{2} + \frac{1}{2}\right] - \frac{V_B}{2} + 0.V_c = 5$$

At node B, $\quad -V_A/2 + \left(\dfrac{1}{2} + \dfrac{1}{2} + \dfrac{1}{2}\right) V_B - V_C/2 = 0$

At node C, $\quad 0.V_A - \dfrac{V_B}{2} + V_C\left(\dfrac{1}{2} + \dfrac{1}{2}\right) = 0$

These equations can be written in matrix form as follows:

$$\begin{bmatrix} \frac{1}{2}+\frac{1}{2}+\frac{1}{2} & -\frac{1}{2} & 0 \\ -\frac{1}{2} & \frac{1}{2}+\frac{1}{2}+\frac{1}{2} & -\frac{1}{2} \\ 0 & -\frac{1}{2} & \frac{1}{2}+\frac{1}{2} \end{bmatrix} \begin{bmatrix} V_A \\ V_B \\ V_C \end{bmatrix} = \begin{bmatrix} 5 \\ 0 \\ 0 \end{bmatrix}$$

Here the first matrix is known as nodal admittance matrix and is denoted by Y and the two column matrices are known as nodal voltage and nodal current matrices denoted by V and I respectively.

Hence, the above matrices can be written in compact form as

$$Y V = I, \text{ where } Y = 1/Z \text{ mho.}$$

It can be seen that the nodal admittance matrix can be written simply by looking at the interconnections of the network using the following procedure:

(i) The diagonal element of each node is the sum of the admittances connected to it.
(ii) The off diagonal element is the negative admittance between the nodes. If there is no connection between the two nodes, the entry in the off-diagonal for both the nodes is zero i.e. if node i is not connected to the node j, then $Y_{ij} = Y_{ji} = 0$. In general every node will not be connected to all other nodes in the network and hence normally nodal admittance matrix is a highly sparse matrix i.e. it will have very small number of non-zero elements. In fact, in a large network consisting of say 100 nodes, these non-zero elements could be as small as 2% of the total elements.

We come back to original problem of finding out V_{AB}

Using Cramer's rule suggested in sec. 1.57,

$$V_A = \frac{\begin{vmatrix} 5 & -\frac{1}{2} & 0 \\ 0 & \frac{3}{2} & -\frac{1}{2} \\ 0 & -\frac{1}{2} & 1 \end{vmatrix}}{\begin{vmatrix} \frac{3}{2} & -\frac{1}{2} & 0 \\ -\frac{1}{2} & \frac{3}{2} & -\frac{1}{2} \\ 0 & -\frac{1}{2} & 1 \end{vmatrix}} = \frac{5 \times \frac{5}{4}}{\frac{13}{8}}$$

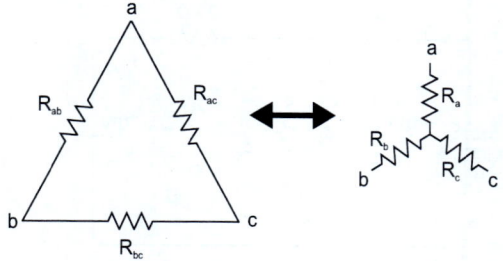
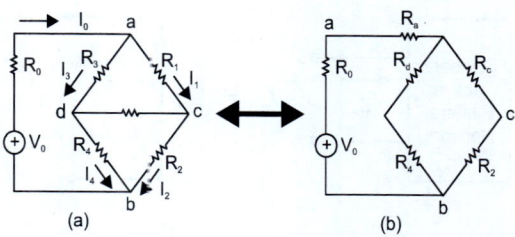

Fig. 1.139. (i) Δ-y transformate

Fig. 1.139. (ii) Example.

Therefore, $V_A = \dfrac{50}{13}$ volts.

Similarly,

$$V_B = \dfrac{\begin{vmatrix} \dfrac{3}{2} & 5 & 0 \\ -\dfrac{1}{2} & 0 & -\dfrac{1}{2} \\ 0 & 0 & 1 \end{vmatrix}}{\Delta}$$

There $V_B = 20/13$ V and $V_{AB} = \dfrac{30}{13}$ V.

Note

• Δ-Y Transformation

The Δ configuration can be converted to star (Y) and vice versa. To convert Δ to Y, the resistance R'_{ab} between terminals a and b of Y should be qual to R_{ab} of Δ, and the same is true for the other two resistances, i.e.

$$\begin{cases} R'_{ab} = R_a + R_b = R_{ab} /\!/ (R_{ac} + R_{bc}) \\ R'_{ac} = R_a + R_c = R_{ac} /\!/ (R_{ab} + R_{bc}) \\ R'_{bc} = R_b + R_c = R_{bc} /\!/ (R_{ab} + R_{ac}) \end{cases}$$

(Refer to Fig. 1.139-i)

Given R_{ab}, R_{ac} and R_{bc}, the three equations can be solved for R_a, R_b and R_c. For example, subtracting the 3rd equation from the sum of the first two, we got expression for R_a. The solutions are:

$$\begin{cases} R_a = R_{ab} R_{ac} / (R_{ab} + R_{ac} + R_{bc}) \\ R_b = R_{ab} R_{bc} / (R_{ab} + R_{ac} + R_{bc}) \\ R_c = R_{ac} R_{bc} / (R_{ab} + R_{ac} + R_{bc}) \end{cases}$$

Reversely, given R_a, R_b and R_c, the three equations can also be solved for R_{ab}, R_{ac} and R_{bc} to get:

$$\begin{cases} R_{ab} = R_a + R_b + R_a R_b / R_c \\ R_{ac} = R_a + R_c + R_a R_c / R_b \\ R_{bc} = R_b + R_c + R_b R_c / R_a \end{cases}$$

The conversion from Δ to Y more useful as Y is easier to analyze than Δ. For example, the circuit in Fig. 1.139(ii)-a, below can be converted to that in (b) to find all the currents in the circuit.

1.59 THEVENIN'S THEOREM (OR HELMHOLTZ'S THEOREM)

The Thevenin's theorem, basically gives the equivalent voltage source corresponding to an active network.

If a linear, active, bilateral network is considered across one of its ports, then it can be replaced by an equivalent voltage source (Thevenin's voltage source) and an equivalent series impedance (Thevenin's impedance), Fig. 1.140

112 Basic Electrical Engineering

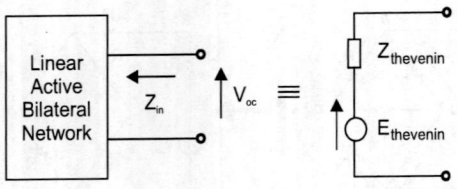

Fig. 1.140. Representation of Thevenin theorem.

Since the two sides are identical, they must be true for all conditions. Thus if we compare the voltage across the port in each case under open circuit conditions, and measure the input impedance of the network with the sources removed (voltage sources short-circuited and current sources open-circuited), then

$E_{thevenin} = V_{oc}$, and

$Z_{thevenin} = Z_{in}$

Let us consider the example of Fig. 1.141 to illustrate Thevenin's Theorem.

Let us aim to calculate the current in the 160 Ω resistor. Let the Thevenin's equivalent circuit across the terminals after disconnecting (open circuiting) the 160 Ω resistor be found out.

Under open circuit conditions, current flowing is

$= (100 - 70)/40 = 0.75$ A

∴ $V_{oc.AB} = 100 - 0.75 \times 20 = 85$ V

(i.e., $E_1 - I \cdot R = V_{AB}$)

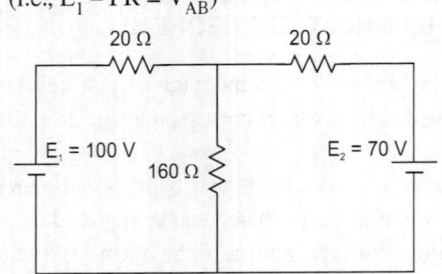

Fig. 1.141-a. Example for Thevenin Theorem

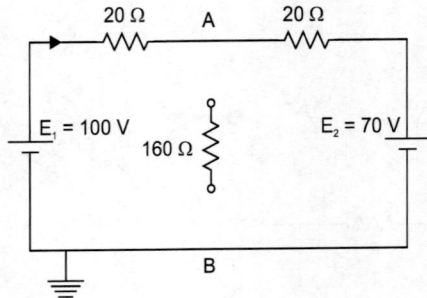

Fig. 1.141-b. 160 Ω resistor is removed and A-B is made OC.

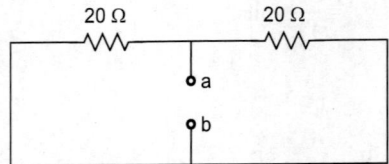

Fig. 1.142.

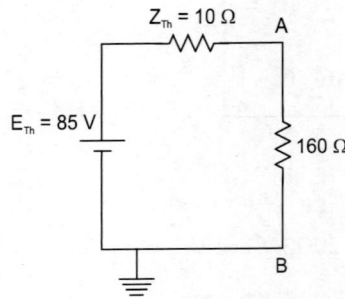

Fig. 1.143. The two 20 Ω resistors are in parallel.

∴ $E_{Th} = 85$ V.

The input impedance across AB (with sources removed) can be found out from Fig. 1.142. So $R_{A \cdot B} = (20 \times 20)/(20 + 20)$

∴ $Z_{Th} = 10 \, \Omega = R_{AB}$

Therefore the Thevenin's equivalent circuit may be drawn with branch AB reintroduced as in Fig. 1.143.

From the equivalent circuit, the unknown current i is determined as

$$i = \frac{85}{10+160} = 0.5 \text{ A}$$

1.60 NORTON'S THEOREM

Norton's theorem is the dual of Thevenin's theorem, and states that any linear, active, bilateral network, considered across one of its ports, can be replaced by an equivalent current source (Norton's current source) and an equivalent shunt admittance (Norton's Admittance), Fig. 1.144.

Since the two sides are identical, they must be true for all conditions. Thus if we compare the current through the port in each case under short circuit conditions, and measure the input admittance of the network with the sources removed (voltage sources short-circuited and current sources open-circuited), then

$I_{norton} = I_{sc}$, and

$Y_{norton} = Y_{in}$

Let the above example of Fig. 1.141 be again considered.

Since we wish to calculate the current in the 160 Ω resistor, let us find the Norton's equivalent circuit across the terminals after short-circuiting the 160 Ω resistor.

The short circuit current I_{sc} is given by

$I_{sc} = 100/20 + 70/20 = 8.5$ A

∴ $I_{norton} = 8.5$ A

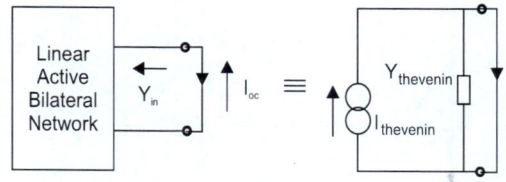

Fig. 1.144. Representation of Norton's theorem.

When shorted across A-B the 160 Ω will become negligible.

Norton's admittance = 1/20 + 1/20 = 0.1 mho (Refer to Fig. 1.146).

∴ Norton's equivalent circuit is, as shown in Fig. 1.147, and the current in the unknown resistor is

$$8.5 \times \frac{0.00625}{0.1+0.00625} = 0.5 \text{ A}$$

which is the same result as before

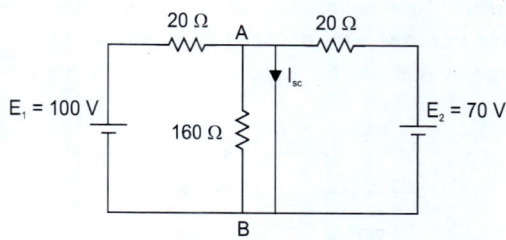

Fig. 1.145

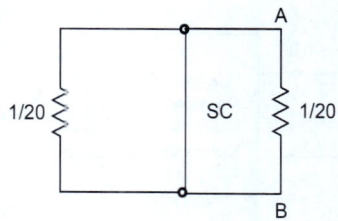

Fig. 1.146. Parallel admittances.

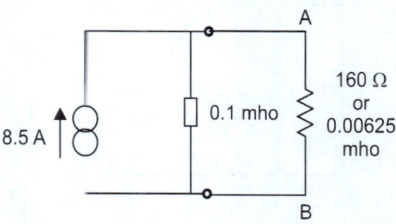

Fig. 1.147. Norton's equivalent circuit.

1.61. RECIPROCITY THEOREM

The reciprocity theorem tells us that in a linear passive bilateral network an excitation and the corresponding response may be interchanged.

In a two-port network, if an excitation e(t) at port (1) produces a certain response r(t) at port (2), then if the same excitation e(t) is applied instead to port (2), then the same response r(t) would occur at the other port (1), Fig. 1.148.

Consider the earlier example of Fig. 1.141, but with only one source as in Fig. 1.149, to study the reciprocity theorem. Determine the current in the 160 Ω branch. Now replace the 160 Ω resistor with the source in series with it and after short-circuit the source at the original location, find the current flowing at the original source location. Show that it verifies the Reciprocality theorem.

For the original circuit, current

$$I_1 = \frac{100}{20 + 160 // 20} \times \frac{20}{20 + 160} = \frac{2000}{37.778 \times 180}$$

$$= 0.294 \, A$$

similarly for the new circuit, current

$$I_2 = \frac{100}{160 + 20 // 20} \times \frac{20}{20 + 20} = \frac{2000}{170 \times 40} = 0.294 \, A$$

It is seen that the identical current has appeared verifying the reciprocality theorem. The advantage of the theorem is when a circuit has already been analysed for one solution, it may be possible to find

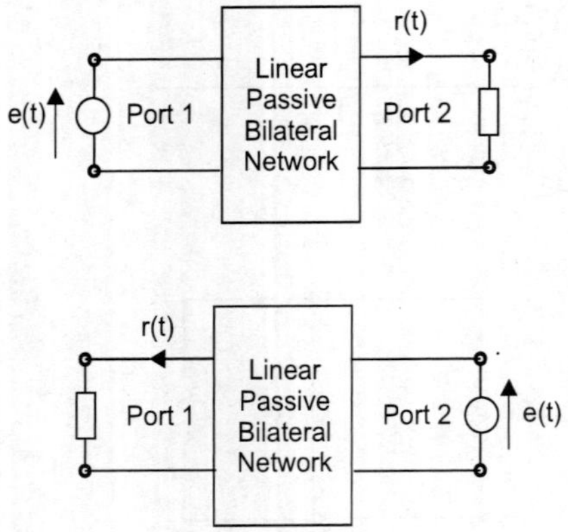

Fig. 1.148. Representation of reciprocity theorem.

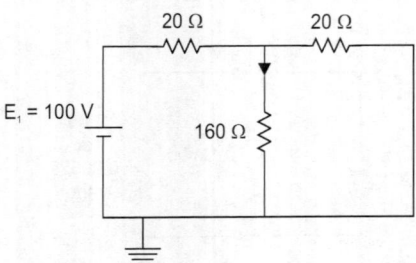

Fig. 1.149.

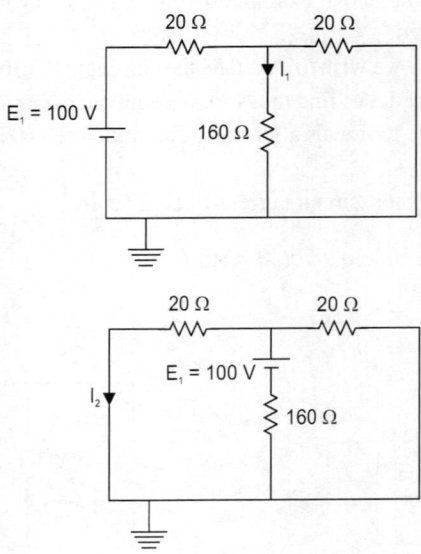

Fig. 1.150. Solution diagram for reciprocity theorem.

a corresponding solution without further work. (// means, parallel).

1.62 COMPENSATION THEOREM

In many circuits, after the circuit is analysed, it is realised that only a small change need to be made to a component to get a desired result. In such a case we would normally have to recalculate. The compensation theorem allows us to compensate properly for such changes without sacrificing accuracy.

In any linear bilateral active network, if any branch carrying a current I has its impedance Z changed by an amount ΔZ, the resulting changes that occur in the other branches are the same as those which would have been caused by the injection of a voltage source of $(-) I \cdot \Delta Z$ in the modified branch.

Consider the voltage drop across the modified branch.

$$V + \Delta V = (Z + \Delta Z)(I + \Delta I)$$

$$= Z \cdot I + \Delta Z \cdot I + (Z + \Delta Z) \cdot \Delta I$$

from the original network, $V = Z \cdot I$

$$\therefore \Delta V = \Delta Z \cdot I + (Z + \Delta Z) \cdot \Delta I$$

Since the value I is already known from the earlier analysis, and the change required in the impedance ΔZ, is also known, $I \cdot \Delta Z$ is a known fixed value of voltage and may thus be represented by a source of emf $I \cdot \Delta Z$.

Using superposition theorem, we can easily see that the original sources in the active network give rise to the original current I, while the change corresponding to the emf $I \cdot \Delta Z$ must produce the remaining changes in the network.

Consider the same example of Fig. 1.141.

Let us say that we want to change the resistor by a quantity ΔR such that the current in the 160 Ω resistor is 0.600 amps. Then the circuit for changes can be written as (Fig. 1.152)

$$\Delta I = 0.6 - 0.5 = 0.1 \text{ A}$$

$$I = 0.5$$

$$\therefore \Delta I = \frac{(-)0.5 \times \Delta R}{160 + \Delta R + 20 // 20}$$

i.e. $0.1 = (-)\dfrac{0.5 \times \Delta R}{170 + \Delta R}$

$$\therefore 17 + 0.1 \Delta R = (-) 0.5 \Delta R$$

i.e. $\Delta R = (-)17/0.6 = (-) 28.333 \ \Omega$

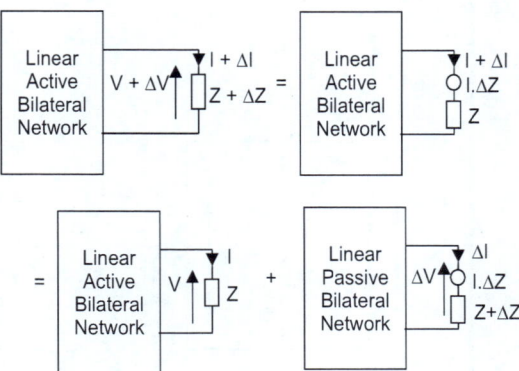

Fig. 1.151. Representation of compensation theorem.

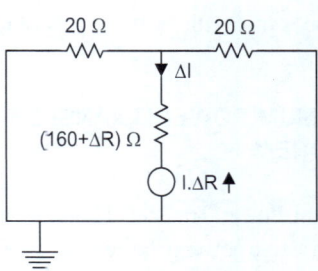

Fig. 1.152. Solution diagram for compensation theorem.

116 Basic Electrical Engineering

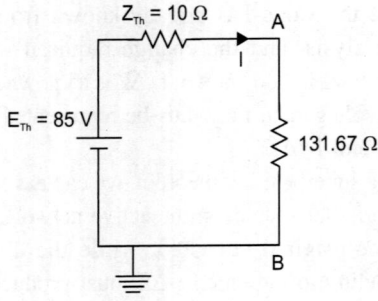

Fig. 1.153.

Therefore the required value of R = 160 – 28.333 = 131.67 Ω

This could have been calculated using Kirchhoff's and Ohm's laws but would have been more complicated.

We can also check this answer with Thevenin's theorem as follows

From Example 2, we had the Thevenin's circuit as shown with the 160 Ω replaced by 131.67 Ω.

The current for this value can be quickly obtained as

$$i = \frac{85}{10 + 131.667} = 0.6 \text{ A}$$

So you can also see that by knowing Thevenin's equivalent circuit for a given network, we can obtain solutions for many conditions with little additional calculations.

The same is true with Norton's theorem.

1.63 MAXIMUM POWER TRANSFER THEOREM

The Maximum Power Transfer theorem states that for maximum active power to be delivered to the load, load impedance must correspond to the conjugate of the source impedance (or in the case of direction quantities, be equal to the source impedance).

Let us analyse this, by first starting with the basic case of a resistive load being supplied from a source with only an internal resistance (this is the same as for D.C.)

Consider a source with an internal emf of E and an internal resistance of r and a load of resistance R, as shown in Fig. 1.154.

$$\text{Current I} = \frac{E}{R+r}$$

$$\text{Load Power P} = I^2 \cdot R = \left(\frac{E}{R+r}\right)^2 \cdot R$$

The source resistance is dependent purely on the source and is a constant, as is the source emf. Thus only the load resistance R is a variable.

To obtain maximum power transfer to the load, let us differentiate with respect to R.

$$\frac{dP}{dR} = \frac{E^2}{(R+r)^4}[(R+r)^2 \cdot 1 - R \cdot 2(R+r)]$$

$$= 0 \text{ for maximum}$$

$$\therefore (R+r)^2 - 2R(R+r) = 0$$

or $R + r - 2R = 0$

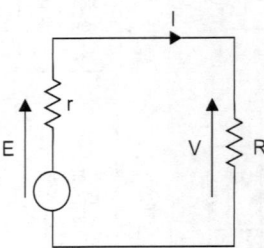

Fig. 1.154. Circuit to demonstrate maximum power transfer theorem.

i.e. R = r for maximum power transfer.

value of maximum power = $P_{max} = \left(\dfrac{E}{r+r}\right)^2 \cdot r = \dfrac{E^2}{4r}$

load voltage at maximum power

$= \dfrac{E}{R+r} \cdot R = \dfrac{E}{r+r} \cdot r = \dfrac{E}{2}$

It is to be noted that when maximum power is being transferred, only half the applied voltage is available to the load, and the other half drops across the source. Also, under these conditions, half the power supplied is wasted as dissipation in the source.

Thus the useful maximum power will be less than the theoretical maximum power derived and will depend on the voltage required to be maintained at the load.

1.64 WORKED ILLUSTRATION VI

Illustration 1.93

Determine the currents i_1, i_2 and i_3 and the voltage V_1 and V_2 in the networks Fig. 1.155 using mesh (or) loop analysis.

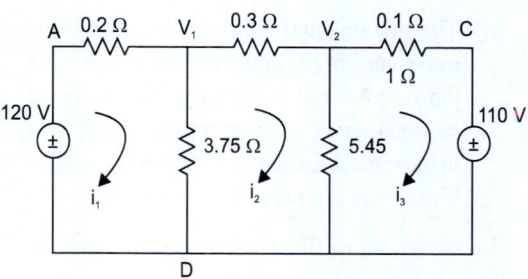

Fig. 1.155. Circuit for illustration 1.93.

Solution

Using KVL we have

$0.2\, i_1 + (i_2 - i_1)\, 3.75 = 120$
Loop I (A – V_1 – D – A)

$3.75\, (i_2 - i_1) + 0.3\, i_2 + 5.45\, (i_2 - i_3) = 0$
Loop II (V_1 – V_2 – D – V_1)

$5.45\, (i_3 - i_2) + 0.1\, i_3 = 110$
Loop III (V_2 – C – V_2)

Rearranging the equations, we have

$(0.2 + 3.75)\, i_1 - 3.75\, i_2 = 120$

$-3.75\, i_1 + (3.75 + 0.3 + 5.45)\, i_2 - 5.45\, i_3 = 0$

$-5.45\, i_2 + (5.45 + 0.1)\, i_3 = 110$

These equations can be written in matrix form as follows

$$\begin{bmatrix} (0.2+3.75) & -3.75 & 0.0 \\ -3.75 & (3.75+0.3+5.45) & -5.45 \\ 0.0 & -5.45 & 5.45+0.1 \end{bmatrix} \begin{bmatrix} i_1 \\ i_2 \\ i_3 \end{bmatrix} = \begin{bmatrix} 120 \\ 0.0 \\ 110 \end{bmatrix}$$

or

Here Z matrix is the loop impedance matrix and is found to be a symmetric matrix along its leading diagonal.

It can be seen that the loop impedance matrix can be written by inspection of the loops.

(i) The diagonal elements are the sum of all the branch impedances in the corresponding loop.

(ii) The off-diagonal elements say Z_{12} is the impedance of the branches common to both the loops 1 and 2. The sign is positive if current i_1 and i_2 in the common branch flow in the same direction and is taken as negative if the two currents oppose each other.

The above set of these equations can again be written in matrix form as follows:

$$\begin{bmatrix} 3.95 & -3.75 & 0.0 \\ -3.75 & 9.50 & -5.45 \\ 0.0 & -5.45 & 5.55 \end{bmatrix} \begin{bmatrix} i_1 \\ i_2 \\ i_3 \end{bmatrix} = \begin{bmatrix} 120 \\ 0.0 \\ 110 \end{bmatrix}$$

$$\Delta = \begin{bmatrix} 3.95 & -3.75 & 0 \\ -3.75 & 9.5 & 5.45 \\ 0 & -5.45 & 5.55 \end{bmatrix}$$

Using Cramer's rule,

$$i_1 = \frac{\begin{vmatrix} 120 & -3.75 & 0.0 \\ 0.0 & 9.60 & -5.45 \\ 110 & -5.45 & 5.55 \end{vmatrix}}{\begin{vmatrix} 3.95 & -3.75 & 0.0 \\ -3.75 & 9.50 & -5.45 \\ 0.0 & -5.45 & 5.55 \end{vmatrix}}$$

Similarly

$$i_2 = \frac{\begin{vmatrix} 3.95 & 120 & 0.0 \\ -3.75 & 0.0 & -5.45 \\ 0.0 & 110 & 5.55 \end{vmatrix}}{\Delta}$$

Similarly expression for i_3 can be written. On solution the approximate values of i_1, i_2 and i_3 are found to be 40A, 10A and –10A. Here –10A in the third loop means the direction of current in the third loop should be opposite to what has been assumed at the beginning of the solution.

The voltage

$$V_1 = (i_1 - i_2) \, 3.75$$
$$= 30 \times 3.75$$
$$= 112.5 \text{ V}$$

and $V_2 = 20 \times 5.45 = 109$ V

Illustration 1.94

Determine the current in the battery, the current in each branch and the p.d. across AB in the network of Fig. 1.56

Solution

Using KVL, as elaborately explained in the previous illustration 1.93, we have the equations in the matrix form.

$$\begin{bmatrix} 9 & -5 \\ -5 & 19 \end{bmatrix} \begin{bmatrix} i_1 \\ i_2 \end{bmatrix} = \begin{bmatrix} 10 \\ 0 \end{bmatrix}$$

From second equation

$$-5 \, i_1 + 19 \, i_2 = 0$$

or $\quad i_1 = \dfrac{19}{5} i_2$

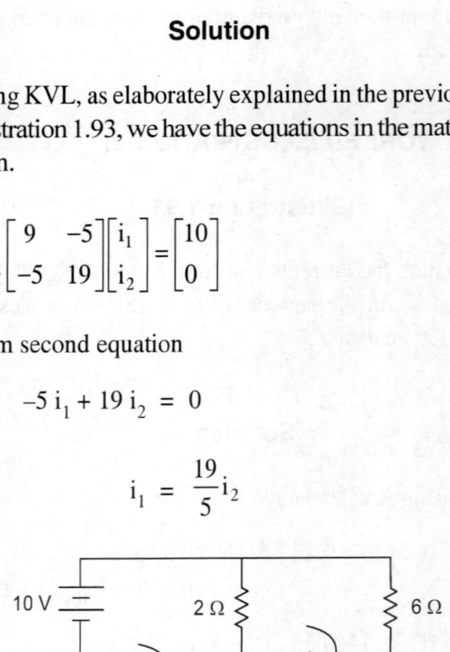

Fig. 1.156. Circuit for illustration 1.94.

From first equation

$$9i_1 - 5i_2 = 10$$

$$\frac{171}{5}i_2 - 5i_2 = 10$$

or $\quad i_2 = \dfrac{10 \times 5}{146}$ or $i_1 = \dfrac{190}{146} = 1.3\,\text{A}$

Potential of A w.r. to C, $0.96 \times 3 = 2.88$ V

and that of B w.r. to C $0.343 \times 8 = 2.744$ V

$$V_{AB} = V_A - V_B = 2.88 - 2.74 = 0.14 \text{ V}$$

Illustration 1.95

Determine the current through the branch AB of the network shown in Fig. 1.157.

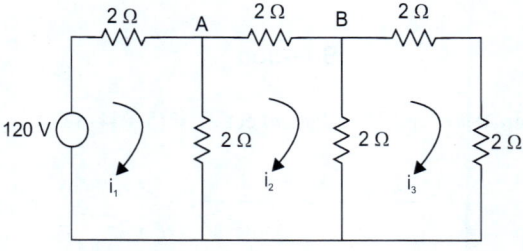

Fig. 1.157. Circuit for illustration 1.95.

Solution

Using KVL we have

$$\begin{bmatrix} 4 & -2 & 0 \\ -2 & 6 & -2 \\ 0 & -2 & 6 \end{bmatrix} \begin{bmatrix} i_1 \\ i_2 \\ i_3 \end{bmatrix} = \begin{bmatrix} 10 \\ 0 \\ 0 \end{bmatrix}$$

Using Cramer's rule

$$i_2 = \frac{\begin{vmatrix} 4 & 10 & 0 \\ -2 & 0 & -2 \\ 0 & 0 & 6 \end{vmatrix}}{\begin{vmatrix} 4 & -2 & 0 \\ -2 & 6 & -2 \\ 0 & -2 & 6 \end{vmatrix}} = \frac{15}{13} \text{ A Ans.}$$

Illustration 1.96

Determine the current through the branch AB of the network of Fig. 1.158 using (i) nodal (ii) Loop analysis.

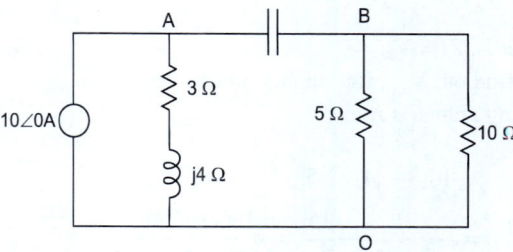

Fig. 1.158. Circuit for illustration 1.98.

Solution

Using Nodal analysis:
Using KCL and writing in matrix form taking O as the reference node

$$\begin{bmatrix} \dfrac{1}{3+j4} + j0.5 & -j0.5 \\ -j0.5 & \dfrac{1}{5} + \dfrac{1}{10} + \dfrac{j}{2} \end{bmatrix} \begin{bmatrix} V_A \\ V_B \end{bmatrix} = \begin{bmatrix} 10 \\ 0 \end{bmatrix}$$

[that is, YV = I]. Using Cramer's rule,

$$V_A = \frac{\begin{vmatrix} 10 & -j0.5 \\ 0 & (3+j5)/10 \end{vmatrix}}{\begin{vmatrix} -1+j1.5 & -j0.5 \\ -j0.5 & (3+j5)/10 \end{vmatrix}}$$

120 Basic Electrical Engineering

Similarly

$$V_A = \frac{\begin{vmatrix} -1+j1.5 & 10 \\ -j0.5 & 0 \end{vmatrix}}{\begin{vmatrix} -1+j1.5 & -j0.5 \\ -j0.5 & (3+j5)/10 \end{vmatrix}}$$

$$V_{AB} = V_A - V_B = \frac{60(3+j4)}{-6+j19}$$

Using loop analysis, we first convert current source into a voltage source, as in Fig. 1.159.

$$\begin{bmatrix} 8+j2 & -5 \\ -5 & 15 \end{bmatrix} \begin{bmatrix} i_1 \\ i_2 \end{bmatrix} = \begin{bmatrix} 10(3+j4) \\ 0 \end{bmatrix}$$

[that is, ZI = V].
To find out V_{AB}, it is enough to find out i_1.
Using cramer's rule:

$$i_1 = \frac{\begin{vmatrix} 10(3+j4) & -5 \\ 0 & 15 \end{vmatrix}}{\begin{vmatrix} 8+j2 & -5 \\ -5 & 15 \end{vmatrix}} = \frac{150(3+j4)}{15(8+j2)-25}$$

$$= \frac{150(3+j4)}{95+j30}, \text{ amps.}$$

Hence,

$$V_{AB} = \frac{150(3+j4)}{95+j30} \cdot (-j2) = \frac{-j300(3+j4)}{95+j30}$$

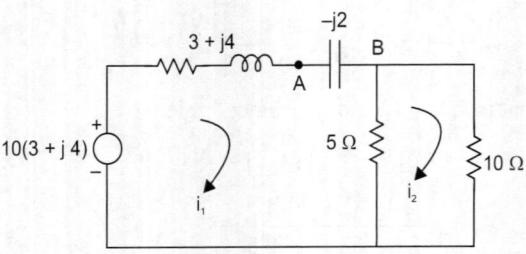

Fig. 1.159.

Dividing numerator and denominator by –j5 we have

$$V_{AB} = \frac{60(3+j4)}{-6+j19} \text{ V}$$

It is found that V_{AB} is same by both the methods i.e. nodal and loop analysis.

Illustration 1.97

Use Thevenin's theorem and find the current through $(5+j4)\,\Omega$ impedance in the Fig. 1.160.

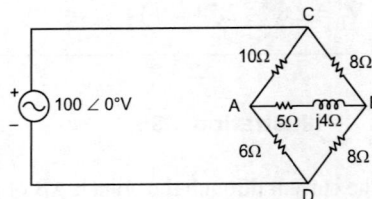

Fig. 1.160

Solution

Step 1: To find V_{Th}: disconnect $Z_L = (5+j4)\,\Omega$

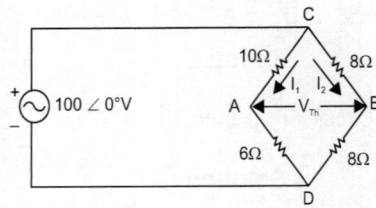

Fig. 1.161

$$I_1 = \frac{100\angle 0}{10+6} = 6.25\angle 0°\,A$$

$$I_2 = \frac{100\angle 0}{8+8} = 6.25\angle 0°\,A$$

Basic Electricity 121

$\therefore V_{Th} = V_{AB} = -6I_1 + 8I_2$

$= -6 \times 6.25 \angle 0° + 8 \times 6.25 \angle 0°$

$= 37.5 \angle 180° + 50 \angle 0°$

$= 12.5 \angle 0° = 12.5$ V

Step 2: To calculate $Z_{Th} = R_{Th}$: In the above circuit, kill the voltage source, by shorting C and D terminals.

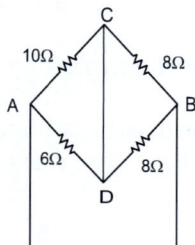

Fig. 1.162

$R_{Th} = \dfrac{10 \times 6}{10+6} \times \dfrac{8 \times 8}{8+8}$

$= 7.75 \ \Omega$

Step 3: Thevenin equivalent ckt

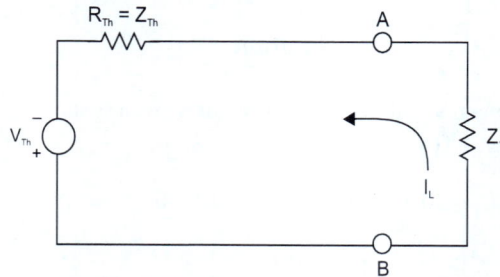

Fig. 1.163

$\therefore \quad I_L = \dfrac{V_{Th}}{Z_{Th} + Z_L}$

$= \dfrac{12.5 \angle 0}{(7.75 + 5 + j4)}$

$= \dfrac{12.5 \angle 0}{13.363 \angle 17.42°}$

$= 0.9354 \angle -17.42°$ A

Illustration 1.98

A loudspeaker is connected across the terminals A and B of the network shown in Fig. 1.164 below. What should be the value of impedance of the speaker to obtain maximum power transferred to it and what is the maximum power?

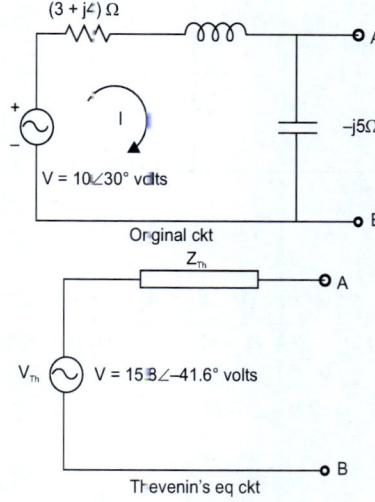

Fig. 1.164

Solution

The Thevenin's equivalent circuit is as shown above.

To find V_{Th}: In the original circuit the current

$= I = \dfrac{10 \angle 30°}{3 + j4 - j5}$

$= \dfrac{10 \angle 30°}{3 - j1}$

$= \dfrac{10 \angle 30°}{3.16 \angle -18.4°} = 3.16 \angle 48.4°$ A

$$\therefore \quad V_{Th} = \text{Voltage across} - j5\Omega$$
$$= 3.16 \angle 48.4° \times 5 \angle -90°$$
$$= 15.8 \angle -41.6° \text{ V}$$

To find Z_{Th}: From the original circuit, killing the voltage source, the following circuit is obtained.

$$Z_{Th} = Z_{AB}$$

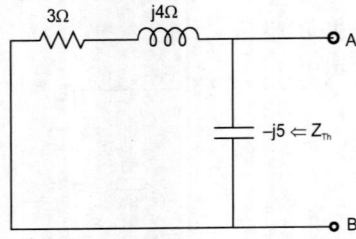

Fig. 1.165A

$$Z_{Th} = \frac{(3+j4)(-j5)}{3+j4-j5} = \frac{5\angle 53° \times 5 \angle -90°}{3.16 \angle -18.4}$$

$$= 7.91 \angle -18.6°$$

$$= (7.5 - j2.52) \, \Omega$$

$$Z_{Th} = R_g - jX_g = (7.5 - j2.52) \, \Omega$$

For maximum power transfer to loudspeaker, let the load impedance be $Z_L = R_L + jX_L$, both variables.

$$\therefore \quad R_L = 7.5 \, \Omega \text{ and } X_L = 2.52 \, \Omega$$

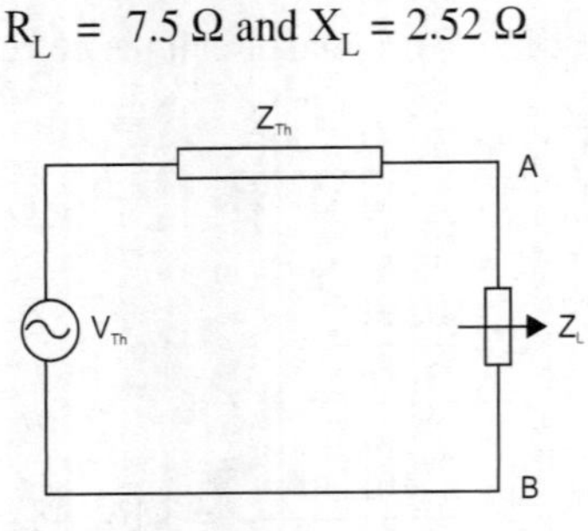

Fig. 1.165B

$$\therefore \quad Z_L = (7.5 + j2.52) \, \Omega$$

$$\text{I at P}_{max} = \frac{V_{Th}}{2R_L} = \frac{15.8 \angle -41.6°}{2 \times 7.5}$$

$$= 1.053 \angle -41.6° \text{ A}$$

$$|I| = 1.053 \text{ A}$$

$$\therefore \quad P_{max} = |I|^2 R_L = (1.053)^2 \times 7.5 = 8.32 \text{ W}$$

Illustration 1.99

Determine the current through the branch AB of the network shown in Fig. 1.166 using Thevenin's equivalent.

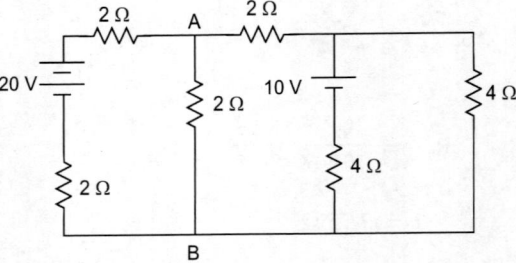

Fig. 1.166. Circuit for illustration 1.99.

Solution

To obtain V_{TH} the equivalent circuit is given as shown in Fig. 1.167.

$$2i_1 - 20 + 4i_1 + 10 + 4(i_1 - i_2) = 0$$

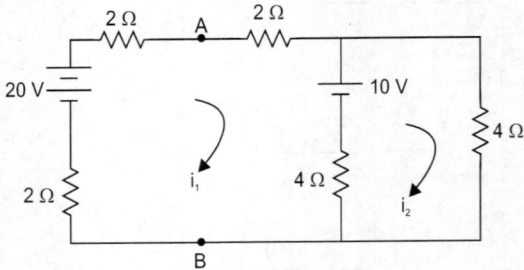

Fig. 1.167. Thevenin circuit reduction

$10i_1 - 4i_2 = 10$ or $5i_1 - 2i_2 = 5$

$4(i_2 - i_1) - 10 + 4i_2 = 0$

$-4i_1 + 8i_2 = 10$

$-2i_1 + 4i_2 = 5$

$-2i_1 + 4i_2 = 5i_1 - 2i_2$

or $6i_2 = 7i_1$ that is $i_2 = \dfrac{7}{6} i_1$

So, $-2i_1 + 4i_2 = 5$ becomes, $-2i_1 + \dfrac{14}{2} i_1 = 5$

So $i_1 = \dfrac{15}{8}$, A

$\therefore V_{AB} + \dfrac{15}{2} = 20$,

$V_{AB} = 12.5$ V

$V_{AB} = V_{TH} = 12.5$ V

To calculate $R_{TH \text{ or }} R_{eq}$, see Fig. 1.168.

Across AB, equivalent resistance is, $(2 + 2) + 2 + (4 // 4)$, which is 2 Ω.

Thus, Thevenin equivalent circuit in drawn as in Fig. 1.169.

Therefore the current through the 2Ω resistor across AB is, 12.5/4 = 3.125 A.

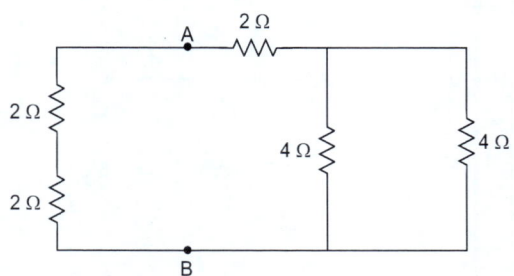

Fig. 1.68.   Circuit to calculate R_{TH}

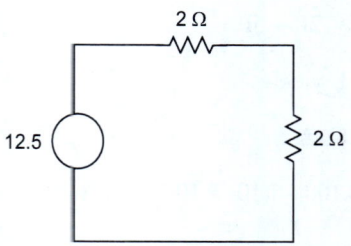

Fig. 1.69.   Thevenin equivalent circuit

Illustration 1.100

Determine the current in the 5-Ω resistor of the circuit shown in Fig. 1.170, by using Thevenin's theorem.

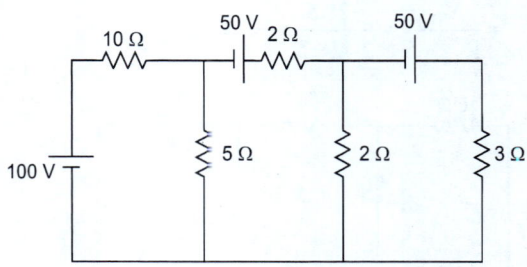

Fig. 1.170.   Circuit for illustration 1.100.

Solution

The circuit to determine the Thevenin voltage is shown in Fig. 1.171, from which we have (for the two meshes)

$100 + 50 = 14I_1 - 2I_2$

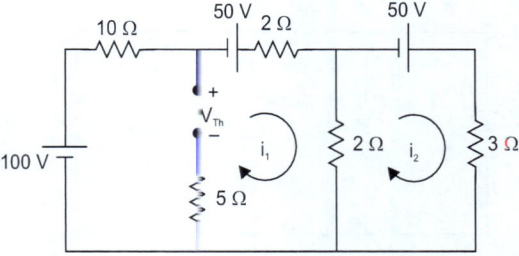

Fig. 1.171.

that is, $50 = -2I_1 + 5I_2$

Solving for I_1 yields

$$I_1 = 12.88 \text{ A}$$

Thus, $V_{Th} = 100 - I(10) = 100 - 12.88(10) = -28.8$ V.

Figure 1.172 shows the circuit to determine the Thevenin resistance. Therefore, $10 \mathbin{/\mkern-5mu/} [2 + 2 \mathbin{/\mkern-5mu/} 3]$

$$R_{Th} = \frac{10(2+1.2)}{10+2+1.2} = 2.43 \text{ } \Omega$$

Hence, current in the 5 Ω resistor is,

$$= \frac{V_{Th}}{R_{Th}+5} = \frac{-28.8}{2.42+5} = -3.87 \text{ A}.$$

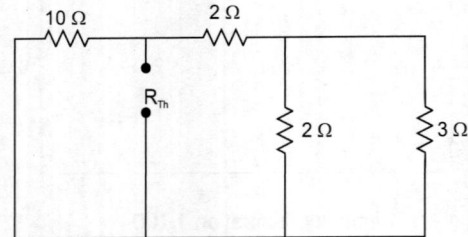

Fig. 1.172.

Illustration 1.101

Using Thevenin' theorem, determine the current in the 2 Ω resistor of the network shown in Fig. 1.173.

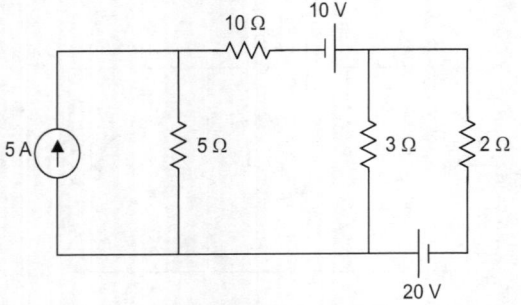

Fig. 1.173. Circuit for illustration 1.101.

Solution

First, we change the 5-A current source to an equivalent voltage source. Consequently, we have the Thevenin equivalent circuit to determine V_{Th} as shown in Fig. 1.174. Writing the mesh equation yields

$$25 - 10 = (5 + 10 + 3)I \text{ or } I = \frac{15}{18} = \frac{5}{6} \text{ A}$$

Hence,

$$20 = -3\left(\frac{5}{6}\right) + V_{Th} \text{ or } V_{Th} = 20 + 2.5 = 22.5 \text{ V}$$

The circuit to determine R_{Th} is shown in Fig. 1.175 from which, RTh is $(5 + 10) \mathbin{/\mkern-5mu/} 3$.

$$R_{Th} = \frac{(10+5)3}{10+5+3} = 2.5 \text{ } \Omega$$

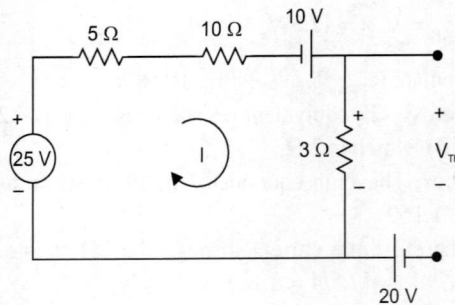

Fig. 1.1.74.

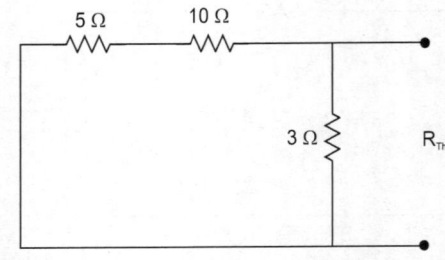

Fig. 1.175.

Thus, $I_{2\Omega} = \dfrac{V_{Th}}{R_{Th}+2} = \dfrac{22.5}{2.5+2} = 5\,A$

Illustration 1.102

Using superposition principle determine the current through 2 Ω resisted between A and B shown in the circuit of Fig. 1.176.

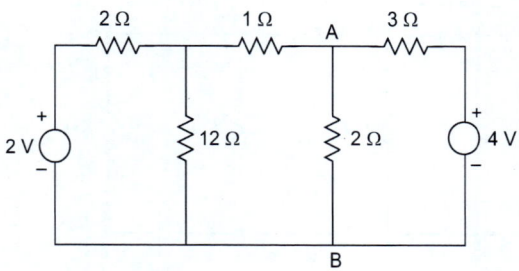

Fig. 1.176. Circuit for illustration 1.102.

Solution

Since there are two sources in the circuit, there will be two circuits for analysis purposes where in one source will be present and the other short circuited being the voltage source. They are shown in Fig. 1.177.

The resistance between CD excluding 12 Ω resistance

$1 + \dfrac{6}{5} = 1 + 1.2 = 2.2\,\Omega$, including 12 Ω resistance

$\dfrac{1}{R} = \dfrac{1}{12} + \dfrac{10}{22} = \dfrac{(11+60)}{132} = \dfrac{71}{132}$

Total resistance $2 + \dfrac{132}{71} = \dfrac{(142+132)}{71} = \dfrac{274}{71}$

Therefore, current through the source is $\dfrac{2 \times 71}{274}$

(i)
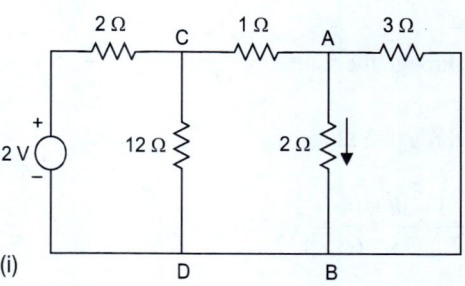

(ii)
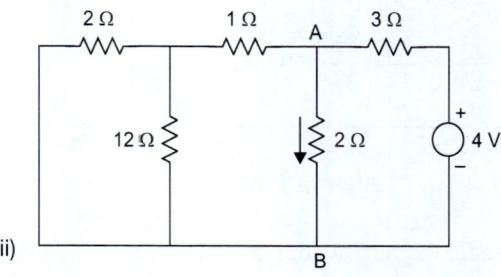

Fig. 1.177.

The current through 1 Ω resistor is

$\dfrac{2 \times 71}{274} \times \dfrac{12 \times 10}{142} = \dfrac{120}{274}$

Therefore current through 2 Ω resistor

$\dfrac{120}{274} \times \dfrac{3}{5} = \dfrac{72}{274}\,A$

(ii)

$1 + \dfrac{24}{14} = \dfrac{38}{14}\,\Omega$

$\dfrac{1}{R} = \dfrac{1}{2} + \dfrac{14}{38} = \dfrac{(19+14)}{38} = \dfrac{33}{38}$

or $R = \dfrac{38}{33}\,\Omega$

Total resistance $3 + \dfrac{38}{33} = \dfrac{137}{33}\,\Omega$

Current through the source $\dfrac{4 \times 33}{137}$ A

Current through 2 Ω resistor

$$\dfrac{4 \times 33}{137} \times \dfrac{38}{14} \times \dfrac{14}{66} = \dfrac{76}{137}$$

Therefore, total current is

$$\dfrac{72}{274} + \dfrac{76}{137} = \dfrac{224}{274} = 0.8 \text{ A}$$

Illustration 1.103

Solve for the current I of Fig. 1.178 by applying the superposition theorem.

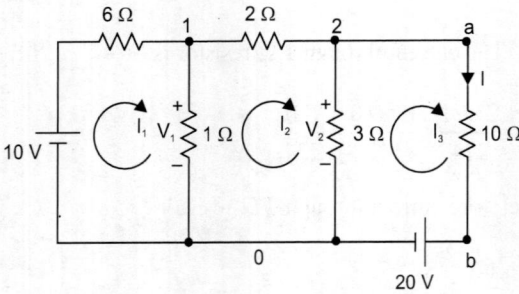

Fig. 1.178. Circuit for illustration 1.103.

Solution

According to the superposition theorem, we determine the current due to each source (in the absence of all other sources). The net current is the sum of all these currents. So let us eliminate the 20 V source by replacing it by a short circuit. The network of Fig. 1.178 then becomes as shown in Fig. 1.179. Combining resistances, we find $I_1 = 0.0636$ A. Next, we eliminate the 10 V source to obtain the circuit of Fig. 1.180. Proceeding as before, we determine $I_2 = -1.744$ A. Consequently,

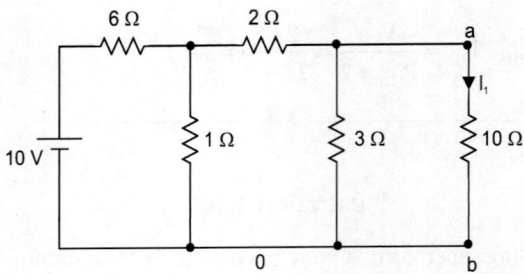

Fig. 1.179. 20 V Source shorted

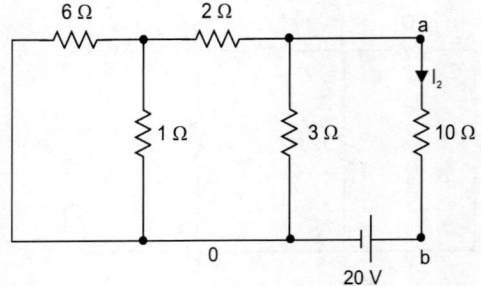

Fig. 1.180. 10 V Source shorted

$$I = I_1 + I_2 = 0.0636 - 1.744 = -1.68 \text{ A}$$

Note that, mesh analysis can also be applied to the respective circuits to determine the currents I_1 and I_2.

Illustration 1.104

Determine the current I_1 in the circuit of Fig. 1.181 by superposition.

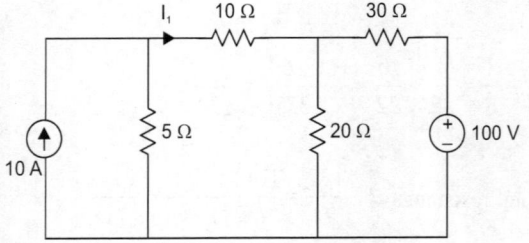

Fig. 1.181. Circuit for illustration 1.104.

Solution

For superposition, we first excite the circuit by the current source only as shown in Fig. 1.182 from which we observe that the 10 A current from the source is divided into a 5 Ω resistor and a 10 + [20 × 30)/(20 + 30)] = 22 Ω resistor. Hence, by current division,

$$I_1' = 10\left(\frac{5}{5+22}\right) = 1.852\,A$$

Next, removing the current source and keeping only the voltage source, from Fig. 1.183 we obtain

$$I_2 = \frac{100}{30 + \{[20(10+5)]/(20+10+5)\}} = 2.593\,A$$

By current division,

$$I_1'' = 2.593\frac{20}{20+10+5} = 1.481\,A$$

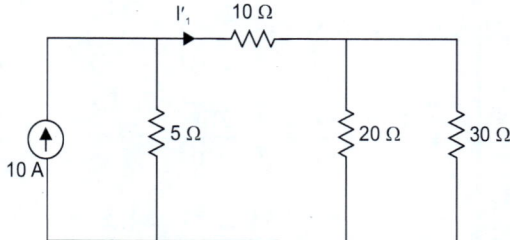

Fig. 1.182. 100 V source shorted.

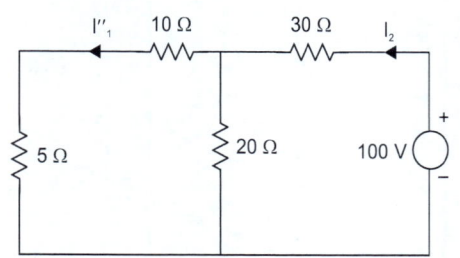

Fig. 1.183. Current source opened.

Hence, $I_1 = I_1' - I_1'' = 1.852 - 1.481 = 0.37\,A$

Illustration 1.105

Calculate the current I in the circuit of Fig. 1.184 by super position.

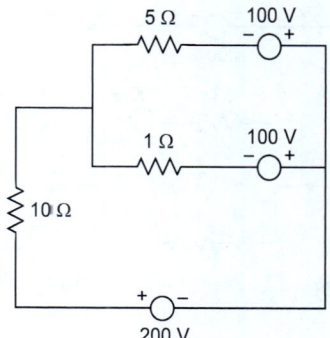

Fig. 1.184. Circuit for illustration 1.105.

We use one source at a time as shown in the circuits of Fig. 1.185a to c. From Fig. 1.185a we have

$$I' = \frac{200}{10 + [(5\times 1)/(5+1)]} = 18.46\,A$$

Fig. 1.185b yields

$$I_1 = \frac{100}{5 + [(10\times 1)/(10+1)]} = 16.92\,A$$

and $I'' = 16.92\dfrac{1}{10+1} = 1.54\,A$

Similarly, from Fig. 1.185c we obtain

$$I_2 = \frac{50}{1 + [(5\times 10)/(5+10)]} = 11.54\,A$$

128 Basic Electrical Engineering

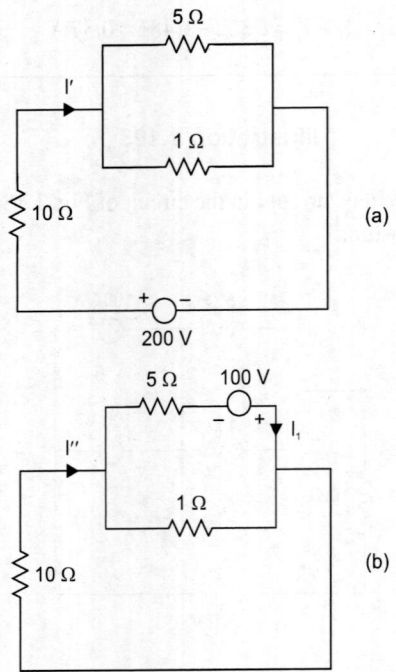

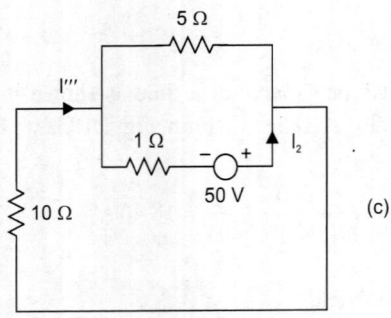

Fig. 1.185. Allowing one source to act at a time.

and $I''' = 11.54 = \dfrac{5}{5+10} = 3.85 \text{ A}$

Hence $I = I' + I'' + I''' = 23.85 \text{ A}$

Illustration 1.106

Find the current in the 2 Ω resistor of the circuit of Fig. 1.186 by superposition.

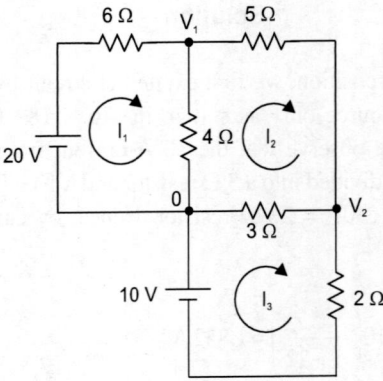

Fig. 1.186. Circuit for illustration 1.106.

Solution

From Fig. 1.187, with the 20 V source removed, we obtain

$$I' = \dfrac{10}{2 + \dfrac{(3\{5 + [(6\times 4)/(6+4)]\})}{3 + 5 + [(6\times 4)/(6+4)]}} = 2.42 \text{ A}$$

From Fig. 1.188 we have

$$I_1 = \dfrac{20}{6 + \dfrac{\{4(5 + [(3\times 2)/(3+2)])\}}{4 + 5 + [(3\times 2)/(3+2)]}} = 2.37 \text{ A}$$

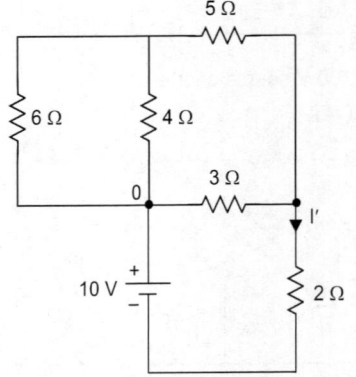

Fig. 1.187. Circuit with 20 V shorted

Basic Electricity

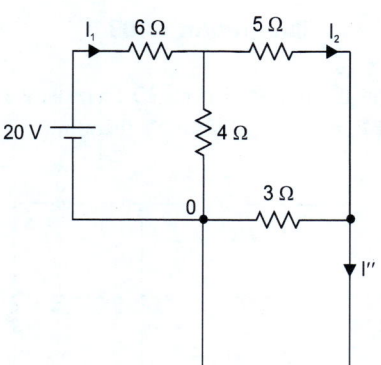

Fig. 1.188. Circuit with 10 V shorted

$$I_2 = 2.37 \frac{4}{\{4+5+[(3\times 2)/(3+2)]\}} = 0.93\,\text{A}$$

$$I'' = 0.93 \frac{3}{3+2} = 0.56\,\text{A}$$

$$I_{2\Omega} = I' + I'' = 2.42 + 0.56 = 2.98\,\text{A}$$

Illustration 1.107

Solve Fig. 1.189 using Norton's theorem

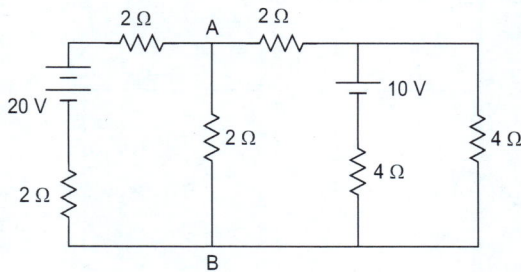

Fig. 1.189. Circuit for illustration 1.107.

Solution

Shorting branch AB and using superposition principle to find current through AB. (Fig. 1.190).
20 V in the circuit and shorting 10 volt source:

The current is $\dfrac{20}{4}\,\text{A} = 5\,\text{A}$

10 volts in the circuit and shorting 20 V:
   Current supplied by the battery

$$\frac{10}{4+4/3} = \frac{3\times 10}{16} = \frac{15}{9}\,\text{A}$$

Current through 2 Ω resistor

$$\frac{15}{8}\times\frac{4}{6} = \frac{5}{4}\,\text{A}$$

Total current through AB is

$$5 + \frac{5}{4} = \frac{25}{4}\,\text{A}$$

The Norton's equivalent circuit, is shown in Fig. 1.191.

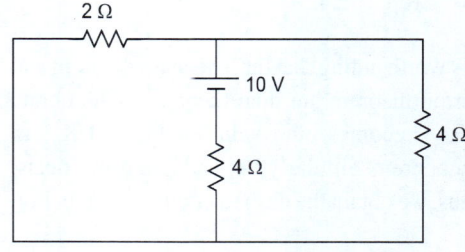

Fig. 1.190. 20 V source shorted.

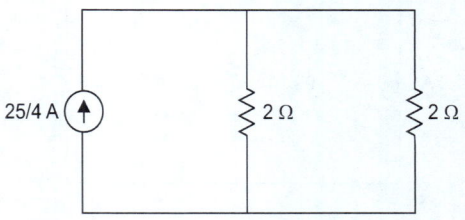

Fig. 1.191. Norton's equivalent circuit.

Therefore, current through AB branch is

$$\frac{25}{4} \times \frac{1}{2} = \frac{25}{8} = 3.125\,\text{A}$$

Illustration 1.108

Obtain a Norton equivalent circuit to determine the current in the 5 Ω resistor of the circuit of Fig. 1.170 of Illustration 1.100.

Solution

From the solution of illustration 1.100,

$$R_{Th} = 2.42\,\Omega \text{ and } V_{Th} = -28.8\,\text{V}$$

Hence, $I_N = \dfrac{V_{Th}}{R_{Th}} = -\dfrac{28.8}{2.42} = -11.9\,\text{A}$

$$G_N = \frac{1}{R_{Th}} = \frac{1}{2.42} = 0.413\,\text{mho}$$

It is worth noting that the Thevenin theorem and the Norton theorem are dual theorems. The ohmic value R_{Th} becomes mho value of $G_N = 1/R_{Th}$ in Norton theorem. Similarly V_{Th} and I_N are also duals.

Thus, we obtain the desired circuit shown in Fig. 1.192.

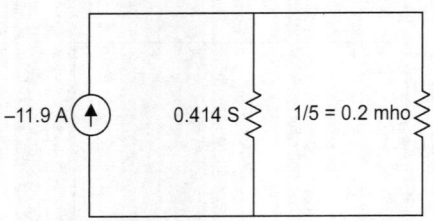

Fig. 1.192. Norton's equivalent.

Illustration 1.109

Determine the current in the 10 Ω resistor of the circuit of Fig. 1.193 by Norton's theorem.

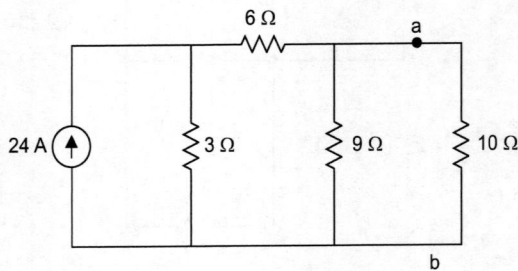

Fig. 1.193. Circuit for illustration 1.109.

Solution

To determine I_N we short-circuit **ab** to obtain the circuit of Fig. 1.194 from which, by current division,

$$I_N = 8\,\text{A}$$

With the source and short-circuit removed from Fig. 1.194 we obtain

$$R_N = \frac{1}{G_N} = \frac{9(6+3)}{9+6+3} = 4.5\,\Omega$$

or $G_N = \dfrac{1}{4.5} = 0.222\,\text{mho}$

(Note that S is siemen or mho)

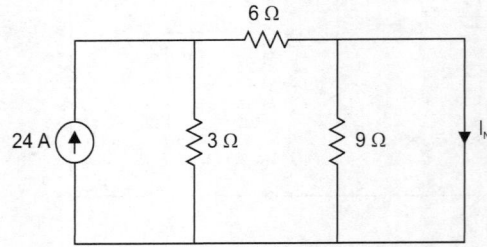

Fig. 1.194. Circuit with a-b short circuited.

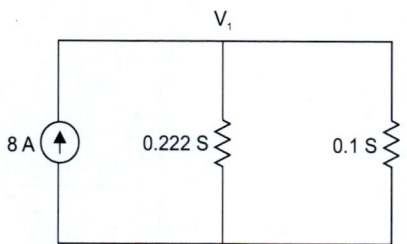

Fig. 1.195.

Hence we obtain the Norton equivalent circuit of Fig. 1.195 which yields

$$V_1(0.222 + 0.1) = 8$$

or $V_1 = 24.84 \text{ V}$ $I_{10\Omega} = I_{0.1S} = 2.484 \text{ A}$

Illustration 1.110

Find the current in the 5 Ω resistor of the circuit of Fig. 1.196 by Norton's theorem.

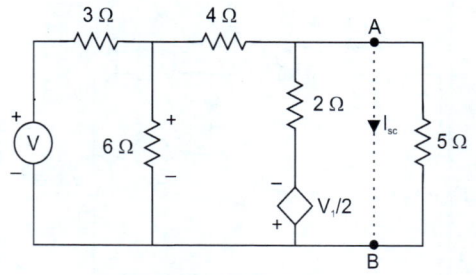

Fig. 1.196. Circuit for illustration 1.110.

Solution

At node A, after short-circuiting the 5 Ω resistor, we obtain

$$\frac{V_1}{4} - \frac{V_1/2}{2} - I_{SC} = 0 \text{ or } I_{SC} = 0$$

Hence no current will flow through the 5 Ω resistor, $I_{5\,\Omega} = 0$.

Basic Electricity 131

Illustration 1.111

Find the current in the 1 Ω resistor of the circuit of Fig. 1.197 by Norton's theorem.

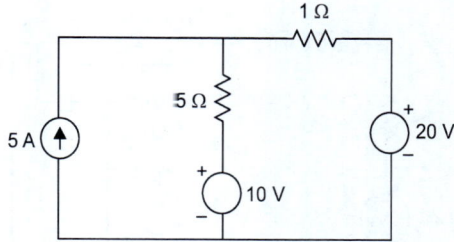

Fig. 1.197. Circuit for illustration 1.111.

Solution

Short-circuiting the 1 Ω resistor yields

$$5 + I_{SC} - \frac{20-10}{5} = 0 \text{ or } I_{SC} = -3\text{A} = I_N$$

By inspection,

$$R_N = 5\Omega$$

Hence we obtain the Norton circuit of Fig. 1.198, from which,

$$\frac{V}{1} + \frac{V}{5} = 3 \text{ or } V = 2.5 \text{ V}$$

and $I_{1\Omega} = \dfrac{V}{1} = 2.5 \text{ A}$

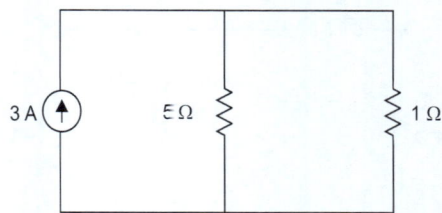

Fig. 1.198

Illustration 1.112

Determine the value of Z_L for maximum power transfer, in Z_L, in the network of Fig. 1.199.

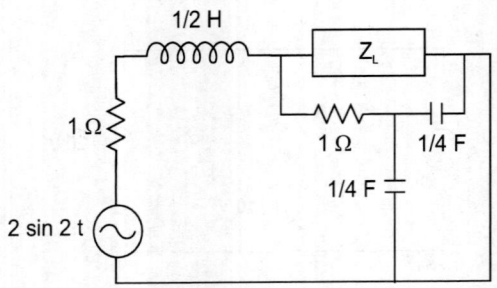

Fig. 1.199.   Circuit for illustration 1.112.

Solution

Since $(\omega) = 2$ rad/sec, the network is drawn as shown in Fig. 1.200.

$$Z_L = \frac{(1+j1)(1-j1)}{2}$$

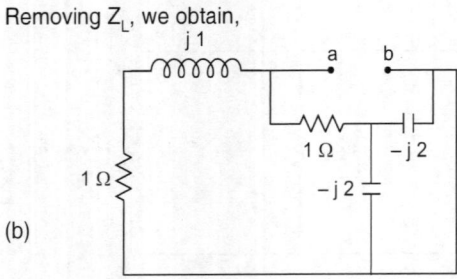

that is,

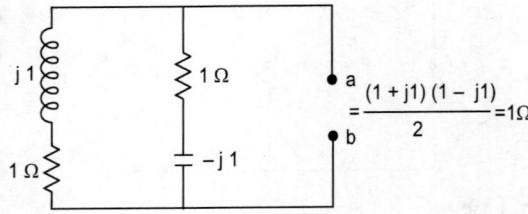

Fig. 1.200.

$$= \frac{(1+j1)(1-j1)}{2} = 1\Omega$$

Hence for maximum power through Z_L, it should be $Z_L = 1\,\Omega$.

Illustration 1.113

In the network shown determine the value of impedance Z_L for maximum power and calculate the maximum power.

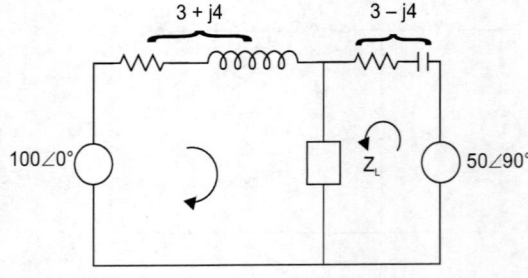

Fig. 1.201.   Circuit for illustration 1.113.

Solution

The equivalent impedance

$$\frac{(3+j4)(3-j4)}{6} = \frac{25}{6} = 4.16\,\Omega$$

To obtain maximum power through this resistance we obtain current through this branch. To find out that we find V_{TH}. The current in the circuit when Z_L is removed is

$$I = \frac{100\angle 0° - 50\angle 90°}{6} = \frac{100}{6} - \frac{50}{6} = \frac{60}{6} \text{ A}$$

Hence

$$V_{TH} = 100\angle 0° - \frac{50}{6}(3+j4) = 100 - 25 - j33.3$$

The current through Z_L

$$I = \frac{V_{TH}}{2 \times 4.16} = \frac{75 - j33.3}{8.32} = \frac{82.06\angle -23.94°}{8.32}$$

$$= 9.86\angle -23.94°$$

The power maximum) through Z_L = is $4.16\, I^2$

$$= 404.6 \text{ W Ans.}$$

Solution

For the Thevenin equivalent, we must have

$$V_{Th} = 450\angle 0° \cdot \frac{3}{3+6} = 150\angle 0° \text{ V}$$

$$Z_{Th} = (2-j4) + \frac{6 \times 3}{6+3} = 6.66\angle 45° \, \Omega$$

For maximum power transfer,

$$Z_L = Z_{Th}^* = 5.66\angle -45° = 4 - j4.$$

Thus

$$I_L = \frac{V_{Th}}{Z_L + Z_{Th}} = \frac{150\angle 0°}{8} = 18.75\angle 0° \text{ A}$$

and maximum power is

$$P_{max} = (I_L)^2 R_L = (18.75)^2 (4) = 1406 \text{ W}.$$

Illustration 1.114

Determine by Thevenin's theorem the load that must be connected across the terminals 'ab' of the circuit of Fig. 1.202, to draw the maximum power from the source. Also find the maximum power.

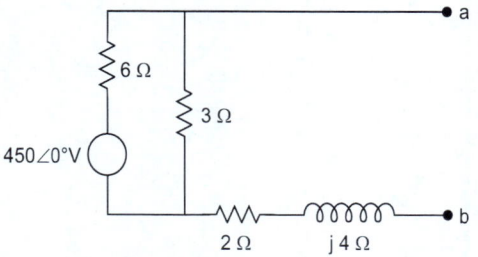

Fig. 1.202. Circuit for illustration 1.114.

Reflection Section

1. What is a linear bilateral element?
2. To find the Thevenin equivalent resistance, what is done to the voltage sources and current sources in the circuit?
3. How Thevenin's equivalent circuit and the Norton's equivalent circuit are related?
4. Nodal analysis is based on what type Kirchhoff's law?
5. In a circuit, what are passive elements and what are active elements? Discuss.
6. Why does a superposition theorem not applicable for power calculations?
7. For a N node network, how many equations are required to solve the circuit using nodal analysis?
8. Differentiate between Thevenin's theorem and Norton theorem.

PART VI: CONCEPT OF 3-PHASE EMF GENERATION

1.65 INTRODUCTION

The complete discussions made in the above three parts are based on 1-phase system. That is, the combination of the circuit constants R, L and C were connected to an alternating source, which has **'single'** current and voltage wave, as found in Fig. 1.52. The Fig. 1.203 shows a single-phase circuit.

If more than such one-phase are joined together it gives raise to a **poly-phase** system. Though it is possible to have any number of phases like this, the increase in power beyond 3-phase is in significant.

An introduction to 3-phase system, is the main aim of this section.

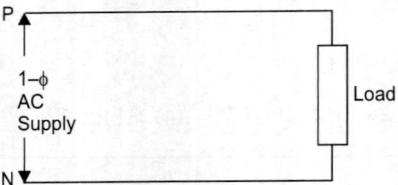

Fig. 1.203. An AC circuit, single-phase

1.66 3-PHASE SYSTEM

Three-phase system, simply denoted as $3 - \phi$ system, is a system in which 3 one-phase systems are merely combined. Each phase of the three-phase system is termed as a **red phase (R) an yellow phase (Y)** and **a blue phase (B)**.

In such system the voltage available in the R-phase, voltage available in the Y-phase and the voltage of the B-phase will be displaced from each other by 120 electrical degrees.

Note: The relation between the electrical degree θ_e and the mechanical (usual mathematical,) degree θ_m is., $\theta_e = \theta_m \dfrac{P}{2}$ where P is the number of poles in the system.

In the text, the degree means, only the electrical degrees.

Consider the Fig. 1.204. It shows three, single-phases without any connection between them.

One end of a phase is taken as start and the other end is taken as end; the ends are marked with a prime (R', Y' and B'). The starts are unprimed.

If these three windings are joined together in some fashion, then such formed single circuit, will be a three-phase system. There are totally two fashions by which a 3-phase system can be formed. They are,

(i) the **delta system** or **mesh system** and
(ii) the **star** or **Y system**

In the delta system, start of one coil and end of the other coil will be joined together; the resultant circuit will be as shown in Fig. 1.205.

The supply will be given to the junction of the joints made.

In the star system the ends of R, Y and B phases will be joined together; the supply will be given between the connected start terminals. It is depicted in Fig. 1.206.

It can be seen in the Fig. 1.206 that it is possible to get a common junction (marked as (N))

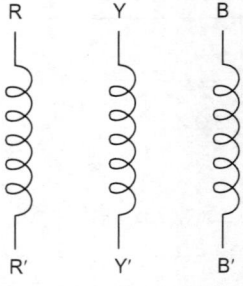

Fig. 1.204. Three, 1-phase coils

Basic Electricity   135

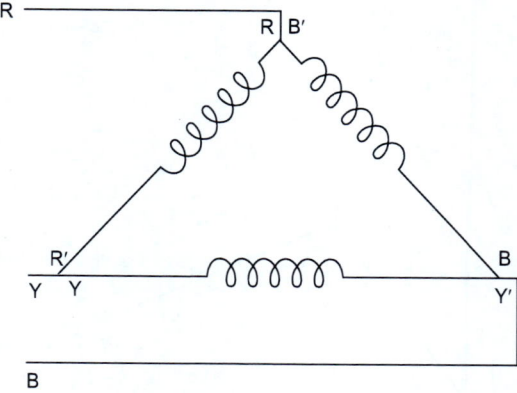

Fig. 1.205. Delta system. Each phase is displaced by 120°

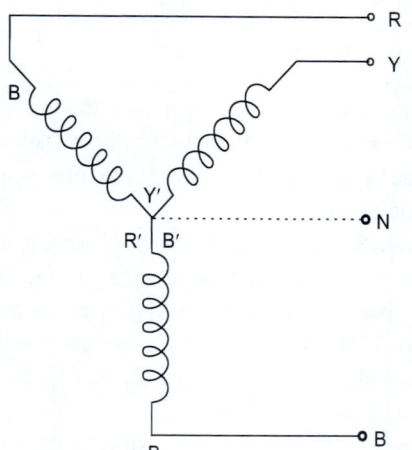

Fig. 1.206. A Star system. Each phase is displaced by 120°

in a star system. This point, is known to be the **neutral** point. If needed a wire can be run from this neutral point also, which will be known as the '**neutral wire**'. Usually this neutral wire will carry only a small amount of current because only the '**lines**' R, Y and B will carry the total current.

If not needed, the neutral wire (alone) may be avoided.

The star system with the neutral wire is known to be the 3-phase, 4-wire star system; without the neutral wire, it is a 3-phase 3-wire star system.

Of course, as it is not possible to form any common junction in a delta system, neutral point cannot be located here.

It is brought out in the initial sections of part I, chapter 1 that by providing a single-phase supply, voltage waveform as shown in Fig. 1.52, can be generated.

On the other hand, if the three phases R, Y and B are arranged in such a way that they are displaced by 120° (i.e. a star or delta system) then a three-phase voltage can be generated. To say precisely, three single phase voltages will be produced; because the windings producing these voltages are 120° apart, the voltages produced by them will also be so.

Let this be studied.

Concentrate on Fig. 1.207.

If the cylinder is allowed to rotate R phase will also revolve through 360° in a magnetic field. As pointed out clearly in the sec. (1.24), it will generate alternating voltage of the waveform shape Fig. 1.52.

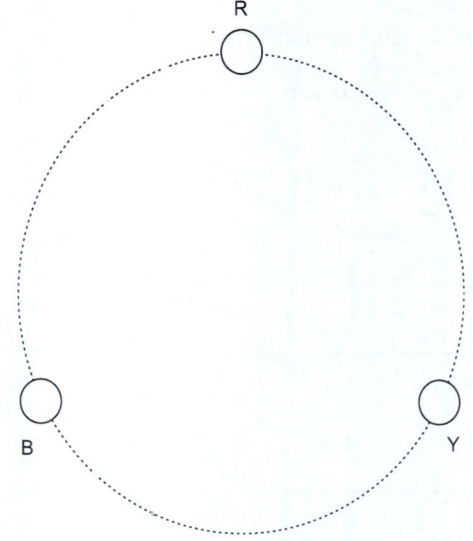

Fig. 1.207. Generation of 3-phase voltage.

As the phase Y is displaced by 120° from the phase R, no doubt that due to the rotation, phase Y will also produce same amount of alternating voltage; but will be displaced by 120°.

Further, phase B also will be in motion when the cylinder is rotating; thus alternating voltage obeying the waveform of Fig. 1.52 will be available in phase B too. Of course, as it is displaced from the phase R by 240°, the voltage of phase B will be displaced from that of R by 240°.

This type of 3-phase voltage production, can be sketched as waveform, as in Fig. 1.208.

Hence, when 3-phase winding is allowed to rotate continuously in a magnetic field, it produces EMF of nature as shown in Fig. 1.208.

If the reader recollects the lagging-leading theory of Chapter 1, the waveform Fig. 1.141 can be drawn in vector from as in Fig. 1.209.

In complex form these voltages are:

$$V_R = V_m \angle 0°$$
$$V_Y = V_m \angle -120°$$
$$V_B = V_m \angle -240°$$

In time domain equation form,

$$v_R = V_m \sin \omega t$$

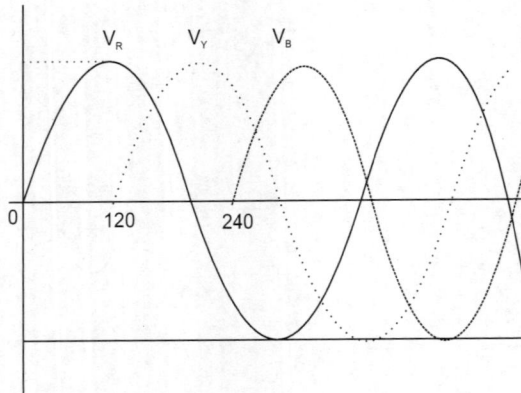

Fig. 1.208. 3-phase voltages

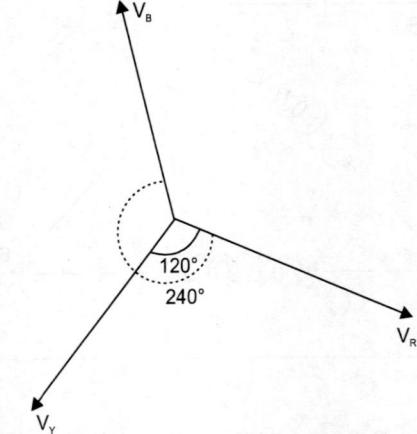

Fig. 1.209. Vector form notation of 3-phase voltages.

$$v_Y = V_m \sin(\omega t - 120)$$
$$v_B = V_m \sin(\omega t - 240)$$

The reader can later experience the use of 3-phase winding for the production 3 fluxes displaced, of course by 120°, in the working principles 3-phase induction motor.

Now referring to Fig. 1.209, we observe that the voltages attain their maximum values in the order V_R, V_Y and V_B. This order is known as the *phase sequence* RYB. A reverse phase sequence will be RBY, in which case the voltages V_B and V_Y lag V_R by 120 and 240°, respectively.

Consider the Fig. 1.210, in which a star (wye) connection is illustrated. The voltage that occurs across the lines or outers—that is between RR' or YY' or BB'—is called the line voltage in a 3-phase system. Whereas the voltage between a line and the neutral—that is, RO, or YO or BO—is called the *phase voltage* in a 3-phase system.

The line voltages are related to the phase voltages such that

$$V_{RO} + V_{RY} = V_{YO}$$

or, $V_{RY} = V_{YO} - V_{RO}$ (i.e., the PD)

Similarly, $(V_{YB} = V_{BO} - V_{YO})$ and $(V_{BR} = V_{RO} - V_{BO})$.

From Eq. (1) and Fig. 1.209 we may write $V_{RY} = V(-0.5 - j0.866) - V(1 + j0) = V(-1.5 - j0.866)$. In polar form,

$$V_{RY} = \sqrt{3}\,V \angle -120°$$

Similar relationships are valid for the phasors V_{YB} and V_{BR}. Because V_{RY} is the voltage across the lines R and Y and V is the magnitude of the voltage across the phase, we may generalize Eq. (2) to $V_L = \sqrt{3}V_p$ where V_L is the voltage across any two lines and V_p is the phase voltage.

A phasor diagram showing the phase and line voltage relationship is shown in Fig. 1.211.

For the wye connection it is clear from Fig. 1.210 that the line currents I_L and phase current I_p are the same. Thus, we may write $I_L = I_p$. The mutual phase relationships of the currents are given in Fig. 1.212.

For the delta connection as shown in Fig. 1.213, the corresponding phasor diagram is Fig. 1.214.

From Fig. 1.214 it follows that $\mathbf{V}_L = \mathbf{V}_p$. See Fig. 1.215 for the phasor diagram.

We show in Fig. 1.213 the phase currents and line currents for the delta-connected system. The phase currents and line currents are related to each other by $I_L = \sqrt{3}\,I_{ph}$.

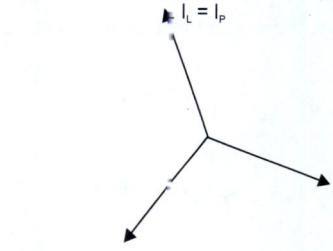

Fig. 1.212.

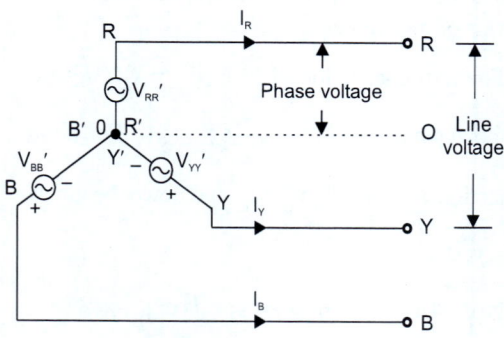

Fig. 1.210. A star system.

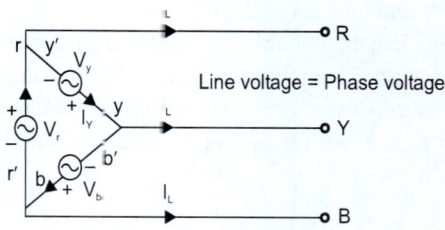

Fig. 1.213. A delta system

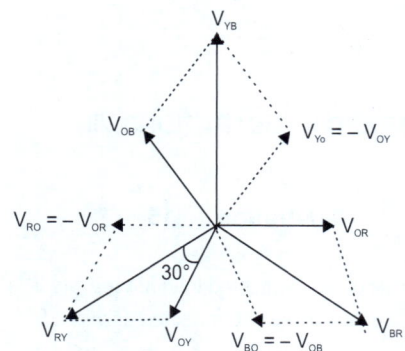

Fig. 1.211. Line & phase voltages.

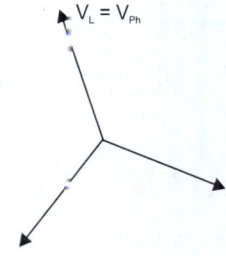

Fig. 1.214.

138 Basic Electrical Engineering

But, $V_L = V_{Ph}$
As a summary, for a star system,

$I_L = I_{Ph}$

$V_L = \sqrt{3} V_{Ph}$

and for a delta system,

$I_L = \sqrt{3} I_{Ph}$

$V_L = V_{Ph}$

In a single-phase system, for $v = V_m \sin \omega t$ is the voltage across and $i = I_m \sin \omega t$ is the current through a resistor, we have the instantaneous and average powers as:

$p = (V_m \sin \omega t)(I_m \sin \omega t)$

$= V_m I_m \sin^2 \omega t = \dfrac{1}{2} V_m I_m (1 - \cos 2\omega t)$

$P = \dfrac{1}{2} V_m I_m = \dfrac{V_m}{\sqrt{2}} \dfrac{I_m}{\sqrt{2}} = VI$

where V and I are rms values.

This can be suitably extended for a 3-phase system, with the knowledge that the three voltage vectors are displaced from each other by 120 degrees. Let an expression for the total average power, be obtained.

$p = v_R i_R + v_y i_y + v_B i_B = (V_m \sin \omega t)(I_m \sin \omega t)$

$+ [V_m \sin(\omega t - 120°)][I_m \sin(\omega t - 120°)]$

$+ [V_m \sin(\omega t + 120°)][I_m \sin(\omega t + 120°)]$

This is because, Y phase lags R by 120°, and B phase lags R phase by 240°.

$= V_m I_m \left(\sin^2 \omega t + \dfrac{1}{4} \sin^2 \omega t + \dfrac{3}{4} \cos^2 \omega t \right.$

$\left. + \dfrac{1}{4} \sin^2 \omega t + \dfrac{3}{4} \cos^2 \omega t \right)$

$= \dfrac{3}{2} V_m I_m (\sin^2 \omega t + \cos^2 \omega t) = \dfrac{3}{2} V_m I_m$

$3 \dfrac{V_m}{\sqrt{2}} \dfrac{I_m}{\sqrt{2}} = 3 V_p I_p$

where V_p and I_p are rms values of phase voltage and current. Since the instantaneous power is constant, the average power is the same as the total instantaneous power; that is, $P_T = 3 V_p I_p$.

The total power in a three-phase circuit having a balanced load with a phase power factor angle ϕ, is:

Power per phase = $V_p I_p \cos \phi$.

Total power $P_T = 3 V_p I_p \cos \phi$

The above total power can also be a expressed in terms of line values, as:

For wye connection, $I_p = I_L$, $V_p = V_L / \sqrt{3}$, so that power becomes $P_T = \sqrt{3} V_L I_L \cos \theta$. For delta connection $I_p = I_L \sqrt{3}$, $V_p = I_L$, and power becomes $P_T = \sqrt{3} V_L I_L \cos \phi$.

Note that the power is $\sqrt{3} V_L I_L \cos \phi$ W, irrespective of star or delta system.

1.67 WORKED ILLUSTRATIONS VII

Illustration 1.115

A three-phase delta-connected load having a $(3 + j4)$–Ω impedance per phase is connected across a 220 V three-phase source. Calculate the magnitude of the line current.

Solution

$V_{line} = V_{phase} = 220\,V$

$Z_{phase} = 3 + j4 = 5\angle -53.2°\,\Omega$

$I_{phase} = \dfrac{220}{5} = 44\,A$

$I_{line} = \sqrt{3} \times 44 = 76.21\,A$

Note that the problems are to be solved by considering phase quantities only.

Illustration 1.116

A 220 V three-phase source supplies a three-phase wye-connected load having an impedance of (3 + j4) Ω per phase. Calculate the phase voltage across each phase of the load.

Solution

$V_{phase} = \dfrac{220}{\sqrt{3}} = 127\,V$

$Z_{phase} = 3 + j4 = 5\angle 53.2°\,\Omega$

$I_{phase} = \dfrac{127}{5} = 25.4\,A = I_{line}$

Illustration 1.117

Calculate the line current I_A in the delta-connected system shown in Fig. 1.215.

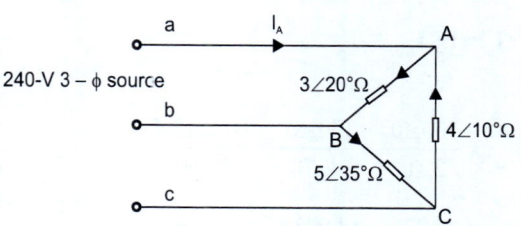

Fig. 1.215 Circuit for illustration 1.117.

Solution

In solving a delta system assume the branch currents as a closed loop, as shown in Fig. 1.215. Then,

$I_A = I_{AB} - I_{CA} = \dfrac{240\angle 0}{3\angle 20} - \dfrac{240\angle 120}{4\angle 10}$

$= 127.16\angle -41.19°\,A$

Illustration 1.118

Determine the current I_a for the three-phase system shown in Fig. 1.216.

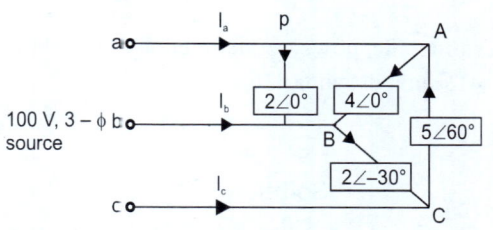

Fig. 1.216. Circuit for illustration 1.118.

Solution

Note that the nodes 'p' and 'a' are the same. Now applying KCL to 'a'.

$$I_a = I + I_{AB} - I_{CA}$$

$$= \frac{100\angle 0°}{2\angle 0°} + \frac{100\angle 0°}{4\angle 0°} - \frac{100\angle 120°}{5\angle 60°}$$

$$= 67.27\angle -14.92° \, A$$

Illustration 1.119

In the circuit of Fig. 1.216, verify that I_c is unaffected by the $2\angle 0°$-Ω impedance. Also calculate I_c.

Solution

$$I_C = I_{CA} - I_{BC} = \frac{100\angle 120°}{5\angle 60°} - \frac{100\angle -120°}{2\angle 30°}$$

$$= 68.55°\angle 81.55° \, A$$

Illustration 1.120

Determine the power consumed by the load $(3 + j4)\,\Omega$ in illustration 116.

Solution

$3 + j4 = 5\angle 53.2\,\Omega$. This means $\phi = 53.2$. Also given that,

$$V_L = 200\,V \text{ and } I_L \text{ was found out as } 25.4;\, A$$

Hence, $P_T = \sqrt{3}(220)(25.4)\cos 53.2° = 5.8\,kW$

Illustration 1.121

Three impedances, $Z_1 = 6\angle 20°\,\Omega$, $Z_2 = 8\angle 40°\,\Omega$, and $Z_3 = 10\angle 0°\,\Omega$, are connected in wye and are supplied by a 480 V three-phase source. Solve for the line currents.

Solution

The currents are

$$I_a = \frac{(480/\sqrt{3})}{6\angle 20°} = 46.19\angle -20°\,A$$

$$I_b = 34.64\angle -150\,A$$

$$I_c = 27.71\angle 120°\,A$$

Illustration 1.122

A load of $(4 - j3)\,\Omega$ is in all the 3-phases of a star system, supplied at 220 V. Calculate the power consumed by the load.

Solution

$$V_p = \frac{220}{\sqrt{3}} = 127\,V$$

$$Z_p = 4 - j3 = 5\angle -36.87°\,\Omega$$

$$I_L = I_p = \frac{127}{5\angle -36.87} = 25.4\angle 36.87°\,A$$

Hence, $P = \sqrt{3}\times 220\times 25.4\cos 36.87 = 7.74\,kW.$

Illustration 1.123

A three-phase balanced load has a 10 Ω resistance in each of its phases. The load is supplied by a 220 V three-phase source. Calculate the power absorbed by the load if it is connected in wye; calculate the same if it is connected in delta.

Solution

In the wye connection,

$$V_p = \frac{220}{\sqrt{3}} 127 \text{ V}$$

$$I_p = \frac{127}{10} = 12.7 \text{ A} = I_L$$

$\cos \phi = 1$ (load being purely resistive)

Hence,

$$P = \sqrt{3} V_L I_L \cos \phi = \sqrt{3} \times 220 \times 12.7 \times 1 = 4.48 \text{ kW}.$$

In the delta connection,

$$V_L = 220 \text{ V}$$

$$I_p = \frac{220}{10} = 22 \text{ A} \quad I_l = \sqrt{3} \times 22 = 38.1 \text{ A}$$

Hence,

$$P = \sqrt{3} \times 220 \times 38.1 \times 1 = 14.52 \text{ kW}.$$

Notice that the power consumed in the delta connection is three times that of the wye connection.

Illustration 1.124

A three-phase 450 V 25 Hz source supplies power to a balanced three-phase resistive load. If the line current is 100 A, determine the active and reactive powers drawn by the load.

Solution

The power factor of a resistor load is 1.0

$$P = \sqrt{3} V_{line} I_{line} \cos 0°$$

$$P = \sqrt{3}(450)(100)(1) = 77.94 \text{ kW}$$

Note that reactive power $Q = VI \sin \phi$

$$Q = \sqrt{3}(450)(100) \sin 0° = 0$$

Illustration 1.125

A three phase 600 V 25 Hz source supplies power to a balanced three-phase motor load. If the line current is 40 A and the power factor of the motor is 0.80, determine the active and reactive powers drawn by the motor.

Solution

$$P = \sqrt{3} V_{line} I_{line} \cos \phi$$

$$P = \sqrt{3}(600)(40)(0.8) = 33.26 \text{ kW}$$

Since $\cos \phi = 0.8$, $\sin \phi = 0.6$;

hence, $Q = \sqrt{3}(600)(40)(0.6) = 24.945 \text{ kvar}$.

1.68 TUTORIAL PROBLEMS V

TV-1: Use Y-Δ and Δ-Y transformation to find the input resistance of the network shown in the figure.

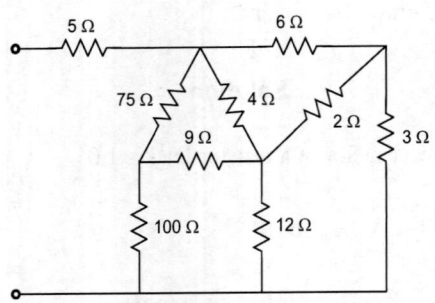

Ans: 9.09 Ω

TV-2: Reduce the network of the figure into a single resistance between A and B terminals by suitable star delta conversion

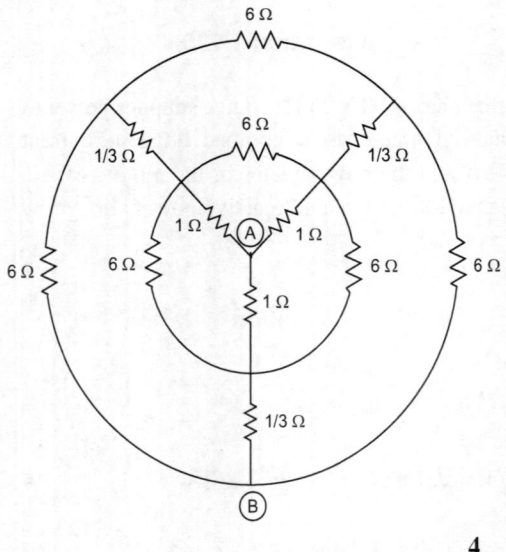

Ans: $\dfrac{4}{3}$ Ω

TV-3: Three impedances each of 10 Ω resistance and 5 Ω inductive reactance are connected in delta to a 400 V 3ϕ supply. Determine the current in each phase and in each line. Calculate also the total power drawn from the supply and p.f. of the load.

Ans: 62 A, 38.23 kW, 0.89 lag

TV-4: Determine the line currents for the unbalanced delta connected load of figure given. Assume the phase sequence in RYB. E = 200 V.

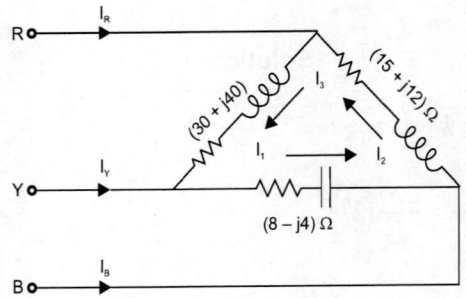

Ans: $I_1 = 2.4 - j3.2$ A
$I_2 = 6.26 - j10.71$ A
$I_3 = 1.58 + j\,10.3$ A
$I_R = 13.53\,\angle -86.5°$ A; $I_Y = 8.42\,\angle -63.2°$ A
$I_B = 21.51\,\angle 102.4°$ A

TV-5: A 3-phase, 4 wire, 208 V, ABC system supplies a star connected load in which $Z_A = 10\,\angle 0°$ Ω $Z_B = 15\,\angle 30°$ Ω and $Z_c = 10\,\angle -30°$ Ω. Find the line currents, the neutral currents and load power.

Ans: $I_A = 12\,\angle 90°$ A
$I_B = 8\,\angle -60°$ A $I_c = 12\,\angle -120°$ A
$I_N = 5.69\,\angle -110.6°$ A
P = 3519 W.

TV-6: The currents in RY, YB and BR branches of a mesh connected system with symmetrical voltages are 20 A at 0.8 power factor lagging 25 A at 0.7 power factor leading and 20 A at unity power factor respectively. Determine the line currents assuming that the phase sequence is RYB.

Ans: $I_R = 39.2\,\angle -48.4°$ A
$I_Y = 15.17\,\angle -127.7°$ A
$I_B = 44.6\,\angle 112°$ A.

TV-7: A balanced 3-phase system supplies an unbalanced 3-phase delta connected load made up of two resistors 100 Ω and 200 Ω and a reactor having an inductance of 0.3 H with negligible resistance. $V_c = 100$ V at 50 Hz. Calculate (a) total power in the systems and (b) the total volt ampere reactive.

Ans. 150 W, 106 VAR (lag)

TV-8: Find the current through a load impedance of 5 ∠30° ohm connected across terminals AB in the circuit shown below using Narton's theorem.

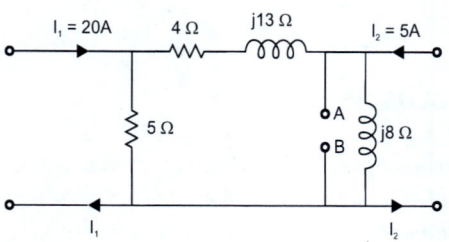

Ans: $I_L = 8.2 \angle 0.93°$ A

TV-9: If superposition is used on the circuit below. find V_1 with (a) only the 20 ∠0° mA source operating (b) only 50 ∠–90° mA source operating

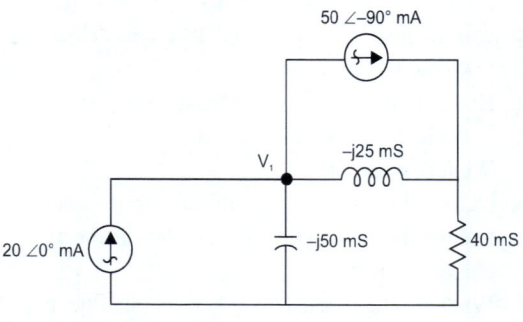

Ans: (a) 0.1951 – j0.556
(b) 0.780 + j0.976

TV-10: Use superposition to find the voltages $V_1(t)$ and $V_2(t)$ in the circuit

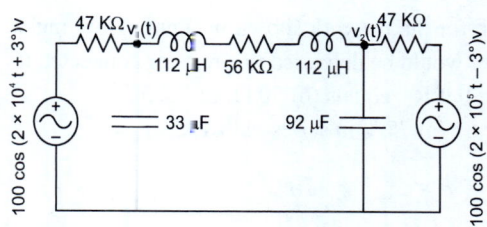

Ans:

$V_1(t) = 3.227 \cos(10^4 t - 83.62°)$
$+ 2.006 \times 10^{-4} \cos(10^5 t + 127.1°)$ V

$V_2(t) = 31.13 \cos(10^4 t - 179.3°)$
$+ 115.7 \cos(10^5 t - 92.91°)$ mV

TV-11: A 1 kHz generator has an internal impedance containing a resistance of 50 Ω in series with and inductance of 0.01 H. It is supplying a load of 1000 Ω. Resistance. A capacitance is connected in parallel with the load and a circuit of inductance L and negligible resistance is put in series with the generator. Find the values of C and L which will give the maximum power in the load.

Ans: C = 0.694 micro farad
L = 24.7 mH

TV-12: The circuit shown in the circuit depicts a circuit in two stages. Select R_1 so that maximum power is transferred from stage 1 to stage 2.

Ans. 8 kΩ.

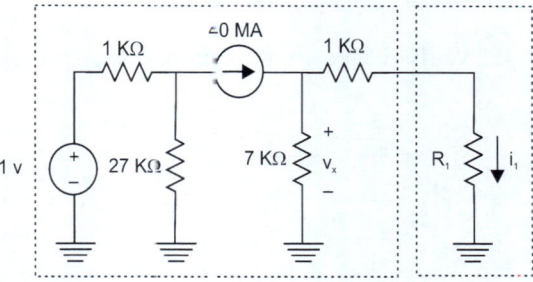

TV-13: Find the Thevenin equivalent at terminals a

144 Basic Electrical Engineering

and b for the network shown in figure. How much power would be delivered to a resistor connected to a and b if R_{ab} equals (b) 50 Ω; (c) 12.5 Ω?
Ans: (a) 75 V in series with 12.5 Ω

shown in the figure and also calculate the transfer resistance.

Ans (2.169 A)

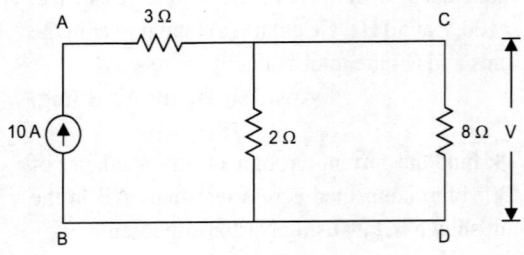

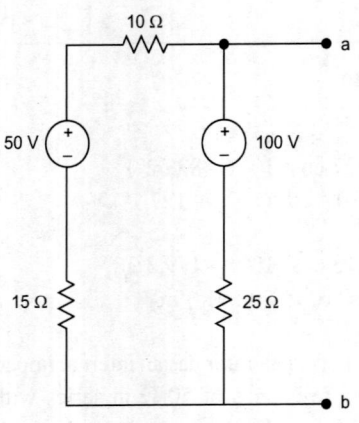

(b) 72 W
(c) 112.5 W

TV-14: Determine the thevenin equivalent seen by j10 Ω of figure and use this to compute V_1
Ans: $V_1 = 1 - j2$ V

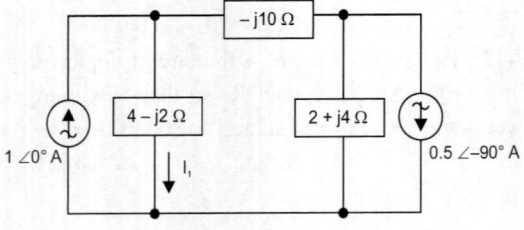

TV-15: Verify reciprocity theorem in the circuit

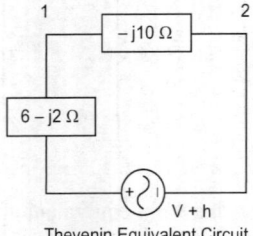
Thevenin Equivalent Circuit

CONCLUSION

This chapter presented in detail the fundamentals of DC and AC circuits with complete theory of the topics with numerous illustrated examples are given. Additional topics such as resonance, network theorems and introduction to 3-phase circuits are also added.

QUESTIONS

1. Define the following terms and their units: current, resistance, potential difference, electric power and electric energy.
2. Discuss on what factors does the resistance of a conductor depend upon.
3. What is a circuit?
4. Define Ohm's law. Apply this to a series circuit and parallel circuit and discuss their respective characteristics.
5. What are the two Kirchhoff's laws? Define and discuss them.
6. How to solve a circuit using KVL and KCL? Elaborate on the sign convention procedure in doing so.
7. Differentiate between a DC circuit and an AC circuit.

8. What is an alternating supply? Explain.
9. Discuss how an alternating sinusoidal voltage is produced.
10. With reference to an alternating quantity, define: cycle, time period, amplitude and instantaneous equation.
11. How do you evaluate the average value and the rms value of an alternating quantity?
12. What is the physical significance of 'phase difference' in alternating quantities? Explain this for a pure resistive circuit, a pure inductive circuit and a pure capacitive circuit.
13. What is phase angle? Define power factor of a circuit.
14. Define power factor in terms of circuit constants.
15. Explain the significance of the 'j' operator in analyzing AC circuits.
16. Assume two alternating quantities displaced at an angle. Denote them in polar form, rectangular form, as time domain equations and as sinusoidal waveforms.
17. Define reactances. State their values for a pure L circuit, pure C circuit and for a circuit with C-L-both in polar and rectangular forms.
18. Write a short note on impedance for R-L-C AC circuits.
19. Prove that the electric power in pure L or pure C alternating circuit is zero.
20. Prove that the electric power in pure R-L-C alternating circuit is VI × power factor.
21. What is reactive power? Differentiate between real power and reactive power.
22. What is resonance? What is its notable application?
23. Derive an expression for the resonant frequency of a series RLC circuit.
24. What is Q-factor? What is its application?
25. What bandwidth? State its ceiling limit frequencies.
26. Prove that at resonance, the voltage across the inductance and the voltage across capacitance are 180 degree away.
27. What is voltage magnification and current magnification in resonant circuit?
28. Define conductance, susceptance and admittance.
29. What is a 3-phase circuit? What are the two famous types of 3-phase connections? Discuss their features.
30. What are line and phase quantities? Bring out a relation of these two quantities for current and voltage in (a) a star system and (b) in a delta system.
31. Prove a relation for electric power in a 3-phase system.
32. Write short notes on a balanced 3-phase circuit and an unbalanced 3-phase circuit.
33. What are the applications of network theorems?
34. Express that the mesh theorem is $V = IZ$ and nodal theorem is $I = VY$.
35. State, and explain with an example, the Thevenin's theorem and its dual theorem.
36. Prove the condition for a maximum power to be transferred to the load in a circuit.
37. State, and explain with an example, the reciprocity theorem and superposition theorem.

INTRODUCTION

This chapter, in its initial parts introduces the fundamentals of magnetic circuits. Various terms associated with magnetic circuits are presented. Series and parallel magnetic circuits are discussed. Worked examples are presented. In the second part of the chapter, properties of magnetic materials are introduced—with particular reference to the hysteresis loop. Applications of magnetic materials are also presented.

CHAPTER OBJECTIVE

At the completion of this chapter, the reader will be able to:
* understand the terms associated with magnetic circuits;
* solve numerical examples on the fundamental magnetic circuit problems; and
* understand the rudiments of magnetic materials.

KEYWORDS

Magnetic circuit, Electromagnetism, Flux and flux density, MMF, Magnetizing force, Leakage flux, Fleming's rules, Dynamically induced EMF, Statically induced EMF, Inductance and Magnetic materials.

2

Magnetism and Electromagnetism

PART I : BASIC TERMS IN ELECTROMAGNETISM

2.1 INTRODUCTION

The current, the voltage, the resistance are three quantities of an electric circuit. This, of course, as discussed in the earlier chapter 1, is an 'electrical theory'.

Similar to this, in the beginning, another theory was in existence and was the 'magnetic theory'. As an electric circuit, magnetic circuit also was defined with several quantities. (The main scope of this chapter is to see all these quantities in depth.) But in those days, **electricity** and **magnetism** were treated entirely as *separate* theories, altogether.

In the beginning of 1800's Oersted and Ampere (whose name later kept as the unit of electric current, as a mark of respect to his uncomparable contribution to the electric science) postulated then an **electric current** could deflect a **magnetic compass** needle; hence **all magnetic phenomena are fully due to the electric charges in motion**. (Statement 1)

After that, few decades later, Faraday experienced in his workshop that, a moving magnet generates an electric current. This indeed states that, **all electrical phenomena are influenced proportionally by the magnetic field** (Statement 2)

As a summary, statement 1 says that magnetic phenomena are due to electricity; Statement 2 says that, electrical phenomena are due to magnetism;

Quite interesting!!

With this conclusion, Maxwell released a final theory, in which he strongly stated that electricity and magnetism are proportional, intertwined, one and only, inseparable—and they are twofolds of a single subject known as **electromagnetism**.

The chapter 1 dealt with the first fold of the same **subject**—electricity; this chapter enters now into its other fold, magnetism.

2.2 MAGNETIC CIRCUIT

It is known that the like poles repel and the unlike poles attract, each other, when they are kept at a distance, closer together. But then, how these two *separate* magnetic materials know that the nearby sides are like or unlike so that they can repel or attract? That is the "Field", which makes these poles to identify whether or not, the respective nearby poles are like or unlike. This means nothing but the two poles are surrounded by this field (refer to Fig. 2.1), without which the magnets could not have been **informed** about their polarities.

This is the fullest explanation of a (magnetic) field.

Hence, any space in which a magnetic effect can be experienced or detected is known as the **magnetic field**. The space is shown in Fig. 2.1.

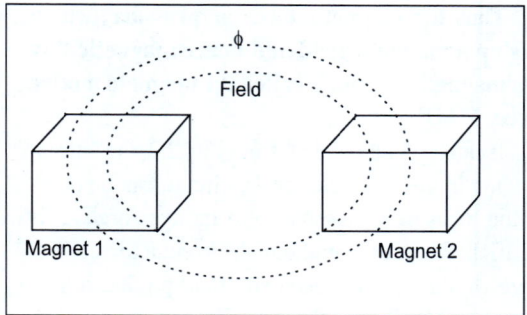

Fig. 2.1. Field demonstration.

This magnetic field is made up of series of fictitious magnetic lines. These imaginary lines are otherwise known as, **lines of magnetic flux** or **magnetic lines of force**, and they always will form a closed non-intersecting path. (See Fig. 2.1). Thus, the total number of magnetic lines in a field is called the magnetic flux, with a symbol ϕ.

Therefore, the closed path in a magnetic field is a magnetic circuit. (2.1)

or

The closed path followed by magnetic flux is called a magnetic circuit. (2.2)

or

The region around the magnet(s) where the poles exhibit a force of attraction or repulsion is called a magnetic field. (2.3)

2.3 DEFINITIONS OF TERMS CONCERNING MAGNETIC CIRCUITS

Flux

As stated in sec 2.2, the magnetic lines of force (or simply the *flux*) are imaginary and they really "do not flow" in the magnetic circuit. To deal the magnetic circuit mathematically, they are introduced and are "assumed" to be available in the field.

Hence **The quantity—that is the total number—of magnetic lines of force produced by a magnet and which are available in the pole's field, is called the magnetic flux.**

This is generally given the symbol ϕ (phi). Unit is webers 1 wb = 10^8 lines the force or Maxells.

A way of defining flux is by considering the **flux density**.

Flux density

The magnetic *flux* per square-metre at right angles to the magnetic field is called the **magnetic flux density** or **magnetic inductions**.

By this, it is seen that ϕ/a is the magnetic flux density, where **a** is the area is square-metre. Giving symbol 'B' to the density,

$$B = \phi/a, \text{ wb/m}^2 \text{ or Tesla.} \qquad (2.4)$$

Thus, $\phi = Ba$; webers (2.4-A)

The product of flux density at a point, and the area in square-metres of this considered point at right angles to the field is the total flux emerging through the point.

The inter-relations between a magnetic field and an electric field

Magnetic field and electric field are inseparable (Refer Sec 2.1). This is brought out clearly by a rule known as cork screw rule, which states that the direction of the magnetic field produced by a current carrying conductor will be in the direction of rotation of a cork screw, which is, threaded, in the direction of the current.

Observe Fig. 2.2.

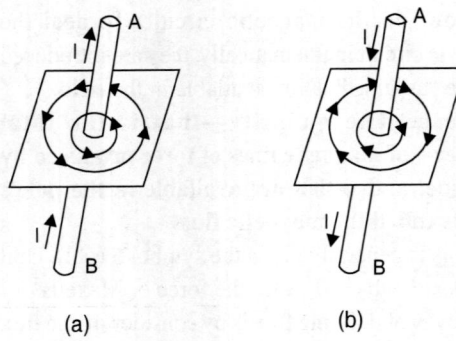

Fig. 2.2. Magnetic field set-up around a current-carrying conductor.

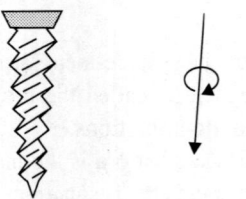

Fig. 2.3.

In Fig. 2.2(a) the conductor carries current upwards. If it is imagined that the screw tip denotes the arrowhead of the current flow (refer to Fig. 2.3) and if the screw is to go in the direction of the current, it must be screwed in, which clearly indicates that the magnetic field set up by the current is in clockwise direction, by cork screw rule.

By the same analogy, see Fig. 2.2(b), downward direction of current produces a magnetic field in anticlockwise direction.

Hence when a current flow in there, invariably, inherently, there will be a magnetic field too! (Refer Sec. 2.1).

The magneto motive force (MMF)

The force which sets up the above discussed magnetic field, (hence, of course, the total magnetic flux) is known as the **magneto motive force**.

Thus **the magnetic force or pressure, which sets up a magnetic field and hence magnetic flux in a magnetic circuit, is called the magneto motive force, MMF.**

It can be declared that, from Fig. 2.2, the current flowing in the conductor or the circuit, can be taken as the measure of the magneto motive force. It is justifiable because, without the conductor current there would not have been any field produced, and hence the MMF.

$$\therefore \text{MMF} \propto I$$

$$= I, \text{ for 1 conductor}$$

$$\text{MMF} = NI, \text{ for N conductors} \qquad (2.5)$$

Its unit is, ampere—turns, symbolised by AT or $\overline{A}$.

Magnetising force or magnetic field strength

So it is clear that the available MMF produces an amount of total flux. Thus the MMF **remains** the same for all the flux lines.

But, MMF per metre length will vary. For instance, considering the Fig. 2.1, the field (hence the flux) near the conductor, will have **larger** MMF per meter length whereas the field (hence the flux) at the extremity will have **lesser** MMF per unit length, because, in both the cases the numerator (that is, the MMF) is constant and the denominator only changes; in the former a lesser denominator and in the latter, a larger denominator.

A term is defined to introduce this aspect and is the **magnetising force** or **magnetic field strength**, symbolised as H.

Hence, **the MMF per unit length of flux path is the magnetising force**.

Letting, l, as the length of the flux path in metre, and using equation (2.5)

$$H = \frac{NI}{l} \quad (2.6)$$

Note: In the Fig. 2.2, the field is shown as circular in nature; otherwise also it is so. For such circular field (or flux path), the length is $2\pi r$. substituting this in equation (2.6),

$$H = \frac{NI}{2\pi r} \quad (2.6\text{-A})$$

which implies that the H is inversely proportional to 'r', the radius. Thus, lesser the 'r', larger will be the 'H'. It is of course, brought out in the above paragraphs.

Relation between B and H

At any point in a magnetic field the magnetising force H is the force maintaining some flux and thus producing a particular flux density B at that point. This means that, without H, there cannot be B. Thus H is the main cause producing B, and their ratio, B/H will remain almost a constant in any point in the entire field

So in a magnetic field, at any point,

$$\frac{B}{H} = \text{a constant} = \mu_0 \quad \text{in vacuum} \quad (2.7)$$

$$= \mu_0 \mu_r \quad \text{in any space,} \quad (2.8)$$

where μ_r is its relative permeability.

[*Note:* It roots out from the fact that, $B \propto H$ in a magnetic field, at any point. Introducing a proportional constant μ_0,

$$B = \mu_0 H \quad (2.9)$$

Thus invariably, μ_0 is a constant]

μ_0 is known to be the **permeability** or the *magnetic space constant*. It holds the value, in MKS system, $4\pi \times 10^{-7}$.

Ohm's law in magnetic circuits

Quickly recollecting the electric circuits discussion, in a closed circuit, an electric force of 'V' volts force a current of 'I' amperes to flow, against a resistance R ohms. The Ohm's law in electric theory states that, I = V/R.

In magnetic circuits, also, equivalently such phenomena occurs. A MMF of NI amp-turns, forces a total flux of ϕ wb, in the magnetic circuit.

Thus, $\phi \propto NI$ \quad (2.10)

As in the case of electric circuits where the flow of electric current was opposed by the resistance of the conductor (or circuit), in magnetic circuits also, the flow of magnetic flux is frictioned or opposed by, what is known as, the reluctance of the magnetic circuit. Thus the flux ϕ set-up by the MMF, of NI ampere turns, should always flow in the magnetic circuit against the **reluctance**. Reluctance is symbolised by S.

Larger the reluctance, more will be the opposition to the flow magnetic flux thus causing lesser flux, ϕ. So, the flux is inversely proportional to the reluctance.

Incorporating this is equation 2.10.

Flux = MMF/reluctance

i.e., $$\phi = \frac{NI}{S} \quad (2.11)$$

This is the Ohm's law of magnetic circuit.

Reluctance in terms of physical quantities

In sec. 1.8, electrical resistance value was given in terms of the physical structure of the resistor. An attempt now is made in a similar way, to define the magnetic resistance known as the reluctance.

The reluctance, depends upon the length (l) of flux travel in metres, the area of cross-section of the magnetic material through which the flux passes in square-metres, and the permeability $\mu = \mu_0 \mu_r$ of the magnetic media and all are related by the below relation.

$$S = \frac{l}{\mu_0 \mu_r a} \quad \ldots (2.12)$$

where μ_r is the relative permeability of the magnetic core material.

Note: (i) μ_r can be read from the magnetic material's B-H. Curve. The B-H curve is available for different materials.

(ii) Permanence is the reciprocal of reluctance, and it measures how easily flux can travel in a magnetic circuit.

$$\text{Permeance} = \frac{1}{\text{reluctance}}$$

$$\text{Permeance} = \frac{\mu_0 \mu_r a}{l} \quad (2.13)$$

The ampere turn formula

Note: Work's law: By work's law, the work done in moving an unit pole around a magnetic circuit is equal to the ampere-turn enclosed in the magnetic circuit.

Refer to Fig. 2.4. It is considered to define the ampere-turn formula.

By fundamentals, work done is the product of force and displacement. In magnetic circuit the force which magnetises the circuit is the magnetising force,

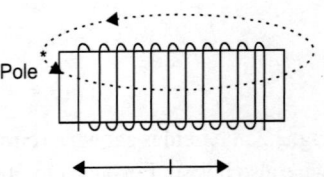

Fig. 2.4. Circuit to define the ampere-turn formula.

H. Hence the work done in a magnetic field is Hl. By work's law,

$$Hl = NI \quad (2.14)$$

Substituting the value of H from equation (2.8) into the equation (2.14)

$$\frac{B}{\mu_0 \mu_r} l = NI \quad (2.14\text{-A})$$

Substituting the value of B from equation (2.4), the above equation becomes,

$$\frac{\phi}{a \mu_0 \mu_r} l = NI$$

or

$$\phi = \frac{NI a \mu_0 \mu_r}{l}$$

or

$$\phi = \frac{NI}{\left(\dfrac{l}{\mu_0 \mu_r a}\right)} \quad (2.15)$$

which is nothing but the magnetic Ohm's law of equation (2.11)!

2.4 SERIES MAGNETIC CIRCUITS

In practice a magnetic circuit may consist of several parts in series of different lengths, cross-sectional areas, and permeabilities.

In such series magnetic circuit all the reluctances of several parts will get summed up together (as resistors in electric circuit) to form the net reluctance of the circuit. For N series paths,

$$S_{net} = S_1 + S_2 + S_3 + \ldots + S_N \quad (2.16)$$

Using equation (2.12),

$$S_{net} = \frac{l_1}{\mu_0 \mu_{r_1} a_1} + \frac{l_2}{\mu_0 \mu_{r_2} a_2} + \frac{l_3}{\mu_0 \mu_{r_3} a_3} + \ldots + \frac{l_n}{\mu_0 \mu_{r_n} a_n}$$

$$S_{net} = \frac{1}{\mu_0} \left[\frac{l_1}{\mu_{r1} a_1} + \frac{l_2}{\mu_{r2} a_2} + \frac{l_3}{\mu_{r3} a_3} \ldots + \frac{l_N}{\mu_{rN} a_N} \right]$$

In series magnetic circuits the flux passing through each part will be the same (as current in the series electric circuits). Let the flux in the series magnetic circuit be ϕ, wb. By finding the product of ϕS_1, ϕS_2 etc., (refer the equation 2.11) the different MMF of different parts of the series circuit can be found out. Then the net MMF is,

The net circuit MMF = $\phi [S_{net}]$ (2.17)

where $S_{net} = S_1 + S_2 + \ldots S_N$

Thus in a series magnetic circuit,
— flux remains the same in each path
— the reluctance of each path changes
— the MMF of each path will be different

Note: The magnetic circuits, like in the circuit of Fig. 2.5-A are said to be **composite** magnetic circuit because the magnetic flux produced passes through **more than one** part (in this case, through iron and air gap)—whereas—circuit as in Fig. 25-B is said to be '**simple**' magnetic circuit as it has **only one** media as the path. (in this case iron).

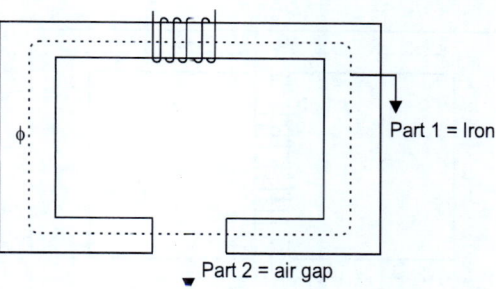

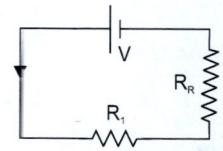

Fig. 2.5-A. A series magnetic circuit.

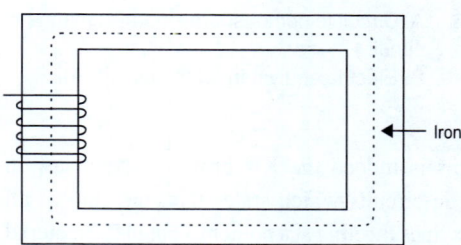

Fig. 2.5-B. Simple magnetic circuit.

2.5 SERIES PARALLEL MAGNETIC CIRCUITS

A magnetic circuit which has two or more than two paths for the magnetic flux to pass through, is a series parallel magnetic circuit.

Consider Fig. 2.6.

The current carrying coil in central portion produces the magnetic flux ϕ, which gets divided into two parallel paths ϕ_2, and ϕ_3.

It is clear that,

$$\phi_1 = \phi_2 + \phi_3 \quad (2.18)$$

154 Basic Electrical Engineering

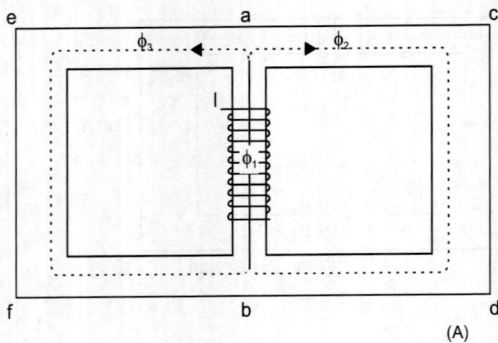

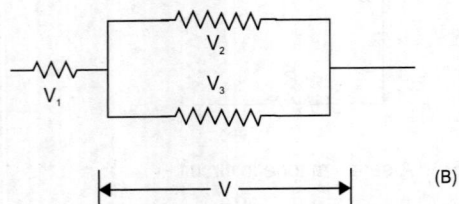

Fig. 2.6. A. Circuit to demonstrate a parallel magnetic circuit
B. Electric equivalent of the circuit (A).

This is indeed the Kirchroff's current law in magnetic circuits which states that, the sum of all the flux lines meeting at a point in a magnetic material is equal to the sum of the flux lines leaving that junction.

It is so easy to observe that the existing two parallel paths of the Fig. 2.6 are:
— the path followed by ϕ_2 (acdba) and
— the path followed by ϕ_3 (aefba)

It is said in Sec (1.20-B) that in parallel electric circuits the EMF remains the same. In analogy to this, in a magnetic circuit, the MMF will remain the same.

Thus the MMF (or the AT's) needed for the above parallel circuit is equal to the MMF (or AT's) needed for any one of the two paths.

Letting the suffixes 1, 2, 3 to denote the respective paths, the different magnetic quantities for this series parallel circuit of Fig. (2.6), using the equation derived in See (2.3) are:

Reluctance of path 1 → that is ad

$$= \frac{l_1}{\mu_0 \mu_{r1} a_1}$$

Reluctance of path 2 → that is acdba

$$= \frac{l_2}{\mu_0 \mu_{r2} a_2}$$

Reluctance of path 3 → that is aefba

$$= \frac{l_3}{\mu_0 \mu_{r3} a_3}$$

∴ The total MMF = MMF needed for path ab + MMF needed for any one of the parallel paths, [as the MMF for them is the same.]

$$\phi_1 S_1 + \phi_2 S_2 = \phi_1 S_1 + \phi_3 S_3 \quad (2.19)$$

[For better understanding, refer to Fig. 2.6B—an electrical equivalent circuit of this magnetic circuit—in which, total volts is $V_1 + V_2$ or $V_1 + V_3$ and both these sums will be the same; this is what, in magnetic circuits, given by equations (2.19)]

2.6 LEAKAGE FLUX

The magnetic flux, that is not following the intended path is known as the leakage flux. (2.20)

Or

The total number of flux lines, that do not contribute to the energy transformation is the leakage flux.

Consider the Fig. 2.7, in which the intended path is abcda.

Flux ϕ is the only flux travelling in this intended path, whereas ϕ_1, does not. So out of the total flux

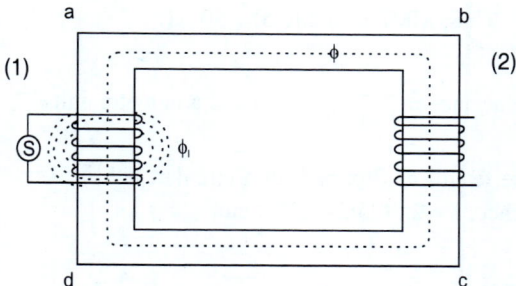

Fig. 2.7.

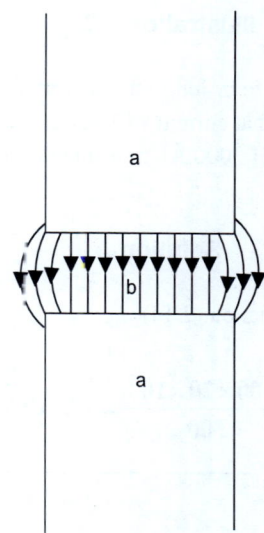

Fig. 2.8. Sketch to demonstrate the magnetic fringing.

produced, say ϕ_T due to MMF at the source, most of flux ϕ will follow the specified path. However a few flux lines (ϕ_1) may not follow this specified or intended path, thus forming the leakage flux.

[It can be seen in Fig. 2.7 that the flux ϕ_1 does not link with side 2. The energy will be transferred to side 2 only if the flux produced by side 1 links with side 2. ϕ_1 is the flux which do not contribute to energy transformation, thus forming leakage flux.]

$$\phi_T = \phi_1 + \phi. \qquad (2.21)$$

Note: The ratio between the total flux ϕ_T and the useful flux ϕ is known as the leakage factor, LF.

$$LF = \phi_T/\phi \qquad (2.22)$$

2.7 MAGNETIC FRINGING

When the magnetic flux lines enter a non-magnetic media suddenly from a magnetic media, these flux lines try to move out or bulge out as indicated in Fig. 2.8.

The reason for the "**bulging out**" process is that the flux lines repel each other when they pass through a non-magnetic material.

The effect is the "**Magnetic Fringing**".

2.8 WORKED ILLUSTRATIONS VIII

Note: All the units of magnetic quantities must be general. That is, flux in webers, flux density in wb/m², length in meters, area in square meters, etc. If it is not given so in the problem, convert them to this general form.

Illustration 2.1

A solenoid 25 cm long is wound with 200 turns. What is the value of the field strength inside the solenoid, when carrying a current of 2 amps.

Solution

From the equation (2.14), the field strength, H is,

$$H = \frac{NI}{l} = \frac{200 \times 2}{25 \times 10^{-2}} = 1600 \text{ AT/m (Ans)}.$$

Illustration 2.2

A solenoid is 20 cm long and is wound with 500 turns of wire. What current will be needed to set up a field strength of 3000 AT/m inside the solenoid.

Solution

By using the equation (2.14)

$$I = \frac{Hl}{N} = \frac{3000 \times 20 \times 10^{-2}}{500} = 1.2 \text{ A (Ans)}.$$

Thus, MMF = (500(0.5)=250 AT.

Reluctance = $\dfrac{l}{\mu_0 \mu_r a}$ with **l** and **a** in metre units.

The length of flux path in a circular path is, $2\pi r$. Hence, $l = 2\pi (25/2) = 25\pi$ cm.

$$\therefore \text{Flux} = \frac{250}{\left(\dfrac{25\pi \times 10^{-2}}{4\pi \times 10^{-7} \times 500 \times 400 \times 10^{-6}}\right)}$$

$$= 80 \ \mu \text{ wb. (Ans)}$$

Illustration 2.3

An iron ring has a cross-section area of 400 mm² and a mean diameter of 25 cm. It is wound with 500 turns. If the value of relative permeability is 500, find the total flux set-up in the ring. The coil resistance is 400 Ω and the supply voltage is 200 V.

Solution

The equation (2.4) (2.11) and (2.12) are used.

$$\text{Flux} = \frac{\text{MMF}}{\text{reluctance}}$$

and

$$\text{MMF} = NI$$

From the given data, the current in the ring I is found out using the Ohm's law, as:

$$I = \frac{V}{R} = \frac{200}{400} = 0.5 \text{ A}.$$

Illustration 2.4

An iron ring of mean length 1 m has an air gap of 1 mm and a winding of 200 turns. If the relative permeability of iron is 500, when a current of 1 amp flows through the coil, find the flux density.

Solution

The problem is depicted in Fig. 2.9. It is series magnetic circuit.

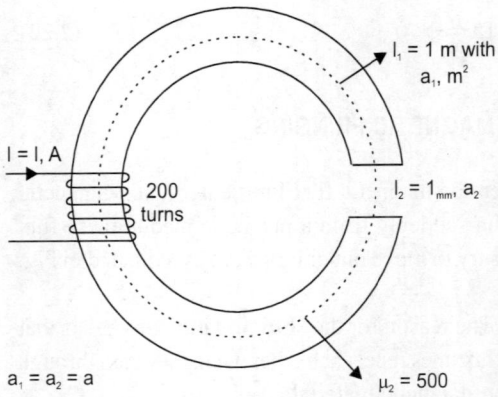

Fig. 2.9. Circuit for illustration 2.4.

Airgap:
For air, $\mu_r = 1$. Thus for the air gap, the reluctance is

$$S_2 = \frac{l_2}{\mu_0 \mu_{r2} a_2} = \frac{1 \times 10^{-3}}{4\pi \times 10^{-7} \times 1 \times a_2} = \frac{795.17}{a}$$

Iron:

$$S_1 = \frac{l_1}{\mu_0 \mu_{r1} a_1} = \frac{1}{4\pi \times 10^{-7} \times 500 \times a_1} = \frac{1591.55}{a}$$

Net MMF = N I = 200 × 1 = 200 AT.
By equation (2.18), net reluctance

$$S_{net} = S_1 + S_2 = \frac{1}{a}(795.77 + 1571.55) = \frac{2387.32}{a}$$

$$\text{Flux} = \frac{\text{MMF}}{\text{Reluctance}} \Rightarrow \frac{200}{\left(\frac{2382.32}{a}\right)} = 0.08378 a$$

and flux density = flux/area = 0.08378a/a

∴ **B = 0.08378 wb/m² or Tesla (Ans).**

Illustration 2.5

A circular ring has a mean circumference of 1.5 m and a cross-sectional area of 100 cm². A saw-cut of 0.4 cm is made in the ring. Calculate the magneting current needed to produce a flux of 0.8 mwb in the air gap of the ring which is round with a coil of 350 turns.

Assume relative permeability of 400 and leakage factor of 1.25.

Solution

The circuit of Fig. 2.10 depicts the circuit.

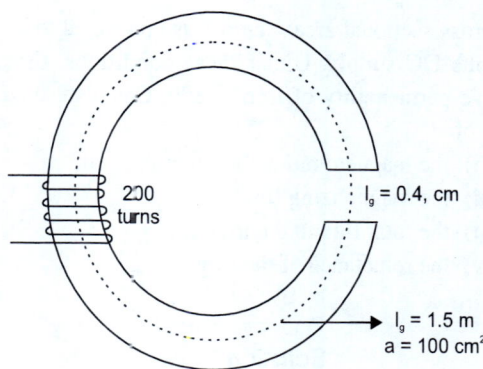

Fig. 2.10. Circuit for illustration 2.5.

Given that the flux in the air gap is 0.8 mwb. This must be the flux in the iron also, because it is a series magnetic circuit. This flux is the useful flux and some flux might have linked with the coil itself without reaching the gap; they, of course, are the leakage flux. (Fig. 2.10).

Note: The leakage factor is not accounted for in the gap, as the leakage flux is present in the iron near winding, and not in the gap.

Thus the total flux ϕ is found out, by using the equation (2.22).

$$\phi_T = (\phi) \, LF = (0.8) \, 1.25 = 1 \text{ mwb.}$$

Using the equation (2.11), are (2.12),

$$NI = \phi S$$

$$(350 \, I = (1 \times 10^{-3})$$

$$\left[\frac{I}{4\pi \times 10^{-7} \times 100 \times 10^{-4}} \left(\frac{1.5}{400} + \frac{0.4 \times 10^{-2}}{1} \right) \right]$$

∴ I = 1.762 A

Illustration 2.6

A coil of 500 turns and resistance 20 Ω wound uniformly on a ring of mean circumference 50 cm

and cross-sectional area 4 cm². It is connected to a 24-volts DC supply. Under these conditions, the relative permeability of iron is 800. Calculate the value of
(i) the magneto motive force of the coil
(ii) the magnetizing force
(iii) the total flux in the iron
(iv) the reluctance of the ring

Solution

Given data: $N = 500$ turns; $a = 40$ cm² $= 4 \times 10^{-4}$ m²

$R = 20\ \Omega$; $25 - V$, DC supply.

$l = 50$ cm (ring);

$\quad = 50 \times 10^{-2}$ m.

$\mu_r = 800$.

(i) The magneto motive force (MMF) is the product of ampere and turns. Current in the coil which sets up the MMF is $V/R = 24/20 = 1.2$ A. Thus,

MMF $= 500 \times 1.2 = 600$ AT. **(Ans)**

(ii) By equation (2.14), the magnetizing force, H, is NI/l

$\therefore H = (500 \times 1.2)/(50 \times 10^{-2})$

$\quad = 1200$ AT/m (Ans).

(iii) By equation (2.11), the flux = MMF/reluctance, where reluctance,

$$S = \frac{l}{\mu_0 \mu_r a}. \quad \text{(By equation 2.12)}$$

$\quad = 50 \times 10^{-2}/(4\pi \times 10^{-7} \times 800 \times 4 \times 10^{-4})$

$\quad = 1.244 \times 10^6$

Thus, flux $= 600/1.244 \times 10^6 = \mathbf{4.82 \times 10^{-4}\ wb}$ **(Ans)**

(iv) The reluctance, as found out by part (iii), is,

1.244×10^6 (Ans).

Illustration 2.7

A sheet steel circular ring of rectangular section 2 cm × 3.5 cm, has inside diameter of 10 cm and outside diameter of 14 cm. A coil of 250 turns is uniformly wound amount it. Calculate the current required to produce a flux of 0.91 mwb in the circuit. Calculate the permeability, reluctance and permanence of the magnetic circuit at this flux density.

Solution

The circuit is depicted in Fig. 2.12.

Needed: I, S, μ_r.

Usually, the mean diameter is considered in case the magnetic media is circular. [It is assumed that the flux is 'much' concentrated on the path provided by its mean diameter.]

$$D_m = (D_i + D_0)/2 = \frac{10 + 14}{2} = 12\ \text{cm}.$$

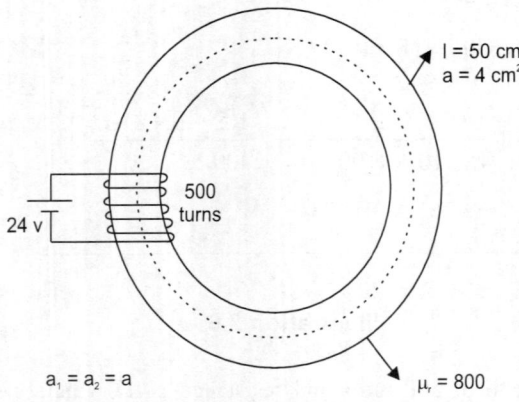

Fig. 2.11. Circuit of illustration 2.6.

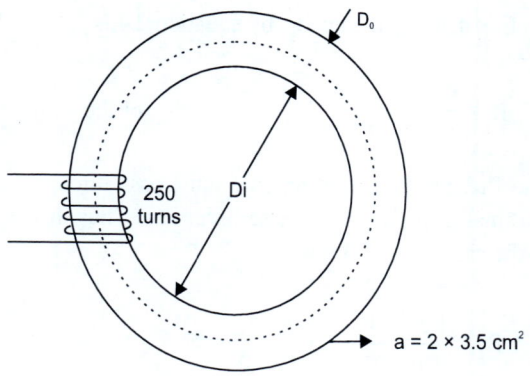

Fig. 2.12. Circuit for illustration 2.7.

and the mean circumference is

$$l = 2\pi r = 2\pi \left(\frac{12}{2}\right)$$

$$= 37.7 \text{ cm} = 37.7 \times 10^{-2} \text{ m}.$$

The area of the magnetic media is $2 \times 3.5 = 7 \text{ cm}^2$

$$= 7 \times 10^{-4} \text{ m}^2$$

[By equation (2.12) permeability $S = \dfrac{1}{\mu_0 \mu_r a}$]

$$S = \frac{37.7 \times 10^{-2}}{4\pi \times 10^{-7}} = 3 \times 10^5$$

(The further steps are left to the reader as an exercise.)

Illustration 2.8

An electromagnet is of the form shown in Fig. 2.13. Mean length of iron path is 48 cm, area of cross-section of core is 10 cm². It is excited by two coils each having 500 turns. When the current is 0.8 amps, the resulting flux density gives a permeability of 1250. Find:

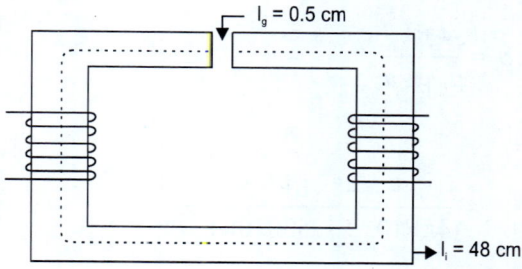

Fig. 2.13. Circuit for illustration 2.8.

(a) reluctance of iron part of the magnetic circuit.
(b) reluctance of air gap.
(c) total reluctance.
(d) total flux.
(e) flux density in gap. Neglect leakage and fringing.

Solution

It is a series magnetic circuit.
The reluctance of iron path of the magnetic circuit, by equation of (2.12), is

$$S = \frac{1}{\mu a}$$

where l = length of path

μ = permeability

a = area of cross-section

The mean length given is 48 cm = 48×10^{-2}.

area, a = 10 cm²

$= 10 \times 10^{-4} \text{ m}^2$

μ_r = 1250

$$\therefore S_i = \frac{l_i}{\mu_0 \cdot \mu_{ri} \cdot a}$$

$$= \frac{48 \times 10^{-2}}{(4\pi \times 10^{-7} \times 1250 \times 10 \times 10^{-4})}$$

$$= 306 \times 10^3 \text{ AT/wb approx, (Ans)}$$

(b) Reluctance of air gap $= \dfrac{l_g}{\mu_0 \cdot \mu_{rg} \cdot a}$,

where $l_g = 0.5 \times 10^{-2}$ m and 'a', of course, the same.

$$\therefore S_g = \frac{0.5 \times 10^{-2}}{4\pi \times 10^{-7} \times 1 \times 10 \times 10^{-4}}$$

$$\therefore S_g = 3980 \times 10^3 \text{ At/Wb Ans.}$$

(c) Total reluctance is the sum of both reluctances (by equation 2.18)

$$S_{net} = S_i + S_g$$

$$306 \times 10^3 + 3980 \times 10^3$$

$$= 4286 \times 10^3 \text{ AT/Wb. (Ans.)}$$

(d) Total flux in this series magnetic circuit is produced by the coil in the core. Hence the total flux, by equation 2.11, can be said as MMF of the coil/ (reluctance of the iron)

By equation (2.5),

MMF = NI = 500 × 0.8 = 400 AT

$$\therefore \phi_{total} = \frac{400}{306 \times 10^3} = 1.3 \times 10^{-3} \text{ Wb}$$

or $\qquad = 1.3$ m Wb.

(e) flux density in gaps B_g, by equation 2.4 is

$$B_g = \frac{\phi}{a} \qquad \text{(of the gap)}$$

The flux produced by the core must pass through the small gap also, as it is a series circuit. Thus flux in the gap is also 1.3 mwb.

$$\therefore B_g = \frac{1.3 \times 10^{-3}}{10 \times 10^{-4}}$$

$$= 1.3 \text{ Wb/m}^2 \text{ or } 1.3 \text{ Tesla.}$$

Illustration 2.9

A soft iron ring of 25.4 cm mean diameter and circular cross-section 5.1 cm diameter, is wound with a magnetising coil. Four amperes flowing in the coil produce a flux of 2.5 m Wb in the air gap which is 0.254 cm wide. Taking μ_r to be 1000 at this flux density and allowing for a leakage coefficient of 1.2, find the number of turns of the coil.

Solution

The mean diameter = 25.4 cm so the mean length of the ring is,

$$\pi D = (\pi \times 25.4) \text{ cm}$$

$$= 0.79796 = 0.798 \text{ m} = l$$

The area of cross-section, $a = \pi D^2/4$

$$= \frac{\pi}{4}[5.1 \times 10^{-2}]^2 = 0.002043 \text{ m}^2.$$

The flux in air gap = 2.5 m Wb = 2.5×10^{-3} Wb.

MMF or AT required for air gap $= \dfrac{B l_g}{\mu_0 \mu_r}$

In the air gap, B_g, the flux density

$$= \frac{\phi}{a} = \frac{2.5 \times 10^{-3}}{0.002043}$$

$= 1.224$ Wb/m² (by equation, 2.4)

Now

$$AT_g = \frac{(1.224 \times 0.252 \times 10^{-2})}{(4\pi \times 14^{-7} \times 1)} (\therefore \mu_r = 1 \text{ for air})$$

$= 2474$ AT

The MMF required for iron path:
Total flux = Leakage factor × useful flux

i.e. $\phi_T = (LF) \phi$, (by equation 2.22)

$= 1.2 \times 2.5 \times 10^{-3} = 3 \times 10^{-3}$ wb

Thus, MMF (or AT) for the iron part is, (by using equation 2.14-a).

$$AT_i = \frac{Bl}{\mu_0 \mu_r} \text{ (of the iron)}$$

$$= \frac{(\phi/a)l}{\mu_0 \mu_r} \text{ (of the iron)}$$

$$= \frac{(3 \times 10^{-3}/0.002043)0.798}{4\pi \times 10^{-7} \times 1000}$$

$= 933$ AT

Now total AT required = $AT_{gap} + AT_{iron}$ (Q it is a series magnetic circuit)

$= 2474 + 933 = 3421$ AT

Here current given = 4 amp.

$$\therefore \text{Turns} = \frac{AT}{I} = \frac{3421}{4} = 856 \text{ AT (Ans)}$$

Illustration 2.10

An electromagnet is of the form and dimension as shown in Fig. 2.14. It is made of iron of square section 5.08 cm side. A flux of 1.2 mwb is required in the air gap. Neglecting leakage and fringing. Find the number of AT required. Take μ_r to be 2000 at this flux density.

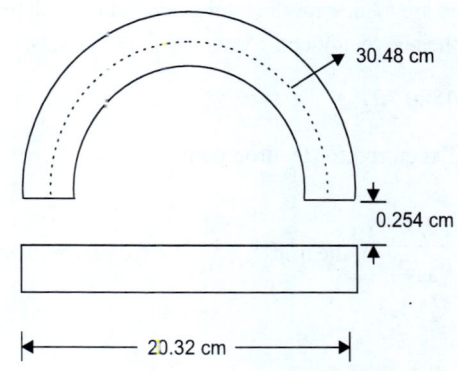

Fig. 2.14. Specimen for illustration 2.10.

Solution

It is a series magnetic circuit problem, in which, two air gap and two iron parts are in series.
Here the gap is given as = 0.254 cm

$= 0.254 \times 10^{-2}$ m

The cross-section area

$= 5.08 \times 5.08 = 25.8064$ cm² $= 25.8064 \times 10^{-4}$ m²

162 Basic Electrical Engineering

flux density

$$= \frac{\text{flux}}{\text{area}} = \frac{1.2 \times 10^{-3}}{25.8064 \times 10^{-4}} = 0.465 \text{ Wb/m}^2,$$

(by equation 2.4)

The ampere-turn for air gap $= \dfrac{Bl_g}{\mu_0 \mu_r}$ (by equation 2.14-a)

$= 0.465 \times 0.254 \times 10^{-2}/(4\pi \times 10^{-7} \times 1)$

$= 940$ AT

As there are to air gaps the total ampere turns will be twice the above amount, for the air gaps. That is,

$= 1880$ AT.

The AT required for the iron path.

$= \dfrac{Bl}{\mu_0 \cdot \mu_r}$ of the iron $+ \dfrac{Bl}{\mu_0 \mu_r}$ of the rectangular iron

i.e., $\text{AT}_{\text{iron}} = \dfrac{0.465 \times 30.48 \times 10^{-2}}{4\pi \times 10^{-7} \times 2000}$

$+ \dfrac{0.465 \times 20.32 \times 10^{-2}}{4\pi \times 10^{-7} \times 2000}$

$= 94$ AT

so total ampere turns = amp. turns (for iron + gap)

$\text{AT}_{\text{total}} = \text{AT}_{\text{iron}} + \text{AT}_{\text{gap}}.$

$= 94 + 1880$

$= \mathbf{1974}$ **AT (Ans.)**

Illustration 2.11

An iron ring of mean circumference 0.8 meter is uniformly wound with 500 turns of wires. When current of 1 amp, is passed through the coil a flux density of 1.1 Wb/m², is produced in the iron. Find the permeability of the iron under these circumstances.

Solution

The situation is depicted in the Fig. 2.15.
The flux thus produced = flux density × area.
Let us consider the area to be A m²

∴ total flux $= (1.1 \times A)$ Wb.

Total magnetising amp-turn $= I \times N = $ mmf

$= 1 \times 500 = 500$ AT (by equation 2.5)

Reluctance in the iron.

$$S = \frac{1}{\mu_0 \mu_r a} \quad \text{(by equation, 2.12)}$$

$$= \frac{0.8}{4\pi \times 10^{-7} \mu_r A} = \frac{636620}{\mu_r A}$$

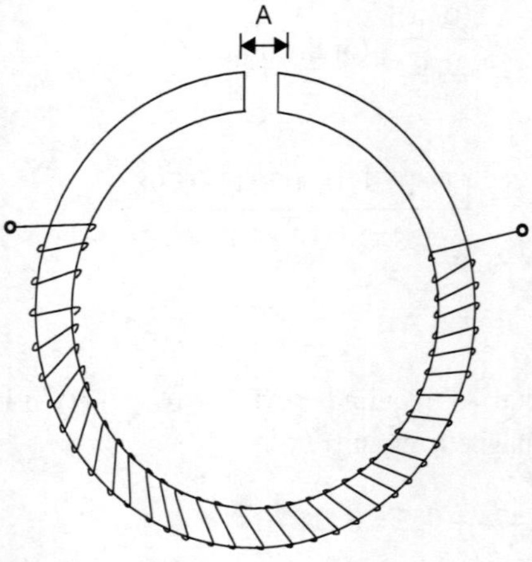

Fig. 2.15. Circuit for illustration 2.11.

flux thus produced = $\dfrac{\text{MMF}}{\text{Reluctance}}$

$$= \dfrac{500}{\left(\dfrac{636620}{\mu_r A}\right)} = \dfrac{500 A \mu_r}{636620}$$

So, $1.1 \times A = \dfrac{500 A \mu_r}{636620}$

$\Rightarrow \mu_r = 1400$.

so relative permeability, $\mu_r = 1400$ (Ans.)

Illustration 2.12

A magnetic core made up of steel has the dimensions as in Fig. 2.16. The cross-sections everywhere, is 25 cm². The flux in branches A and B is 3500 μ Wb, but that in the branch C is Zero. Find the needed ampere turns for coil A. The relative permeability of Steel is 100.

Solution

Refer Section 2.5.

Let the coil A be energised. Thus the total flux is produced here, and that flux will get divided into

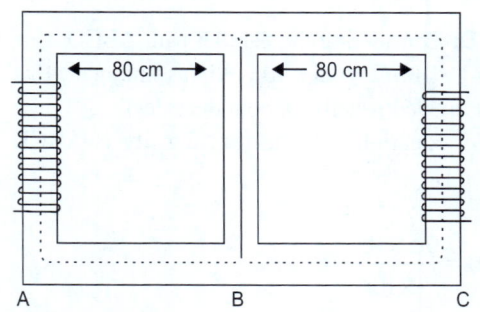

Fig. 2.16. Problem circuit for illustration 2.12.

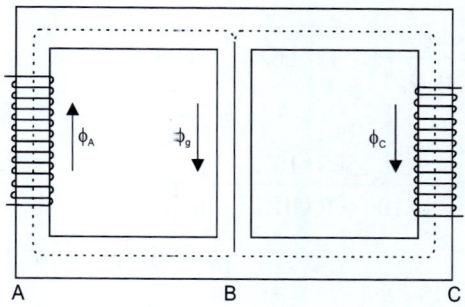

Fig. 2.17. Solution circuit for illustration 2.12.

two parallel parts in the section B and C, giving raise, to two parallel flux paths ϕ_B and ϕ_C. Let total flux produced in section A be ϕ_A. The flux division is in Fig. 2.17 and $\phi_A = \phi_B + \phi_C$.

As in a parallel magnetic circuit the MMF's are the same, using equation 2.11 it can be written that,

$$\phi_B S_B = \phi_C S_C$$

Using the equation (2.12) for S,

$$3500 \times 10^{-6} \times 30 \times 10^{-2} = \phi_C \, 80 \times 10^{-2}$$

$$\phi_C = \dfrac{3500 \times 10^{-6} \times 30 \times 10^{-2}}{80 \times 10^{-2}} = 1.3125 \text{ mWb}.$$

Thus total flux, $\phi_A = \phi_B + \phi_C$

$$= 3500 \times 10^{-6} + 1.3125 \times 10^{-3}$$

$$= 4.8125 \times 10^{-3} \text{ Wb}.$$

Ampere-turns needed for coil A: By equation (2.19), it is Ampere turns for path A + Ampere turns for path B.

i.e., $\phi_A S_A + \phi_B S_B$

$$\dfrac{\phi_A \cdot l_A}{\mu_0 \mu_r a} + \dfrac{\phi_B \cdot l_B}{\mu_0 \mu_r a}$$

$$= \frac{\phi_A}{\mu_0 \mu_r a}[l_A + l_B] \; (Q \; \phi_A = \phi_B)$$

$$= \frac{3500 \times 10^{-6}}{4\pi \times 10^{-7} \times 1000 \times 25 \times 10^{-4}} \times [80 \times 10^{-2} + 30 \times 10^{-2}]$$

= 1225.5 AT (Ans.)

Illustration 2.13

The magnetic circuit, made of iron is arranged as in Fig. 2.18. The central limb has a cross-sectional area of 8 cm², and each side has 5 cm². Calculate the ampere-turn needed to produce a flux of 1 mWb in the control limb.

The B-H relation for iron is given by

B—Telsa	1	1.25
H—AT/M	200	500

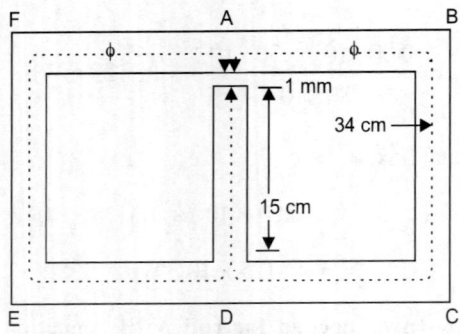

Fig. 2.18. Circuit for illustration 2.13.

Solution

(Refer to section 2.5)

It is a composite-parallel magnetic circuit. As discussed in section 2.5, the flux in one portion is getting divided into two other portions. Here, according to Fig. 2.8, the flux in the central portion (where flux is produced) is getting divided into two side portions. Thus total mmf would be the sum of central portion MMF and either of the side portions.

But unlike Fig. 2.6, here in Fig. 2.18, another portion—i.e. the air gap—is also available, whose mmf must also be added to get final total circuit MMF.

Central portion:

Flux, $\phi = 1 \text{ mWb} = 1 \times 10^{-3}$ Wb.

Cross-sectional area = $a = 8 \text{ cm}^2 = 8 \times 10^{-4} \text{ m}^2$

By equation 2.4 the flux density here is,

$$B = \frac{\phi}{a} = \frac{1 \times 10^{-3}}{8 \times 10^{-4}} = 1.25 \text{ T}$$

From the given tables, for 1.25 T, the AT/M in 500. Thus the total ampere-turn = ampere-turn/m × m

= 500 × 0.15 = 75 AT

Air gap

Using equation (2.14-a), the mmf NI is,

$$NI = \frac{B}{\mu_0 \mu_r} \cdot I = \frac{1.25}{4\pi \times 10^{-7}} \times 1 \times 10^{-3}$$

= 994.7 = 995 approx.

ABCD: Considering the parallel path ABCD, flux ϕ is = $1 \times 10^{-3}/2 = 0.5 \times 10^{-3}$ Wb (assuming the flux to get divided equally on both the sides).

By equation 2.4, the flux density is ABCD portion is

$$B = \frac{0.5 \times 10^{-3}}{5 \times 10^{-4}} = 1 \text{ Tesla}.$$

From the given table for 1 tesla, AT/M = 200.

Thus the ampere-turn needed for the entire region—ABCD—is, 200 × 0.34 = 68 AT.

NET: The net ampere turn is, 75 + 994.7 + 68 = 1137.7

= 1138 AT (Ans.)

Illustration 2.14

A torroidal air-core with 2000 turns has a mean radius of 25 cm, the diameter of each turn being 5 cm. If the current is 10A, find (i) MMF, (ii) flux and (iii) flux density.

Solution

By equation (2.5). The MMF can be written as, MMF = NI

= 2000 × 10 = **20000 AT (Ans.)**

Reluctance of the coil, by equation (2.17), is

$$S = \frac{l}{\mu_0 \mu_r a}$$

As the coil is circular, its cross-sectional area will be $\pi D^2/4$.

∴ $a = \pi(5)^2/4 = 19.635$ cm^2

$= 19.635 \times 10^{-4}$ m^2

$l = 2\pi r_{mean} = 2\pi (25)$

$= 157.08$ cm $= 1.5708$ m.

∴ $S = \dfrac{1.5708}{4\pi \times 10^{-7} \times 1 \times 19.635 \times 10^{-4}}$

(Q air core, $\mu_r = 1$)

$= 636.62 \times 10^6$

By equation (2.11), flux - MMF/reluctance.

i.e., $\phi = \dfrac{20000}{636.62 \times 10^6}$

$= \mathbf{3.14 \times 10^{-5}}$ **wb (Ans.)**

Using equation (2.4), flux density B is,

$$B = \frac{\phi}{a} = \frac{3.14 \times 10^{-5}}{19.631 \times 10^{-4}}$$

$= 0.158$ Tesla.

Illustration 2.15

Estimate the total MMF required to produce a flux of 1 milli weber round an iron ring of 6 cm^2 cross-section and 90 cm mean diameter, having an air gap 9 mm wide across it. Relative permeability is 1200.

Solution

Figure 2.10 may be consider, with the data of this illustration.

$\phi = 1$ mWb $= 1 \times 10^{-3}$ Wb

$a = 6$ cm$^2 = 6 \times 10^{-4}$ m^2

$l_g = 3$ mm $= 9 \times 10^{-3}$ m

$\mu_r = 1200$.

$l_i = \pi D_{mean} = \pi (90) = 287.74$ cm $= 2.87$ m.

This is a series magnetic circuit having iron and the air gap. Thus, the net reluctance will be, as given by equation (2.18), the sum of the two individual reluctances,

Air gap

$$S_g = \frac{l}{\mu_0 \mu_r a} \quad \text{(air gap)}$$

$$= \frac{9 \times 10^{-3}}{4\pi \times 10^{-7} \times 1 \times 6 \times 10^{-4}}$$

$(\therefore \mu_r = 1 \text{ for air})$

$= 11.93 \times 10^6$

Iron

$$S_i = \frac{l}{\mu_0 \mu_r a} \quad \text{(of iron)}$$

$$= \frac{2.87}{4\pi \times 10^{-7} \times 1200 \times 6 \times 10^{-4}} = 3.17 \times 10^6$$

$S_{net} = S_g + S_i = (11.93 + 3.17) \, 10^6 = 15.1 \times 10^6$

Total MMF $= \phi \, S_{net}$ (by equation 2.11)

$= (1 \times 10^{-3})(15.1 \times 10^6)$

$= 15100 \text{ AT (Ans.)}$

Illustration 2.16

An iron ring 10 cm mean diameter is made up of 1.5 cm diameter round iron of permeability 900 has an air gap of 5 mm wide. If the number of turns are 400 and current flowing is 3.5, calculate magneto motive force, the reluctance and the airgap flux density.

Solution

The problem is exactly similar to the above illustration 2.17, with different data. Reader is thus expected to proceed the problem similarly. Answers:

Magneto motive force = 1400 AT

Flux = 58.2 μ Wb

Flux density = 0.328 Tesla.

2.9 TUTORIAL ILLUSTRATIONS VI

Illustration TVI-1

A soft iron ring is 0.6 m in mean circumference and 2.5 cm² in cross-section. Find the number of AT (ampere-across) needed to produce a total flux of 0.25 m Wb, if at 1 tesla at AT is 240. **[240 AT]**.

Illustration TIII-2

A coil of 1000 turns is wound on a laminated core of sheet steel having a cross section of 25 sq. cm and mean length of 50 cm. What is the current required to produce a core flux of 3 mWb.

2.10 ELECTROMAGNETIC INDUCTION

An electro-magnetic induction, as the name itself implies, describes about the 'electrical-magnetic' dependence. It has been pointed out in section 2.1 that electricity and magnetism are inseparable. This section justifies this statement. Here we say as 'induction' because a live magnetic field will **induce** some electrical quantity in the electric circuit it travels or the electric current will **induce** a magnetic field around the circuit through which it passes.

2.11 FARADAY'S POSTULATIONS

The electromagnetic induction is found out and formulated properly by Faraday.

Faraday postulated that 'an electric voltage' can certainly be produced in a coil by the magnets flux

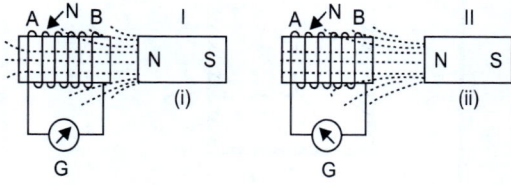

Fig. 2.19. Demonstration of induced EMF and flux.

linking with it. The relation he brought out was: 'an EMF is generated or induced in a circuit whenever the magnetic flux linking with the circuit is changed, and is proportional to the rate of change of flux linkage. In simple words, **whenever the flux linking a coil charges, an emf is induced in the coil (First law). Its magnitude is proportional to the rate of change of flux linkage. (Second law)**.

In can the observed that, when the magnet is moved from position (i) to (ii) in Fig. 2.19 obviously the flux is altered, which will alter the flux lines falling on the coil; hence the total number of flux lines 'linking' the N number of turns of the coil is changed. Let this change occur in dt sec.

Then, by Faraday, the EMF induced (e) in the coil, is

$$e = \frac{d}{dt}(\phi) = \frac{d\phi}{dt} \quad \text{with one turn} \quad (2.23)$$

$$e = N\frac{d\phi}{dt} \quad \text{with N number of turns} \quad (2.24)$$

These are the Faraday's laws of electromagnetic induction equations. From, equation (2.23), $d\phi = edt$

or $\phi = et$ \quad (2.25)

i.e., 1 weber = 1 V × 1 sec.

It is usual practice to place a '–' sign in the equations (2.23) and (2.24) for the fact that, this induced emf (and hence the respective induced current), will always be in such a direction as to oppose the very first cause producing it. This is the Lenz's law. The very first cause, of course, is the flux change in the coil, without which no emf would have been induced. Hence it will be opposed.

2.12 FLAMING'S RIGHT HAND RULE (FRH)

The magnitude of the emf induced in the coil due to a change in the flux linking with it is given by the equation (2.24); the Flaming's right hand rule, gives its **direction**. It states that, if the thumb, forefinger and middle finger of the right hand are kept at right angles to each other (refer to Fig. 2.20) in such a way that the thumb points in the direction of the motion of the conductor, the forefinger in the direction of the flux (from north to south) then the middle finger will denote the direction of the induced EMF in the conductor.

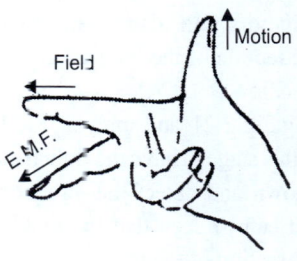

Fig. 2.20. FRH illustration.

2.13 DEMONSTRATION OF FLAMING'S RIGHT HAND RULE (FRH)

The flux direction (i.e., the magnetic field) is assumed to flow from N → S pole. The direction of rotation of the iron body known as the armature—which is discussed in depth in the chapter 3) and hence the connected conductor is also chosen as indicated.

168 Basic Electrical Engineering

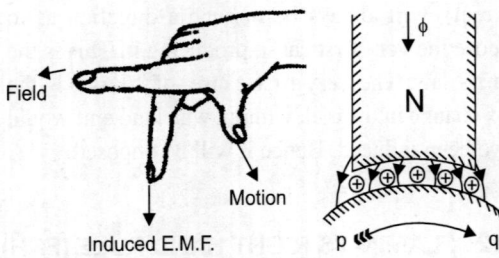

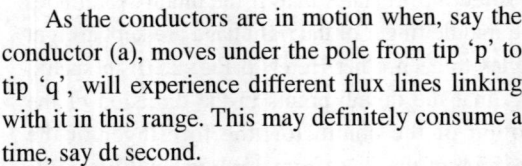

Fig. 2.21. Circuit to demonstrate FRH.

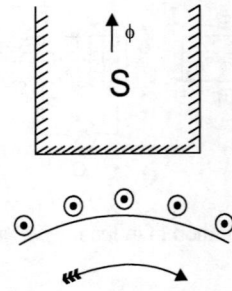

Fig. 2.22.

2.14 ELECTROMAGNETIC FORCE

Statement: "**When a current carrying conductor is placed in a magnetic field the conductor experiences a mechanical force which acts in a direction perpendicular to both the directions of current and the magnetic field.**"

The magnitude of the mechanical force is given by the expression,

$$F = BI\, l, \text{ newtons.}$$

where B is the flux density of the magnetic field in tesla; I, the current through the conductor in amps; l, the active length of the conductor in meters.

As the conductors are in motion when, say the conductor (a), moves under the pole from tip 'p' to tip 'q', will experience different flux lines linking with it in this range. This may definitely consume a time, say dt second.

This phenomena is nothing but, Flaming's first law, which states that rate of change of flux linkage, induces an emf in the concerned conductor. Hence an emf of magnitude $d\phi/dt$ will be induced in the conductor 'a'. (refer to equation 2.23) [Similarly, all the other conductors will also be induced with a voltage]. The direction of the induced emf in found out as detailed below.

Considering Fig. 2.21, and applying FRH (reader is suggested at this stage to have the right hand finger arranged as shown and detect the direction of the induced emf) it can be seen that the middle finger will point the direction of the induced emf '**into the paper**'—'**into the conductor**'. Hence the induced current also will flow into the conductor.

Usually a '+' is kept in the conductor if the current flows '**into**' it, on the other hand, a dot (.) will be kept in the conductor, if the current comes out.

Note: had n been a S pole, then the flux direction would have been assumed into the pole. (Refer to Fig. 2.22). Maintaining the same direction of rotation, it is readily seen, by applying FRH that, the middle finger point **out side the paper**—i.e. **outside the conductor**. Hence a 'dot' is kept in the conductor.

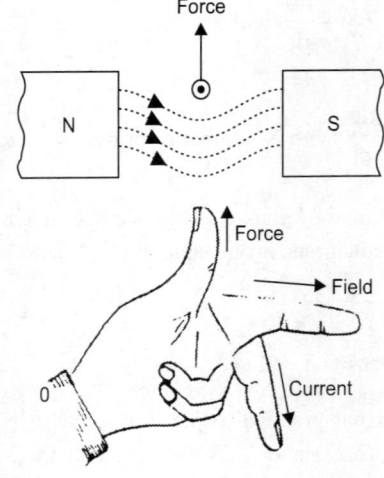

Fig. 2.23. Arrangement of fingers to demonstrate FLH.

The direction of the force acting on the conductor is given by the Flaming's left hand rule (FLH), which states that, "**if the thumb, the forefinger and the middle finger of left hand are kept at right angles to each other (refer to Fig. 2.23), in such a way that the forefinger points the direction of flux (from N to S) and the middle finger points the direction of the current in the conductor, then the thumb will indicate the direction in which the force will act on the conductor.**"

Note: The reader may test the validity of FLH, as guided in sec. (2.13)

2.15 TYPES OF MAGNETIC INDUCTIONS

When the flux linking with a coil changes an EMF is induced in the coil. This change in flux linkage (i.e., the change in the product of number of conductors N and the flux cutting it ϕ) can be obtained in two ways viz,
  (i) Dynamically (i.e., by motion)
  (ii) Statically (i.e., without motion)

— giving the two different induced EMF's viz (i) dynamically induced EMF and (ii) statistically induced EMF respectively.

2.16 DYNAMICALLY INDUCED EMF

It is brought out in Fig. 2.19 that the movement of the electric conductors in a stationary magnetic field system induces an EMF in the conductors. Here, the conductor (i.e., **the electric circuit**) was in **motion** and the flux (i.e., **the DC excited magnetic field system**) was **stationary**, thus producing a continuously changing flux linkage of Nϕ as long as the electric circuit is in motion. An EMF will be induced in the moving electric conductors.

This is the basic principle on which a DC generator works.

The set-up can be otherwise also, in which, a moving magnetic field system will induce an EMF in the stationary electric conductors. In such case the conductors (i.e. **the electric circuit**) will be **stationary** and the magnets (i.e., **the DC excited magnetic field system**) will be **in motion**. This will produce a continuously changing flux linkage of Nϕ, thus inducing on EMF in the stationary conductors. This is the basic principle of an AC generator.

Such induced EMF's is known as the DYNAMICALLY INDUCED EMF. Observe this in Fig. 2.24).

Thus it can be concluded that either the electric circuit or the magnetic field system must be in motion to have EMF induced in the conductors of the electric circuit; such a motion in which either of the system is moving is turned as the "**relative motion**".

Hence, **whenever there is a relative motion between a magnetic field and an electric circuit, an EMF is induced in the conductors of the electric circuit.**

Stating as 'relative motion' will take into account the two aspects, viz (i) either of the system is in motion and (ii) the change in flux linkage (i.e., Nϕ)

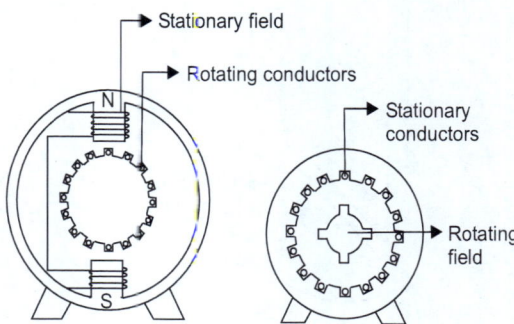

Fig. 2.24. Methods of inducing EMF (dynamically) in a conductor a: Electric conductors in motion; magnetic field system stationary. b: magnetic field system in motion; electric conductors stationary.

is taking place for a definite time as long as the motion exists.

Magnitude of the Dynamically induced EMF

The procedure of getting the dynamically induced EMF is discussed above. Its magnitude is determined as stated below:

Let the conductor move in a magnetic field of strength B Wb/m².

Let the conductor have a length of 'l' metres in the magnetic field and an area of cross-section of 'a' m².

Consider in Fig. 2.25.

Let the conductor move with a velocity of v m/sec.

By using the equation, (2.23) the EMF induced in the conductor is found out.

The flux linking the conductor is, by equation (2.4)

$\phi = B a$

The area covered by the conductor of 1 metre, when it moves through a distance of dx metre is l dx.

i.e., $a = l dx$

$\therefore \phi = B [l\, dx]$

The dynamically emf, $e_d = N \dfrac{d\phi}{dt}$.

As only the conductor is there, $N = 1$. Thus,

$$e_d = \dfrac{d\phi}{dt} = \dfrac{Bl\, dx}{dt}$$

But the rate of change of distance, dx/dt, is the velocity v of the moving conductor. So,

$$e_d = Bl\, v, \text{ volts} \qquad (2.26)$$

Note: When the conductor is at an angle $\theta°$ measured along the direction of the magnetic field, then the magnitude of the dynamically induced voltage will be,

$$e_d = Bl\, v \sin\theta \text{ V} \qquad (2.27)$$

The equations (2.26) and (2.27) gives the magnitude of the induced emf, e_d, whose direction is found out by the Flaming's right hand rule.

2.17 STATICALLY INDUCED EMF

It is an EMF induced in the conductors when both the magnetic field system and the winding (electric circuit) are stationary.

For easy understanding refer to Fig. 2.26.

It is brought out in Sec. 2.3, Fig. 2.2., that a current carrying conductor will always be associated with an amount of flux produced by it.

Let a current of flow through the coil thus producing a flux of ϕ, Wb. (Fig. 2.26). If this current is changed, no doubt that, the flux produced by it will also change, proportionally, the total conductors, N_1 being a constant, the flux linkage $N_1\phi$, is now

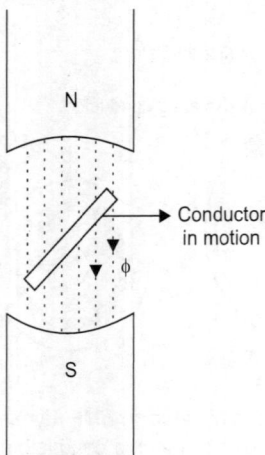

Fig. 2.25. The electric conductor in a magnetic field to demonstrate the dynamically induced EMF.

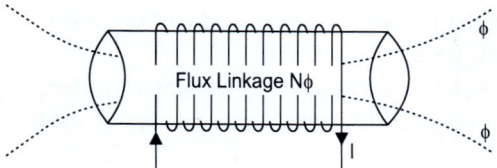

Fig. 2.26. Circuit for illustrating statically induced EMF (self-induced type) N1: Number of turns in the coil.

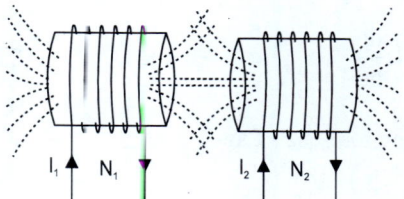

Fig. 2.27. Illustration circuit to demonstrated mutually induced EMF.
N_1: Number of turns in coil 1.
N_2: Number of turns in coil 2.

changed. The stationary electric conductors (i.e., coil) is experiencing a change in the flux linkage in it.

Indeed, this is what the conditions to induce an emf in a coil; thus an emf—known as 'STATICALLY INDUCED EMF'—will be induced in the conductor.

The transformer works entirely on this types of induced emf.

The statically induced emf is of two types.
 (i) Self-induced emf
 (ii) Mutually induced emf.

Self-Induced EMF

EMF induced in a coil or conductor due to its own current change (and hence the own flux linkage change) is known as the **self-induced EMF.**

This is presented in the above discussion.

Mutually Induced EMF

EMF induced in a coil or conductor due to a change in the current in the nearby coil is known as the **mutually induced emf**.

The magnetic flux lines are un-insulatable and they can pass through anywhere, emerging upto infinity.

Consider Fig. 2.27. A change in current in coil 2 will definitely cause a change in its original flux produced. As flux lines have no magnetic insulation, a part of the total flux ϕ_2 produced by coil 2 can link with the coil one kept near the coil 2.

Let ϕ_2', the portion of flux linking with coil 1 before the current change and ϕ_2'' be the portion of the flux linking with coil 1 after the current change. Respectively, the flux linkages are $N_1\phi_2'$ and $N_1\phi_2''$. thus there is a change in flux linkage in coil 1, from $N_1\phi_2'$ to $N_1\phi_2''$ and hence an emf—known as 'mutually induced emf'—will be induced in the coil 1. This can be instantly observed by the pointer movement in the voltmeter, if connected.

Such EMF gets the name as mutually induced EMF because, the flux ϕ_2 (of coil 2) is mutually shared by coil 1 to get an EMF induced it it.

Note carefully that, obviously, there will be a flux change hence flux linkage change in the coil 2 too; but this change is due to its own flux. Hence, as far as coil 2 is concerned, the EMF induced in it due to this action will be, of course, a self induced EMF.

2.18 SELF AND MUTUAL INDUCTANCES

Self inductance is the property of a system, by which an emf is induced in a coil, when the current through that coil is changed in a time duration of dt, seconds. It is given by symbol L. Thus the induced emf is proportion to the rate of change of current;

i.e., $e \propto di/dt$.

Introducing a proportionality constant, known as the self-inductance of the coil,

$$e = L\frac{di}{dt} \tag{2.28}$$

accounting for Lenz's Law.

$$e = -L\frac{di}{dt} \tag{2.29}$$

However, mutual inductance is the property of a circuit in which a change in current in a coil in the vicinity can induce an EMF is this circuit. It is given the symbol M.

The self-inductance is measured by the unit Henry which can be defined from the above discussions as: A COIL HAS AN INDUCTANCE OF ONE HENRY, IF AN EMF OF ONE VOLT IS INDUCED IN IT, WHEN THE CURRENT CHANGES AT A RATE OF ONE AMPERE PER SECOND.

Self-inductance can otherwise also be stated as, the total number of flux linkage per ampere flow through the coil. That is,

$$L = \frac{\text{Flux linkage}}{\text{ampere}} = \frac{N\phi}{I} \tag{2.30}$$

It is shown that a portion of the total flux ϕ_2 produced by coil 2, links with all the N_1 turns of coil 1, to induce mutually an EMF in the coil 1. Moreover, the flux ϕ_2 of coil 2 is produced due to its current I_2.

Taking all this into account, placing appropriate suffixes, the mutual inductance, for the circuit setup of Fig. 2.27 is, (using equation 2.30)

$$M = \frac{N_1\phi_2}{I_2} \tag{2.31}$$

That is, in simple words, the flux ϕ_2 of coil 2 produced by its own current I_2, links with N_1 turns of coil 1 to induce mutually and emf in the coil 1; hence, the equation (2.31)

Note: If the flux ϕ_1 due to its current $(I)_1$ change, causes e_2, then the mutual inductance can be written as,

$$M = \frac{N_2\phi_1}{I_1} \tag{2.32}$$

where M is the mutually inductance; its unit is henry.
Reader may try this.

It is brought out in equation (2.30) that, the inductance of a coil is

$$L = \frac{N\phi}{I}.$$

But, by equation (2.11), the flux ϕ is given by, $NI/(l/\mu_0 \mu_r a)$. This is substituted in the equation (2.23), to give L as,

$$L = \frac{N}{I} \cdot \frac{NI}{l/\mu_0 \mu_r a}$$

$$L = \frac{N^2}{l/\mu_0 \mu_r a} \tag{2.33}$$

From this it is seen that, the inductance L of a coil:

(a) Inversely depends on the length of the media and directly depends on the area of cross-section of the coil. That is it depends on the **shape** of the conductor.
(b) Directly depends on the relative permeability of the magnetic media. Higher the permeability, higher will be inductance.
(c) Also depends directly on the square of the number of turns of the coil.

2.19 ENERGY STORED IN A MAGNETIC FIELD

Consider the Fig. 2.28, a coil supplied by a source.
Applying KVL of equation (1.21),

$$v = L\frac{di}{dt} + Ri$$

where iR is voltage drop across the resistor. Similarly, the voltage across an inductor is given by L di/dt.

By equation (1.5), the energy is VI times t. Performing the product in the above equation.

$$v\, i\, dt = R\, i(i\, dt) + L\frac{di}{dt}(i\, dt)$$

i.e., $v\, i\, dt = Ri^2\, dt + Li\, dt$

where V i dt = Energy supplied in time dt

$i^2 R\, dt$ = energy dissipated by the resistor is the form of heat.

Li di = Energy observed by the inductor in building up the magnetic field.

[*Note:* Thus it is to be noted that as long as the current change (to say precisely, current raise) is there in a circuit, the inductor of the circuit will store, energy in the magnetic field. Of course, the inductance of the inductor will 'oppose' continuously the change in the current.]

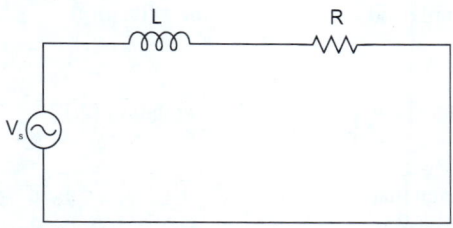

Fig. 2.28. Circuit to discuss energy stored in an inductor.

It can now be concluded that, the inductance is the property of a coil by which it stores energy due to current change in it; it will also continuously oppose the change in current.

As the main aim is to find the energy stored in an inductor, let the term Li di be concentrated.

As the current raises from zero to the final value I, the total energy stored in the field is the sum of Li di from 0 to I. Mathematically,

$$\text{Energy stored} = \int_0^I Li\, dt;\ \left[L\frac{i^2}{2}\right]_0^I$$

∴ Energy stored in an inductor = $\frac{1}{2}LI^2$ (2.34)

2.20 WORKED ILLUSTRATION IX

Illustration 2.17

A coil is wound with 500 turns. When a current of 4 amps flows the total flux threading the coil is found to be 0.06 mWb. What is the inductance of the coil in henries?

Solution

It is a problem involving one coil; so it is, self-inductance category. Refer to Sec. 2.18 and equation (2.30).

Inductance, $L = \frac{N\phi}{I}$

Given that N = 500 turns, $\phi = 0.06 \times 10^{-3}$ Wb and I = 4 A. Thus,

$L = 500 \times 0.06 \times 10^{-3}/4 = \mathbf{7.5 \times 10^{-3}\ H}$

(Ans.)

Illustration 2.18

A conductor 50 cm long is moved at right angles to a uniform magnetic field strength of 0.12 Wb/m² at a speed of the 20 cm/sec. Find the value of the emf induced.

Solution

As the motion is involved, the emf induced will be a dynamically induced emf. Refer to sec. 2.16.

Given that, $l = 50$ cm $= 50 \times 10^{-2}$ m; $B = 0.12$ Wb/m²; velocity $= 20$ cm/sec $= 20 \times 10^{-2}$ m/sec.; $\theta = 90°$ [at right angles]. Thus,

$e_d = Bl\, v \sin \theta$

$= 0.12 \times 50 \times 10^{-2} \times 20 \times 10^{-2} \times \sin 90$

$= 0.12$, volts.

Illustration 2.19

An inductive circuit is carrying current of 4 amps. If its inductance is 0.15 henry find the value of self-induced emf when the current is reduced to zero in 0.01 seconds.

Solution

It is a classic example which involves both the inductance and induced emf definition in proper proportions.

Using the equations (2.30) the total flux linkage can be found out.

$L = \dfrac{N\phi}{I}$

$\therefore\ N\phi = LI = 0.15 \times 4 = 0.6$ Wb-turns.

Now, the induced EMF can be found out.

$e = \dfrac{d}{dt}(N\phi)$.

Here, dt, the change in time is 0.01 sec. so,

$= 0.6/0.01 = 60$ V (Ans.)

Illustration 2.20

An iron-cored coil is wound with 400 turns. It is 40 cm long and 5 cm² in cross-section. When carrying 0.5 A the relative permeability of the coil may be taken as 2500. Find

(a) Inductance of the coil
(b) Value of self-induced EMF when current is reduced to zero in 0.01 seconds.

Solution

[*Note:* the coils can be wound on a magnetic material. It can be steel, iron etc. and are respectively known as steel-cored and iron-cored.]

Reader is suggested to develop technical steps to solve this classic problem, before the below solution is scanned.

(a) As far as part (a) is concerned the inductance is to be found out using the equation (2.30), which needs the value of flux ϕ; the equation (2.12) and (2.11) are used together to get the value of ϕ.

reluctance, $S = \dfrac{1}{\mu_0 \mu_r\, a}$ by equation (2.12)

Given that, $l = 40$ cm $= 40 \times 10^{-2}$ m; $a = 5$ cm² $= 5 \times 10^{-4}$ m²; $\mu_r = 2500$; $I = 0.5$ A; $N = 400$; $t = 0.01$ sec; Thus,

$$S = \frac{40 \times 10^{-2}}{4\pi \times 10^{-7} \times 2500 \times 5 \times 10^{-4}} = 254648.$$

and

$$\phi = \frac{NI}{S}$$

$$= \frac{400 \times 0.5}{254648} = 7.8 \times 10^{-4} \text{ Wb}.$$

Now using the equation (2.30), the inductance is,

$$L = \frac{N\phi}{I} = \frac{400 \times 7.8 \times 10^{-4}}{0.5} = 0.628 \text{ henry. (Ans).}$$

(b) The induced emf is $N\phi/t$.

$$\therefore e = \frac{400 \times 7.8 \times 10^{-4}}{0.01} = 3.12 \text{ V (Ans)}$$

Illustration 2.21

Two coils having 50 and 500 turns respectively are wound side by side on a closed iron circuit of section 50 cm² and mean length 12 cm. Estimate the mutual inductance between the coils if the permeability of iron is 1000. Also find the self-inductance of 2nd coil. If the current in one coil grows from 0 to 5A in 0.01 sec., find the EMF induced in the other coil.

Reference: Sec. 2.17.

Given that, $N_1 = 50$, $N_2 = 500$ turns.

$a = 50$ cm² $= 50 \times 10^{-4}$ m²

$l = 12$ cm $= 1.2$ m

Magnetism and Electromagnetism 175

$\mu_r = 1000$; $I_2 = 5$ amps; $t = 0.01$ Sec.

The mutual inductance is given by $M = N_1 \phi_2/I_2$
The reluctance of the magnetic circuits, by equation (2.12),

$$S = \frac{1.2}{4\pi \times 10^{-7} \times 1000 \times 50 \times 10^{-4}} = 190986.$$

The flux in coil 2, ϕ_2, by equation (2.11) is,

$$\phi_2 = \frac{N_2 I_2}{S} = \frac{500 \times 5}{190986} = 0.01309 \text{ Wb}.$$

$\therefore$ $M = 50 \times 0.01309/5 =$ **0.1309 H (Ans)**.

The Self-inductance is given by the formula, of equation (2.30). For the 2nd coil,

$$L_2 = \frac{N_2 \phi_2}{I_2} = \frac{500 \times 0.01309}{5} = 1.32 \text{ H}$$

(Ans).

The EMF induced in the coil, is

$$e = N\frac{dI}{dt} = \frac{50 \times (0.5)}{0.01} \text{ V}.$$

Illustration 2.22

A coil having an inductance of 75 mH is carrying a current of 100 A. What is the self-induced emf when the current is (a) reduced to zero in 0.02 sec. (b) Reverses in 0.02 sec.

Solution

Given $L = 75$ mH $= 75 \times 10^{-3}$ H; $I = 100$ A $e = ?$ when

176 Basic Electrical Engineering

(a) current is reduced to zero in 0.02 Sec (= dt)
It is given that the initial current is 100; and it reaches finally to zero. Therefore, the change in current, di, is 100 – 0 = 100 A.

$$\therefore e = L\frac{di}{dt} = 75 \times 10^{-3} \times \frac{100}{0.02} = \textbf{375 V (Ans)}.$$

(b) Current is reversed is 0.02 sec. (= dt).

The total current is reversed means nothing but, the current has reached –100 A. Thus the change in current is, 100 – (–100) = 200 A.

$$\therefore e = L\frac{di}{dt} = 75 \times 10^{-3} \times \frac{200}{0.02} = \textbf{750 V (Ans)}.$$

Reflection Section

1. When can EMF be induced dynamically?
2. What happens to the current in a coil (like solenoid) when accelerating a magnet inside it? Explain.
3. Conceptually differentiate between flux and flux density.
4. Give the formula for induced EMF when the magnetic field, length and the velocity related to conductor are mutually perpendicular to each other.
5. If a conductor 0.2 m long moves with a velocity of 0.3 m/s in a magnetic field of 0.5 T, calculate the EMF induced when the magnetic field, length and the velocity related to conductor are mutually perpendicular to each other.
6. Find the strength of the magnetic field in a conductor 0.5 m long moving with a velocity of 10 m/s, inducing an EMF of 20 V if magnetic field, velocity and length of conductor are mutually perpendicular to each other.
7. What is the force that tends to set up a magnetic field? Abbreviate it. Differentiate this force from that of the force that tends establish current in an electric circuit.
8. Differentiate between a magnetic induction and an electromagnetic induction.
9. What is the name for the point in a magnet where the intensity of magnetic lines of force is maximum?
10. Write about the application of Lenz's law.

PART II: MAGNETIC CHARACTERISTICS OF MATERIALS

2.21 MAGNETIZATION CURVES

Any discussion of the magnetic properties of a material will include the type of graph known as a *magnetization* or *B-H curve*. Various methods are used to produce B-H curves. Fig. 2.29 shows how the B-H curve varies according to the type of material within the field.

The 'curves' here all are straight lines and have *magnetic field strength* (H) as the horizontal axis and the *magnetic flux density* (B) as the vertical axis. Negative values of H aren't shown but the graphs are symmetrical about the vertical axis.

Fig. 2.29(a) is the curve in the absence of any material: a vacuum. The gradient of the curve is $4\pi.10^{-7}$ which corresponds to the fundamental physical constant μ_0. More on this later. Of greater interest is to see how placing a specimen of some material in the field affects this gradient. Manufacturers of a particular grade of ferrite material usually provide this curve because the shape reveals how the core material in any component made from it will respond to changes in applied field.

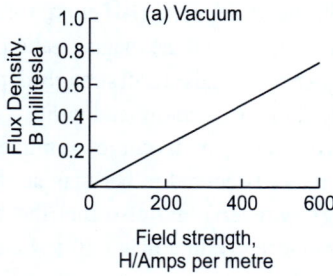
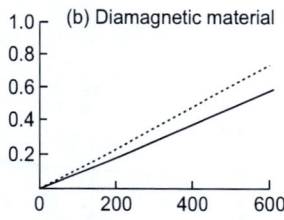
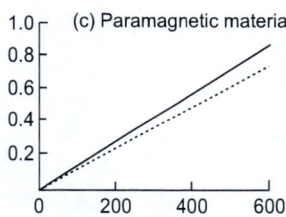

Fig. 2.29. Magnetization in different materials

2.22 DIAMAGNETIC AND PARAMAGNETIC MATERIALS

Imagine a hydrogen atom in which a nucleus with a single stationary and positively charged proton is orbited by a negatively charged electron. Can we view that electron in orbit as a sort of current loop? The answer is yes, and you might then think that hydrogen would have a strong *magnetic moment*. In fact ordinary hydrogen gas is only very weakly magnetic. Recall that each hydrogen atom is not isolated but is bonded to one other to form a molecule, giving the formula H_2—because that has a lower chemical energy (for H by a whopping 218 kJ mol^{-1}) than two isolated atoms. It is not a coincidence that in these molecules the angular momentum of one electron is opposite in direction to that of its neighbour, leaving the molecule as a whole with little by way of magnetic moment. This behaviour is typical of many substances which are then said to lack a *permanent magnetic moment*.

When a molecule is subjected to a magnetic field those electrons in orbit planes at a right angle to the field will change their momentum (very slightly). This is predicted by *Faraday's Law* which tells us that as the field is increased there will be an induced E-field which the electrons (being charged particles) will experience as a force. This means that the individual magnetic moments no longer cancel completely and the molecule then acquires an *induced magnetic moment*. This behaviour, whereby the induced moment is opposite to the applied field, is present in all materials and is called *diamagnetism*. Hydrogen, ammonia, bismuth, copper, graphite and other *diamagnetic substances*, are **repelled** by a nearby magnet (although the effect is extremely feeble). Think of it as a manifestation of Lenz's law. Diamagnetic materials are those whose atoms have only *paired electrons*.

In other molecules, however, such as oxygen, where there are unpaired electrons, the cancellation of magnetic moments belonging to the electrons is incomplete. An O_2 molecule has a net or *permanent magnetic moment* even in the absence of an externally applied field. If an external magnetic field is applied then the electron orbits are still altered in the same manner as the diamagnets but the permanent moment is usually a more powerful influence. The 'poles' of the molecule tend to line up parallel with the field and reinforce it. Such molecules, with permanent magnetic moments are called *paramagnetic*.

Although paramagnetic substances like oxygen, tin, aluminium and copper sulphate are attracted to a magnet the effect is almost as feeble as diamagnetism. The reason is that the permanent moments are continually knocked out of alignment with the field by thermal vibration, at room temperatures anyway (liquid oxygen at −183 °C can be pulled about by a strong magnet).

Particular materials where the magnetic moment of each atom can be made to favour one direction are said to be *magnetizable*. The extent to which this happens is called *magnetization*. Fig. 2.29(b) above is the *magnetization curve* for diamagnetic materials. In diamagnetic substances the flux grows slightly more slowly with the field than it does in a vacuum. The decrease in gradient is greatly exaggerated in the figure—in practice the drop is usually less than one part in 6,000.

Fig. 2.29(c) is the curve for paramagnetic materials. Flux growth in this case is again linear (at moderate values of H) but slightly faster than in a vacuum. Again, the increase for most substances is very slight.

Although neither diamagnetic nor paramagnetic materials are technologically important (geophysical

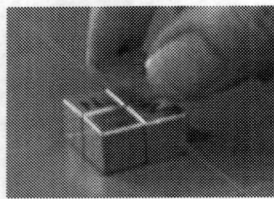

surveying is one exception), they are much studied by physicists, and the terminology of magnetics is enriched thereby.

2.23 FERROMAGNETIC MATERIALS

The most important class of magnetic materials is the *ferromagnets*: iron, nickel, cobalt and manganese, or their compounds (and a few more exotic ones as well). The *magnetization curve* looks very different to that of a *diamagnetic* or *paramagnetic* material. We might note in passing that although pure manganese is not ferromagnetic the name of that element shares a common root with magnetism: the Greek *mágnes lithos*—"stone from Magnesia" (now Manisa in Turkey).

Fig. 2.30 shows a typical B-H curve for iron. It is important to realize that the magnetization curves for ferromagnetic materials are all strongly dependent upon purity, heat treatment and other factors. However, two features of this curve are immediately apparent: it really is curved rather than straight (as with *non-ferromagnets*) and also that the vertical scale is now in teslas (rather than milliteslas as with Fig. 2.29).

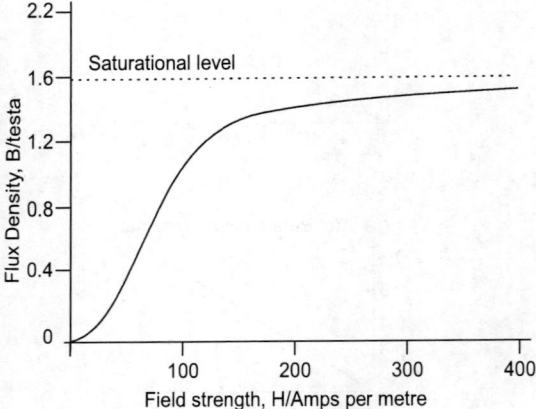

Fig. 2.30. Magnetization in iron

Fig. 2.30 is a *normal magnetization curve* because it starts from an unmagnetized sample and shows how the *flux density* increases as the *field strength* is increased. You can identify four distinct regions in most such curves. These can be explained in terms of changes to the material's magnetic 'domains':

1. Close to the origin a slow rise due to 'reversible growth'.
2. A longer, fairly straight, stretch representing 'irreversible growth'.
3. A slower rise representing 'rotation'.
4. *An almost flat region corresponding to paramagnetic behaviour and then* μ_0—the core can't handle any more flux growth and has *saturated*.

At an atomic level ferromagnetism is explained by a tendency for neighbouring atomic *magnetic moments* to become locked in parallel with their neighbours. This is only possible at temperatures below the *curie point*, above which thermal disordering causes a sharp drop in *permeability* and degeneration into *paramagnetism*. Ferromagnetism is distinguished from paramagnetism by more than just permeability because it also has the important properties of *remnance* and *coercivity*. (Discussed in sec. 2.26).

2.24 FERRIMAGNETIC MATERIALS

Almost every item of electronic equipment produced today contains some *ferrimagnetic* material: loudspeakers, motors, deflection yokes, interference suppressors, antenna rods, proximity sensors, recording heads, transformers and inductors are frequently based on *ferrites*. The market is vast.

What properties make *ferrimagnets* so ubiquitous? They possess *permeability* to rival most *ferromagnets* but their *eddy current losses* are far lower because of the material's greater electrical resistivity. Also it is practicable (if not straightforward) to fabricate different shapes by pressing or extruding—both low cost techniques.

What is the composition of ferrimagnetic materials? They are, in general, oxides of iron combined with one or more of the transition metals such as manganese, nickel or zinc, e.g. $MnFe_2O_4$. Permanent ferrimagnets often include barium. The raw material is turned into a powder which is then fired in a kiln or *sintered* to produce a dark gray, hard, brittle ceramic material having a cubic crystalline structure.

At an atomic level the magnetic properties depend upon interaction between the electrons associated with the metal ions. Neighbouring atomic *magnetic moments* become locked in anti-parallel with their neighbours (which contrasts with the *ferromagnets*). However, the magnetic moments in one direction are weaker than the moments in the opposite direction leading to an overall magnetic moment.

2.25 SATURATION

Saturation is a limitation occurring in inductors having a *ferromagnetic* or *ferrimagnetic* core. Initially, as current is increased the *flux* increases in proportion to it (see Fig. 2.30). At some point, however, further increases in current lead to progressively smaller increases in flux. Eventually, the core can make no further contribution to flux growth and any increase thereafter is limited to that provided by μ_0—perhaps three orders of magnitude smaller. Iron saturates at about 1.6·T while ferrites will normally saturate between about 200·mT and 500·mT.

It is usually essential to avoid reaching saturation since it is accompanied by a drop in *inductance*. In many circuits the rate at which current in the coil increases is inversely proportional to inductance (I = V * T / L). Any drop in inductance therefore causes the current to rise faster, increasing the *field strength* and so the core is driven even further into saturation.

Core manufacturers normally specify the saturation *flux density* for the particular material used. You can also measure saturation using a *simple circuit*. There are two methods by which you can calculate flux if you know the number of turns and either:

1. The current, the length of the magnetic path and the B-H characteristics of the material.
2. The voltage waveform on a winding and the cross-sectional area of the core—see Faraday's Law.

Although saturation is mostly a risk in high power circuits it is still a possibility in 'small signal' applications having many turns on an ungapped core and a DC bias (such as the collector current of a transistor).

If you find that saturation is likely then you might:

- Run the inductor at a lower current
- Use a larger core
- Alter the number of turns
- Use a core with a lower permeability
- Use a core with an air gap

or some combination thereof—but you'll need to re-calculate the design in any case.

When a *ferromagnetic* material is magnetized in one direction, it will not relax back to zero magnetization when the imposed magnetizing field is removed. It must be driven back to zero by a field in the opposite direction. If an alternating magnetic field is applied to the material, its magnetization will trace out a loop called a *hysteresis loop*. The lack of retraceability of the magnetization curve is the property called hysteresis and it is related to the existence of *magnetic domains* in the material. Once the magnetic domains are reoriented, it takes some energy to turn them back again. This property of ferrromagnetic materials is useful as a magnetic "memory". Some compositions of ferromagnetic materials will retain an imposed magnetization indefinitely and are useful as "permanent magnets". The magnetic memory aspects of iron and chromium oxides make them useful in audio *tape recording* and for the magnetic storage of data on computer disks

2.26 THE HYSTERESIS LOOP AND MAGNETIC PROPERTIES

A great deal of information can be learned about the magnetic properties of a material by studying its hysteresis loop. A hysteresis loop shows the relationship between the induced magnetic flux density (**B**) and the magnetizing force (**H**). It is often referred to as the B-H loop. An example hysteresis loop is shown in Fig. 2.31.

The loop is generated by measuring the magnetic flux of a ferromagnetic material while the magnetizing force is changed. A ferromagnetic material that has never been previously magnetized or has been thoroughly demagnetized will follow the

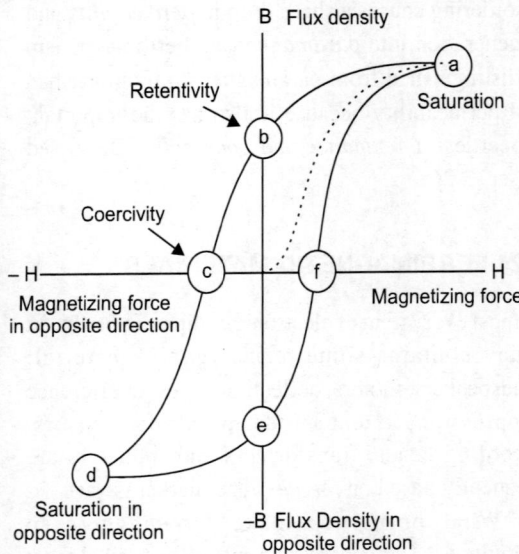

Fig. 2.31. A complete hysteresis loop.

dashed line as **H** is increased. As the line demonstrates, the greater the amount of current applied (**H+**), the stronger the magnetic field in the component (**B+**). At point "**a**" almost all of the magnetic domains are aligned and an additional increase in the magnetizing force will produce very little increase in magnetic flux. The material has reached the point of magnetic saturation. When **H** is reduced to zero, the curve will move from point "**a**" to point "**b**." At this point, it can be seen that some magnetic flux remains in the material even though the magnetizing force is zero. This is referred to as the point of **retentivity** on the graph and indicates the **remanence** or level of **residual magnetism** in the material. (Some of the magnetic domains remain aligned but some have lost their alignment.) As the magnetizing force is reversed, the curve moves to point "**c**", where the flux has been reduced to zero.

This is called the point of **coercivity** on the curve. (The reversed magnetizing force has flipped enough of the domains so that the net flux within the material is zero.) The force required to remove the residual magnetism from the material is called the coercive force or coercivity of the material.

As the magnetizing force is increased in the negative direction, the material will again become magnetically saturated but in the opposite direction (point "**d**"). Reducing **H** to zero brings the curve to point "**e**." It will have a level of residual magnetism equal to that achieved in the other direction. Increasing **H** back in the positive direction will return **B** to zero. Notice that the curve did not return to the origin of the graph because some force is required to remove the residual magnetism. The curve will take a different path from point "**f**" back to the saturation point where it will complete the loop.

From the hysteresis loop, a number of primary magnetic properties of a material can be determined.

1. **Retentivity**—A measure of the residual flux density corresponding to the saturation induction of a magnetic material. In other words, it is a material's ability to retain a certain amount of residual magnetic field when the magnetizing force is removed after achieving saturation. (The value of **B** at point **b** on the hysteresis curve.)
2. **Residual Magnetism** or **Residual Flux**—the magnetic flux density that remains in a material when the magnetizing force is zero. Note that residual magnetism and retentivity are the same when the material has been magnetized to the saturation point. However, the level of residual magnetism may be lower than the retentivity value when the magnetizing force did not reach the saturation level.
3. **Coercive Force**—The amount of reverse magnetic field which must be applied to a magnetic material to make the magnetic flux return to zero. (The value of **H** at point "**c**" on the hysteresis curve.)
4. **Permeability, μ**—A property of a material that describes the ease with which a magnetic flux is established in the component.
5. **Reluctance**—Is the opposition that a ferromagnetic material shows to the establishment of a magnetic field. Reluctance is analogous to the resistance in an electrical circuit.

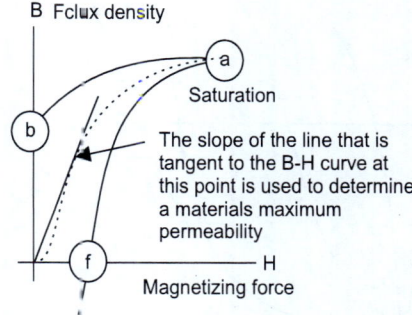

Fig. 2.32. Value of μ.

Permeability

As previously mentioned, permeability is a material property that describes the ease with which a magnetic flux is established in a component. It is the ratio of the flux density to the magnetizing force and is represented by the following equation:

$$\mu = B/H \text{ (Fig. 2.32)}$$

It is clear that this equation describes the slope of the curve at any point on the hysteresis loop. The permeability value given in papers and reference materials is usually the maximum permeability or the maximum relative permeability. The maximum permeability is the point where the slope of the B/H curve for the unmagnetized material is the greatest. This point is often taken as the point where a straight line from the origin is tangent to the B/H curve.

The relative permeability is arrived at by taking the ratio of the material's permeability to the permeability in free space (air).

$$\mu_{(relative)} = \mu_{(material)}/\mu_{(air)}$$

where: $\mu_{(air)} = 1.256 \times 10^{-6}$ H/m

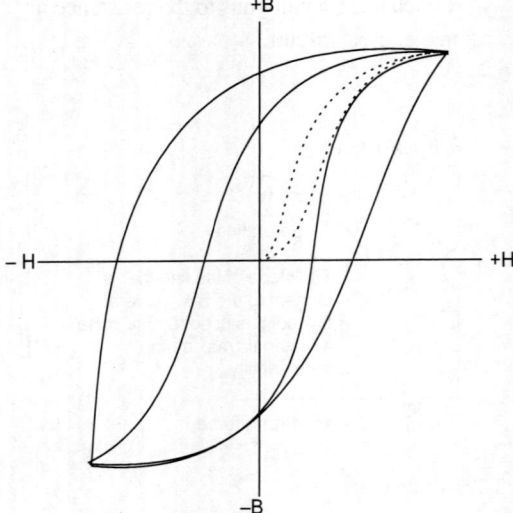

Fig. 2.33.

The shape of the hysteresis loop tells a great deal about the material being magnetized. The hysteresis curves of two different materials are shown in the graph, Fig. 2.33.

Relative to other materials, a material with a wider hysteresis loop has:

- Lower Permeability
- Higher Retentivity
- Higher Coercivity
- Higher Reluctance
- Higher Residual Magnetism

Relative to other materials, a material with the narrower hysteresis loop has:

- Higher Permeability
- Lower Retentivity
- Lower Coercivity
- Lower Reluctance
- Lower Residual Magnetism.

In magnetic particle testing, the level of residual magnetism is important. Residual magnetic fields are affected by the permeability, which can be related to the carbon content and alloying of the material. A component with high carbon content will have low permeability and will retain more magnetic flux than a material with low carbon content.

Reflection Section

1. Which type of magnetic material which can be magnetized on both directions?
2. Name the materials that have very high permeabilities (hundreds and even thousands times of that of free space).
3. Name the materials whose permeabilities are slightly greater than that of free space.
4. What does the area of hysteresis loop indicate?
5. Give the role of hysteresis co-efficient in minimizing the hysteresis loss.
6. Discuss the fact that the hysteresis loss is caused because of the work required for the magnetising the material.
7. According to Steinmetz hysteresis law, hysteresis loss in a material is proportional to Bx. What is x?

2.27 HYSTERESIS IN MAGNETIC RECORDING

Refer to Fig. 2.34 and Fig. 2.35. It is customary to plot the *magnetization* M of the sample as a function of the *magnetic field strength* H, since H is a measure of the externally applied field which drives the magnetization.

Because of *hysteresis*, an input signal at the level indicated by the dashed line could give a magnetization anywhere between C and D, depending upon the immediate previous history of the tape (i.e., the signal which preceded it). This clearly unacceptable situation is remedied by the *bias*

Magnetism and Electromagnetism 183

2.28 VARIATIONS IN HYSTERESIS CURVES

There is considerable variation in the *hysteresis* of different magnetic materials.

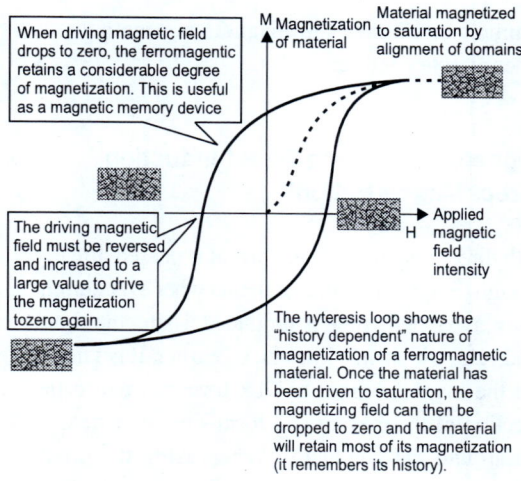

Fig. 2.34.

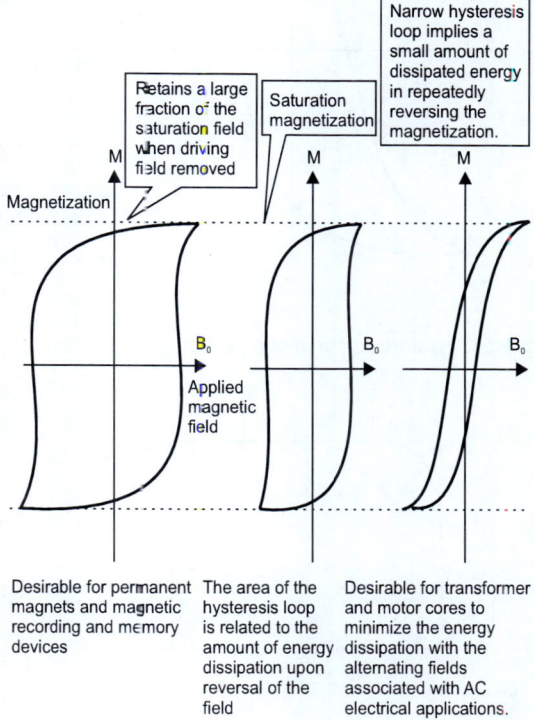

Fig. 2.36. Various hysteresis curves.

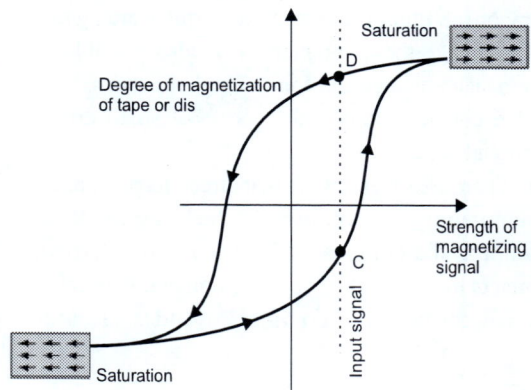

Fig. 2.35. An example to describe the application of hysteresis loop.

signal which cycles the oxide grains around their hysteresis loops so quickly that the magnetization averages to zero when no signal is applied. The result of the bias signal is like a magnetic eddy which settles down to zero if there is no signal superimposed upon it. If there is a signal, it offsets the bias signal so that it leaves a remnant magnetization proportional to the signal offset.

2.29 THE HYSTERESIS CURVE OF ALLOYS

The hysteresis curve of soft alloys

The hysteresis curve of soft alloys is thin and therefore the coercive force is small. These alloys are mostly used in electromechanical machines and transformers.

The hysteresis curve of hard alloys

The hysteresis curve of hard alloys is wide and therefore the coercive force is high.

184 Basic Electrical Engineering

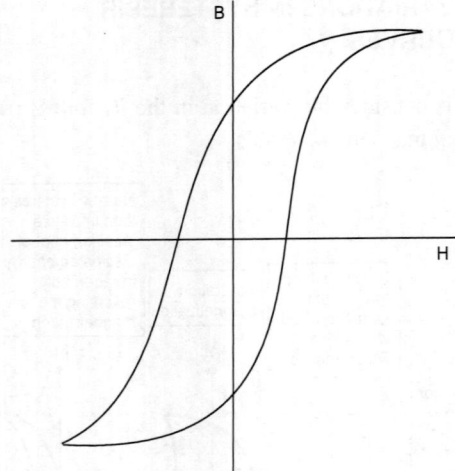

Fig. 2.37. Hysteresis curve of soft alloys.

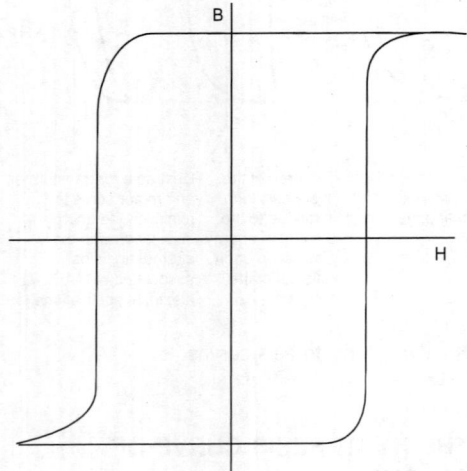

Fig. 2.38. Hysteresis curve of hard alloys.

2.30 MAGNETIZATION OF FERROMAGNETIC MATERIALS

There are a variety of methods that can be used to establish a magnetic field in a component for evaluation using magnetic particle inspection. It is common to classify the magnetizing methods as either direct or indirect.

Magnetization Using Direct Induction (Direct Magnetization)

With direct magnetization, current is passed directly through the component. Recall that whenever current flows, a magnetic field is produced. Using the right-hand rule, which was introduced earlier, it is known that the magnetic lines of flux form normal to the direction of the current and form a circular field in and around the conductor. When using the direct magnetization method, care must be taken to ensure that good electrical contact is established and maintained between the test equipment and the test component. Improper contact can result in arcing that may damage the component. It is also possible to overheat components in areas of high resistance such as the contact points and in areas of small cross-sectional area.

There are several ways that direct magnetization is commonly accomplished. One way involves clamping the component between two electrical contacts in a special piece of equipment. Current is passed through the component and a circular

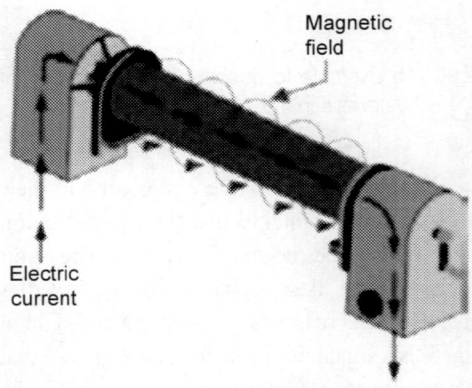

Fig. 2.39.

magnetic field is established in and around the component. When the magnetizing current is stopped, a residual magnetic field will remain within the component. The strength of the induced magnetic field is proportional to the amount of current passed through the component.

A second technique involves using clamps or prods, which are attached or placed in contact with the component. Electrical current flows through the component from contact to contact. The current sets up a circular magnetic field around the path of the current.

Magnetization Using Indirect Induction (Indirect Magnetization)

Indirect magnetization is accomplished by using a strong external magnetic field to establish a magnetic field within the component. As with direct magnetization, there are several ways that indirect magnetization can be accomplished.

The use of **permanent magnets** is a low cost method of establishing a magnetic field. However, their use is limited due to lack of control of the field

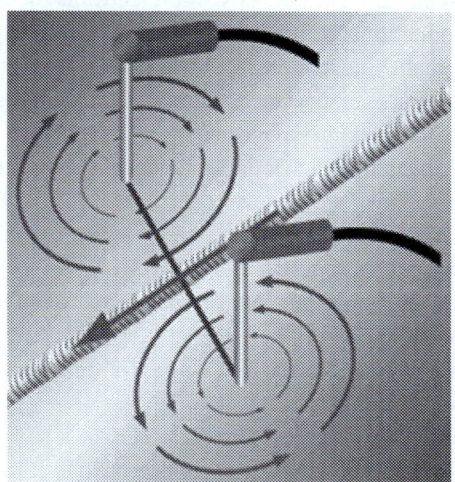

Fig. 2.40.

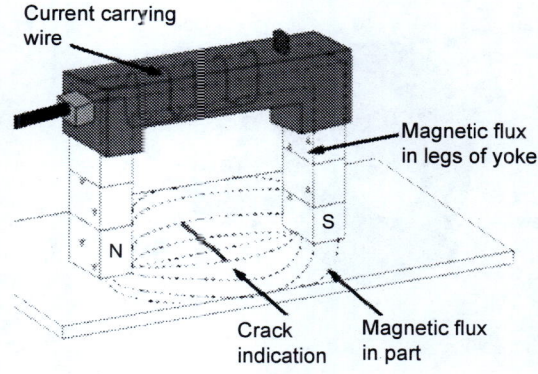

Fig. 2.41.

strength and the difficulty of placing and removing strong permanent magnets from the component.

Electromagnets in the form of an adjustable horseshoe magnet (called a yoke) eliminate the problems associated with permanent magnets and are used extensively in industry. Electromagnets only exhibit a magnetic flux when electric current is flowing around the soft iron core. When the magnet is placed on the component, a magnetic field is established between the north and south poles of the magnet.

Another way of indirectly inducting a magnetic field in a material is by using the magnetic field of a current carrying conductor. A circular magnetic field can be established in cylindrical components by using a **central conductor.** Typically, one or more cylindrical components are hung from a solid copper bar running through the inside diameter. Current is passed through the copper bar and the resulting circular magnetic field establishes a magnetic field within the test components.

The use of **coils** and **solenoids** is a third method of indirect magnetization. When the length of a component is several times larger than its diameter, a longitudinal magnetic field can be established in the component. The component is placed

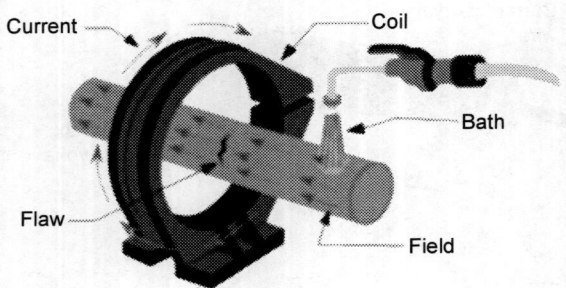

Fig. 2.42.

longitudinally in the concentrated magnetic field that fills the center of a coil or solenoid. This magnetization technique is often referred to as a "coil shot."

CONCLUSION

This chapter presented in detail the fundamentals of magnetic circuits. Terms associated with and the induction of magnetic circuits are given. Complete theory of the topics with numerous illustrated examples are given.

QUESTIONS

1. What is the concept of electromagnetism?
2. Tabulate the similarities of an electric circuit with that of a magnetic circuit.
3. What is a magnetic circuit?
4. What is a field?
5. Explain the concept of flux in a magnetic circuit.
6. What is the difference between a permanent magnet and an electromagnet?
7. What is meant by excitation?
8. Discuss the cork-screw rule.
9. What is MMF? What is its unit? Explain it.
10. Explain Ampere's law.
11. Relate mmf and the magnetizing force, H and define it.
12. Define flux density. Bring out a relation between B and H.
13. Define reluctance of a magnetic circuit.
14. What is permeance?
15. Explain the Ohm's law of a magnetic circuit.
16. Write a note on (a) series magnetic circuit and (b) parallel magnetic circuit. Explain the method of solving such circuits.
17. What is a leakage flux? Why does it take place in magnetic circuit? Is it avoidable?
18. How the magnetic leakage is accounted in magnetic circuits?
19. Discuss the leakage factor.
20. What is magnetic fringing?
21. Write short note on the electromagnetic induction.
22. What are the Faraday's postulations on electromagnetic induction? Explain them with suitable sketches.
23. Discuss the Fleming's right hand rule and Fleming's left hand rule. Tell the applications, respectively.
24. Write short notes on how a mechanical force could be produced in a magnetic circuit. Derive any suitable formula.
25. What are the two possible electromagnetic inductions? Explain them thoroughly using suitable illustrations.
26. Define self inductance. With a suitable example, define a mutual inductance?
27. Explain the behaviour of an inductance when it is supplied with (a) DC and (b) AC.
28. Discuss the coefficient of coupling in the mutually arranged coils.
29. Develop an expression for the energy stored in a coil in terms of its L.
30. What are the various types of magnetic materials?
31. What is hysteresis?
32. Draw a B-H curve. Explain its growth in steps. Indicate various regions of it. Explain them.
33. What are the major applications of magnetic materials. Explain.

INTRODUCTION

This chapter introduces the concept of DC machines and transformers. In the first part, a detailed study on the fundamentals of DC machines is presented. The construction and types of DC machines are first presented in general, and then focus is made in particular on DC generator and on DC motor. The principles of operation of these two machines, governing equations, electric equivalent circuits, accessories for the operation of these two machines and numerical examples to understand the principle of DC motors and DC generators are presented.

In part II, fundamentals of transformers are presented. The construction, principles of operation and the governing equations of transformers are presented. The behaviour of transformer when it is loaded is given with relevant figures. The equivalent circuit of a transformer, evaluation of its components using suitable tests and the importance of voltage regulation are then presented, which are followed by a thorough discussion on 3-phase transformers. Various illustrated examples will help to understand the machines in a better way.

CHAPTER OBJECTIVE

At the completion of this chapter, the reader will be able to:

* solve questions related to the performance of DC machines and transformers;
* solve typical numerical examples on DC machines and transformers; and
* understand the concept of 3-phase transformers.

KEYWORDS

DC generator, Types of DC generators, Current and voltage equations, EMF induced, Building up of induced EMF, Critical field resistance, Characteristics.

DC motor, Concept of torque, Back EMF, Type and characteristics and Concept of speed control.

1-phase transformers, Construction and principle of operation, EMF equation, Phasor diagram, Equivalent resistances, Equivalent reactances, Equivalent circuit, Voltage regulation and Concept of 3-phase transformers.

3

Electric Machines

PART I: D C MACHINES

3.1 INTRODUCTION

An electromechanical energy conversion device—simply, an electric machine—is a bridge between electrical energy and mechanical energy systems. Conversion of energy takes place in two ways viz, electrical to mechanical or from mechanical to electrical, whenever a change of flux linkage takes place (Sec. 2.16).

When a device converts a mechanical energy into an electrical energy, such an electrical device is said to be a **generator**; on the other hand, when a device converts an electrical energy into a mechanical energy, such device is a **motor**. This clearly indicates that the electromechanical process is a **reversible** process and thus a **same** device can be used either as a generator or a motor.

Thus, in general, the device which undertakes either of the above convertion is termed to be an **electric machine**.

When the operation is completely DC—that is mechanical energy to DC electrical energy and vice-versa—then it is further pointed out as a **DC electric machine**; of course, if the operation is completely an AC, it is an **AC electric machine**.

In such machines, there will be a moving body which always will be set into motion and another stationary body. Because there is **always** motion in such electric machines. They are otherwise known as **dynamic devices** or **dynamic machines**.

There is yet another electric device in which an electric-to-electric energy transformation takes place **without** any rotating action; such a device is also known to be a **static device** (because there is **no** rotating part); it is a **transformer**.

These main classifications of electric machines is picturised in a flow diagram of Fig. 3.1.

The main objective of this chapter is to bring out the construction, the working principles and the energy equation aspects of the electric machines.

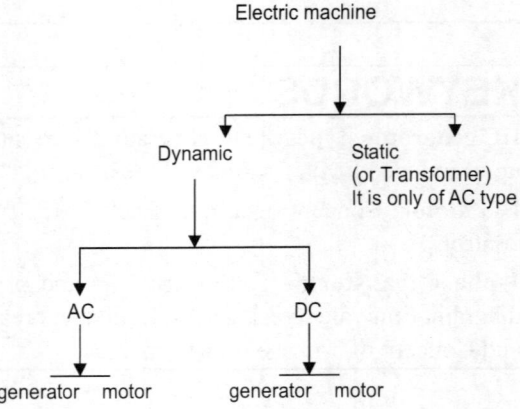

Fig. 3.1. Flow diagram for electric machine classifications.

3.2 DC MACHINES

As mentioned in the previous section, a rotating machine (whether AC or DC) can be either of a generator or a motor. These classifications are based on the individual energy transformation modes; but basically their construction is the same. In fact a DC machine can very much be used as a generator and also as a motor.

The general construction and various parts of a DC machine is discussed at first.

The essential parts of DC machines are (i) the **'field'** (a stationary part) ii) the **'armature'** (a moving part) (iii) a small air gap between these two and (iv) commutators and brushes.

The field is an electromagnet assembly which will be provided with a separate winding wound over the magnets known as the **field winding**. The winding will be excited by a separate DC source. This excitation provided to the field windings of the electromagnets will **excite the magnets**; this excitation will make the electromagnets to act as N and S poles. Thus the field system will provide the working field in the airgap. (Refer to Fig. 3.2)

The stationary part of the DC machine is this field system. The poles will be held in position with the help of rivets, attached to a part known as the **yoke**. (Fig. 3.3)

The yoke thus keeps or holds the poles in stationary position; Apart from this the yoke's main job is to provide a return path for the flux lines from the neighbouring poles as shown.

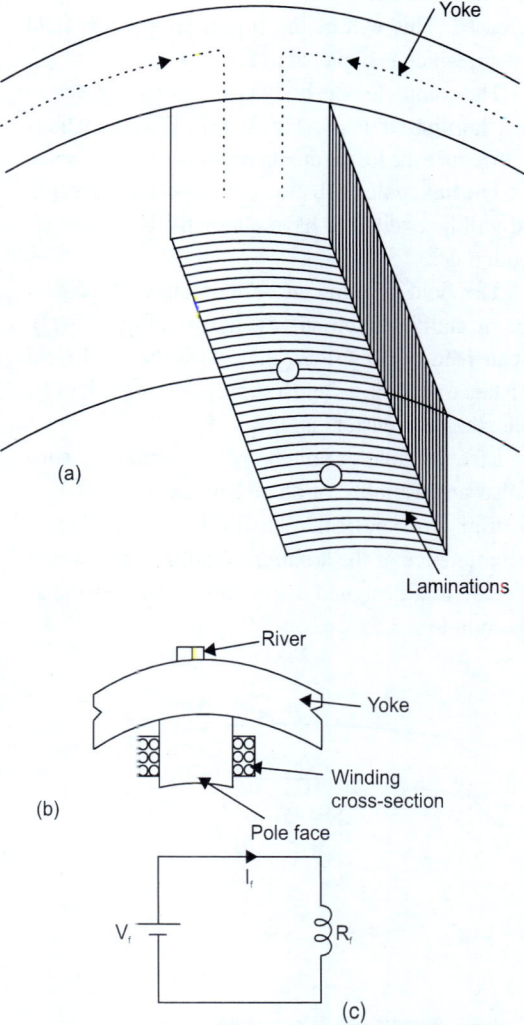

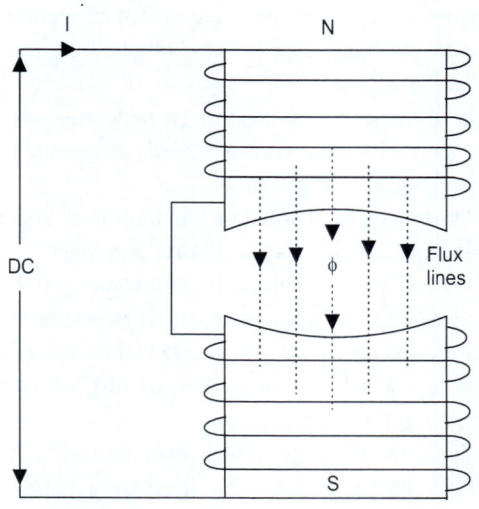

Fig. 3.2. Field system.

Fig. 3.3. (a) Pole assembly. (b) Front view (c) Field circuit diagram.

As seen in Fig. 3.3(a) if the field winding is wound around the pole, their is no 'base' to hold it. It will fall down due to gravity. Thus the poles are further provided with an arrangement known to be the '**pole shoe**'. This actually provides a hold for the field windings; moreover, it distributes the flux lines produced, on the surface of the armature more uniformly over a wider area, as the pole shoes form a curved bottom surface. Because of this arrangement, the useful flux for the energy transformation gets increased. This will be the complete practical field system, which is depicted in Fig. 3.4.

The complete pole body is not a single structure but a '**laminated**' body. (Fig. 3.3(a)). The laminations are to reduce the losses produced due to eddy currents.

The field poles will always be of even numbered and will be excited to have alternatively North and South poles.

The field system can be thus represented as a electric equivalent circuit, as shown in Fig. (3.3(c)). If can readily be written that, with the total field resistance of R_f ohms and a supply voltage to it of V_f volts, the field current is, $I_f = V_f/R_f$.

Like the field or pole body, the **armature core** is also an assembly of steel laminations, but it is circular in section. Throughout the outer circumference of the armature small grooves would be made to a designed depth, at equal intervals, as shown in Fig. 3.5.

Fig. 3.5. Armature with its winding. Note the commutator assembly in the shaft extension.

These grooves are the slots, the total number and the depth of which are design factors. The slots are provided to receive the armature winding.

The integral part of the armature is the shaft. In case of a motor the torque developed (discussed in the next section) will rotate the armature, in turn the connected shaft will also rotate giving the mechanical energy output to other machine coupled to this shaft.

The armature winding provided in the slots of the armature is the heart of the DC machine; it is where the voltage is generated in case of a generator or the torque is developed in case of a motor.

The armature winding consists of a group of insulated copper coils—that is the conductors—kept in the slots with the coil ends properly connected to the commutators. (Fig. 3.6)

The two types of armature windings used in the DC machines are the **Lap** and **Wave** windings.

A small air gap, in the order of few mm, will be provided between the stationary field system and the rotating armature; the flux available in this air gap is highly responsible for the generation of EMF in a generator or torque in a motor.

The commutator is another main part of a DC machine which converts the alternating voltage generated in the armature of a DC generator to a direct

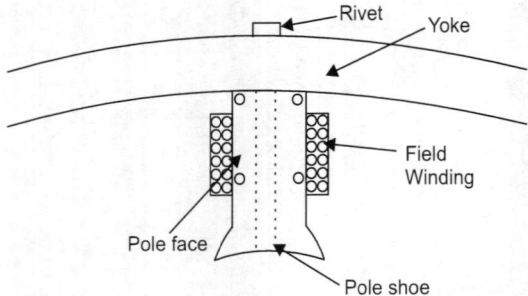

Fig. 3.4. A complete pole system.

Electric Machines 191

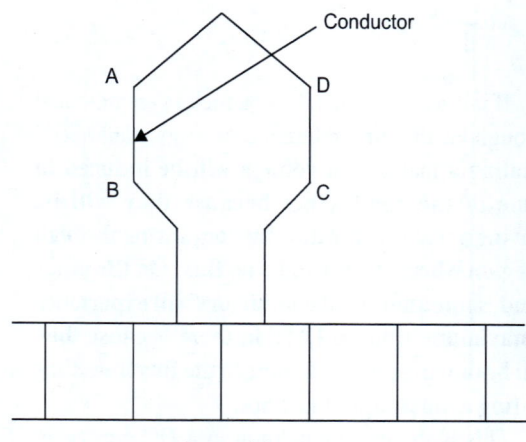

Fig. 3.6. Commutator with winding

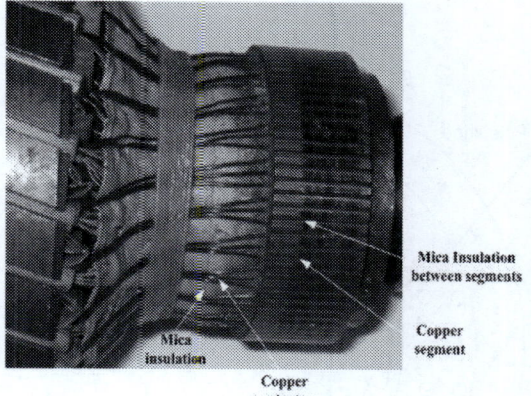

Fig. 3.8. V-grooved, commutator assembly.

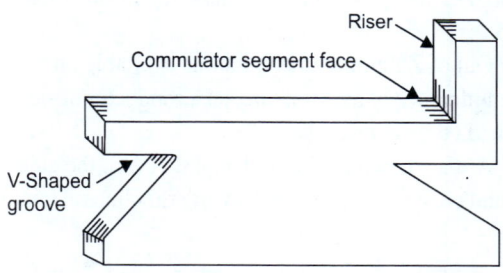

Fig. 3.7. Wedge-shaped commutator segment.

voltage in the load circuit where it is used. (Discussed in depth in Sec. 4.3)

The commutator is a built-up group of hard-drawn copper bars, wedge-shaped in section when viewed on end, and having V-shaped grooves at each end. Fig. 3.7 is a sketch of one such commutator with parts noted.

It is made up of high conductivity material. There will be as many commutator segments as the number of armature coils and they will be insulated from each other by thin mica sheets.

Each armature coil will be connected to a single commutator segment at the riser (or lug). The coil end will electrically be soldered here.

The V-grooves are provided, to prevent these flexible commutator segments from flying out under the action of centrifugal forces when the armature is rotating. As in Fig. 3.8, the V-grooves will be perfectly fitted in the end of the armature core; the armature coil ends will be brought and soldered to this riser.

A 4-pole DC machine with field, yoke armature etc., is shown in Fig. 3.9.

3.3 DC GENERATOR

It is seen in Sec (2.16) that the relative motion between a magnetic field and electric conductors produces an EMF in the electric conductors. This is the fundamental principle behind the working of the DC generators.

The field system will be stationary and the armature of the DC generator will be set into motion by a **prime mover** (may be a motor or a turbine etc.) coupled to the same shaft of this generator. This exhibits the **relative motion** conditions.

The DC excited poles will produce magnetic lines of force (flux) which will leave the

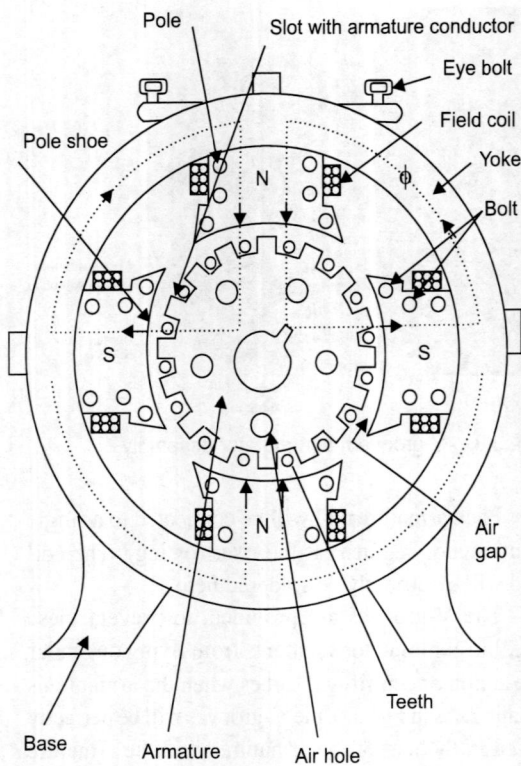

Fig. 3.9. D.C. Machine.

$$e \alpha \frac{d}{dt}(N\phi)$$

If it is assumed that the conductors are mounted throughout the circumference of a constant-speed **rotating** armature, **no voltage will be induced in some of the conductors because they will be moving parallel to the flux lines or passing through a region where there will be no flux. On the other hand, some amount of conductors will experience a maximum induced EMF in them because they will be moving perpendicular to the flux lines thus cutting a maximum flux-lines**.

This is the exact condition in a DC generator, too! Thus, the EMF generated inside the armature is not a pure direct voltage and it will be of varying magnitude, Hence it will be of alternating nature.

As all the conductors of the armature slots are going to cover 360°, let a conductor be examined from 0 to 360°, which will be the same for the other also.

Out of Z number of conductors available on the circumference of the armature, let a conductor of side **ab** and **cd** be considered.

At the shown position of Fig. 3.10(A), the side **ab** and **cd** are parallel to the flowing flux lines. Thus,

cylindrically—curved pole shoes and pass across the air gap; immediately they fall over the rotating armature surface. It is clear that because of this action, the moving copper conductors located in the slots of the armature will cut the flux lines. This flux-cutting action of the armature conductors is responsible for the generated voltage is them.

Thus to produce voltage a generator needs the presence **of unvarying (DC) magnetic lines of force** and **motion of the conductors cutting these flux,** before the voltage is generated.

It is brought out in the equation (2.23) that the magnitude of the generated voltage in a conductor is directly proportional to the rate at which the conductor cuts the magnetic lines of force. That is,

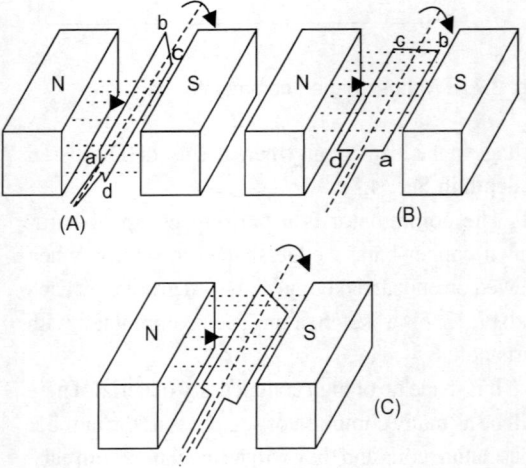

Fig. 3.10. Generator principle

both the sides cannot cut any flux lines, hence there is no EMF induced in the coil. Let his be at 0°.

Figure 3.10(B) shows the same coil after 90° movement. Readily the reader can see that the coil side **ab** and **cd** are moving **perpendicular** to the flux lines. The induced EMF in the conductor will be at a maximum at this position. So the conductor from zeroth degree would have started cutting the flux line, thus getting induced EMF in it increased. AT 90th degree, the generated EMF will reach the maximum due to the maximum flux-cutting active here. After another 90 degrees the position of the conductor is in Fig. 3.10(C) (the reader should absorb the coil sides). This is exactly as the zeroth degree case. The sides are parallel to the flux lines thus having zero induced EMF. So the maximum induced EMF from 90th degree would have gradually reduced to a zero value at this 180th degree, because the flux cutting action also would have reduced respectively.

The induced EMF curve upto this situation is in Fig. 3.11.

A Keen observation will reveal the fact that the coil sides **ab**, which was revolving under the south pole (which was receiving the flux) will be now under the north pole (which was **giving** out the flux); similarly the coil sides **cd** was under the north pole, and will revolve now under the south pole as in Figs. 3.3(D) and 3.3(E); because of this, the **direction** of the induced EMF **will change**. But note that, **still the flux-cutting action in going to be the same**. This simply implies that the amount of flux lines the sides may cut will remain unchanged. Thus the magnitude of the induced EMF will remain **the same**. Hence, the variations in the magnitude of the induced EMF will be similar to the 0° to 180° rotation but in the negative side. It is depicted in Fig. 3.12.

As a summary, in a conductor,

0–90°: induced EMF reaches to positive maximum.

90°–180°: it reaches a zero from maxmimum in the positive side.

180°–270°: Induced EMF reaches negative maximum.

270°–360°: it approaches a zero, in the negative side.

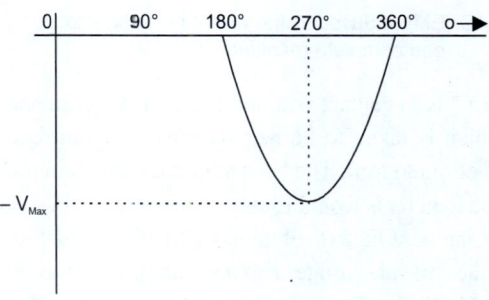

Fig. 3.12. EMF from 180° to 360°.

For one revolution the EMF induced in a conductor is in Fig. 3.13; it will be the same in all the conductors also.

Thus, **the EMF induced in the armature conductors of a DC generator is alternating**.

This alternating induced EMF is made direct in the output circuit with the help of brushes and commutators.

Consider, for this discussion, the Fig. 3.14.

It is seen in Fig. 3.14(A) that in the first half revolution current flows along **a b l m c d a**. Also note that the brush no. 1 is in contact with the commutator segment 'p' which is taken to be positive (carrying current from top to bottom) and brush

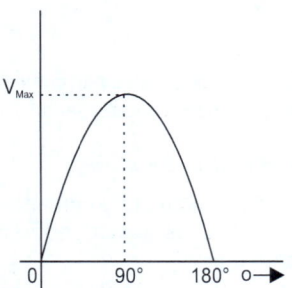

Fig. 3.11. From 0° to 180°

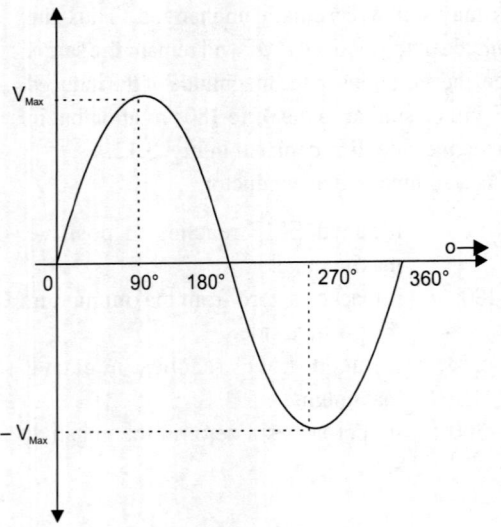

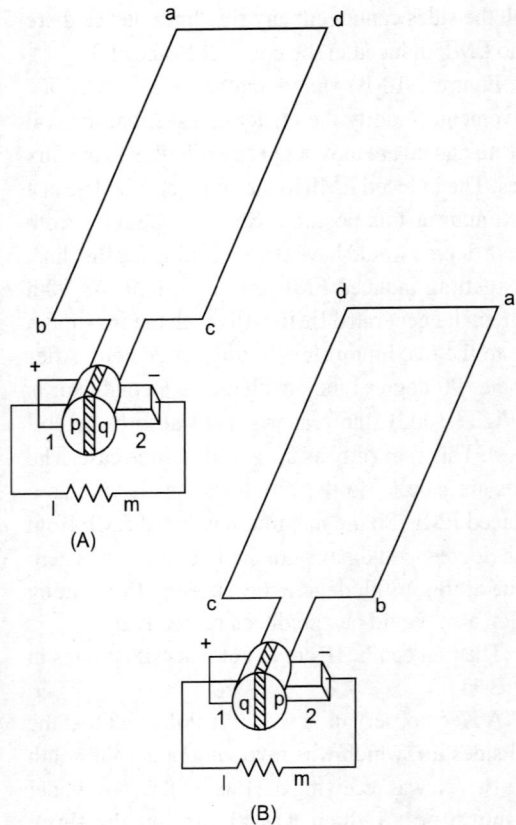

Fig. 3.13. EMF induced in the armature conductor for one complete revolution.

number 2 is in contact with the commutator segment 'q' which is taken to be negative (carrying current from bottom to top). The current in the output circuit (i.e. the load R) is from **l** to **m**.

In the next half revolution, that is from 180° to 360°, the coil sides rotates and occupies position as in Fig. 3.14(B); the direction of the induced EMF in the coil will thus be reversed. But **at the same time**, the position of commutator segments 'p' and 'q' would have also been reversed with the result that the brush no. 1 comes in touch with the segment 'q' which is now positive and the brush no. 2 is in contact with the segment 'p' which is negative. Here again the current direction in the load R is from **l** to **m**.

From the above two paragraphs it is clear that the brush no. 1 is always in contact with commutator segment which is positive (i.e.; the one which carries current from top to bottom, according to our convention) and brush no. 2 will always be in contact with the commutator segment which is negative (carrying current from bottom to top), thus maintaining in the **output circuit** unidirectional current. Refer to Fig. 13.15.

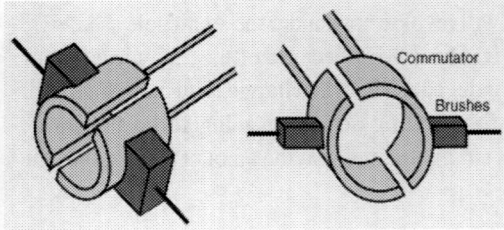

Fig. 3.14. (a) The action of brushes and commutators.
(b) Diagrammatic representation.

Reflection Section

1. In connecting commutator segments to the armature conductors, describe the role of commutator lugs.
2. Explain the role of split rings in making the AC induced EMF to be DC EMF in the output circuit of a generator.

3. How the alternative poles of a DC machine are made north and south?
4. Why the magnets of a DC machines are called electromagnets? Can they be replaced with permanent magnets. Discuss.

3.4 TYPES OF DC GENERATORS

There are broadly two types of DC generators classified with reference to their manner of field

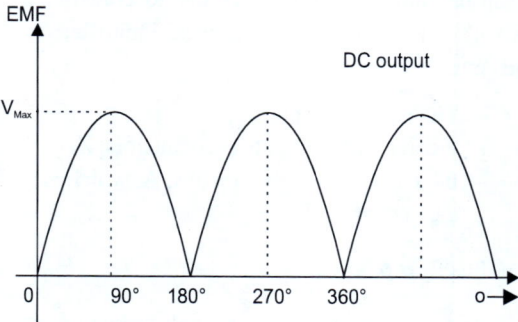

Fig. 3.15. Current in the output circuit

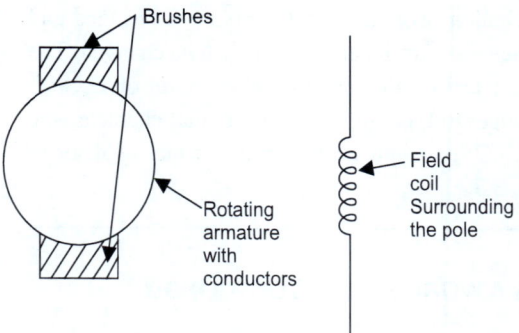

Fig. 3.16. (a) An armature symbol (b) Field coil symbol.

excitations. It is brought out that the field coils (coils surrounding the poles, kept to magnitize them) is to be excited by a DC source.

If the field coil excited separately from an outside DC source, then such machine is grouped under the separately excited DC generators; on the other hand the field coil may be excited by the current supplied by its own armature without a separate source of supply; this way excited generators are known to be the self-excited DC generators, and they are of three types.

In self-excised DC generators, the field winding may be connected in parallel with the armature; then it is known to be the **self-excited DC shunt generators.**

The field may also be connected in series with the armature; then the machine is known to be the **self-excited series generator**.

It is possible to insert both the shunt field winding and the series field winding together, forming a **self-excited DC compound generator**.

In general, diagrammatically, the armature and the field winding are denoted by the symbols, as shown in Fig. 3.16(a) and (b).

These symbols are used to denote the above-mentioned DC generators as shown in Fig. 3.17.

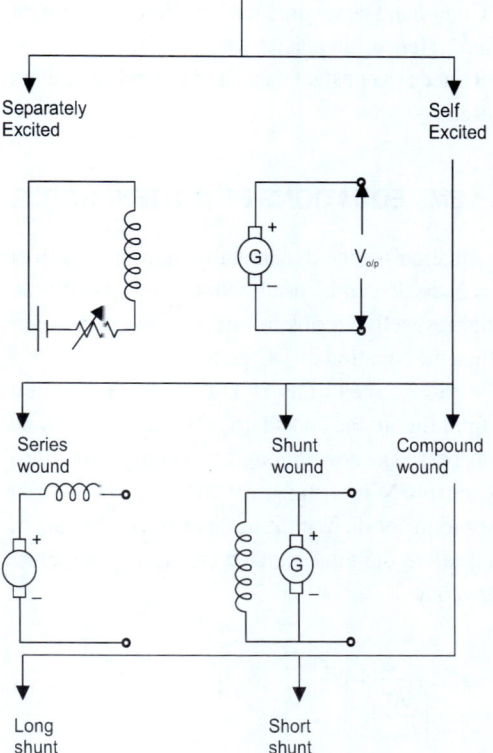

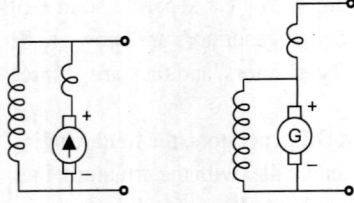

Fig. 3.17. Types of DC generators

Applications of DC generators

Shunt generators are used for supplying nearly constant loads. They are used for battery charging, as a supply source for field winding of synchronous machines and separately excited DC machines.

A classic application of the series generators is their use as **boosters**. When the series generators are used as boosters, they 'boost' (or increase) the voltage of the circuit. Series generator as boosters are used widely used in electric transmission systems.

Compound generators fairly maintain a constant voltage. Hence, at places where voltage is to be maintained at a constant value, compound generators are used.

3.5   EMF EQUATIONS OF DC GENERATOR

The equation for the dynamically induced emf, that is $e = N\, d\phi/dt$, can be used with proper substitution for the respective terms to arrive at the expression for the emf equation of DC generator.

With P number of poles and a flux per pole of ϕ wb, total flux in the considered DC machine will be $P\phi$, webers. The revolutions for second, N/60, will become 60/N in time units which will be the substitution for dt. With Z number of conductors in the armature the total induced emf in Z conductors will become,

$$e = Z\frac{P\phi}{\left(\dfrac{60}{N}\right)} = \left(\frac{P\phi ZN}{60}\right)$$

With a number of parallel paths for armature winding the final expression will be,

$$e = \frac{P\phi NZ}{60A} \text{ V} \qquad (3.1)$$

The armature winding in a DC machine is either wound in lap connections (where A=P) or wave connection (where A=2), which determines the parallel points, which is a very detailed study, and has no scope here.

In the above expression for the generated emf the following terms can readily be identified as constants :

- The number of poles, P
- the number of armature conductors, Z
- the number of parallel paths, A, which is a dependent on Z

$$\text{Thus } E_g \propto \phi N \qquad (3.2)$$

which is a very valuable proportionality equation of a DC generator.

ϕ, the flux in the pole can be varied if the field excitation (excitation to the poles) is changed and hence the induced emf E_g can also be changed. The characteristic that shows the variation of E_g with changes in I_f is called the open circuit characteristic (O.C.C) or magnetization curve, which is dealt in Sec 3.7.

3.5-A WORKED ILLUSTRATIONS X

Illustration 3.1

An eight pole lap wound armature has 960 conductors and is run at 500 rpm. If the flux per pole is 40 m Wb. find the e.m.f generated.

Solutions

Here poles are 4, speed 500 rpm, conductors 960 and flux 40 m Wb.

$$E_g = \frac{\phi.ZNP}{60.A}$$

$$= \frac{8 \times 40 \times 10^{-3} \times 960 \times 500}{60 \times 8}$$

$$= 320 \text{ V}.$$

Illustration 3.2

A six pole, wave wound armature has 640 conductors. If the flux per pole is 16 m Web at what speed must be driven in order to generator 256 V.

Solution

Using equation (3.1),

$$E = \frac{P\phi ZN}{60A} \text{ V}$$

$$256 = \frac{6 \times 16 \times 10^{-3} \times 640 \times N}{60 \times 2}$$

$$N = 500 \text{ rpm}$$

Illustration 3.3

A six pole, wave wound armature has 250 conductors when driven at 400 rpm, the e.m.f generated is 260 volt. What is the useful flux per pole.

Solutions

Useful flux per pole

$$= \frac{\text{E.m.f. generated} \times 60 \times \text{No. of parallel path}}{\text{No. of Conductors} \times \text{speed} \times \text{poles}}$$

$$= \frac{260 \times 60 \times 2}{250 \times 400 \times 6}$$

$$= 0.053 \text{ Wb}.$$

$$= 53 \text{ mWb}$$

Illustration 3.4

A four pole, lap wound armature when driven at 600 rpm generates 120 V. If the flux per pole is 25 m Wb. find the number of armature conductors.

Solutions

According to the e.m.f. equation the value of armature conductors, Z is:

$$Z = \frac{E \times 60 \times A}{\phi.N.P.} = \frac{120 \times 60 \times 4}{25 \times 10^{-3} \times 600 \times 4}$$

$$= 480 \text{ conductos}$$

3.6 CURRENT AND VOLTAGE EQUATIONS OF A DC MACHINE

In this section current division in DC machine (generator and as a motor) with the respective voltage equation are presented. Fig. 3.17 indicated, in general, the types of DC machines, in which, the current distribution in armature winding and field winding will vary depending on whether the DC machine is used as a generator or motor. Even though the principles of DC generator or motor is presented in Sec 3.9 the current division is presented here.

The current cirections of the armature current I_a, field current I_f and the line current (in case of motor) or load current (in case generator), I_L will be marked in the respective circuits based on the fact whether the DC machine is a DC generator or DC motor.

Consider the circuit for the shunt wound machine in the Fig. 3.18. If this DC machine is a DC generator then the armature will 'supply' the current (or power) to the load, which thus becomes the source. The reader should immediately understand that I_a is the total current which then branches out as I_f and I_L at the node 'a' as per KCL. But, had this been a DC motor, then the input supply is total power for the entire circuit, which thus becomes the source. The reader should immediately understand that I_L is the total line current which then branches out as I_f and I_a as per KCL. These respective cases are represented in equivalent circuits as shown in Fig. 3.18(a) and Fig. 3.18(b). The knowledge of current division is of utmost important in analysing a DC machine.

As similar analysis can be done an DC series moter also. As given in Fig. 3.19, if it is a generater, the total current is supplied from the armature, whereas if it is a motor, the total current is suplied from the source. However, in both the cases, $I_a = I_L = I_{se}$, where I_L is load current for generator and is supply current for motor.

A similar exercise can be made for long shunt DC machines (Fig. 3.20) and for short shunt DC machine. (Fig 3.21).

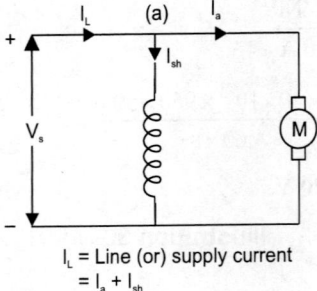

I_L = Line (or) supply current
= $I_a + I_{sh}$

Fig. 3.18b.   Equivalent circuit of shunt motor.

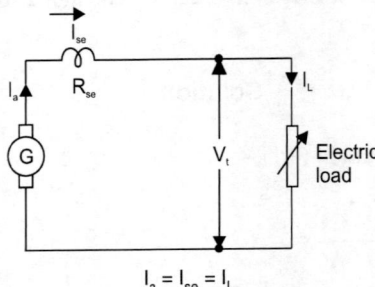

$I_a = I_{se} = I_L$

Fig. 3.19a.   Equivalent circuit of series generator.

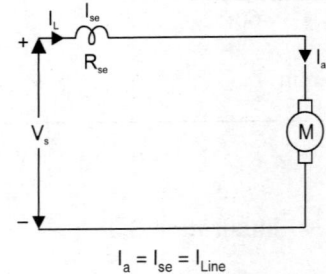

$I_a = I_{se} = I_{Line}$

Fig. 3.19b.   Equivalent circuit of series motor.

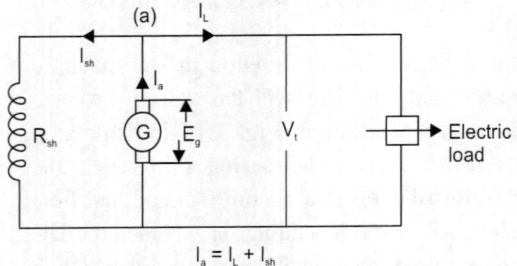

$I_a = I_L + I_{sh}$

Fig. 3.18a.   Equivalent circuit of shunt generator.

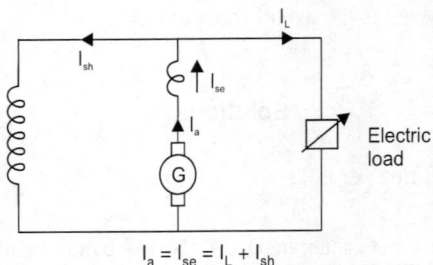

$I_a = I_{se} = I_L + I_{sh}$

Fig. 3.20a.   Equivalent circuit of long shunt compound generator.

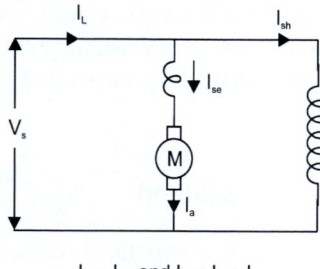

$I_a = I_{se}$ and $I_L = I_a + I_{sh}$

Fig. 3.20b. Equivalent circuit of long shunt compound motor.

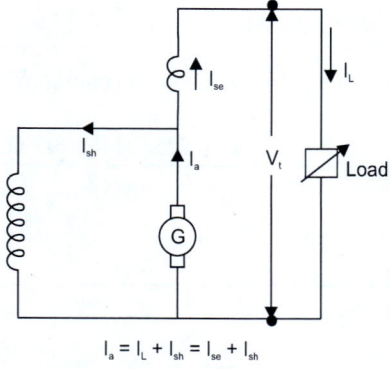

$I_a = I_L + I_{sh} = I_{se} + I_{sh}$

Fig. 3.21a. Equivalent circuit of short shunt compound generator.

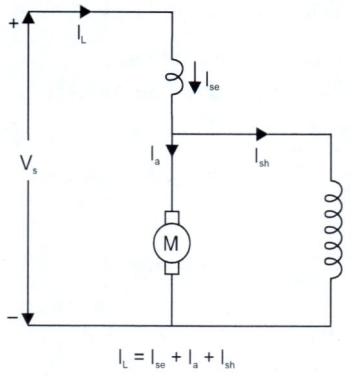

$I_L = I_{se} + I_a + I_{sh}$

Fig. 3.21b. Equivalent circuit for short shunt compound motor.

Electric Machines

Voltage Equations

The application of KVL to any of the equivalent circuits of DC machine will involve the three major terms viz, the supply voltage, the final voltage at the receiving side terminal, and the voltage drop that occur in between in the circuit elements like armature resistance etc.

In case of a DC generator the supply voltage is the generated voltage in the armature and the final terminal voltage is the voltage at the load. The difference between these two voltages forms, of course, the circuit voltage drop. So,

$$E_g = V_t + I_a R_a \qquad (3.3)$$

In case of a DC motor the supply is the total voltage and the voltage across the armature terminals is the final voltage. Thus

$$V_s = E_g + I_a R_a \qquad (3.4)$$

The reader is required to self analyse the significance of E_g, V_t, V_s and $I_a R_a$, at this juncture.

Illustration 3.5

The armature of a four pole shunt generator is lap wound, has 128 slots, four conductor per slot. Shunt field resistance 48 Ω. armature resistance 0.04 ohm. The flux per pole is 48 mWb. Find the speed of the marchine when supplying 400 A at a terminal voltage of 240 V.

Solutions

The shunt generator is supplying 400 A at 240 V. So the shunt field current

$$= \frac{V_{sh}}{R_{sh}} = \frac{240}{48} = 5 \text{ A}$$

∴ Armature current

$$= I_L + I_{sh} \text{ (Fig. 3.18(a))}$$

$$= 400 + 5$$

$$= 405 \text{ A}$$

Now armature drop

$$= I_a R_a \text{ V}$$

Armature drop

$$= 405 \times 0.04$$

$$= 16.2 \text{ V}$$

∴ e.m.f. generated

$$= 240 + 16.2 = 256.2 \text{ V}$$

Now the number of conductor

$$= \text{No. of Slots} \times \text{cond. / slot}$$

$$= 128 \times 4$$

$$= 512 \text{ conductor}$$

Poles are 4, parallel paths are 4 and flux = 0.048 Wb

According to emf equation, the speed

$$N = \frac{E \times 60 \times A}{\phi \times Z \times P}$$

$$= \frac{256 \times 60 \times 4}{0.048 \times 512 \times 4} = 625 \text{ rmp}$$

Illustration 3.6

The armature of a four pole shunt generator is lap wound and generates 216 V when running at 600 rpm. The armature has 144 slots, with six conductor per slot. If this armature is rewound, wave connected, find the emf generated at the same speed and flux per pole.

Solution

From the generator conditions, the flux is found out.

$$\phi = \frac{E \cdot 60 A}{PZN} = \frac{216 \times 60 \times 4}{4 \times (144 \times 6) \times 600}$$

$$= 0.025 \text{ Wb}.$$

Using this value, for wave wound conditions,

$$E = \frac{P\phi ZN}{60A} = \frac{4 \times 0.025 \times (144 \times 6) \times 600}{60 \times 2}$$

$$= 432 \text{ V}$$

Illustration 3.7

A four-pole shunt generator with lap connected armature supplies a load of 200 A at 100 V. Shunt field resistance 50W. Ra = 0.05 Ω. Calculate;

a. Total armature current
b. Current per armature path
c. emf generated

Allow a brush contact drop of 2 V

Solution

(a) The shunt field current

$$\frac{V_{sh}}{R_{sh}} = \frac{100}{50}$$

$$= 2 \text{ A}$$

Armature current = $I_{sn} + I_L$ = 200 + 2

= **202 A Ans**

(b) The armature is lap wound where the number of parallel paths are equal to the number of poles. So here being four pole generator there are four parallel paths.

$$= \frac{200}{4}$$

= **50.5 A Ans.**

(c) The current in armature

= 202 A

∴ armature drop = $I_a R_a$ = 202 × 0.05

= 10.10 V

E.M.F. generated = P.d + drops + contact drop

= 100 + 10.1 + 2

= **112.1 V Ans**

Illustration 3.8

A shunt generator supplies a load of 5.5 kW at 110V through a pair of feeder, total resistance = 0.042 Ω. R_a = 0.1 Ω and R_{sh} = 40 Ω. Find the V_t and E_g.

Solution

The load as given is 5.5 kW at 110 V. So the load current is,

$$= \frac{\text{Power}}{\text{Voltage}}$$

$$\frac{5.5 \times 1000}{1100} = 50 \text{ A}.$$

Now voltage drop in feeders

= (I × R)

= 50 × 0.042

= 2.1 V

Voltage across the armature is the sum of terminal voltage plus the drop in feeder.

= 110 + 2.1 = 112.1 V

Now shunt field current

$$= \frac{112.1}{40} = 2.8025 \text{ A}$$

So the current in armature

= $I_L + I_f$

= 50 + 2.8 = 52.8 A

Armature drop = 52.8 × 0.1

= 5.28 V

So the e.m.f. generated

= Voltage at brush + Armature drop

= 112.1 + 5.21 = 117.38

= 117.4 V

∴ e.m.f = 117.4 V

and terminal voltage

= **112.1 V Ans.**

Note: The reader can understand from this problem that, in the case of a generator, the total supply voltage is the voltage generated in the armature windings.

Thus, all kinds of drops when added to the load terminal voltage, will yield the armature voltage.

Illustration 3.9

The e.m.f generated in the armature of a shunt generator is 625 V, when delivering its full load current of 400 A to the external circuit. The field current is 6 A and the armature resistance is 0.06 Ω. What is the terminal voltage.

Solution

The emf generated is 625 V. The armature drop is the product of current in armature and armature resistance.

Here
$$I_a = I_L + I_f$$
$$= 400 + 6 = 406 \text{ A}$$

The armature drop $= 406 \times 0.6$
$$= 24.36 \text{ V}$$

So terminal voltage = EMF − drop in armature.
$$= 625 − 24.36$$
$$= 620.64 \text{ V}$$

∴ terminal voltage = **620.64 Ans.**

Illustration 3.10

The output of a shunt generator is 500 A, at a terminal voltage of 225 V. Armature resistance is 0.02 ohm. shunt field resistance 50 Ω. What is the emf generated?

Solution

The current in the external circuit is 500 A at 225 V.

Here the current in shunt field is,

$$I_{Sh} = \frac{V_{sh}}{R_{sh}}$$

$$= \frac{225}{25} = 4.5 \text{ A}$$

The current in armature
$$= I_L + I_{sh}$$
$$= 500 + 4.5$$
$$= 504.5 \text{ A}$$

The armature drop $= I_a R_a$
$$= 504.5 \times 0.02$$
$$= 10.090 \text{ V}$$

e.m.f. generated = load P.D + drop in side the source
$$= 225 + 10.09$$
$$= 235.09 \text{ V}$$

Illustration 3.11

Short shunt compound generator supplies a load at 100 V. through a pair of feeders of total resistance 0.05 Ω. The load consists of four motors each taking 40 A and a lighting load of 200, 60 W lamps. Armature resistance 0.02 Ω. series field resistance 0.05 Ω. Shunt field resistance 50 Ω. Find

(a) Load current
(b) Terminal voltage
(c) Emf generated

Solution

Each motor is taking 40 A so total motor current is

= 160 A

Current taken by one lamp of 60 W is,

$$= \frac{60}{100} = 0.6 \text{ A}$$

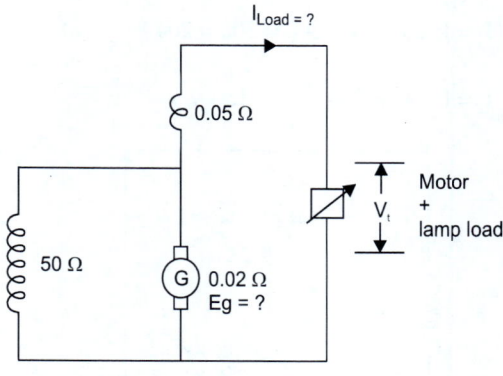

Fig. 3.22. Circuit for illustration 3.11

Total current in 200 lamps is:

= 0.6 × 200

= 120 A

Now total current supplied

= 160 + 120

= 280 A

Now current 280 A is flowing through the feeders having resistance 0.05Ω.
So voltage drop in feeders is:

= 280 × 0.05

= 14 V

The voltage at the terminals is, thus:

= Voltage at load + feeder drops

= 100 + 14

= **114 V**

The current 280 A is flowing through the series field. So the series field drop is

$$= I_{se} R_{se}$$

$$= 280 \times 0.05 = 14 \text{ V}$$

∴ Voltage across shunt field

= Voltage at brushes

= 114 + 14 = 128 V

Shunt field current

$$= \frac{V_{sh}}{R_{sh}} = \frac{128}{50}$$

= 2.48 A

Total armature current

$$= I_L + I_{sh}$$

= 280 × 2.48

= 282.48 V

so the armature drop $= I_a R_a = 282.48 \times 0.02$

= 5.65 V

Now e.m.f. generated = Load voltage + feeder drops + series field drop + armature drop

= 100 + 14 + 14 + 5.65

= 133.65 V

Illustration 3.12

A 150 kW, 250 V shunt generator has a field circuit of 50 Ω and an armature circuit resistance of 0.05 Ω. Calculate:

a. The full load line current flowing to the load
b. The field current
c. The armature current
d. The full load generator voltage

Solutions

(a) $I_L = \dfrac{kW \times 1000}{V_1} = \dfrac{1500 \times 1000}{250} = 600 A \left(= \dfrac{VI}{V} \right)$

(b) $I_f = \dfrac{V}{R_f} = \dfrac{250}{50} = 5 A$

(c) $I_a = I_f + I_L = 5 + 600 = 605\ A$

(d) $E_g = V_a + I_a R_a$

   $= 250\ V + 605 \times 0.05 = 280.25\ V$

Illustration 3.13

A long-shunt compound generator, rated at 100 kW and 500 V DC, has an armature resistance of 0.03 Ω, a shunt field resistance of 125 Ω, and a series field resistance of 0.01 Ω. The diverter carries 54 A. Calculate

a. The diverter resistance at full load
b. The generated voltage at full load.

Solutions

Refer to Fig. 3.23.

a. $I_L = \dfrac{kW \times 1000}{V} = \dfrac{100 \times 1000\ W}{500\ V} = 200\ A$

   $I_f = \dfrac{V_f}{R_f} = \dfrac{500}{125} = 4\ A$

   $I_a = I_f + I_L = 4 + 200 = 200 = 204 A$

   $I_s = I_a - I_d = 204 - 54 = 150 A$

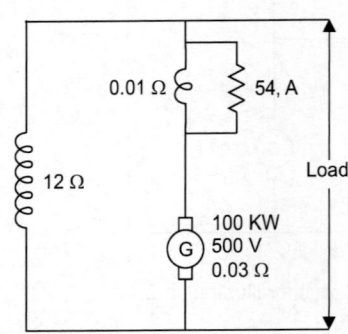

Fig. 3.23. Circuit for illustration 3.13

Since diverter and series fields are in parallel,

$I_d R_d = I_s R_s$ and

$R_d = \dfrac{I_s R_s}{I_d} = \dfrac{150 \times 0.01}{54} = 0.0278\ \Omega$

b. $E_g = V_1 + I_a R_a + I_s R_s$

   $= 500 + (240 \times 0.03) + (150 \times 0.01)$

   $= 507.62\ V$

Illustration 3.14

Calculate the voltage induced in the armature widing of a 4-pole, lap-wound, DC machine having 728 active conductors and running at 1800 rpm. The flux per pole is 30 mWb.

Because the armature is lap wound P = A and

$$E = \frac{P\phi ZN}{60A} = \frac{(30\times10^{-3})(1800)(728)}{60} = 655.2 \text{ V}$$

Illustration 3.15

What voltage would be induced in the armature of the machine of Illustration 3.14 if the armature were wave wound?

For a wave-wound armature, A = 2. Thus,

$$E = \frac{(30\times10^{-3})(1800)(728)}{60}\left(\frac{4}{2}\right) = 1310.4 \text{ V}$$

Illustration 3.16

If the armature in illustration 3.14 is designed to carry a maximum line current of 100 A, what is the maximum electromagetic power developed by the armature?

Solution

Because there are 4 parallel paths (A = P = 4) in the lap-wound armature, each path can carry a maximum current of

$$\frac{I_a}{A} = \frac{100}{4} = 25 \text{ A}$$

The power developed by the armature is
$$P_d = EI_a = (655.2)(100) = 65.5 \text{ kW}$$

Illustration 3.17

Calculate the electromagetic torque developed by the armature described in Illustration 3.14.

Solution

From the energy-conversion equation, $EI_a = T_e\omega_m$, and the result of illustration 3.16,

where $\omega_m = \frac{2\pi N}{60}$

Illustration 3.18

A 4-pole, lap-wound armature has 144 slots with two coil sides per slot, each coil having two turns. If the flux per pole is 20 mWb and the armature rotates as 720 rpm, what is the induced voltage?

Substitute P = A = 4, N = 720, ϕ = 0.020, and Z = 144 × 2 × =576 in the emf equation to obtain

$$E = \frac{(0.020)(720)(576)}{60}\left(\frac{4}{4}\right) = 138.24 \text{ V}$$

Note that, Z = 2T.

Illustration 3.19

A 100 kW, 230 V, shunt generator has R_a = 0.05 Ω and R_f = 57.5 Ω. If the generator operates at rated voltage, calculate the induced voltage at (a) full-load and (b) half full-load. Neglect brush-contact drop.

Solution

I_f = 230/57.5 = 4 A.

(a) $I_L = \dfrac{100 \times 10^3}{230} = 434.8 \text{ A}$

$I_a = I_L + I_f = 434.8 + 4 = 438 \text{ A}$

$I_a R_a = (438.8)(0.05) = 22 \text{ V}$

$E = V + I_a R_a = 230 + 22 = 252 \text{ V}$

(b) $I_L = 217.4$ A, because at half load, current will also be halved.

$I_a = 217.4 \text{ A}$

$I_a R_a = 11 \text{ V}$

$E = 230 + 11 = 241 \text{ V}$

Illustration 3.20

A 50 kW, 250 V, short-shunt, compound generator has the follwoing data: $R_a = 0.06 \; \Omega$, $R_{se} = 0.04 \; \Omega$, and $R_f = 125 \; \Omega$. Calculate the induced armature voltage at rated load and terminal voltage. Take 2 V as the total brush-contact drop.

Solution

Refer to Fig. 3.21a.

$I_L = \dfrac{50 \times 10^3}{250} = 200 \text{ A}$

$I_L R_{se} = (200)(0.04) = 8 \text{ V}$

$V_f = 250 + 8 = 258 \text{ V}$

$I_f = \dfrac{258}{125} = 2.06 \text{ A}$

$I_a = 200 + 2.06 = 202.06 \text{ A}$

$I_a R_a = (202.06)(0.06) = 12.12 \text{ V}$

$E = 250 + 12.12 + 8 + 2 = 272.12 \text{ V}$

Illustration 3.21

A separately-excited DC generator has a constant loss of P_c (W), and operates at a voltage V and armature current I_a. The armature resistance is R_a. At what value of I_a is the generator efficiency a maximum?

$\text{output} = VI_a$

$\text{input} = VI_a + I_a^2 R_a + P_c$

$\text{efficiency } \eta = \dfrac{VI_a}{VI_a + I_a^2 R_a + P_c}$

For η to be a maximum, $d\eta/dI_a = 0$, or

$V(VI_a + I_a^2 R_a + P_c) - VI_a(V + 2I_a R_a) = 0$

or $I_a = \sqrt{P_c / R_a}$

In other words, the efficiency is maximised when the armature loss, $I_a^2 R_a$, equals the constant loss, P_c.

Illustration 3.22

The generator of illustration 3.19 has a total mechanical and core loss of 1.8 kW. Calcualte (a) the generator efficiency at full-load, and (b) the horsepower output from the prime mover to drive the generator at this load.

Solution

From illustration 3.19, $I_f = 4$ A and $I_a = 438$ A.

There are three major losses that occur in a DC machine. They are (i) the copper loss, that occur in copper, that is, windings of a DC machine. It is found out as I^2R. Therefore the two windings, viz., armature winding and field winding cause the respective copper losses as $I_a^2 R_a$ and $I_f^2 R_f$, both in watts. (ii) the core loss that occur in iron parts due to flux changes and (iii) friction loss, a mechanical loss occurring at bearing.

$$I_f^2 R_f = (16)(57.5) = 0.92 \text{ kW}$$

$$I_a^2 R_a = (428.8)^2(0.05) = 9.63 \text{ kW}$$

Total losses $= 0.92 + 9.63 + 1.8 = 12.35$ kW

output $= 100$ kW

input $= 100 + 12.35 = 112.25$ kW

$$\text{efficiency} = \frac{100}{112.35} = 89\%$$

(b) prime mover output $= \dfrac{112.35 \times 10^3 \text{ W}}{746 \text{ W/hp}} = 150.6$ hp

Illustration 3.23

(a) At what load does the generator of illustration 3.19 and 3.22 achieve maximum efficiency?
(b) What is the value of this maximum efficiency?

Solution

(a) From Illustration 3.22, the constant losses are $P_c = 920 + 1800 = 2720$ W. Hence, by Illustration 3.21,

$$I_a = \sqrt{\frac{2720}{0.05}} = 233.24 \text{ A}$$

and $I_L = I_a - I_f = 233.24 - 4 = 229.24$ A

(b) output power $= (229.24)(230) = 52.72$ kW

$I_a^2 R_a = P_c = 2.72$ kW

(by Illustration 3.21)

input power $= 52.72 + 2(2.72) = 58.16$ kW

$$\text{Maximum efficiency} = \frac{52.72}{58.16} = 90.6\%$$

Illustration 3.24

A 20 hp, 250 V, shunt motor has an armature-circuit resistance of 0.22 Ω and a field resistance of 170 Ω. At no-load and rated voltage, the speed is 1200 rpm and the armature current is 3.0 A. At full-load and rated voltage, the line current is 55 A, and the flux is reduced 6% (due to the effects of armature reaction) from its value at no-load. What is the full-load speed?

$$E_{\text{no-load}} = 250 - (3.0)(0.22) = 249.3 \text{ V}$$

$$I_f = \frac{250}{170} = 1.47 \text{ A}$$

$$E_{\text{full-load}} = 250 - (55 - 1.47)(0.22) = 238.2 \text{ V}$$

Since speed is proportional to E/ϕ, we have

$$N = 1200 \left(\frac{238.2}{249.3}\right)\left(\frac{1}{0.94}\right) = 1220 \text{ rpm}$$

Illustration 3.25

A 10 hp, 230 V, shunt motor takes a full-load line current of 40 A. The armature and field resistances are 0.25 Ω, and 230 Ω, respectively. The total brush-contact drop is 2 V and the core and friction losses are 380 W. Calculate the efficiency of the motor. Assume that stray-load loss, is 1% of output.

input = (4)(230) = 9200 W

field-resistance loss = $\left(\dfrac{230}{230}\right)^2 (230)$ = 230 W

armature-resistance loss = $(40-1)^2(0.25)$ = 380 W

core loss and friction loss = 380 W

brush-contact loss = (2)(39) = 78 W

stray-load loss = $\dfrac{10}{100} \times 746$ = 75 W

total losses = 1143 W

power output = 9200 – 1143 = 8057 W

efficiency = $\dfrac{8057}{9200} = 87.6\%$

Illustration 3.26

A 10 kW, 250 V, shunt generator, having an armature resistance of 0.1 Ω and a field resistance of 250 Ω, delivers full-load at rated voltage and 800 rpm. The machine is now run as a motor while taking 10 kW at 250 V. What are induced volts in each case?

Solution

As a generator:

$I_f = \dfrac{250}{250} = 1\,A$ $I_L = \dfrac{10 \times 10^3}{250} = 40\,A$

$I_a = 40 + 1 = 41\,A$ $I_a R_a = (41)(0.1) = 4.1\,V$

$E_z = 250 + 4.1 = 254.1\,V$

As a motor:

$I_L = \dfrac{10 \times 10^3}{250} = 40\,A$ $I_f = \dfrac{250}{250} = 1\,A$

$I_a = 40 - 1 = 39\,A$ $I_a R_a = (39)(0.1) = 3.9\,V$

$E_m = 250 - 3.9$

$= 246.1$, volts.

3.7. CHARACTERIZATION OF DC GENERATOR

3.7-A. NO-LOAD VOLTAGE CHARACTERISTICS OF A DC GENERATOR

The circuit of Fig 3.24 is commonly used in machinery laboratories to investigate both no-load and load characteristics of shunt generators. With switch S open, the generator is driven by a prime mover at an approximately constant rate of speed.

For a given generator, the number of poles P, the total number of armature conductors Z, and the number of paths A may be determined from the armature winding data. Thus, for a given armature, P, Z and A in the EMF equation are fixed. Equation (3.2) may be written as.

$$E_g = KN\phi \qquad (3.5)$$

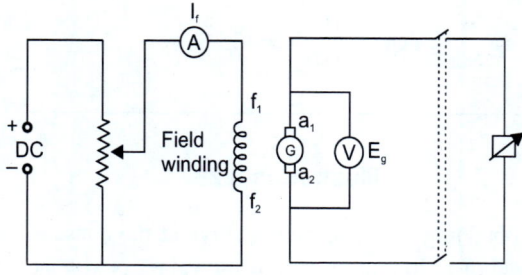

Fig. 3.24. Circuit connections to load a DC generator.

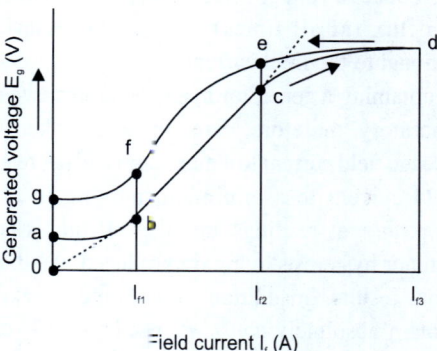

Fig. 3.25.

where $K = (PZ/60A)$

Since the prime mover is being driven at an approximately constant speed, the generated EMF E_g is, $Eg_1 = K\phi$ itself.

It would appear, on the basis of above equation that the voltmeter reading in Fig. 3.24 is solely and purely a function of the mutual air-gap flux produced by the field winding. If the potentiometer is adjusted for zero field current, we assume that E_g would be zero. Such is not the case, however, and even when the field MMF ($I_f N_f$) is zero the air-gap flux is not zero. A small voltage is recorded across the armature by the voltmeter when the field current is zero. This voltage is indicated as point 'a' on the curve of Fig. 3.25 where the field current is zero and the generated voltage E_g is some small value of a few volts. The voltage of, '**o-a**' is due to the retentivity of the field poles and is proportional to the amount of residual magnetism that was left in the generator iron when the generator was last turned off.

If the field current is increased by means of the potentiometer so that a field current is recorded of I_{f1}, the voltage will rise to point *b* in Fig. 3.25. If the current is increased in the same direction so that the ammeter records a field current of I_{f2}, the generated voltage will rise to point **c** in Fig. 3.25. Thus, the generated induced voltage rises in proportion to the air-gap MMF produced by the field current ($I_f N_f$).

It should be noted that the portion of the curve **a** to **b** is nonlinear since it is composed of a *fixed* residual MMF and a *variable* field-current MMF. The portion **b** to **c** is *linear*, however, since the residual MMF is now negligible in comparison to the MMF produced by the field current, and the generated voltage varies directly with the variation in field current. Beyond point **c** (the knee of the curve), an increase in field current does *not* produce a proportional increase in generated voltage. Here, the iron of the field poles and the surrounding core of the magnetic circuit approaches *saturation*. Beyond point c, therefore, any increase in MMF above the knee of the saturation curve will fail to produce a proportionate increase in flux, and the magnetization curve from c to d is again nonlinear, this time because of the effect of *magnetic saturation*.

It can be seen, then, that the shape of the magnetization curve (E_g versus I_f) is no different from the shape of the saturation curve (B versus H) obtained for any ferromagnetic material. As a matter of fact, even if the machine were not rotating, and if measuremetns were made of the air-gap flux versus the magnetizing force, the B–H curve would be identical to that shown in Fig. 3.25. Since $E_g = K\phi N$, rotation of the armature conductors at a constant speed produces a voltage directly proportional to the air-gap flux (at all times) and *not* necessarily proportional to the field current!

In obtaining a generator magnetization curve in the laboratory, therefore, care should be taken to increase the field current to a maximum and decrease the field current to a minimum, moving in one direction *only*, as readings are taken. If this is not done, minor hysteresis loops are produced, yielding erroneous results. In addition, care should be taken to maintain absolutely *constant* speed, since $E_g \alpha \phi$ assumes that the speed is, in fact, constant. The magnetization curve of Fig. 3.25 is a graphic representation of this equation. If the speed is recorded at the same instant that the field current and voltage readings are taken, then it is a simple matter to *correct for any variations* of speed that may occur, using the *ratio* method.

Illustration 3.27

Assuming constant field excitation calculate the no-load voltage of separately excited generator whose armature voltage is 150 V at a speed of 1800 rpm when

a. The speed is increased to 2000 rpm
b. The speed is reduced to 1600 rpm

Solution

From Eq. (3-7), $E_g = K'' S$ at constant field excitation, and therefore $E_{final}/E_{orig} = S_{final}/S_{orig}$ from which

(a) $E_{final} = E_{orig} \left(\dfrac{S_{final}}{S_{orig}} \right)$

$= 150 \text{ V} \left(\dfrac{200}{1205} \right) = 166.\overline{6} \text{ V}$

(b) $E_{final} = 150 \text{ V} \left(\dfrac{1600}{1800} \right) = 133.\overline{3} \text{ V}$

Illustration 3.28

In obtaining a magnetization curve at the constant speed of 1200 rpm, the following values of voltage were recorded when the speed was simultaneously noted as varying from 1200 rpm:

a. 64.3 V at 1205 rpm
b. 82.9 V 1194 rpm
c. 162.3 V at 1194 rpm

What corrections must be made in the data before the curve is plotted?

Solution

$E_1 = 64.3 \text{ V} \left(\dfrac{1200}{1205} \right) = 64.0 \text{ V at 1200 rpm}$

$E_2 = 82.9 \text{ V} \left(\dfrac{1200}{1194} \right) = 83.3 \text{ V at 1200 rpm}$

$E_3 \ 162.3 \text{ V} \left(\dfrac{1200}{1202} \right)$

$= 162.0 \text{ V at 1200 rpm}$

3.7-B BUILDUP OF A SELF-EXCITED SHUNT GENERATOR

The magnetization curve for the separately excited generator of Fig. 3.25 and a particular shunt field resistance line R_f are shown in Fig. 3.26 for the identical generator connected as a self-excited shunt generator. Since the field circuit is connected directly across the armature, the ordinate of the field resistance line R_f is the terminal voltage of the generator V_a. The manner in which a self-excited generator manages to excite its own field and build a DC voltage across its armature is described with reference to Fig. 3.26 in the following steps:

1. Assume that the generator starts from rest, i.e., prime mover speed is zero. Despite a residual magnetism, the generated EMF E_g is zero.
2. As the prime mover rotates the generator armature and the speed approaches rated speed, the voltage due to residual magnetism and speed ($E = K\phi N$) increases.
3. At rated speed, the voltage across the armature due to residual magnetism is small (E_1 = as shown in the figure). But this voltage is also across the field circuit whose resistance is R_f. Thus the current that flows in the field circuit (I_1) is also small.
4. When I_1 flows in the field circuit of the generator, an increase in MMF results (due to $I_f N_f$) which acids the residual magnetism in increasing the induced voltage to E_2 as shown in Fig. 3.26.
5. Voltage E_2 is now impressed across the field, causing a larger current I_2 to flow in the field circuit. I_2 to flow in the field circuit. $I_2 N_f$ is an increased MMF, which produces generated voltage E_3.
6. E_3 yields I_3 in the field circuit, producing E_4 But E_4 cause I_4 to flow in the field, producing E_5; and so on up to E_4. But E_4 cause I_4 flow in the field producing E_5; and so on, up to E_8 the maximum value.
7. The process continues until that point where the field resistance line crosses the magnetization curve in Fig. 3.26. Here the process a current flow that in turn produces an induced voltage of the same magnitude (E_8 as shown in the figure).

3.7-C CRITICAL FIELD RESISTANCE

The preceding description for the buildup of a self-excited shunt generator, shown in Fig. 3.26 used particular value of field resistance R_f. If the field resistance were reduced by means of adjusting the field rheostat to a lower value, say R_{f1} shown in Fig. 3.26 the buildup process would take place along field resistance line R_{f1} and build up a somewhat higher value than E_8, i.e., the point where R_{f1} interacts the magnetization curve, or E_9.

Since the curve is extremely saturated in the vicinity of E_9 reducing the field resistance (to its limiting field winding resistance) will not increase the voltage appreciably. Conversely, increasing the field rheostat resistance and the field circuit resistance (to a value having a higher slope than R_f

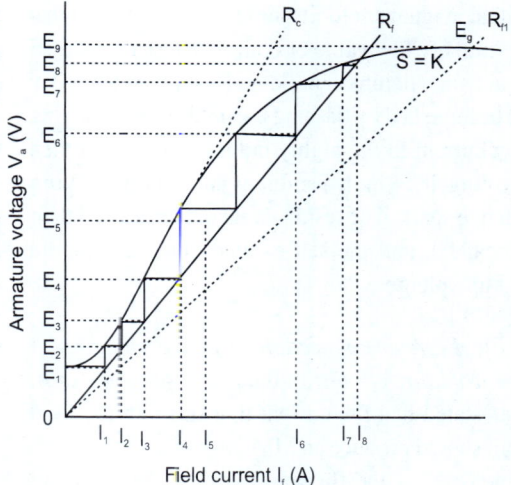

Fig. 3.26.

in the figure) will cause a reduction of the maximum value to which buildup can possibly occur.

The resistance of the field reostat may be increased until the field circuit reaches a critical field resistance. Field circuit resistances above the critical field resistance will fail to produce buildup. The critical field circuit resistance R_c is shown as a tangent to the saturation curve passing through the origin (0) of the axes of the curve of Fig. 3.26. Thus a field circuit resistance higher than R_c will produce an armature voltage of E_1 –approximately and no more.

3.7-D REASONS FOR FAILURE (OF SELF-EXCITED SHUNT GENERATOR) TO BUILD UP VOLTAGE

There are four specific (electrical) reasons for the failure of a self-excited unload shunt generator to build up voltage. They are as follows:

1. *Lack of (or low) residual magnetism:* Since the buildup process as described above requires some residual magnetism for its initiation, it is evident that an extremely low value or complete loss of residual magnetism will inhibit buildup. Residual magnetism may be remedied by flashing the field i.e., by applying direct current to the highly inductive field and then removing it, which produces an accompanying inductive spark. The residual magnetism should then be regained, and the self-excited generator should build up voltage.

2. *Field circuit connection reversed with respect to the armature circuit:* In the above section step 4, it was stated that the current that flows in the field circuit should produce an MMF that aids the residual magnetism i.e., the flux produced by the field coil must be of the same magnetic polarity as the residual MMF. If the field connections are reversed with respect to the armature, the resulting field flux will tend to buck or decrease the residual flux, thus decreasing the net flux and the generated EMF E_g when the field circuit is closed.

A simple test for this condition is to open the field circuit of a running generator and observe the voltmeter across the armature. If the voltage across the armature rises when the field circuit is opened, then the field circuit connections are reversed with respect to the armature. This failure may be remedied merely by reversing $f_1 - f_2$ with respect to $a_1 - a_2$ as in Fig. 3.24.

It should be noted that reversing the prime mover direction of rotation will accomplish the same purpose, where possible, since it reverses the polarity of the armature. A possible and really identical cause of failure is the use of the wrong direction of rotation, which produces reversed armature connections with respect to the field.

3. *Field circuit resistance higher than critical field resistance:* An open connection in the field circuit windings, rheostat, or connections will result in a resistance higher than critical field resistance. This will prevent buildup. This trouble may be checked by means of an ohmmeter.

4. *Open connection or high resistance in the armature circuit:* An open connection in the armature circuit, a dirty commutator, a loose brush, or the lack of brushes will tend to act in the same manner as a high resistance in the field circuit because they reduce field current and tend to prevent a voltage higher than the voltage due to residual magnetism. The resistance of connections $a_1 - a_2$ to the armature should also be checked by means of an ohmmeter (for the desirable comparatively low resistance). A high armature circuit resistance indicates an open connection in the armature circuit.

Reflection Section

1. Give the need of lap winding and wave winding in the armature winding of a generator.
2. Describe the influence of critical field resistance on the induced EMF.
3. When does critical armature resistance occur?
4. 'When a DC motor operates, it generates EMF also.' Discuss on this statement.
5. State and explain various torques in a DC motor.

3.8 LOAD VERSUS VOLTAGE CHARACTERISTIC OF A SHUNT GENERATOR

It was stated in an earlier section that the effect of application of load across the armature terminals is to reduce the generated and armature voltage. When a generator is loaded, its armature current will increase to handle with increase in load. This means that the amature circuit drop, $I_a R_a$, will also increase. As this influences E_g, an increase in it eventually reduces E_g. There are three reasons for this drop in voltage:

1. An internal armature voltage drop produced by the armature circuit resistance R_a.
2. The effect of armature reaction on the air gap flux.
3. The reduction in field current caused by the two preceding factors.

Let us consider each of these factors in turn

1. For a separately exited generator, shown in Fig. 3.17 a which is not supplying load current, I_a is zero and V_a equals E_g. For a loaded self-excited shunt generator, as the load current I_1 increases, the armature current I_a also increases as does the armature circuit voltage drop, or $I_a R_a$. Thus, the terminal voltage across the shunt generator armature V_a decreases with application of load, because $V_t = E_g - I_a R_a$. This is shown in Fig. 3.27.
2. *Armature reaction.* The individual current-carrying conductors of the armature furnish a load current in the same direction as the induced voltage. Since these conductors are embedded in an iron armature, an armature MMF is produced in proportion to the load current. The effect of this armature flux is to distort and reduce the air-gap flux produced by the field. The reduction in mutual field flux ϕ_m reduces the generated and terminal armature voltages, E_g and V_{a1} respectively. Thus, as the armature current I_a increases, the effect of armature reaction is progressive reduction of ϕ and V_{a1}.
3. *Reduction in field current.* The terminal voltage V_a drops as a function of the load current as a result of (1) armature reaction and (2) the internal armature circuit voltage drop. This drop in V_a results in a decreased field current and excitation ($I_f N_f$) results in decreased air-gap flux and reduced E_g and V_a. If the generator field current and speed are such that the field poles are unsaturated, the machine will rapidly unbuild (see Fig. 3.27) this cause of voltage drop does not

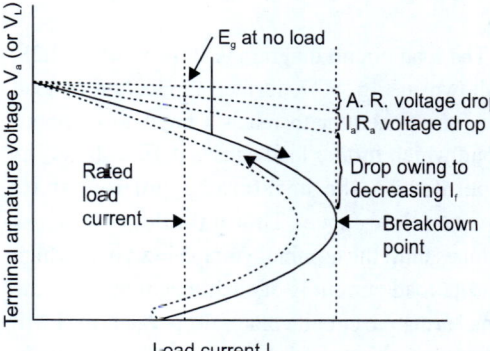

Fig. 3.27.

occur in a separately exited DC generator; and, for this reason, the same generator operated with separate excitation always has improved regulation.

The effect of each of the preceding three factors is shown in Fig 3.27, which shows the (external) load-voltage characteristic of a shunt generator. For the circuit of Fig. 3.28, the readings of the voltage across the armature V_a is the same as E_g at no load (neglecting the $I_a R_a$ and armature reaction drop produced by the field current). The effects of armature reaction (in reducing the mutual air-gap flux), armature circuit voltage drop, and decrease in field current are all shown with progressive increases in a load.

Note that both the armature reaction and the $I_a R_a$ drop are shown as dashed straight lines, representing theoretically linear voltage decreases directly proportional to the increase in load current. The drop owing to decreased field current is a curved line, since it depends on the degree of saturation existing in the field at that value of load.

In general, the external load-voltage characteristic decreases with application of load only to a small extent up to its rated load (current) value. Thus, the shunt generator is considered having a fairly constant output voltage with application of load and, in practice, is rarely operated beyond the rated load current value continuously for any appreciable time.

The load circuit diagram is shown in Fig. 3.28. As shown in Fig. 3.27, further application of load causes the generator to reach a breakdown point, beyond which further load causes it to "unbuild" as it operates on the unsaturated portion of its magnetization curve. This unbuilding process continues until the terminal voltage is zero, at which point the load current is of such magnitude that the internal armature circuit voltage drop equals the EMF generated on the unsaturated or linear portion of its magnetization curve, as illustrated below.

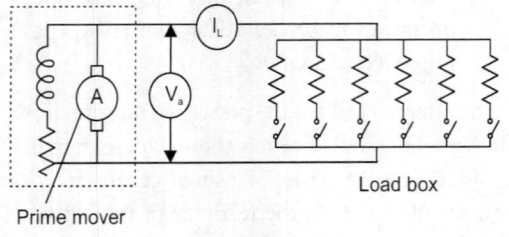

Fig. 3.28.

Illustration 3.29

A 125 V shunt generator having an armature resistance of 0.15 Ω is loaded progressively until the terminal voltage across the load is (practically) zero. If the load current is 96 A and the field current is 4 A, what is the voltage generated in the armature?

Solution

$$I_a = I_f + I_1 = 4 + 96 = 100 \text{ A}$$

$$E_g = V_a + I_a R_a = 0 + (100 \times 0.15) = 15 \text{ V}$$

This example serves to indicate the possible extent of unbuilding of the generator. It also serves to indicate that the generated EMF at no load is not the same as the generated EMF at any given load condition, since the generator is no longer operating on the same portion of the magnetization curve, due primarily to the unbuilding of the generator.

3.9 DC MOTOR

An electric motor is an electric machine which converts electrical energy into mechanical energy.

It is seen in Sec. (2.14) that when a current-carrying conductor is kept in a magnetic field experiences a mechanical force whose magnitude is given by the product BIl sin θ and the direction is given by Fleming's left hand (FLH) rule. It is the basic principle behind the working of a DC motor.

As far as the construction is concerned there is no difference between the DC generator and a motor; motor too has an armature, a field region, commutator, brushes etc. In fact, as mentioned in the beginning of Sec. 3.2, a DC machine can be used as a generator or motor.

Concentrate on the Fig. 3.29, to discuss the working principle of a DC motor.

It is a multipolar DC motor, a part of which is shown, the action occurring here will also be the same in other pole-pairs.

The armature windings and the field winding are given the rated DC supply, and hence there will be a current flow in both the windings.

It is assumed that the armature conductors carry current out when they are under the north pole (marked with a '·' to indicate this); the armature conductors carries current 'in' when they are under the south pole (marked with a '+' to indicate this); the north poles give out flux lines as shown in Fig. 3.29, and the south poles receive the flux lines.

It is evident that the current carrying (armature) conductors are kept in the magnetic field produced by the field magnets. Thus the conductors will experience a mechanical force. Note that the armature is still not rotating.

Let the FLH be applied now (reader may try this). The direction of the force on each conductor is found to be clockwise. This direction is marked above each conductor.

Constructionally, each conductor of the armature has a definite perpendicular distance from the centre of the armature. It is shown in Fig. 3.30.

The force at a definite perpendicular distance from a common reference point, is nothing but, the twisting moment, the torque. Thus, the force in all the armature conductors will collectively produce a torque, which will finally rotate the armature in the same clockwise-direction.

The function of the commutator in the motor is the same as in a generator; here, by reversing the current in each conductor as it pass from one pole to another, it helps to develop a **continuous** and **unidirectional torque**.

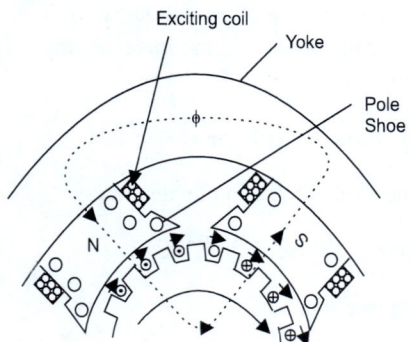

Fig. 3.29. Illustration to discuss the working principle of DC motor.

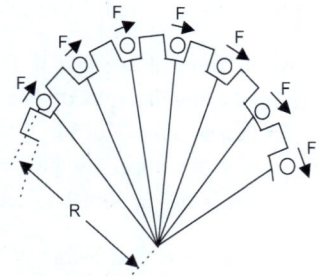

Fig. 3.30. To Illustrate the production of torque.

3.10 THE BACK EMF

It is brought out in chapter 2 under Sec. (2.3) that, a current carrying conductor always surrounded by the magnetic flux produced by it; and, by Sec (2.16), whenever there is a relative motion between the magnetic field and the electric coil, an EMF is induced in the conductors of the coil.

Both these magnetic phenomena join together and make the motor generate a voltage in its armature. This generated voltage is the back emf or the counter emf of the motor.

Let this be examined.

It was shown in the previous section that the motor action result when the armature is placed in a magnetic field and the armature winding is supplied with a DC supply; a direct current hence flows in the armature conductors.

Under this condition, magnetic field will be set up by the current carrying armature conductors around them. Note that, due to the excitation of the field coils, the main field is already set up. Refer to Fig. 3.31.

This interaction, i.e., **The effect of the armature field over the main field produced by the poles** is known to be the **Armature Reaction**.

Due to the mechanical force exerted in the armature conductors, the armature rotates. In this course of operation of rotation armature conductors also continually cut through the stationary resultant magnetic field, and because of such flux-cutting action, the 'relative motion' condition is exhibited between the rotating armature conductors and the stationary resultant field; hence voltage is generated in the very **same** conductors that experience force action.

This can only mean that when a motor is operating, it is simultaneously acting as a generator.

Obviously the motor action will be stronger than the generator action; hence the generator action of the motor will be comparatively small and, for the motoring discussion purpose, not considered; however, the generated voltage, does oppose the applied voltage, thus reducing it.

Because the generated voltage of the motor is in opposition to the applied voltage it is reported as the **back emf** or the **counter emf**; symbol E_b. Diagrammatically it is shown in Fig. 3.32.

E_b, the back EMF, is a voltage, opposing which, the motor should **always** produce torque.

The net voltage in the armature is:

$$V - E_b = I_a R_a \qquad (3.6)$$

3.11 TORQUE DEVELOPED IN A DC MOTOR

Consider a pulley of radius R metres acted upon by a circumferential force of F newton which causes it to rotate at N. rps (Fig. 3.33).

By the term torque is meant the turning or twisting moment of a force about an axis. It is measured by the product of the force and the radius at which this force acts.

Then, torque T = FR, newton-metre

Work done by this force in one revolution

= Force × distance = F × 2πR, J

Work done per second

$$W = F \times 2\pi R \times N \text{ J/second}$$
$$= (F \times R) \times 2\pi N \text{ J/second}$$

Now 2π N = angular velocity ω radians / second

It is also obvious from above that if T is the torque in N-m and ω the angular velocity in radian / second, then power developed = T ω W.

Let T_a be the torque developed by the armature of the motor running at N rps, in Nm. Then,

Power developed = T × ω = T_a × 2πN W

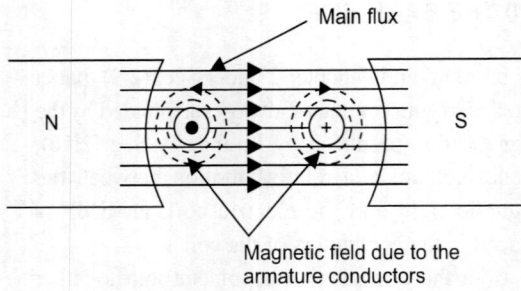

Fig. 3.31. (a) The main field and (b) Magnetic field due to the armature conductors.

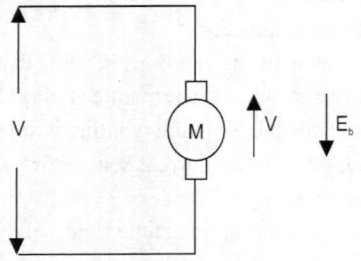

Fig. 3.32. The back EMF.

We also know that electrical power converted into mechanical power in the armature is $E_b I_a$ watt.

Equating the above two powers, $T_a \, 2\pi N = E_b I_a$

Since $E_b = \dfrac{P\phi ZN}{A}$, V

$T_a \, 2\pi N = E_b I_a = \dfrac{P\phi ZN}{60A} I_a$

or $T_a =$

$\dfrac{1}{2\pi} \Phi Z I_a (P/A), N-m = 0.159 \Phi Z I_a (P/A), N-m$

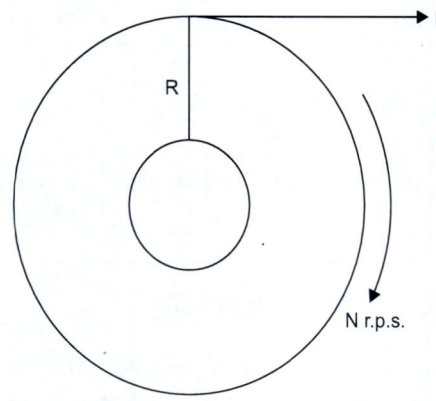

Fig. 3.33.

$= (0.159/9.81) \, \phi Z I_a \times (P/A)$ kg-m

$= 0.0162 \, \phi \, Z I_a \, (P/A)$ kg-m, \hfill (3.7)

From the above equations

$T = \dfrac{1}{2\pi} \dfrac{P\phi Z I_a}{A}$, kg-M

The P, Z, and A being constants, we obtain a very useful proportionality equation for torque as,

$T_a \phi I_a$ \hfill (3.8)

It is worth examining this expression.

There are majorly two torques available in a DC motor. They are the torque developed T_{dev} and the demanded or the load toque, T_D. T_D has to be supplied by T_{dev}, and it is evident from the equation (3.7) that this can happen only if I_a is tuned. Thus, inherently if the demanded torque at the load increases, the armature current increases to develop an increased armature torque, T_{dev}. In case of shunt motor, ϕ is totally independent of I_a and this is a constant. This is because, I_f is decided by V_f/R_f, both of which are almost constants. So, for shunt motors ϕ is practically constant, hence

$T_a \propto I_a$ \hfill (3.9)

In the case of series motor, ϕ is directly proportional to I_a (before saturation) because field windings carry full armature current.

$T_a \propto I_a^2$ \hfill (3.10)

But after saturation of the field magnets, any increase in field current will not produce any additional flux. Thus when the load increases, after saturation, the increase in I_a (hence I_f) will not cause an increase in flux. Hereafter, so, flux will be independent of I_a and hence

$T_a \propto I_a$ \hfill (3.11)

Shaft Torque (T_{sh})

The whole of the armature torque, as calculated above, is not available for doing useful work at the shaft because a certain percentage of this is required for supplying iron and friction losses in the motor.

The torque which is available for doing useful work is known as shaft torque T_{sh}. It is so called because it is available at the shaft, immediately for the load.

218 Basic Electrical Engineering

Output = $T_{sh} \times 2\pi N$, W

where T_{sh} is in N–m and N in rps

$$\therefore T_{sh} = \frac{\text{output in watts}}{2\pi N} \text{ N–m, (N in rps)}$$

$$= \frac{\text{output in watts}}{2\pi N/60} \text{ N–m, here N in r.p.m.}$$

$$T_{sh} = \frac{60}{2\pi} \frac{\text{output}}{N}$$

$$= 9.55 \frac{\text{out-put}}{N} \text{ N–m} \quad (3.12)$$

The difference $T_a - T_{sh}$ is known as lost torque

3.12 TYPES OF DC MOTOR

As was brought out earlier, there are three types of DC motors viz., DC shunt motor, DC series motor and DC compound motor. The armature winding and the field winding are supplied by a DC source; with this knowledge, as shown in Fig. 3.35, all the types of DC motors can be understood.

3.13 WORKED ILLUSTRATIONS XI

Illustration 3.30

A DC motor takes 40 A at 220 V and runs at 800 r.p.m. If the armature and field resistance are 0.2 Ω and 0.1 Ω respectively, find the torque developed in the armature.

Solution

$T_a = 9.55 \, E_b I_a / N$, newton-metre

Now,

$E_b = V - I_a(R_a + R_{se}) = 220 - 40(0.2 + 0.1) = 208$ V

$\therefore T_a = 9.55 \times 208 \times 40/800 = 99.3$ N-m

Illustration 3.31

Determine the value of torque in N-m units established by the armature of a 4 pole motor having 774 conductors, two paths in parallel, 24 mWb per pole when the total armature current is 50 A.

Solutions

Here

$Z = 774; I_a = 50A$

$\Phi = 24 \times 10^{-3}$ Wb, $P = 4$, $A = 2$

$T_a = 0.159 \times 24 \times 10^{-3} \times 774 \times 50 \times 4/2$ N-m
    $= 295.3$ N-m

Illustration 3.32

A DC motor has 6 poles, flux per pole of 0.05 Wb with lap wound armature of 600 conductors. Motor speed is 500 rpm. Determine the applied voltage and back emf given that, armature resistance is 0.25 Ω, armature current 40 A. Also, determine the torque developed by the motor in N-m.

Solution

Here $A = 6$

$$E_b = \frac{\phi Z N}{60}\left(\frac{P}{A}\right) = \frac{0.05 \times 600 \times 500}{60} = 250 \text{ V}$$

(since E_b the back voltage in the armature, which is, in analogy, is E_g)

Armature drop = $I_a R_a$ = 40 × 0.25 = 10V

∴ applied voltage, $V = E_b + I_a R_a$ = 250 + 10 = 260V

$T_a = 0.159 \, E_b \, I_a / N$, where N in rps

∴ T_a = 0.159 × 250 × 40/(500 / 60) = 1908 N-m

Illustration 3.33

A six pole 250 volts series motor is wave connected. There are 240 slots and each slot has four conductors. The flux per pole is 1.75×10^{-3} when the motor is taking 80 A. The field resistance is 0.05 Ω, R_a = 0.1 ohm and the iron and frictional loss is 0.1 kW. Calculate (a) speed (b) total input.

Solutions

$E_b = V - I_a (R_a + R_{se})$ = 250 – 80 (0.05 + 01) = 238V

Now, $E_b = \dfrac{\phi Z N}{60}\left(\dfrac{P}{A}\right)$

or, $238 = \dfrac{17.5 \times 10^{-3} \times (240 \times 4) \times N}{60}\left(\dfrac{6}{2}\right)$ (Q A = 2)

(a) ∴ N = 283.3 rpm

(b) Total input = 250 × 80 = 20,000 W = 20 kW

3.14 SPEED OF A DC MOTOR

From the voltage equation of a motor we have

$E_b = V - I_a R_a$

or $\phi ZN(P/A) = V - I_a R_a$

∴ $N = \dfrac{V - I_a R_a}{\phi} \times \left(\dfrac{A}{ZP}\right)$, rps

As stated earlier, P, Z, and A are constants. Now V- $I_a R_a = E_b$

∴ $N = K \cdot \dfrac{V \cdot I_a R_a}{\phi} = K \dfrac{E_b}{\phi}$

Or $N \alpha \dfrac{E_b}{\phi}$ \hfill (3.13)

It shows that speed is directly proportional to back e.m.f. E_b and inversely to the flux ϕ.

For Series Motor

Let N_1 = Speed in the 1st case, that is at some load

I_{a1} = armature current in the 1st case

ϕ_1 = flux / pole in the 1st case

N_2, I_{a2}, ϕ_2 = corresponding values in the second case, at some other load.

Then, using the above relation, we get,

$N_1 \alpha \dfrac{E_{b1}}{\phi_1}$

$N_2 \alpha \dfrac{E_{b2}}{\phi_2}$

∴ $\dfrac{N_2}{N_1} = \dfrac{E_{b2}}{E_{b1}} \times \dfrac{\phi_1}{\phi_2}$

In a series motor, before saturation ($I_a = I_{se}$). So, prior to saturation of magnetic poles, $\phi \alpha I_a$

∴ $\dfrac{N_2}{N_1} = \dfrac{E_{b2}}{E_{b1}} \times \dfrac{I_{a1}}{I_{a2}}$

For shunt motor

In this case, the same equation applies

$$\therefore \frac{N_2}{N_1} = \frac{E_{b2}}{E_{b1}} \times \frac{\phi_1}{\phi_2}$$

As flux is constant, $\phi_1 = \phi_2$

So, $\dfrac{N_2}{N_1} = \dfrac{E_{b2}}{E_{b1}}$

Speed Regulations

The term speed regulation can now be defined which refers to the change in the speed of a motor with change in applied load torque, other conditions remaining constant. By change in speed here is meant the change which occurs under these conditions due to inherent properties of the motor itself and not those changes which are affected through manipulation of rheostats or other speed controller devices.

The speed regulations is defined as the change in speed when the laod on the motor is reduced from rated value to zero, expressed as per cent of the rated load speed.

$\therefore$ % speed regulations

$$-\frac{\text{N.L.Speed} - \text{F.L.Speed}}{\text{F.L.Speel}} \times 100$$

$$\frac{N_0 - N_N}{N} \times 100 = \frac{dN}{N} \times 100 \qquad (3.14)$$

3.15 RELATIONS BETWEEN TORQUE AND SPEED OF A MOTOR

From the derivations and discussion of the basic speed equation $N = (V - I_a R_a) / K\phi$, the reader may have noticed what may appear to be an obvious inconsistency between this equation and $T = K\phi I_a$. Since torque is defined as a force tending to produce rotation, increasing the field flux would tend to increase the torque and (possibly) the speed. On the other hand, increasing the field flux would reduce the speed. Is there an inconsistency and is it possible to reconcile the two equations?

Actually, there is no inconsistency; and with the help of equation, $I_a = (V_a - E_b) / R_a$, it is possible to give both a qualitative and a quantitative explanation of what happens when the field flux is reduced. Qualitatively, the steps are:

1. The field flux of a shunt motor is reduced by decreasing the field current.
2. The counter emf, $E_b = K\phi N$, drops instantly (the speed remains constant as a result of the inertia of the large and heavy armature).
3. The decrease in E_b causes an increases in the armature current, I_a; refer to equation for I_a cited above.
4. small reduction in field flux produces a large increase in armature current.
5. In $T = K\phi I_a$ the small decrease in flux is more than counterbalanced by a large increase in armature current. Note that the torque has increased more than the flux was reduced.
6. This increase in torque produces an increase in speed.

In summary, decreasing field current (and field flux) results in an increase in speed.

Illustration 3.34

A 100 V shunt motor is taking a current of 220 A. Armature resistance 0.015 Ω. Shunt field resistance 20 Ω, Calculate the back e.m.f and the power spent in turning the armature.

Solutions

The current taken by the shunt field is,

$$\frac{V_{Sh}}{R_{Sh}} = \frac{100}{20}$$

$$= 5 \text{ A}$$

Now current in armature

$$I_a = I_L - I_{sh}$$

$$= 220 - 5 = 215 \text{A}.$$

Now the back e.m.f

$$E_b = V - I_a R_a$$

$$= 100 - 215 \times 0.015$$

$$= 100 - 3.225 = 96.8 \text{ V}$$

The power spent in turning the armature is,

$$= E_b \cdot I_a$$

$$= 96.775 \times 215$$

$$= 20807 \text{ Watts}$$

Illustration 3.35

A 220 V series motor is taking a current of 40 A. Resistance of armature is 0.5 Ω, resistance of series field is 0.25 Ω. Calculate,

(a) Voltage at the brushes
(b) Back emf
(c) Power wasted in armature
(d) Power wasted in series field

Solution

In a series motor, current in the armature and in the series field are same under normal conditions.

a. Now voltage drop in series field

$$= I_{se} R_{se}$$

$$= 40 \times 0.25$$

$$= 10.00 \text{ V}$$

Voltage at brushes

$$= \text{Voltage supply} - \text{drop in series field}$$

$$= 220 - 10 = 210 \text{ V}$$

b. Back e.m.f = Voltage at brushes − armature drop

$$E_b = 210 - 40 \times 0.5$$

$$E_b = 190 \text{ V}$$

c. The power wasted in armature

$$= I_a^2 R_a$$

$$= 40^2 \times 0.5$$

$$= 800 \text{ W}$$

d. Power in series field

$$= I_{se}^2 R_{se}$$

$$= 40^2 \times 0.25$$

$$= 400 \text{ W Ans}$$

Illustration 3.36

A motor is running at 950 rev / min and the torque exerted at the pulley is 1150 N-m. What h.p is being transmitted.

Solution

Let T be the shaft torque and N the speed in r.p.m

$$\text{H.P} = \frac{2\pi NT}{60 \times 746} \left(Q \; \frac{\text{Watts}}{746} = \text{HP} \right)$$

$$= \frac{2\pi \times 950 \times 1150}{60 \times 746}$$

$$= 153.5 \text{ h.p}$$

Illustration 3.37

If the power transmitted by the shaft of a motor is 50 h.p., and the speed being 480 rev/min, what is the torque?

Solution

H.P of motor 50

Speed = 450 rpm

Hence, H.P. developed $= \dfrac{2\pi NT}{60 \times 764}$

or $T = \dfrac{HP. \times 60 \times 746}{2\pi N} = \dfrac{50 \times 60 \times 746}{2 \times 3.14 \times 480}$

$= 742$ Nm

Illustration 3.38

The power transmitted by shaft of a motor is 20 h.p. If the torque is 222 N-m. What is the speed?

Solution

Here h.p $= \dfrac{2\pi N.T.}{60 \times 746}$

$N = \dfrac{60 \times 746 \times h.p}{2\pi T}$

$$= \frac{60 \times 746 \times 20}{2 \times \pi \times 222}$$

$$= 642 \text{ rpm}$$

Illustration 3.39

A series motor takes 45 A at 210 V and runs at 750 rev/min. The horse power output is 10, the series field resistance 0.2 Ω armature resistance, 0.30 Ω. Find (a) Iron and friction losses (b) the lost torque.

Solution

The input = 210 × 45 = 9450 W

The output = 10.h.p = 10 × 746 = 7460 W

Total loss = Input − output

$= 1990$ W

(a) The Cu loss $= I_a^2 (R_{se} + R_a)$

$= 45^2 \times (0.2 + 0.3)$

$= 1012.5$ W

Iron and friction loss + Cu loss = Total losses

∴ Iron and frication loss = 1990 − 1012.5

$= \mathbf{977.5 \text{ W Ans}}$

b. Lost torque $= T_a - T_{sh}$

$= \dfrac{955 \times \text{Iron and friction loss}}{N}$

$= \dfrac{9.55 \times 977.5}{750}$

$= \mathbf{12.45 \text{ N-m. Ans}}$

3.16 APPLICATIONS OF DC MOTORS

DC shunt motors have a constant speed characteristics. So, whatever be the change in the load connected to the motor, if the speed of the motor is to be maintained nearly constant, then the DC shunt motors are used. In driving lathe, wood working machines etc., shunt motors are used.

Series motors are used where, the starting torque needed to drive a load, is high. It finds better application in hoists, lifts, conveyors, and mainly in electric trains.

Compound motors are used for driving heavy loads, printing machines and related areas.

3.17 CHARACTERISTICS OF DC MOTORS

3.17.1 Electromagnetic Torque Characteristic of DC Motors

The fundamental torque equation $T = K\phi I_a$, provides means of predicting how the torque of each of the three types of motors will vary with application of load (ie. with increased armature current). The torque-load characteristic of each motor type will be taken up in turn. It is now assumed that each motor has been properly started and accelerated so that its armature is connected directly across the line terminals, V.

What is the effect of increased load on the torque of DC motors?

Shunt Motor

From Fig. 3.18(b), it is clear that the shunt field current I_{Sh} is V_{Sh}/R_{Sh}. Thus, the starting and the running period, the current in the shunt field circuit is essentially constant for a given setting of the field rheostat, and consequently the flux (for the present) is also essentially constant. As the mechanical load is increased, the motor slows down somewhat, causing decreased counter EMF and increased armature current. In the basic torque equation, therefore, if the flux is essentially constant and if the armature current increases directly with the application of mechanical load, the torque equation for the shunt motor may be expressed as a perfectly linear relation $T = K I_a$, the characteristic shown in Fig. 3.34 for the shunt motor.

Series Motor

If shunt field coils were removed and replaced with a full series field winding, the identical armature and construction would produce the torque curve shown in Fig. 3.35 for the series motor. In a series motor,

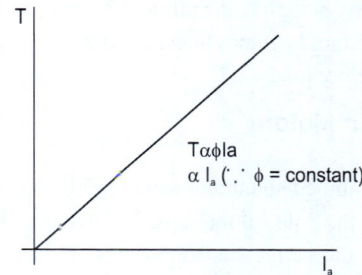

Fig. 3.34.

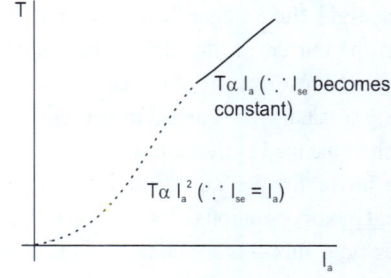

Fig. 3.35.

the armature and series field currents are the same (ignoring the effects of a divertor), and the flux produced by the series field ϕ is at all times proportional to the armature current I_a. The basic torque equation for series motor operation, therefore becomes $T = K I_a^2$. To the extent that the field core is unsaturated (i.e on the linear portion of its magnetizsation curve), the relation between series motor torque and load current is exponential, as shown in Fig. 3.35.

The flux in the field winding will get saturated as the load increases. This is because $I_{se} = I_a$ and a larger I_a occurs at heavy loads which saturates the field system. Here after the flux will practically remain constant, whatever increase in I_{se} be. So, after saturation, in a DC series motor, $T_a \propto I_a$ only. It can be seen in Fig. 3.35 that, the initial exponential shape of the torque thus becomes linear in saturation region, that is, at heavy loads.

Compound Motors

When a combined shunt and series field winding is installed on the poles of the same DC dynamo under discussion, the series field may be cumulative or differentially compounded. Cumulative means the flux produced by the series field will aid the shunt field flux, whereas the differential compound machine means, the flux by the series field will unaid the shunt field flux. Regardless of compounding, however, the current in the shunt field circuit and the field flux ϕ_f during starting or running is essentially constant. The current in the series field is a function of the load current drawn by the armature.

The basic torque equation for cumulative compound motor operation is $T = K(\phi_f + \phi_s) I_a$, where the series field flux ϕ is a function of the armature current I_a. Starting with a flux equal to the shunt field flux at no load and one that increases with armature current, the cumulative compound motor produces a torque curve that is always higher than the shunt motor for the same armature current.

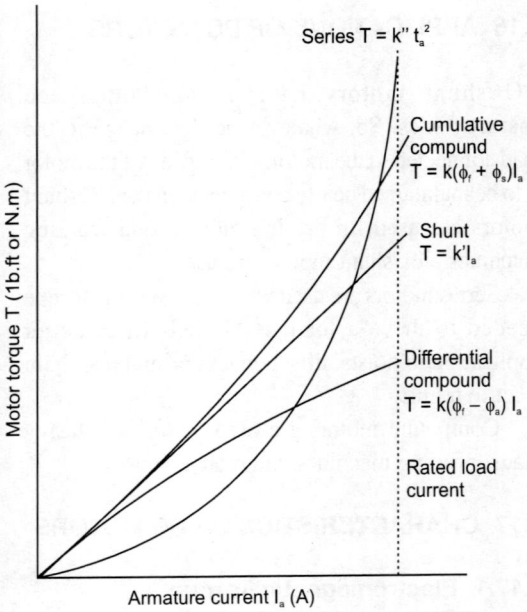

Fig. 3.36.

For the differential compound motor, however, the proceeding torque equation may be written as $T = K(\phi_f - f_s) I_a$, where ϕ_s is still a function of I_a and ϕ_f is (presumably) constant. Starting with a flux equal to the shunt field flux at no load, any value of armature current will produce a series field MMF that reduces the total air-gap flux and hence the torque. Thus, the differential compound motor produces a torque curve that is always less than that of the shunt motor. (Fig. 3.36)

3.17.2 Speed Characteristics of DC Motors

The fundamental speed equations

$$N \propto \frac{V - I_a R_a}{\phi}$$

$$N = K \cdot \frac{V - I_a R_a}{\phi}$$

provides a means of predicting how the speed of each of the motors, will vary with application of load. The speed-load characteristic of each motor will be taken up in turn. To simplify the discussion, it is assumed that the brush volt drop is zero.

Shunt Motor

Assume that the shunt motor of Fig. 3.18(b) has been brought up to rated speed and is operating at no load. Since the field flux of the motor may be considered constant, the speed of the motor may be expressed in terms of the basic speed equation:

$$N \propto V - I_a R_a, \text{ only}$$

As mechanical load is applied to the armature shaft, the counter EMF decreases and the speed decreases proportionately. But since the counter EMF from no load to full load is change of approximately

$$N = K \frac{V - I_a(R_a + R_{se})}{\phi}$$

where V_a is the voltage applied across the motor terminals. Since the air-gap flux produced by the series field is proportional to the armature current only, the speed may be written as

$$N = K \frac{V_a - I_a(R + R_{se})}{\phi}$$

Above equation gives us an indication of the speed-load characteristic of a series motor. If a relatively small mechanical load is applied to the shaft of the armature of a series motor, the armature current I_a is small, making the numerator of the fraction in the above equation large and its denominator small, resulting in an unusually high speed. At no load, therefore, with little armature current and field flux, the speed is extremely excessive. For this reason,

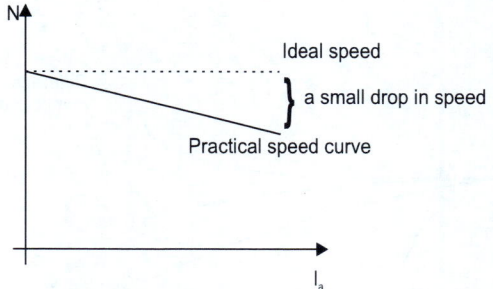

Fig. 3.37.

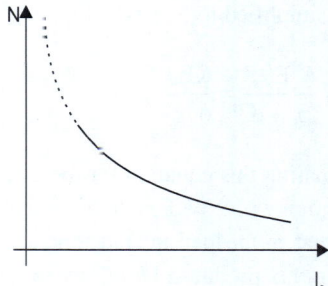

Fig. 3.38.

20 percent (i.e from 0.75 V_a at full load to approximately 0.95 to no load), the motor speed is essentially constant, as shown in Fig. 3.37.

Series Motor

The basic speed equations indicated above as modified for the circuit of the series motor is clearly.

series motors are always operated coupled or geared to a load, as in hoists, cranes, or DC traction (railway) service and never operated without any load. As the load increases, however, the numerator of the fraction in the speed equation decreases faster than the denominator increases (the numerator decreases by a product of I_a, compared to the denominator, which increases directly with I_a), and the speed drops rapidly as shown in Fig. 3.38. The dashed line represents that lightly loaded portion of the characteristics in which series motors are not operated.

As shown in Fig 3.38 excessive speed for a series motor does not results in a high armature current (as with shunt and compound motors) that will open a fuse or a circuit breaker and disconnect the armature from the line. Some other method of protection against runaway must by used. Series motors are usually equipped with centrifugal switches, normally closed in the operating range, that open at speeds of approximately 150 percent of the rated speed or higher.

Cumulative Compound Motor

The basic speed equation for the cumulative compound motor may be written as

$$N = K \frac{V_a - I_a(R_a + R_{se})}{\phi_f + \phi_s}$$

And further simplified to

$$= N = K \frac{E_b}{\phi_f + \phi_s} = \frac{KE_b}{\phi_{total}}$$

By comparing this equation for the cumulative compound motor with $N = KE/\phi$ for the shunt motor, it is evident that, as the load and the armature current increases, the flux produced by the series field also increases, while the counter EMF decreases. The denominator therefore increases while the numerator decreases proportionately more than for a shunt motor. The result is that the speed of the cumulative compound motor will drop at a faster rate than the speed of the shunt motor with application of load, as shown in Fig. 3.39.

Differential Compound Motor

The above equation for the speed of cumulative. compound motor, may be modified slightly to show the effect of the opposing field MMFs, and the speed is

$$N = \frac{KE_b}{\phi_f - \phi_b} = K \frac{V - I_a(R_a + R_{se})}{\phi_f - \phi_s}$$

As the load and I_a on a differential compound motor increase, the numerator of the fraction in the above equations decreases somewhat but the denominator decreases more rapidly. The speed may drop slightly at light loads ; but as the load increases, the speed increases. This condition sets up a dynamic instability.

As the speed increases, most mechanical loads increase automatically (since more work is done at a higher speed), causing an increase in current, a decrease in total flux, and a higher speed, producing still more load. Because of this inherent instability, differential motors are rarely used.

In a machinery laboratory where differential motors are tested, one may occasionally observe a

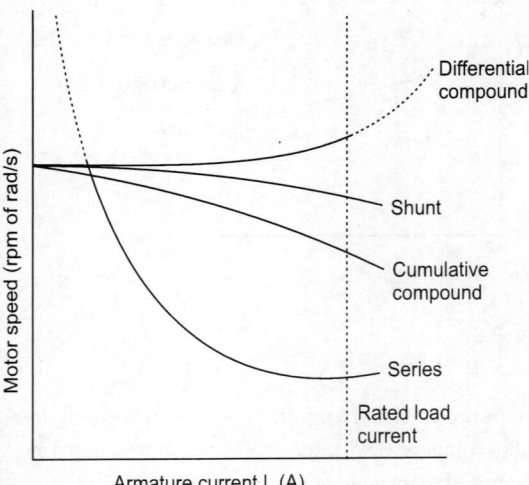

Fig. 3.39.

condition where a differential motor begins to race away and suddenly drops in speed and reverses direction. This may be explained by using the above speed equation and Fig. 3.39. As the counter EMF decreases due to decreased mutual flux, the armature current and torque increase is so excessive that the series field flux exceeds the shunt field flux, and the motor reverse direction.

It is for this reason that, when a differential motor is started for testing purposes in the laboratory, care should be taken to sort out the series field so that the high staring and armature current will not start the motor in the reverse direction.

Refer to Fig. 3.39.

Illustration 3.40

A 230 V, 10 h.p., 1250 rpm cumulative compound motor has an armature resistance of 0.25 W a combined compensating winding and interlope resistance of 0.25, and a brush volt drop of 5 V. The resistance of the series field is 0.15 W, and the shunt field resistance is 230 W. When the motor is shunt connected the line current rated load is 55 A and the no-load line current is 4 A. The no-load speed is 1810 rpm. Neglecting armature reaction at rated voltage, calculate

(a) Speed at rated load
(b) Internal power in watts and internal horse power developed.

Solution

(a) $I_a = I_L - I_f = 4 - 1 = 3$ A

No-load $E_b = V_a - (I_a R_a + BD)$

$= 230 - (3 \times 0.5 + 5) = 223.5$ V, at 1810 rpm

Similarly the full load $E_b = 198$ V

$$N_r = 1810 \left(\frac{198}{223.5} \right) = 1603 \text{ rpm}$$

(b) $P_{dev} = E_b I_a = 198 \times 54 = 10700$ W

$$hp = \frac{10700}{746} = 14.3 \text{ h.p.}$$

Illustration 3.41

The motor of Illustration 3.24 is reconnected as a long shunt cumulative compound motor. At rated load (55 A), the compound winding increases the flux per pole 25 percent. Calculate

(a) The speed at no load (4 A line current)
(b) The speed at rated load (55 a line current)

Solution

No load $E_b = V_a - (I_a R_a + I_a R_{se} + BD)$

$= 230 - ((3 \times 0.5) + (3 \times 0.15) + 5)$

$= 223.05$ V

$$N_{nl} = 1810 \times \left(\frac{223.05}{223.5} \right) = 1806 \text{ rpm}$$

b. Full load $E_b = 230 - ((54 \times 0.5) + (54 \times 0.15) + 5)$

$= 190$ V

$$N_r = K \left(\frac{\delta E_c}{\delta \phi} \right)$$

$$= 1806 \text{ rpm} \left(\frac{190}{223.05} \right) \left(\frac{1.0}{1.25} \right)$$

$= 1231$ rpm

Illustration 3.42

The armature of a D.C. machine has a resistance of 0.1 Ω and is connected to a 230 V supply. Calculate the generated e.m.f. when it is running
    (a) as a generator giving 80 A;
    (b) as a motor taking 60 A.

(a) Voltage drop due to armature resistance = 80 × 0.1 = 8 V. From equation (3.3)

    Generated e.m.f. = 230 + 8 = 238 V.

(b) Voltage drop due to armature resistance = 20 × 0.1 = 6 V. From. equation (3.4)

    Generated e.m.f. = 230 – 6 = 224 V.

Illustration 3.43

A four pole motor is fed at 440 V and takes an armature current of 50 A. The resistance of the armature circuit is 0.28 Ω. The armature winding is wave-connected with 888 conductors and the useful flux per pole is 0.023 Wb. Calculate the speed.

Solution

From expression (3.4) we have

    440 = generated e.m.f. + 50 × 0.28

    Generated e.m.f. = 440 – 14 = 426 V

Substituting in the e.m.f. equation (3.1), we have

$$426 = 2 \times \frac{888}{2} \times \frac{N \times 2}{60} \times 0.023$$

$$N = 626 \, r/min$$

Illustration 3.44

A motor runs at 900 r/min of a 460 V supply. Calculate the approximate speed when the machine is connected across a 200 V supply. Assume the new flux to be 0.7 of the original flux.

If ϕ is the original flux, then from expression (3.12),

$$900 = \frac{460}{k\phi}$$

$$\therefore k\phi = 0.511$$

and

$$\text{new speed} = \frac{\text{new voltage}}{k \times \text{original flux} \times 0.7} \text{ (approx.)}$$

$$= \frac{200}{0.511 \times 0.7} = 559 \, r/min$$

Illustration 3.45

A D.C. motor takes an armature current of 110 A at 480 V. The resistance of the armature circuit is 0.2 Ω. The machine has six poles and the armature is lapconnected with 864 conductors. The flux per pole is 0.05 Wb. Calculate
    (a) the speed;
    (b) the groos torque developed by the armature.

(a) Generated e.m.f. = 480 – (110 × 0.2)

    = 458 V

Since the armature winding is lap-connected, c = 6.

Substituting in expression 38.1, we have

$$458 = 2 \times \frac{846}{6} \times \frac{N \times 3}{60} \times 0.05$$

$$N = 636 \, r/min$$

(b) Mechanical power developed by armature is

110 × 458 = 50 380 W

Substituting in expression [3.11] we have

$$2\pi T \times \frac{636}{60} = 50380$$

T = 756 N-m

Alternatively, using equation (3.11), we have

$$T = 0.318 \times \frac{110}{6} \times 864 \times 3 \times 0.05$$

= 756 N-m

Illustration 3.46

The torque required to drive a DC generator at 15 r/s is 2 kNm. The core, friction and windage losses in the machine are 8 kW. Calculate the power generated in the armature winding.

Driving torque = 2 kNm = 2000 Nm

From expression [3.11], power required to drive the generator is

2π × 2000 [Nm] × 15 [r/s]

= 188 400 W = 18.3 kW

Since core, friction and windage losses are 8 kW,

∴ Power generated in armature widing = 188.4 − 8

= 180.4 kW

Illustration 3.47

A series motor runs at 600 r/min when taking 110 A from a 230 V supply. The resistance of the armature circuit is 0.12 Ω and that of the series widning is 0.03 Ω. The useful flux per pole for 110 A is 0.024 Wb and that for 50 A is 0.0155 Wb. Calculate the speed when the current has fallen to 50 A.

Total resistance of armature and series winding is

0.12 + 0.03 = 0.15 Ω

therefore emf generated when current is 110 A is

230 − 110 × 0.15 = 213.5 V

It was shown that for a given machine:

Generated emf = a constant (say k) × speed × flux

Hence with 110 A

213.5 = k × 600 × 0.024

∴ k = 14.82

With 50 A,

Generated emf = 230 − 50 × 0.12 = 22.5 V

But the new emf generated is

k × new speed × new flux

∴ 22.5 = 14.82 × new speed × 0.0155

∴ Speed for 50 A = 969 r/min

Illustration 3.48

The induced emf in a DC machine while running at 500 rpm is 180 V. Calculate the induced emf while the machine is running at 600 rpm. Assume constant flux.

Solution

We know, $E = K \phi N$

If flux ϕ is assumed constant,

$E = K_1 N$

Thus from the data given,

$180 = K_1 \times 500$

or $K_1 = \dfrac{180}{500}$

Therefore induced emf at 600 rpm is

$E = \dfrac{180}{500} \times 600 = 216 \, V$

Illustration 3.49

The induced emf in a DC machine while running at 750 rpm is 220 V. Calculate

(a) assuming constant flux the speed at which the induced will be 250 and

(b) the percentage increase in the field flux for an induced emf of 250 V and speed of 700 rpm.

Solution

(a) $E_1 = K_1 N_1$

or $220 = K_1 \times 750$

or $K_1 = \dfrac{220}{750}$

$E_2 = K_1 N_2$

$N_2 = \dfrac{E_2}{K_1} = 250 \times \dfrac{750}{220} = 852 \, \text{rpm}$

(b) $E_1 = K \phi_1 N_1$

or $220 = K \phi_1 \times 750 \quad\quad (1)$

and $E_2 = K \phi_2 N_2$

or $250 = K \phi_2 \times 700 \quad\quad (2)$

Dividing Eqn. (2) by Eqn. (1)

$\dfrac{250}{220} = \dfrac{K \phi_2}{K \phi_1} \times \dfrac{700}{750}$

$\dfrac{\phi_1}{\phi_2} = \dfrac{250}{220} \times \dfrac{750}{700} = 1.217$

Therefore the percentage increase in flux is 21.7%.

Illustration 3.50

The induced emf in a DC machine is 200 V at a speed of 1200 rpm. Calculate the electromagnetic torque developed at an armature current of 15 A.

Solution

We know,

$EI = T \omega$

or $T = \dfrac{EI}{\omega}$

$= \dfrac{200 \times 15}{\dfrac{2\pi \times 1200}{60}} = \dfrac{200 \times 15}{6.28 \times 20}$

$= 23.9 \, Nm$

Illustration 3.51

On the armature of a DC machine is placed one coil having 10 turns. The length and diameter of the armature are both 0.2 m. The armature is rotating at a uniform magnetic field density of 1 Wb/m² at 1500 rpm. If the coil is connected to the load, the total resistance of the circuit becomes 4 Ω. Calcualte the torque developed on the armature coil.

Solution

The induced emf in one conductor is given by

$$e = Bl\omega r$$

Each turn of the armature coil has two coil-sides (i.e. two conductors). The coil has 10 such turns. Therefore the total induced emf

$$E = Bl\omega r \, 2 \times 10$$

$$= 1 \times 0.2 \times \frac{2\pi \times 1500}{60} \times 0.1 \times 2 \times 10$$

$$= 62.8 \text{ V}$$

The current through the armature coil when connected to the load,

$$I = \frac{62.8}{4} = 15.7 \text{ A}$$

Torque

$$T = \frac{EI_a}{\omega} = \frac{62.8 \times 15.7}{\frac{2\pi \times 1500}{60}}$$

$$= 6.27 \text{ N-m}$$

Illustration 3.52

The armature supply voltage of a DC motor is 230 V. The armature current is 12 A, the armature resistance is 0.8 Ω, and the speed is 100 rad/s. Calculate (a) the induced e.m.f., (b) the electromagnetic torque, (c) the electrical power input to the armature, (d) the mechanical power developed by the armature, and (e) the armature copper-losses.

(a) $\quad E = V - I_a R_a$

$\qquad = 230 - 12 \times 0.8$

$\qquad = 220.4 \text{ V}$

(b) $\quad EI_a = T_e \omega$

$\therefore \quad T_e = \dfrac{EI_a}{\omega}$

$\qquad = \dfrac{220.4 \times 12}{100}$

$\qquad = 26.448 \text{ N-m}$

(c) $\quad P_{input} = VI_a$

$\qquad = 230 \times 12$

$\qquad = 2760 \text{ W}$

(d) $\quad P_d = T\omega$

$\qquad = EI_a = 220.4 \times 12$

$\qquad = 2644.8 \text{ W}$

(e) $\quad I_a^2 R_a = (12)^2 \times 0.8 = 115.2 \text{ W}$

Illustration 3.53

A DC generator is connected to a 220 V DC mains. The current delivered by the generator to the mains is 100 A. The armature resistance is 0.1 Ω. The generator is driven at a speed of 400 rpm. Calculate (a) the induced emf, (b) the electromagnetic torque, (c) the mechanical power input to the armature neglecting iron, windage and friction losses, (d) the power input and output of the armature when the speed drops to 350 rpm. State whether the machine is generating or motoring. Assume constant flux.

(a) $\quad E = V + I_a R_a$

$\quad\quad = 220 + 100 \times 0.1$

$\quad\quad = 230\ V$

(b) $\quad T_e = \dfrac{E I_a}{\omega} = \dfrac{230 \times 100}{2\pi \times \dfrac{400}{60}} = 549.09\ Nn$

(c) Mechanical power input = Electromagnetic power developed + Iron-loss in the armature + windage and friction-losses

Neglecting iron loss, windage-loss and friction-loss,

Mechanical power input = $\omega T_e = E I_a$

$\quad\quad = 230 \times 100 = 23000\ W$

(d) When the machine runs at 400 rpm

$\quad\quad E = K\phi N$

$\quad\quad 230 = K\phi \times 400$

or $\quad K\phi = \dfrac{230}{400}$

$K\phi$ remains constant when the machine runs at 350 rpm also.

Therefore,

$\quad\quad E = K\phi N$

or $\quad E = \dfrac{230}{400} \times 350 = 201.25\ V$

As induced e.m.f. E is less than the terminal voltage V (remains constant as the machine is connected across a 220 V mains), the machine will now work as a motor. Thus it will draw current from the mains. Current drawn from the mains can be calculated as, $E = V - I_a R_a$.

$\quad I_a = \dfrac{E - V}{R_a} = \dfrac{220 - 201.25}{0.1}$

$\quad\quad = 187.5\ A$

$P_{input} = 220 \times 187.5 = 41.25,\ kW$

P_{output} (neglecting iron friction and windage-losses)

$\quad\quad = E I_a$

$\quad\quad = 201.25 \times 187.5 = 37.73\ kW$

Illustration 3.54

A shunt generator delivers 50 kW at 250 V and 400 rpm. The armature resistance is 0.02 Ω and field resistance is 50 Ω. Calculate the speed of the machine when running as a shunt motor and taking 50 kW input at 250 V.

Solution

$\quad I_f = \dfrac{250}{50} = 5\ A$

$\quad I_L = \dfrac{50000}{250} = 200\ A$

$\quad I_a = 200 + 5 = 205\ A$

$\quad E_G = V + I_a R_a = 250 + 205 \times 0.02$

$\quad\quad = 254.1\ V$

As a motor,

$\quad E_M = V - I_a R_a$

$$I_L = \frac{50000}{250} = 200 \text{ A}$$

$$I_a = I_L - I_f = 200 - 5 = 195 \text{ A}$$

$$E_M = 250 - 195 \times 0.02 = 246.1 \text{ V}$$

We know

$$\frac{N_M}{N_G} = \frac{E_M}{E_G}$$

Therefore speed of the motor,

$$N_M = N_G \frac{E_M}{E_G}$$

$$N_M = 400 \times \frac{246.1}{254.1} = 387 \text{ rpm}$$

Illustration 3.55

A DC shunt machine, connected to 250 mains, has an armature resistance of 0.12 Ω, and the resistance of the field circuit is 100 Ω. Calculate the ratio of the speed as a generator to the speed as a motor, the line current in each case being 80 A.

Solution

Given V = 250 V, R_a = 0.12 Ω, R_f = 100 Ω, I_L = 80 A
Field current,

$$I_f = \frac{V}{R_f} = \frac{250}{100} = 2.5 \text{ A}$$

When the machine is generating,

$$I_a = I_L + I_f$$
$$= 80 + 2.5$$
$$= 82.5 \text{ A}$$

$$E_G = V + I_a R_a$$
$$= 250 + 82.5 \times 0.12$$
$$= 259.9 \text{ V}$$

When the machine is motoring,

$$I_L = I_a + I_f$$
$$I_a = I_L - I_f$$
$$= 80 - 2.5$$
$$= 77.5 \text{ A}$$

$$E_M = V - I_a R_a$$
$$= 250 - 77.5 \times 0.12 = 240.7 \text{ V}$$

Induced emf, $E \propto N$ (other terms remaining constant)

$$\frac{E_G}{E_M} = \frac{N_G}{N_M}$$

Ratio of speeds,

$$\frac{N_G}{N_M} = \frac{E_G}{E_M}$$

$$= \frac{259.9}{240.3} = 1.081$$

Illustration 3.56

A 1500 kW, 550 V, 16 pole, generator runs at 150 rpm. What must be the useful flux per pole if there are 2500 conductors in the armature and the winding is lap connected and full-load armature copper loss is 25 kW? Calculate the area of the pole shoe if the air-gap flux density has a uniform value of 0.9 wb/m². Also find the no-load-terminal voltage. Neglect change in speed.

Solution

Given V = 550 V, P = 16, N = 150 rpm, Z = 2500, A = 16.

Power = 1500 kW

Full-load copper loss = 25 kW

Flux density in the pole B = 0.9 Wb/m^2

Full load current (I_a) is calculated as

$$= \frac{1500 \times 1000}{550} = 2727.3 \text{ A}$$

We have

$$I_a^2 R_a = 25 \times 1000$$

$$R_a = \frac{25 \times 1000}{(2727.3)^2}$$

$$= 3.36 \times 10^3 \, \Omega$$

Induced emf,

$$E = V + I_a R_a$$

$$= 550 + 2727.3 \times 3.36 \times 10^{-3}$$

$$= 559.2 \text{ V}$$

We have,

$$E = \frac{\phi ZNP}{60 A}$$

Substituting the values,

$$559.2 = \frac{\phi \times 2500 \times 150 \times 16}{60 \times 16}$$

or, $\phi = 0.0895$ Wb

To calculate the area of the pole shoe, we take,

Flux = Flux density × Area of the pole shoe

Thus,

$$\text{Area of pole shoe} = \frac{\text{flux}}{\text{flux density}}$$

$$= \frac{0.0895}{0.9} \text{ m}^2$$

$$= 994.4 \text{ cm}^2$$

Illustration 3.57

A commutator having a diameter of 76 cm rotates at 600 rpm. Calculate the approximate time of commutation if the width of brush is 1.5 cm.

Solution

Given commutator diameter = 76 cm = 0.76 m.

Commutator radius, = 0.38 m, speed N = 600 rpm, speed in rps, n = 10

Brush width = 1.5 cm = 1.5×10^{-2} m and peripheral speed of commutator,

$$V = r \cdot \omega = r \times 2\pi n, \text{ m/sec}$$

Substituting the data,

$$V = 0.38 \times 2 \times 3.14 \times 10$$

$$= 23.86 \text{ m/sec}$$

Time of commutation,

$$T_c = \frac{\text{Brush width}}{\text{peripheral speed, V}}$$

Substituting the data,

$$T_c = \frac{15 \times 10^{-2}}{23.86} \text{ Sec}$$

$$= 0.628 \times 10^{-3} \text{ Sec}$$

3.18 STARTERS FOR DC MOTORS

At the instant of applying a voltage V across the armature terminals in order to cause a motor to rotate, the motor armature is not producing any counter EMF since the speed is zero. The only current-limiting factors are the armature brush volt drop and the resistance of the armature circuit, R_a. Since neither of these, under normal conditions, amounts to more than 10 or 15 percent of applied voltage V across the armature, the overload is many times the rated armature current. Under this condition, the starting current of the motor is approximately 8 times as great as the normal rated full-load armature current. It shows that severe damage might be done to a motor, whenever it is starting, unless the starting current is limited by means of a commercial starter.

The current is excessive because of a lack of counter EMF at the instant of starting. Once rotation has begun, counter EMF is built up in proportion to speed. What is required then is a device, usually a tapped or variable resistor, whose purpose is to limit the current during the starting period and whose resistance may be progressively reduced as the motor gains speed.

Note that, a progressively decreasing value of starting resistance is required as the motor develops an increased counter emf owing to acceleration. This is the principle of the armature resistance motor starter.

The manner in which a starter is used in conjunction with the three basic types of DC motors, is shown in Fig. 3.40. The techniques shown here for motor starting are schematic diagrams only; as stated previously commercial forms of manual and automatic starters and controllers differ somewhat from these.

The shunt and compound motors are started with full field excitation (i.e., the full line voltage is impressed across the field circuit) in order to develop maximum starting torque ($T=K \phi I_a$). In all three types of motor, the armature starting current is limited by a high-power series – connected variable starting resistor. In commercial practice, the initial inrush of armature current is generally limited to a higher value than the full-load current, again to develop greater starting torque, particularly in the case of large motors that have great inertia and that come up to speed slowly.

With the starting arm at position 1 in Fig. 3.40, the maximum series resistance will limit the armature current on starting to about 150 percent of its rated value. As the motor slowly increases its speed, the armature develops counter emf and the armature current drops to approximately full load. If the starting arm were left at position 1, the armature current would drop somewhat and the speed would stabilize at some value well below the rated speed. In order to accelerate the motor armature once more, it is necessary to move the arm to position 2. Again, there is an inrush of armature current, and the motor rises in speed. This process is continued until the position 5, at which, all the external starter resistances will be cut, leaving only the R_a to function during the running conditions.

With this major arrangement, accessories like overload relay and no voltage relay are provided,

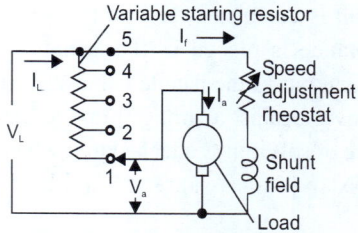

(a) Shunt motor starter (schematic form)

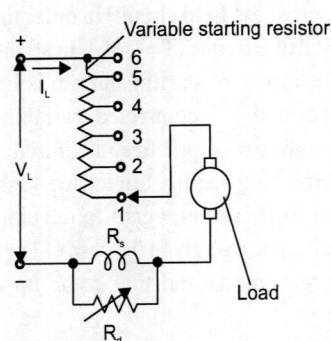

(b) Series motor starter (schematic form)

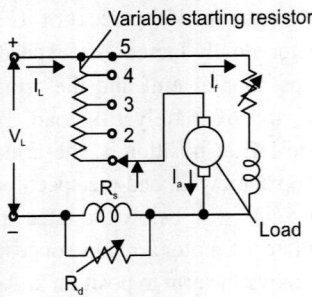

Fig. 3.40. (a) and (b)

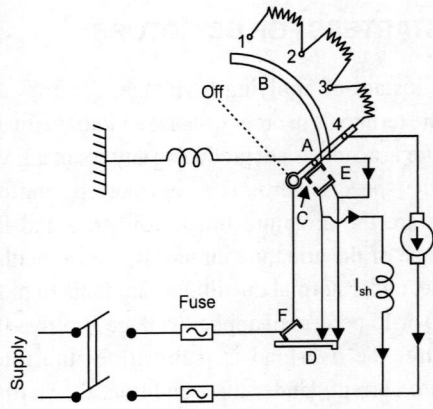

Fig. 3.41.

respectively, to stop the motor when it is overloaded and stop it if the supply is off, as shown in Fig. 3.40b.

Figure 3.41 denotes the circuit diagram for a commerical starter.

In consists of an arm of handle A which moves over the studs. When the arm touches the first stud, field circuit is completed through brass are B and full resistance is placed in the armature but is gradually cut out as the handle is moved over. The handle moves against a strong spring as shown. It has a piece of soft iron C attached to it which in the 'FULL-ON' position is attracted and held by the electromagnet E which is energised by shunt field current. This is known as *'hold-on'* coil or *'low-voltage'* (formerly *NO-voltage*) release. The action of this protective device is, in case of a failure or disconnection of the supply or a break in the field circuit, to release the arm and allow the spring to bring it back to 'OFF' position. This prevents the fuses from blowing, as they otherwise would if the supply were restored with the handle in the 'FULL-ON' position.

An *over-current* (or *overload*) release is also fitted in the starter. This consists of an electromagnet F which is connected in the supply line. If the machine becomes overloaded beyond a certain predetermined value, then D is lifted and short-circuits E. Hence, the handle is released and returns to 'OFF' position.

3.19 SPEED CONTROL OF DC MOTORS

The speed of the DC motor depends upon Eb and φ as already stated in equation (3.12),

$$N = K\frac{E_b}{\phi}$$

$$= K\frac{V - I_a R_a}{\phi}$$

Thus the speed of the motor can be controlled by
i. field control (control of ϕ)
ii. Armature control (control of I_a hence E_b)
iii. Voltage control (control of V)

Field Control

In case of field control method, a series resistance of high value is connected in series with the shunt field as shown in Fig. 3.42. The flux produced is directly proportional to current in field winding, and the speed is inversely proportional to the flux. An increase in the externally added shunt field resistance will reduce the shunt field current and thus the value of ϕ. The speed increases. On the other hand a decrease in the externally added shunt field resistance will increase the shunt field current and hence the flux, ϕ. Thus the speed reduces.

Armature Control

In this method a variable resistance of low value and more current capacity is introduced in the armature circuit. As we know that speed of a motor is directly proportional to the back emf of the motor.

$$E_b = V - I_a R_a$$

Consider the Fig. 3.43. If now the external resistance in the armature circuit (R_{ext}), is increased, $I_a R_a$ in the above equation will also increase; but E_b will decrease. So, the speed N (α Eb) will decrease. Similarly we can find that a decrement in R_{ext} will decrease $I_a R_a$, but increase E_b. Thus the speed increases. In this method the speed is controlled

Electric Machines 237

below normal and cannot be above normal. The other disadvantage is that more current (as compared to the field current) causes more power loss (I^2R) in it and causes heat development in the resistance.

Voltage Contro

In this method the voltage given to the motor is controlled and the speed is controlled below normal.

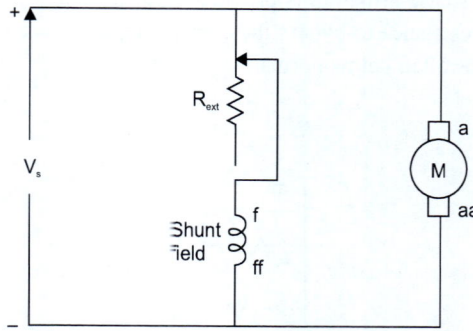

Fig. 3.42-A. Circuit for speed by field control method.

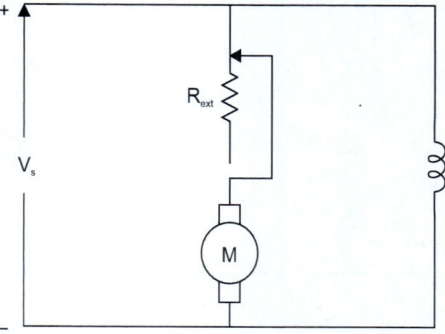

Fig. 3.42-B. Circuit for speed control by armature control method.

(a) Multiple Voltage Control

In this method the shunt field of the motor is permanently connected across the definite voltage but the armature can be connected by means of a suitable switchgear across one of several different voltages, the speed will be approximately proportional to these voltages.

In this case a field divertor is used to bypass the current to armature thus the flux weakens and speed accelerates.

Armature Divertor

In this case armature is provided with the shunt of low resistance to bypass the current and thus speed is controlled below normal.

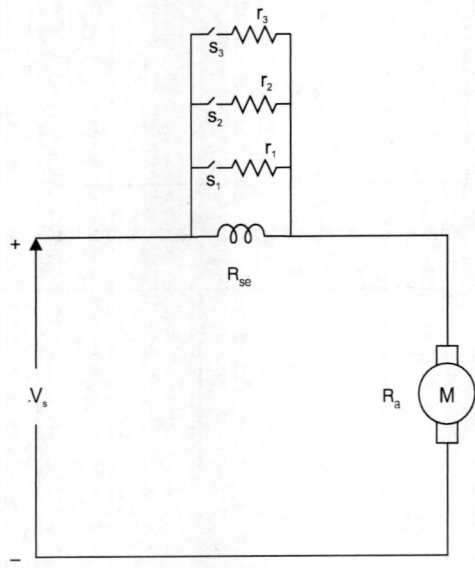

Fig. 3.43.

Consider Fig. 3.43. There are three parallel paths for field current. When all the switches S_1, S_2 and S_3 are open, the normal operating condition is not disturbed, and $I_{se} = I_a$. But when S_1 is closed, some current of I_a flows through r_1 and thus I_{se} is reduced. Therefore ϕ is reduced and speed increases. If further S_2 is closed, I_{se} will proportionally further be reduced. Therefore ϕ is further reduced, and thus the speed.

a. Tapped Field Control

Generally used in traction purpose. The field tappings are used. The full field winding results less speed and if it is cut then the higher speed is obtained.

b. Paralleling Field Windings

The speed can be controlled by connecting two field coils in series or in parallel generally in fan motors etc.

c. Variable Supply Voltage

Here also the variable voltage is given to the motor. Less voltage less speed and thus we get below normal speed.

Reflection Section

1. Which DC motor can be used for the following applications? Justify your answer for each case. (1) Weaving machines and lathe machines, (2). Hoists, (3). Electric traction, (4). JCBs, and (5). Heavy duty loads.
2. What is the role of No-voltage coil and over-current relays in starters?

PART II: TRANSFORMERS

3.20 TRANSFORMER

A transformer is a simple, efficient and comparatively inexpensive device used only in AC circuits for the purpose of changing the voltage of one value from one circuit to another voltage value in the other circuit, maintaining the **same frequency** at both the places.

There are **no moving parts** in the transformer, and hence it is a **static device**; because of this, the mechanical forces which are always present in the rotating machines, are entirely absent in the transformers. This makes the total losses to get reduced (by the mechanical losses amount) thus increasing the total efficiency.

Essentially a conventional transformer consists of two electric circuits which are insulated from each other and **joined** by a common magnetic path. The electric circuit which is connected to the source of AC supply is called the **primary**, while the other electric winding that feeds or supplies the load is known as the **secondary**.

As said earlier, in a transformer, voltage of one value one circuit will be transformed to another value in another electric circuit; this is achieved in two ways. Some transformers are designed **to raise** the primary applied voltage to a higher value. This raised higher value of the voltage will be available in the secondary side, in which case the transformer is known to be a **step-up** transformer as it 'steps up' the voltage from primary to secondary. Other types is constructed in such a way that the higher input voltage of the primary side will be lowered in the secondary side. Such transformers are designated as the **step-down** transformer as the primary applied voltage is **stepped down** in the secondary side. The voltage which is lesser out of the two winding voltages in called as the low voltage (LV) and the other greater voltage is called as the high voltage (HV).

An important point worth noting is that the voltage in either side of the winding depends directly on the number of turn in the respective windings.

3.21 TYPES AND CONSTRUCTION

The Fig. 3.44 shows a very simple transformer in which two electrically insulated coils are **linked** by a common magnetic flux path ϕ. For the flow of the flux from the primary side to the secondary side and most of all, to provide a magnetic path so that no flux (ϕ) produced will be **wasted out**, a laminated magnetic steel block is used known as the **core**. It is visible in Fig. 3.44.

The core and the two windings viz, the primary and the secondary, forms the essential parts of a transformer.

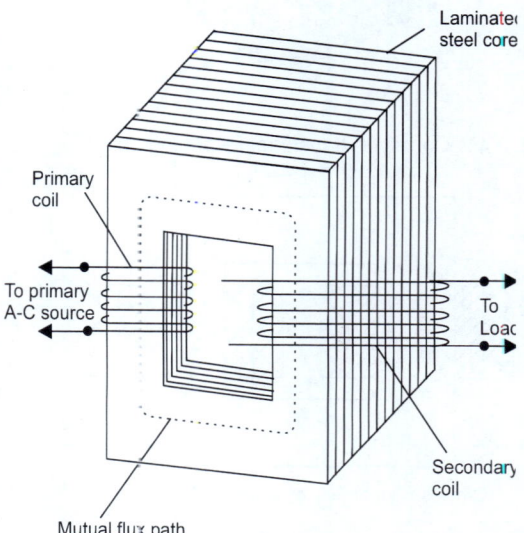

Fig. 3.44. Simple transformer diagram.

The laminations are provided to reduce the eddy current losses in the core.

There are two types of transformers, distinguished from each other by the manner in which the primary and secondary winding are placed along the laminated steel core.

In one type, (known as the core type transformer), the coils will surround the core. The photograph of such a transformer is in Fig. 3.45 and the diagrammatic representation in Fig. 3.46.

Note that the primary and secondary windings are wrapped or wound around the laminated core sides. Thus the core is kept **inside** the windings.

In the second type of transformer construction (known as the 'shell type') the core will surround the coils. Refer to the Figs. 3.47 and 3.48.

Note in the Fig. 3.47 and 3.48 that the primary and secondary windings are assembled (insulated from each other, of course), **inside** the laminated steel core. In both the cases, near to the core, the LV winding will be placed.

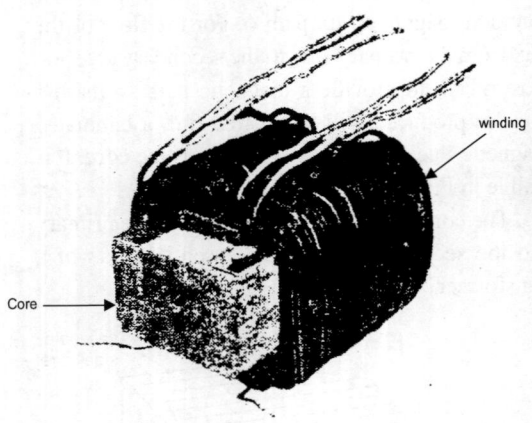

Fig. 3.45. Core-type transformer—photograph

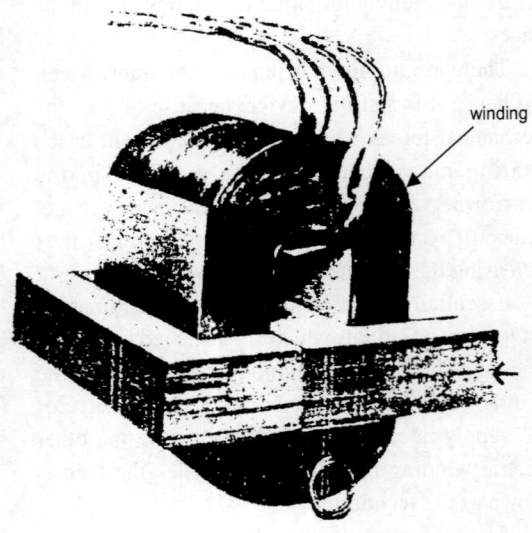

Fig. 3.47. Shell-type transformer

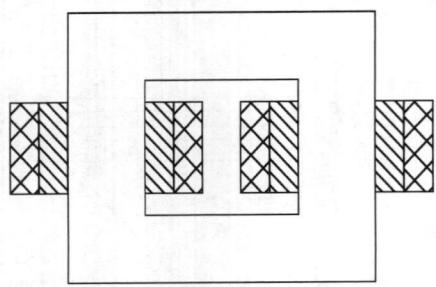

Cross-sectional view of windings are shown

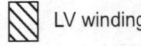 LV winding 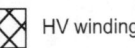 HV winding

Fig. 3.46.

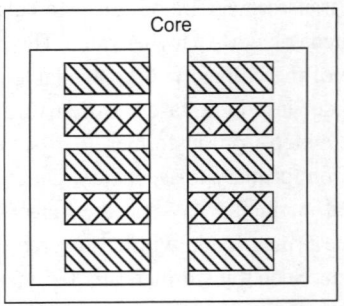

Cross-sectional view of windings are shown

 LV-winding   HV-winding

Fig. 3.48.

Note: In some special cases the electric circuits are joined together, and the so formed one single coil will serve as both primary and secondary; it is the **auto-transformer**. It is symbolically shown in Fig. 3.49. Auto transformer is basically intended to provide a variable voltage to a circuit. The output is taken across (or the secondary is), the points **PQ**. The point **P** is movable; as **P** moves up more number of turns is included in the secondary and hence more voltage could be got in the secondary, as the voltage in a transformer winding is directly proportional to its number of terms. Movement of P downwards will cause reduction in number of secondary turns thus causing the secondary voltage to get reduced.

Anyway, the principle of operation still remains the **same** for both the conventional transformer and an auto-transformer, which forms the next section. Reader is suggested to refer back to the mutually induced emf and the related topics of Sec (2.18) of chapter 2, because they are very essential to understand the working principle of a transformer.

Consider a transformer as shown in Fig. 3.50, in which the secondary winding is let open.

The primary is connected to an AC supply of suitable voltage V_1 volts. This applied voltage will

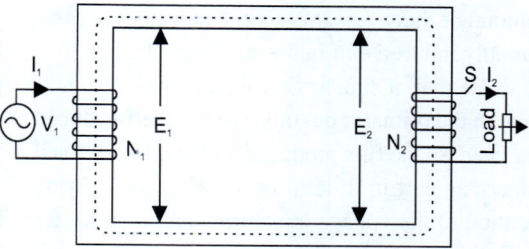

Fig. 3.50. Demonstrative circuit for transformer principle

cause an alternating current to flow in the primary side; let this current be I_1 amps.

The laminated core (made up of magnetic material such as steel etc.) will now be **magnetised** by the primary current I_1, of MMF $N_1 I_1$, thus releasing flux of ϕ_1 webers. As this flux flux ϕ_1 is produced by an alternating current I_1, obviously the flux will also be of alternating in nature and, at all times, proportional to the current I_1.

Such produced flux will be offered a less reluctance closed path by the core. So the flux lines will have their path completed in the core as shown in Fig. 3.50.

It can be readily seen that the flux ϕ_1 produced by the primary current I_1, passes through the secondary winding too.

Because this flux ϕ_1 is an AC flux, its magnitude will continuously change sinusoidally with respect to time.

Note carefully that the number of turns in the secondary (and the primary) are fixed; only the linking flux ϕ_1 changes. Thus the value of flux linkage $N\phi$ is continuously changing. In primary the continuously changing flux linkage will be $N_1 \phi_1$ and in secondary $N_2 \phi_2$.

This change is linkage of flux in the secondary winding will induce an emf in it. This of course, is a mutually induced emf because the flux produced by the primary coil induces an emf in the other coil (here the secondary coil), the magnitude of which is given by the equation, $M \, di/dt$, where M is the mutual

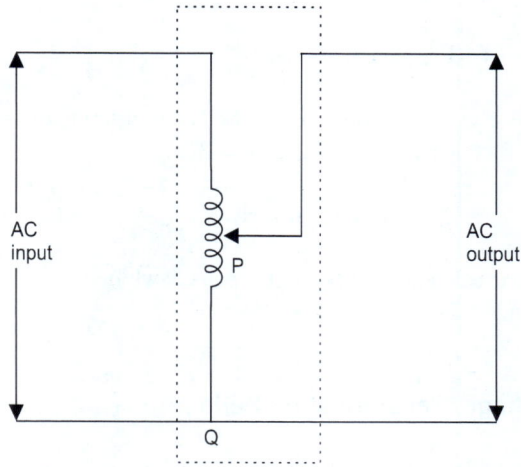

Fig. 3.49. An auto-transformer

inductance between this two windings. Let this mutually induced emf in the secondary be E_2.

Note that, a similar flux linkage will also take place in the primary; but this will be a **self-induced emf** because the flux produced by the primary **itself** induces an emf in it; let it be E_1. However, paying attention to the voltage transformation, after all for which the transformer is intended, E_2 is given much more importance for the analysis.

Thus the transformer transforms voltage from one circuit (here, the primary) to another circuit (here, the secondary) without any moving part. As there is no moving part, the 'relative motion' conditions will totally be absent in transformers which is the main cause for the frequency change; so frequency of operation of transformer as both the places will be the "**same**".

If the reader carefully focus his attention on the circuit of Fig. 3.50 the E_2 can be treated as the **supply voltage source** to the secondary.

The closing of the switch S will obviously cause a current I_2 to flow in the load circuit, thus causing the energy to get transformed from one circuit to other; that is, primary to secondary.

From Fig. 3.50 it is clear that V_1 is supply voltage for the primary and thus $V_1 > E_1$. But it is also equally clear that, for the secondary E_2 will serve as the supply, air thus,

$$E_2 > V_2. \quad (3.15)$$

3.22 EMF EQUATION OF A TRANSFORMER

Let N_1 and N_2 be the number of turns on the primary winding and secondary winding, respectively

ϕ_m = maximum flux density in the core in webers = $B_m \times A$. Where a is the area of the core through which it passes.

f = the supply frequency, cps

As the supply is of sinusoidal in nature, the core flux increase from zero crossing time instant to the positive maximum value of ϕ_m in one quarter of the cycle; i.e., (1/4)f, seconds. Then, the average rate of change of flux is, $d\phi/dt$, $\phi_m/(1/4f) = 4f\phi_m$, volts.

Using the form factor relation, the rms value can be found out from the maximum value, as, rms value of the voltage = 1.11 × average value of the voltage. As, all the voltage and current values are substituted in rms value, this conversion becomes essential.

So, rms value of the emf = 1.11 (4fϕ_m)

$$= 4.44 \, f\phi_m \text{ V}$$

For N_1 number of primary winding terms, this is equal to,

$$= 4.44 \, f\phi_m \, N_1 \text{ V}$$

Thus, for the primary,

$$E_1 = 4.44 \, f\phi_m \, N_1 \text{ V} \quad (3.16\text{-a})$$

and for the secondary,

$$E_2 = 4.44 \, f\phi_m \, N_2 \text{ V} \quad (3.16\text{-b})$$

N_2 = Number of secondary winding turns

f = Frequency in c/s.

This e.m.f. equation can also be given in terms of the flux density and area, as ($\phi = B_m A$)

$$E_1 = 4.44 \, B_m \, A.f. \, N_1. \text{ V} \quad (3.17)$$

The usual values for B_m is between 0.9 wb/m² to 1.4 wb/m².

Voltage Transformation Ratio

It is the ratio of secondary voltage to primary voltage

$$\frac{E_2}{E_1} = \frac{4.44 f \phi N_2}{4.44 f \phi N_1}$$

or $\quad \dfrac{E_2}{E_1} = \dfrac{N_2}{N_1} = K \qquad\qquad (3.18)$

Here N_2/N_1 is known as turn transformation ratio and both voltage and turn ratio are proportional to each other.

In a transformer, it is a fact that the MMF on both the sides are the same at any instant of operating time. That is, $N_1 I_1 = N_2 I_2$. This gives that, $I_1/I_2 = N_2 N_1 = K$. The ratio I_1/I_2 is the current transformation ratio. Comparing this ratio with equation (3.18), it is equivalent to $1/K$.

3.23 WORKED ILLUSTRATIONS XII

Illustration 3.58

A 1-phase transformer has 400 primary and 1000 secondary turns. The net cross-sectional area of the core is 60 eq-cm. If the primary winding is connected to a 50 cps supply at 500 V, calculate (a). The peak value of the flux density in the core, and (b) the voltage inducted in the secondary winding.

Solution

$E_1 = 4.44 f \phi_m T_1$

$500 = 4.44 \times 50 \times 400 \, \phi_m$

$\therefore \quad \phi_m = \dfrac{1}{4.44 \times 40} = 5.63 \times 10^{-3}$ Wb

$\therefore$ The flux density $= \dfrac{\phi_m}{\text{area in m}^2} = \dfrac{5.63 \times 10^{-3}}{60 \times 10^{-4}}$

$= 0.94 \text{ Wb/m}^2$

Voltage induced in the secondary

$= 500 \times \dfrac{1000}{400} = 1250 \text{ V}$

Illustration 3.59

A 1-phase, 50 Hz, core type tranformer has square cores of 20 cm side. The permissible flux density is 1 Wb/m². Calculate the number of turns per limb on the high and low voltage sides for a 3000/220 V ratio.

Solution

Volts per turn

$E_1 = 4.44 \, f \phi_m$

Assuming a staking factor 0.9.

$\phi_m = B_m \times$ net iron area in m²

$= 1 \times 20 \times 20 \times 0.9 \times 10^{-4}$

$= 0.036$ Wb

$\therefore \quad E_1 = 4.44 \times 50 \times 0.036 = 8.$ emf per turn will be same for both sides.

$\therefore$ The total number of secondary turns is given by EMF/turn.

$= \dfrac{220}{8} = 28$

Turns per limb $= \dfrac{28}{2} = 14$

The number of primary turns per limb

$$= 14 \times \frac{3000}{220} = 191$$

Illustration 3.60

A 1-phase, 25 Hz shell-type transformer with sandwich coil has the dimensions in cm as shown in Fig. 3.51. The ratio is 20,000/4000 V and the flux density is not to exceed 1.2 Wb/m². Find the number of turns in the several sections in the alternative arrangements.

Solution

Allowing a standing factor of 0.9
Net area

$$= 0.9 \times 34 \times 120 \text{ cm}^2 = 0.3672 \text{ m}^2$$

$$\phi_m = 0.3672 \times 1.2 = 0.4406$$

The voltage per turn

$$E_1 = 4.44 \, f \, \phi_m,$$
$$= 4.44 \times 25 \times 0.4406 = 48.9$$

$$\therefore \quad T_2 = \frac{4000}{48.9} = 82 \text{ make this } 84$$

Then $T_1 = 84 \times \frac{20,000}{4000} = 5 \times 84 = 420$

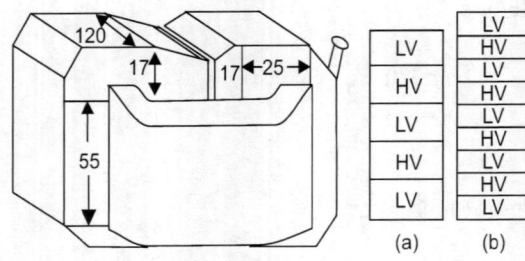

Fig. 3.51. Arrangement of coils

(a) The arrangement will be

$$T_1 = 420 = 2 \times 210$$
$$T_2 = 84 = 2 \times 21 + 42$$

(b) Make $T_2 = 88$; then $T_1 = 88 \times 5 = 440$

The arrangement will be

$$T_1 = 440 = 4 \times 110$$
$$T_2 = 2 \times 11 + 3 \times 22 = 88$$

Illustration 3.61

The area of the core is 0.085 8 sq-m in a 200 kVA 6600/440 V, 25 Hz, 1φ transformer. Find the number of turns for a flux density of about 1.2 Wb/m² and suitable conductor sections for a current density of about 2 A/mm².

Solution

$$A_i = 0.085 \text{ m}^2$$

Assuming a stacking factor of 0.9 the net iron area is

$$A_i = 0.9 \times 0.085 = 0.0764 \text{ m}^2$$

$$\phi_m = EA_i = 1.2 \, A_i = 0.092 \text{ Wb}$$

Volts per turn

$$E_i = 4.444 \, f\phi_m = = 4.44 \times 25 \times 0.092$$
$$= 10.2$$

∴ The number of secondary turns

$$= \frac{440}{10.2} = 44$$

The number of primary turns

$$= 44 \times \frac{6600}{440} = 660$$

$$I_1 = \frac{200 \times 1000}{6600} = 30.3 \text{ A}$$

∴ Primary conductor section

$$= \frac{30.3}{\text{current density}} = \frac{30.3}{2} = 15 \text{ mm}^2$$

(that is current/(current/mm²))

$$I_2 = 30.3 \times \frac{6600}{440} = 455 \text{ A}$$

∴ Secondary conductor section

$$= 445/2 = 230, \text{ mm}^2$$

Illustration 3.62

The core of a 230/260 V 25 Hz continuously variable auto-transformer consists of an area of 57.6 × 10⁻⁴ sq.m.

Allowing 10% for insulation and a B_{max} of 1.2 Tesla, find the required number of turns and size of wire.

Solution

The net iron area

$$A_1 = 57.5 \times 10^{-4} \text{ m}^2$$

The flux $= BA_1 = 1.2 \times 57.6 \times 10^{-4}$

$$= 69.12 \times 10^{-4} \text{ Wb}.$$

∴ The voltrage per turn $= E_t$

$= 4.44 \text{ f} \phi = 4.44 \times 25 \times 69.12 \times 10^{-4}$

$= 0.7672$

∴ The number of secondary turns

$$= \frac{260}{0.7672} = 339$$

$$T_1 = \frac{339 \times 230}{260} = 300$$

(i.e.) There are 339 turns with a tapping at 300

External circumference $= 14 \pi$ cm $= 140 \pi$ mm

∴ Size of wire $= \dfrac{140\pi}{339} = 1.3$ mm (overinsulation)

Illustration 3.63

Determine the number of turns per phase in each winding of three phase transformer with a ratio of 20,000/2000 V, to work on a 50 Hz network. The high voltage winding is delta connected and the low voltage winding star connected. Each core has a gross section of 560 cm². Assume a flux density of about 1.2 Wb/m².

Solution

Assuming standing factor of 0.9

Net iron area $A_i = 0.9 \times 560 = 504 \text{ cm}^2$

∴ The flux $\phi_m = B_m A_i = 1.2 \times 504 \times 10^{-4}$

$$= 0.0605 \text{ Wb}$$

The voltage/turn

$E_i = 4.44 \text{ f}\phi_m = 4.44 \times 50 \times 0.0605$

= 13.4 V

The phase voltage ratio (Δ/Y)

$$= \frac{20{,}000}{2000/\sqrt{3}} = \frac{20{,}000}{1155} = 17.3$$

∴ Secondary turns per phase

$$= \frac{1155}{13.4} = 86$$

Primary turns/phase

$$= 17.3 \times 86 = 1490$$

Illustration 3.64

A 3-phase 50 Hz transformer of the shell type has an iron cross-section of 400 cm² (gross). If the flux density be limited to 1.2 Wb/m², find the number of turns per phase on high and low-voltage windings. The voltage ratio is 11000/550 V, the high voltage side being connected in star and the low-voltage in delta.

Solution

With a stacking factor of 0.9

The net iron area

$$A_i = 400 \times 0.9 = 360 \text{ cm}^2$$

The flux $\phi_m = B_m A_i = 1.2 \times 360 \times 10^{-4}$

$$= 0.0432 \text{ Wb}$$

Volts per turn $E_i = 4.44 f \phi_m$

$$= 4.44 \times 50 \times 0.0432 = 9.59 \text{ V}$$

The phase turn ratio (Δ/Y):

$$= \frac{11000/\sqrt{3}}{550} = 11.55$$

The secondary turns per phase

$$= \frac{550}{9.59} = 58$$

The primary turns per phase

$$= 58 \times 11.55 = 670$$

Note that the voltages are always kept in per phase.

Illustration 3.65

A single-phase transformer has 200 turns on the primary and 100 on secondary. The load draws a current of 20 A from the secondary. If the primary winding is connected to a 200 V supply, determine:
(i) Primary current
(ii) Secondary voltage.

Solution

$K = N_2/N_1 = 100/200 = 1/2$; $I_2 = 20$ A, $V_1 = 200$ V
(i) $I_1 = KI_2 = (1/2) \times 20 = 10$ A
(ii) $E_2 = V_2 = KV_1 = (1/2) \times 200 = 100$ V

e.m.f./turn = 200/200 = 1 V

∴ $E_2 = 100 \times 1 = 100$ V

Illustration 3.66

A single-phase transformer has 400 primary and 1000 secondary trns. The net cross-sectional area of the core is 60 cm². If the primary winding be connected to a 50 Hz supply at 520 V, calculate (i) the peak value of flux density in the core (ii) the voltage induced in the secondary winding.

Solution

$K = N_2/N_1 = 100/400 = 2.5$

(i) $E_2/E_1 = K \therefore E_2 = KE_1 = 2.5 \times 520 = 1300$ V

(ii) $E_1 = 4.44 \, f \, N_1 \, B_m \, A$

$520 = 4.44 \times 50 \times 400 \times B_m \times (60 \times 10^{-4})$

$\therefore B_m = 0.976$ Wb/m^2

Illustration 3.67

A single-phase 50 Hz transformer is required to step down from 2200 V to 250 V. The cross-sectional area of the core is 36 cm^2 and the maximum value of the flux density is 6 Wb/m^2. Determine the number of turns of the primary and secondary windings and also the turn ratio.

Solution

$E_1 = 4.44 \, f \, \phi_m \, N_1$ volt

$E_1 = 2200$ V; $f = 50$ Hz;

$\phi_m = B_m \times A = 6 \times (36 \times 10^{-4}) = 0.0216$ Wb

$2200 = 4.44 \times 50 \times 0.0216 \times N_1$

$\therefore N_1 = 459$

Similarly,

$250 = 4.44 \times 50 \times 0.0216 \times N_2$

$\therefore N_2 = 52$

Turn ratio

$= N_1/N_2 = 459/52.$

Illustration 3.68

A 25 kV A transformer has 500 turns on the primary and 50 turns on the secondary winding. The primary is connected to 3000 V, 50 Hz supply. Find the full-load primary and secondary currents, the secondary e.m.f. and the maximum flux in the core.

Solution

$K = N_2/N_1 = 50/500 = 1/10$

Now, primary current is, KVA/V$_p$.

$I = 25,000/3000 = 8.33$ A,

So, full load current is,

$I_2 = I_1/K = 10 \times 8.33 = 83.3$ A

E.M.F. per turn on primary side = 3000/500 = 6 V

$\therefore$ Secondary e.m.f. = $6 \times 50 = 300$ V (or $E_2 = KE_1 = 300 \times 1/10 = 300$ V)

Also, $E_1 = 4.44 \, f \, N_1 \, \phi_m$;

$3000 = 4.44 \times 50 \times 500 \times \phi_m = 27$ mWb

$\therefore \phi_m = 27$ mWb

3.24 EQUIVALENT RESISTANCE

In Fig. 3.52 shown a transformer whose primary and secondary windings have resistance of R_1 and R_2 respectively. The resistance have been shown external to the windings.

It would now be shown that the resistance of the two windings can be transferred to any one of the two windings. The advantage of concentrating both the resistances in one winding is that it makes calculations very simple and easy because one has then to work in one winding only. It will be proved that a resistance of R_2 in secondary is equivalent to R_2/K^2 in primary. The value of R_2/K^2 will be denoted by R'_2 – the equivalent secondary resistance as referred to primary.

The copper loss in secondary is $I_2^2 R_2$. This loss is supplied by primary which takes a current of I_1. Hence if R'_2 is the equivalent resistance in primary which would have caused the same loss as R_2 in secondary then

$$I_1^2 R'_2 = I_2^2 R_2 \text{ or } R'_2 = (I_2/I_1)^2 R_2 = R_2/K^2$$

Similarly, equivalent primary resistance as referred to secondary is $R'_1 = K^2 R_1$

In Fig. 3.53 secondary resistance has been transferred to primary side leaving secondary circuit resistance less. The $R_1 + R'_2 = R_1 + R_2/K^2$ is known as the equivalent or effective resistance of the transformer as referred to primary and may be designated as R_{01}.

$$R_{01} = R_1 + R'_2 = R_1 + R_2/K^2 \quad (3.19)$$

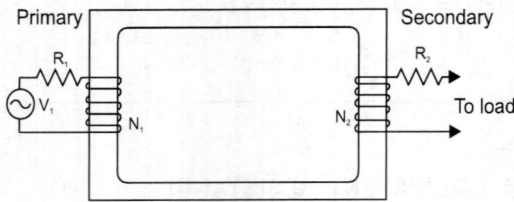

Fig. 3.52.

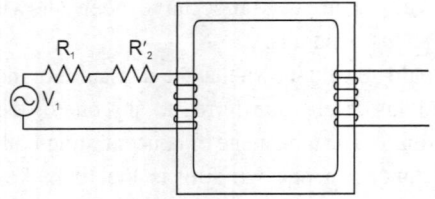

Fig. 3.53.

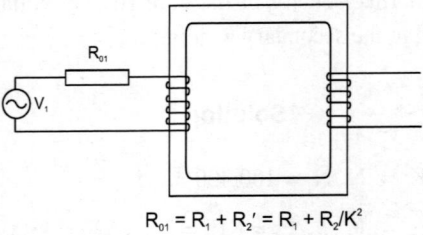

$R_{01} = R_1 + R'_2 = R_1 + R_2/K^2$

Fig. 3.54.

Similarly, the equivalent resistance of the transformer as referred to secondary is

$$R_{02} = R_2 + K^2 R_1 \quad (3.20)$$

This fact is shown in Fig 3.54 where all the resistance of the transformer has been concentrated in the secondary winding.

It will be noted that

1. a resistance of R_1 in primary is equivalent to $K^2 R_1$ in the secondary. Hence it is called the equivalent primary resistance as referred to secondary i.e R'_1
2. a resistance of R_2 in secondary is equivalent to R_2/K^2 in primary. Hence, it is called the equivalent secondary resistance as referred to primary i.e R'_2
3. Total or effective resistance of the transformer as referred to primary is

   R_{01} = primary resistance + equivalent secondary resistance referred to secondary
   $R_1 + R'_2 = R_1 + R_2/K^2 \quad (3.21)$

4. Similarly total transformer resistance as referred to as secondary is

   R_{02} = Secondary resistance + equivalent primary resistance as referred to secondary
   $R_2 + R'_1 = R_2 + K^2 R_1 \quad (3.22)$

3.25 MAGNETIC LEAKAGE

It was usual to assume that all the flux linked with primary winding also links with the secondary winding. But, in practice, it is impossible to realize this condition. It is found, in practice, that all the flux linked with primary does not link with the secondary but part of it i.e., ϕ_{L1} completes its magnetic circuit by passing through air rather than around the core, as shown in Fig. 3.55. This leakage flux is produced when the m.m.f. due to primary ampere-turns existing between points a and b, acts along the leakage paths. Hence, this flux is known as primary leakage flux and is proportional to the primary ampere-turns alone, because the secondary turns do not link the magnetic circuit of ϕ_{L1}. The flux ϕ_{L2} which is linked with secondary winding alone (and not with primary turns). This flux ϕ_{L1} is in time phase with I_1 and produces a self-induced e.m.f. e_{L1} in primary (and not in secondary). Similarly across CD the leakage flux ϕ_{L2} will be in time phase with I_2 producing e_{L2} in secondary (and not in primary).

At no load and light-loads the primary and secondary ampere-turns are small hence leakage fluxes are negligible. But when load is increased both primary and secondary windings carry more currents. Hence large m.m.f.s are set up which, while acting on leakage paths, increase the leakage flux.

As said earlier, the leakage flux linking with each winding, produces a self-induced e.m.f. in that winding. Hence, in effect, it is equivalent to a small

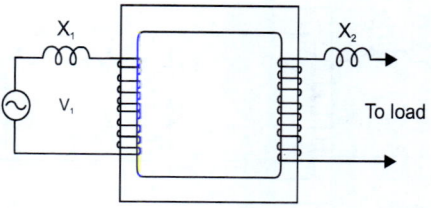

Fig. 3.56.

choke or inductive coil in series with each winding such that voltage drop in each series coil is equal to that produced by leakage flux. In other words, a transformer with magnetic leakage is equivalent to an ideal transformer with an extra inductive coil connected in both primary and secondary circuits as shown in Fig. 3.56, such that the internal-internal e.m.f. in each inductive coil is equal to that due to the corresponding leakage flux in the actual transformer.

The terms X_1 and X_2 are known as primary and secondary leakage reactances respectively accounting for the above said process. The leakage flux links one or the other windings but not both, hence it in no way contributes to the transfer of energy from the primary to the secondary winding. Also the primary voltage V_1 will have to supply reactive drop $I_1 X_1$ in addition to $I_1 R_1$ and similarly, E_2 will have to supply $I_2 R_2$ and $I_2 X_2$.

3.26 TRANSFORMER WITH RESISTANCE AND LEAKAGE REACTANCE

In Fig. 3.57 are shown the primary and secondary windings of a transformer with resistance and leakage reactances taken out of the windings. The primary impedance is given by

$$Z_1 = R_1 + jX_1 = \sqrt{R_1^2 + X_1^2}$$

Similarly, secondary impedance is given by

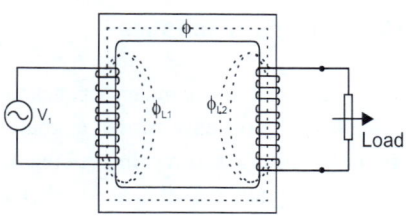

Fig. 3.55.

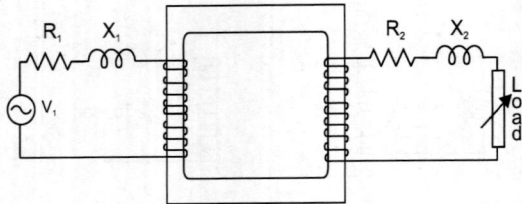

Fig. 3.57.

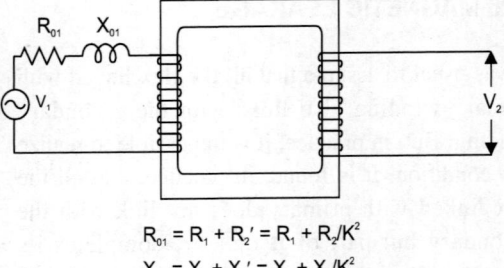

$R_{01} = R_1 + R_2' = R_1 + R_2/K^2$
$X_{01} = X_1 + X_2' = X_1 + X_2/K^2$

Fig. 3.58.

$$Z_2 = R_2 + jX_2 = \sqrt{R_2^2 + X_2^2}$$

The resistance and leakage reactance of each winding is responsible for some voltage drop in each winding. In primary, the leakage reactance drop is $I_1 X_1$ (usually 1 or 2% of V_1). Hence,

$$V_1 = E_1 + I_1(R_1 + jX_1) = E_1 + I_1 Z_1 \quad (3.23)$$

Similarly, there are $I_2 R_2$ and $I_2 X_2$ drops in secondary. Note that, in primary $V_1 > E_1$ but in secondary, $E_2 > V_2$. Thus,

$$E_2 = V_2 + I_2(R_2 + jX_2) = V_2 + I_2 Z_2 \quad (3.24)$$

By adopting the similar approach which was used in tranforming the resistances from one winding to the other we can write,

$$X'_1 = K^2 X_1 \quad (3.25)$$

Similarly

$$X'_2 = X_2 / K^2 \quad (3.26)$$

$$X_{01} = X_1 + X'_2 = X_1 + X_2/K^2 \quad (3.27)$$

$$X_{02} = X_2 + X'_1 = X_2 + K^2 X_1 \quad (3.28)$$

It is obvious that total impedance of the transformer as referred to primary is given by

$$Z_{01} = R_{01} + jX_{01} \text{ or}$$

$$Z_{01} = \sqrt{R_{01}^2 + X_{01}^2} \quad (3.29)$$

and $Z_{02} = R_{02} + jX_{02}$ or

$$Z_{02} = \sqrt{R_{02}^2 + X_{02}^2} \quad (3.30)$$

Reflection Section

1. Why the core structure is laminated? Justify.
2. Can an cir-cored transformer exist? Discuss.
3. Can there be a third winding (that is 2nd secondary winding)? If so, how will it act?
4. What is the significance of indicating resistance R and resistance X in the circuit of a transformer? What do R, X and Z imply?

3.27 CHARACTERISTICS OF A TRANSFORMER

3.27.1 No-load Characteristics

On no-load, i.e. with the secondary open and no current through it, the primary carries a small no-load current of I_0, because this winding is connected to the primary supply and thus it is a closed circuit. I_0 has two components viz.,

(i) A magnetizing component I_m (or I_μ) which lags 90° behind E_1. This is the current whose function is to produce and maintain the magnetic flux.

(ii) A working component I_w (or I_c) which is necessary to take care of the total core or iron loss (= hysteresis loss + eddy current loss) of the transformer. I_w produces the necessary real power to supply the core loss. This component will be in phase with E_1.

Note also that the total no-load current is made up of two vector currents, which are I_m and I_w. So, $\bar{I}_0 = \bar{I}_m + \bar{I}_w$. Thus an electrical equivalent circuit can now be made with this vector sum as shown in Fig. 3.60.

As I_m is the total current responsible for the production of magnetic flux ϕ in the transformer, it is taken to flow against a magnetic coil of reactor X_0, whereas, as I_w is responsible for core losses, it is taken to flow against no-load resistance of R_0.

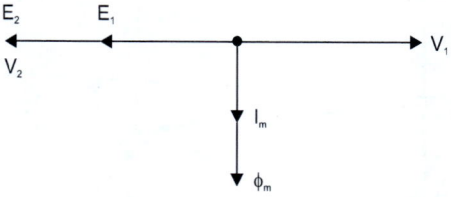

Fig. 3.59. No-load vector diagram of transformer.

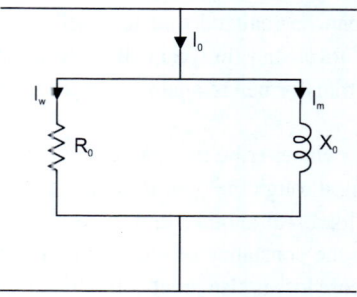

Fig. 3.60. No-load equivalent circuit.

With this two essential components I_m and I_w, the no-load vector diagram can be drawn as illustrated in Fig. 3.59. For high efficiency and good regulation, I_m and I_w must both be small. I_w is kept small by using low-loss iron and not allowing the transformer to work at higher flux-densities. I_m is kept small by having a closed magnetic circuit, that is a magnetic circuit with no air-gap and no leakage.

Examination of the no-load vector diagram, reveals the following important relations:

1. E_1 is equal to and opposite to V. This is because E_1 can be taken as the back e.m.f. of the primary, thus opposing V_1.
2. E_2 is in-phase with V_2, but opposite in-phase by 180° to V_1.
3. The magnetizing current I_m lags V_1 by 90° and produces ϕ_m, in phase with I_m.
4. E_1 and E_2 lags ϕ_m by 90°, and are produced by ϕ_m.
5. V_2 is equal to magnitude to E_2 and opposite to V_1 by 180°.

3.27.2 Transformer On-load

Consider, at first, a non-inductive load (that is, a pure resistive load) connected to the secondary terminals; because the secondary load circuit is closed, a secondary current I_2 will flow now, and as this load is a pure resistive load, E_2 and I_2 will be in-phase with each other. Because of the introduction of the secondary current I_2, an amp-turns of $N_2 I_2$ will now be set-up in the secondary winding. This will produce a flux, ϕ_2, in the secondary. Already there was an established flux, ϕ, available in the transformer established due to I_m; and now we are facing a new flux getting established, called ϕ_2. This secondary flux if allowed to exist, would disturb the initial flux conditions existing at no-load by I_m. That is, the ϕ_2 will be in opposition with ϕ, and thus the opposing secondary flux ϕ_2 will, momentarily weaken the existing transformer flux, ϕ. Thus the primary back emf E_1 will tend to reduce.

252 Basic Electrical Engineering

Now, as soon as the primary back emf E_1 reduces, V_1 gains the upper hand over E_1 and thus will cause the primary current I_1 to flow large. So, there is always an additional in-rush of current in the primary, when the secondary of a transformer is loaded. Let this aditional in-rush primary current be I'_2. This is called the load component of the primary current. This current will be anti-phase with I_2.

Now, the new in-rush current I'_2 will establish an additional m.m.f. at the primary, $N_1 I'_2$. This will set up, a new flux in the primary, and let it be called as ϕ'_2. It can be realized that this will be in just opposition to ϕ_2 and thus cancelling its existence. Thus, the transformer is again left with the initial flux, ϕ.

Hence, whatever be the load condition, the net flux passing through the core is almost the same as that at no load. An another important deduction is that due to the constancy of the core flux at all the loads, the core loss is also practically the same under all load conditions. So, core loss or iron loss, is a constant.

3.27.3 Secondary Vector Diagram of a Transforms On-load

We have, From eqn. (3.22) That,

$$\bar{E}_2 = \bar{V}_2 + \bar{I}_2\bar{R}_2 + \bar{I}_2\bar{X}_2$$

or $\bar{E}_2 = V_2 + jI_2\bar{Z}_2$.

Thus, when I_2R_2 and $I_2 X_2$ are vectorially added to V_2 (the load terminal voltage), we get the secondary induced voltage, E_2.

- Let V_2 be taken as the reference vector
- Let an inductive load be connected to the secondary of the transformer which makes the secondary current I_2 to lag its terminal voltage by a phase angle of ϕ_2.

- I_2R_2 is vectorially added to V_2. This is done by adding I_2R_2 vector to V_2 by placing it in-phase with I_2.

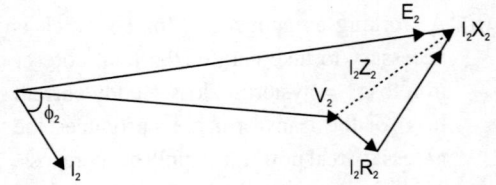

Fig. 3.61. $\bar{E}_2 = \bar{V}_2 + \bar{I}_2\bar{Z}_2$

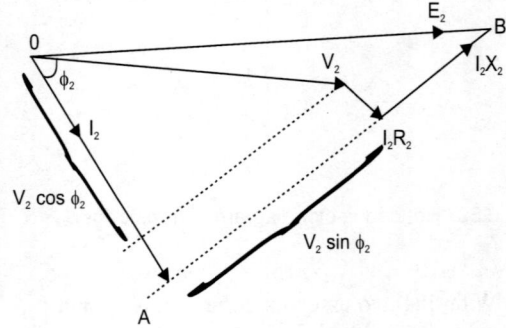

Fig. 3.62

- To this summation, $I_2 X_2$ is vecorially added. The vector $I_2 X_2$ is added to $\bar{V}_2 + \bar{I}_2\bar{R}_2$, by placing this vector is quadrature with $\bar{I}_2$.
- $\bar{I}_2\bar{Z}_2$ is then obtained from the sum, $\bar{I}_2\bar{R}_2 + \bar{I}_2\bar{X}_2$ and $\bar{E}_2 = \bar{V}_2 + j\bar{I}_2\bar{Z}_2$.
- This is illustrated in Fig. 3.61.
- The value of E_2 is obtainable from the above vector diagram. Refer now to the same vector diagram as extended in Fig. 3.62.
- From the right-angle triangle OAB

$$E_2 = \sqrt{(V_2 \cos\phi_2 + I_2R_2)^2 + (V_2 \sin\phi_2 \pm I_2X_2)^2}$$

(3.31)

3.28 VOLTAGE REGULATION OF A TRANSFORMER

When a transformer is loaded, its secondary terminal voltage falls (for a lagging p.f) provided the applied primary voltage V_1 is held constant. As a DC generator, the output of a transformer is voltage (that is electrical energy), which drops as the transformer is graudually loaded. We know that, $V_2 = E_2 - I_2 Z_2$. The secondary current I_2 will obviously increase when the transformer is loaded, and thus the total voltage drop $I_2 Z_2$ in the secondary increase, making V_2 to fall. This drop in secondary terminal voltage from no-load (N.L) to full load (F.L), that is from E_2 to a new value of V_2 is called the voltage regulation of the transformer and is usually expressed as percentage of the secondary no-load voltage.

Percentage regulation

$$\frac{\text{Sec.Volt (N.L.)} - \text{sec.volt (F.L.)}}{\text{Sec. Volt (N.L.)}} \times 100 \quad (3.32)$$

Let V_L = Secondary terminal voltage on no load

V_0 = Secondary terminal voltage on load

then % regulation = $\dfrac{V_0 - V_L}{V_0} \times 100$

Since the value of $V_0 - V_L$ is the voltage drop given by

$V_0 - V_L = I_2 (R_{02} \cos \phi_2 \pm X_{02} \sin \phi_2)$, where

plus for lagging power factor and minus for leading power factor, the percentage voltage regulation is:

$$\% \text{ VR} = \frac{I_2 (R_{02} \cos \phi_2 \pm X_{01} \sin \phi_2)}{V_2} \times 100 \quad (3.33)$$

3.29 LOSSES IN A TRANSFORMER

In a static transformer, there are no friction or windage losses because there are no rotating parts. Hence, the other losses occurring are:

Core or Iron Loss

It includes both hysterics loss and eddy current loss. Because the core flux in a transformer remains practically constant for all loads (its variation being 1 to 3% from no-load to full load) the core loss is practically the same at all loads.

These losses are minimized by using steel of high silicon content the core and by using very thin laminations.

(ii) Copper Loss: This loss is due to the ohmic resistance of the transformer windings. Total Cu los = $I^2 R_1 + I^2_2 R_2 = I^2_1 R_{01} = I^2_2 R_{02}$. It is clear that Cu loss is proportional to $(\text{current})^2$ or $(kVA)^2$. In other word, Cu loss at that Cu oss is proportional to $(\text{current})^2$ or $(kVA)^2$

3.30 EFFICIENCY OF A TRANSFORMER

As is the case with other types of electrical machines, the efficiency of a transformer at a particular load and power factor is defined as the output divided by the input – the two being measured in the same units (either watts or kilowatts).

$$\text{Efficiency} = \frac{\text{Output power}}{\text{Input power}}$$

But a transformer being a highly efficient piece of equipment. has very small losses, hence, it is impractical to try to measure transformer efficiency by measuring its input and output. These quantities are nearly of the same values. A better method is to determine the losses and then to calculate the efficiency from

$$\text{Efficiency} = \frac{\text{output power}}{(\text{output power}) + \text{losses}}$$

$$= \frac{\text{output power}}{(\text{output power}) + \text{Cu loss} + \text{Iron loss}} \quad (3.34)$$

Thus the total losses = constant loss + variable loss

= Iron loss + copper loss

The copper loss is variable because it is proportional to the (load current)^2, which is a variable.

It may be noted here that efficiency is based on power output in watts and not on volt-amperes, although losses are proportional to VA. Hence, at any volt-ampere load, the efficiency depends on power factor, being maximum at a power factor of unity.

Also note that the core losses and copper losses are found out in watt units, whereas the transformer rating (that is the output) will be VA units. By knowing the output (or load) power factor, the equivalent watt (that is, VA × Pf) is found out and used in the expression of the efficiency.

Thus,

$$\text{efficiency} = \frac{(V_2 I_2)\text{pf}, \text{W}}{(V_2 I_2 \text{Pf} + \text{Iron loss} + I_1^2 R_1 + I_2^2 R_2), \text{W}}$$

Note that, as the copper loss is proportional to the (current)^2, if P_{cu} is the copper loss at full load, by the above said proportionality constraint, it is $(1/2)^2 P_{cu}$ in half-load, and $(2)^2 P_{cu}$ at twice the full load. It can be seen in the table below.

	P_{cu} = Copper loss at full-load
Fraction of load	Copper loss is terms of P_{cu}
1/4	$(1/4)^2 P_{cu}$
1/2	$(1/2)^2 P_{cu}$
3/4	$(3/4)^2 P_{cu}$
1	P_{cu}
1 1/2	$(1.5)^2 P_{cu}$
2	$2^2 P_{cu}$

Reflection Section

1. What is the significant difference between $\cos\theta_0$ and $\cos\theta_2$?
2. Derive the vector diagram of a transformer on load.
3. There is a transformer with a % voltage regulation of 4% and another transformer with % voltage regulation of 40%. Which is good and why?
4. List the losses in a transformer and discuss how do they vary with the applied load?

3.31 WORKED ILLUSTRATIONS XIII

Illustration 3.69

(a) A 2,200/200 V transformer draws a no-load primary current of 0.6 A and absorbs 400 W. Find the magnetising and iron loss currents.

(b) A 2,200/250 V transformer takes 0.5 A at a p.f. of 0.3 on open circuit. Find magnetising and working components of no-load primary current.

Solution

(a) Iron-loss current

$$= \frac{\text{No-load input in watts}}{\text{Primary voltage}} = \frac{400}{2,200} = 0.182 \text{ A}$$

Now $I_0^2 = I_w^2 + I_\mu^2$

Magnetising component $I_\mu = \sqrt{(0.6^2 - 0.182^2)} = 0.572$ A

(b) $I_0 = 0.5$ A, $\cos \phi_0 = 0.3$

∴ $I_w = I_0 \cos \phi_0 = 0.5 \times 0.3 = 0.15$ A

$I_\mu = \sqrt{(0.5^2 - 0.15^2)} = 0.476$ A

Illustration 3.70

A single-phase transformer has 500 turns on the primary and 40 turns on the secondary winding. The mean length of the magnetic path in the iron core is 150 cm and the joints are equivalent to an air-gap of 0.1 mm. When a p.d of 3,000 V is applied to the primary, maximum flux density is 1.2 Wb/m². Calculate (a) the cross-sectional area of the core (b) no-load secondary voltage (c) the no-load current drawn by the primary (d) power factor on no-load. Given that AT/cm for a flux density of 1.2 Wb/m² in iron to be 5, the corresponding iron loss to be 2 W/kg at 50 Hz and the density of iron as 7.8 gram/cm³.

Solution

(a) $3{,}000 = 4.44 \times 50 \times 500 \times 1.2 \times A$

∴ $A = 0.0225$ m² $= 225$ cm²

This is the net cross-sectional area. However, the gross area would be about 10% more to allow for the insulation between laminations.

(b) $K = N_2/N_1 = 40/500 = 4/50$

∴ N.L. secondary voltage = $KE_1 = (4/50) \times 3000 = 240$ V

(c) AT per cm = 5

∴ AT for iron core = $150 \times 5 = 750$

AT for air-gap = Hl, (using the formula, Hl = NI).

$$= \frac{B}{\mu_0} \times 1 = \frac{1.2}{4\pi \times 10^{-7}} \times 0.0001 = 95.5$$

Total AT for given $B_{max} = 750 + 95.5 = 845.5$

Max. value of magnetizting current drawn by primary = $845.5/500 = 1.691$ A

Assuming this current to be sinusoidal, its r.m.s. value is $I_m = 1.691/\sqrt{2} = 1.196$ A

Volume of iron = length × area = $150 \times 225 = 33750$ cm²

Density = 7.8 gm/cm³.

∴ Mass of iron = $33750 \times 7.5/1000 = 263.25$ kg.

Total iron loss = $263.25 \times 2 = 526.5$ W

Iron loss component of no-load primary current I_0 is $I_w = 526.5/3000 = 0.176$ A

$$I_0 = \sqrt{I_m^2 + I_w^2} = \sqrt{1.196^2 + 0.176^2} = 1.208 \text{ A}$$

(d) Power factor, $\cos \phi_0 = I_w/I_0 = 0.176/1.208 = 0.1457$.

Illustration 3.71

A 50 kVA. 4400/220 V transformer has $R_1 = 3.45$ Ω. $R_2 = 0.009$ Ω. The values of reactance are $X_1 = 5.2$ Ω and $X_2 = 0.015$ Ω. Calculate for the transformer (i) equivalent resistance as referred to primary (ii) equivalent resistance as referred to secondary (iii) equivalent reactance as referred to both primary and secondary (iv) equivalent impedance as referred to both primary and secondary (v) total Cu loss, first using individual resistances of the two windings and secondly, using equivalent resistance as referred to each side.

Solution

Full load

$I_1 = 50{,}000/4400$

$= 11.36$ A (assuming 100% efficiency)

Full-load

$I_2 = 50{,}000/220 = 227$ A

$K = 220/4400 = 1/20$

(i) $R_{01} = R_1 + R_2/K^2 = 3.45 + 0.009/(1/20)^2$

$= 3.45 + 3.6$

$= 7.05\ \Omega$

(ii) $R_{02} = R_2 + K^2 R_1 = 0.009 + (1/20)^2\ 3.45$

$= 0.0176\ \Omega$

(iii) $X_{01} = X_1 + X'_2 = X_1 + X_2/K^2$

$= 5.2 + 0.015/(1/20)^2$

$= 11.2\ \Omega$

Also $X_{02} = X_2 + X'_1 = X_2 + K^2 X_1 = 0.015 + 5.2/20^2$

$= 0.028\ \Omega$

Also $X_{02} = X_{01} K^2 = 11.2/400 = 0.028\ \Omega$

(iv) $Z_{01} = \sqrt{R_{01}^2 + X_{01}^2}$

$= \sqrt{(7.05^2 + 11.2)^2} = 13.23\ \Omega$

$Z_{02} = \sqrt{R_{02}^2 + X_{02}^2}$

$= \sqrt{(0.0176^2 + 0.028)^2} = 0.0331\ \Omega$

also $Z_{02} = K^2/Z_{01} = 13.23/400 = 0.0331\ \Omega$

Cu loss $= I_1^2 R_1 + I_2^2 R_2$

$= 11.36^2 \times 3.45 + 227^2 \times 0.009 = 909\ \Omega$

Also copper loss is $I_1^2 R_{01} = 11.36^2 \times 7.05 = 910\ \Omega$

$= I_2^2 R_{02} = 227^2 \times 0.0176 = 910$ W

Illustration 3.72

A 230 / 460 V transformer has a primary resistance of 0.2 Ω and a reactance of 0.5 Ω and the corresponding values for the secondary are 0.75 Ω and 1.8 Ω respectively. Find the secondary terminal voltage when supplying (a) 10 A at 0.8 power factor lagging (b) 10 A at p.f. 0.8 leading.

Solution

$K = 2$

$R_{02} = 0.75 + (4 \times 0.2) = 1.55\ \Omega$

$X_{02} = 1.8 + (4 \times 0.5) = 3.8\ \Omega$

$I_2 = 10$ A; $\cos \phi = 0.8$, and $\sin \phi = 0.6$

(a) Voltage drop $= I_2 (R_{02} \cos \phi + X_{02} \sin \phi)$

$= 10 (1.55 \times 0.8 + 3.8 \times 0.6)$

$= 35.2$ V

$V_2 = 460 - 35.2 = 424.8$ V

Illustration 3.73

A 50-KVA transformer has primary voltage of 6600 V a secondary voltage of 250 V. It has 52 secondary turns. Find the number of primary turns and the primary and secondary current neglecting losses.

Solution

The capacity of transformer 50 kVA

The primary voltage = 6600 V

Secondary voltage = 250 V

The transformation rations are

$$\frac{V_2}{V_1} = \frac{N_2}{N_1} = \frac{I_1}{I_2}$$

So here as given we know that

$$\frac{250}{6600} = \frac{52}{N_1}$$

$$N_1 = \frac{6600 \times 52}{250} = 1373 \text{ tuns}$$

or

The current in secondary winding

$$\frac{KVA}{KV}$$

$$= \frac{50 \times 1000}{250} = 200 \text{ A}$$

Now the primary current is

$$\frac{250}{6600} = \frac{I_1}{200}$$

$$I_1 = \frac{250 \times 200}{6600} = 7.58 \text{ A}$$

Illustration 3.74

A single-phase 50 c/s transformer is required to have a no load voltage ratio of 3300/440 V. The total maximum flux in the core is to be about 50 m Wb. Find the number of turns in each winding.

Solution

The e.m.f. equation of the transformer is

$$E = 4.44 \, f\phi N \text{ V}$$

where E = the induced voltage

ϕ = flux in Wb

f = Frequency in c/s

N = Number of turns

and

for secondary side is can be calculated by the ratio method or direct calculation.

$$440 = 4.44 \times 50 \times 50 \times 10^{-3} \times N$$

N = 40 turns (Ans)

Illustration 3.75

The maximum flux density in the core of a 240 / 3000 V, 50 c/s transformer is 1.25 Wb/m². If the e.m.f induced per turns in 8 V, find the number of primary and secondary and net cross sectional area of the core.

Solution

The e.m.f induced per turn is 8 volts, i.e

$$E/N = 4.44 \, f BA$$

$$8 = 4.44 \times 1.25 \times A \times 50$$

and $$A = \frac{8 \times 10^4}{4.44 \times 1.25 \times 50} \text{ cm}^2$$

$= 288.2 \text{ cm}^2$ **(Ans.)**

Now, $E = 4.44 \text{ B.A.f N V}$

or $N = \dfrac{240 \times 10^4}{4.44 \times 1.25 \times 50 \times 288.2} \text{ cm}^2$

$= 30$ turns

or $\dfrac{V_1}{V_2} = \dfrac{N_1}{N_2}$

or $\dfrac{240}{300} = \dfrac{N_1}{N_2}$

$N_2 = 374$ turns

So the number of turns in the secondary winding is 375 turn and in primary winding is 30 turns.

Illustration 3.76

A single-phase transformer has 300 primary turns and 750 secondary turns, the net cross sectional area of the core is 64 sq.cm. If the primary voltage is 440 voltage 50 c/s. Find (a) maximum flux density in the core (b) e.m.f induced in the secondary.

Solution

The primary winding voltage = 440 V

The primary turns of 300, and secondary turns are 750

The transformation ratio $\dfrac{V_2}{V_1} = \dfrac{N_2}{N_1}$

So $\dfrac{V_2}{440} = \dfrac{750}{300}$

So $V_2 = \dfrac{750 \times 400}{300} = 1100 \text{ V (Ans)}$

Now the e.m.f. equated is

$E = 4.44 \text{ B.A f. N V}$

$440 = 4.44 \times B \times 64 \times 10^{-4} \times 50 \times 300$

$= 1.03 \text{ Wb/m}^2$ **(Ans)**

Illustration 3.77

A 6600/1000 volts 50 c/s transformer has 640 primary turns. The sectional area of the core is 440 sq. cm of which 90 per cent is iron, mean length of core 2.4 metres. Find the maximum flux density. Assuming a relative permeability of 1600.

Solution

The working voltage 6600 / 1000, at 50 / cs

now the cross sectional area is 440 cm^2

and 90% are will be (440 × 0.9) cm^2

the e.m.f equation

$E = 4.44 \text{ B.A. f. N. V}$

$6600 = 4.44 \times B \, (440 \times 0.9) \times 10^{-4} \times 50 \times 640$

$B = 1.173 \text{ Wb/m}^2$

3.32 APPLICATION OF DIFFERENT TYPES OF TRANSFORMERS

Step-up transformer

As the name states that the secondary voltage is stepped up in tune with the transformation ratio K-E_2/E_1 compared to primary voltage. This can be achieved by increasing the number of windings in the secondary as compared to primary windings. In power plant, this transformer is used as connecting transformer of the generator to the grid.

Step-down transformer

It is used to step down the voltage level from lower to higher level at secondary, again in tune with K, hence, it is called a step-down transformer. The winding turns more on the primary side than the secondary side.

Air core transformer

Both the primary and secondary windings are wound on a non-magnetic strip where the flux linkage between primary and secondary windings is through the air.

Compared to iron core the mutual inductance is less in air core, i.e., the reluctance offered to the generated flux is high in the air medium. But the hysteresis and eddy current losses are completely eliminated in air-core type transformer.

Iron core transformer

Both the primary and secondary windings are wound on multiple iron plate bunch which provides a perfect linkage path to the generated flux. It offers less reluctance to the linkage flux due to the conductive and magnetic property of the iron. These are widely used transformers in which the efficiency is high compared to the air-core type transformer.

Autotransformer

Standard transformers have primary and secondary windings placed in two different directions, but in autotransformer windings, the primary and the secondary windings are connected to each other in series, both physically and magnetically. Here the same winding serves both as a primary winding and also as a secondary winding, which is achieved by a movable tap contactor.

On a single common coil which forms both primary and secondary windings in which voltage is varied according to the position of secondary tapping on the body of the coil windings.

Power transformer

The power transformers are big in size. They are suitable for high voltage (greater than 33 kV) power transfer applications. It is used in power generation stations and transmission substations. It has high insulation level.

Distribution transformer

In order to distribute the power generated from the power generation plant to remote locations, these transformers are used. Basically, it is used for the distribution of electrical energy at low voltage, that is, less than 33 kV for industrial purpose and 440 V-220 for in domestic purpose.

- It works at low efficiency at 50-70%
- Small size
- Easy installation
- Low magnetic losses
- It is not always fully loaded

Measurement transformer

These are used to measure high voltage, high current, power, etc. These are classified as potential transformers, current transformers, etc.

Protection transformers

This type of transformers is used in component protection purpose. The major difference between measuring transformers and protection transformers is the accuracy that means that the protection

3.33 OPEN CIRCUIT AND SHORT CIRCUIT TESTS

These two tests on a transformer help to determine (i) the parameters of the equivalent circuit of Fig. 3.66, (ii) the voltage regulation and (iii) efficiency. The equivalent circuit parameters can also be obtained from the physical dimensions of the transformer core and its winding details. Complete analysis of the transformer can be carried out, once its equivalent circuit parameters are known.

The power required during these two tests is equal to the appropriate power loss occurring in the transformer.

Open Circuit (or No-load) Test

The circuit diagram for performing open circuit test on a single phase transformer is given in Fig. 3.68. In this diagram, a voltmeter, wattmeter and an ammeter are shown connected on the low voltage side of the transformer. The high voltage side is left open circuited. The rated frequency and voltage is applied to the primary, i.e. low voltage side, is varied with the help of a variable ratio auto-transformer. When the voltmeter reading is equal to the rated voltage of the l.v. winding, all the three instrument readings are recorded, respectively V_0, I_0 and W_0.

Let

V_1 = Applied rated voltage on l.t. side,

I_0 = exciting current (or no-load current),

and P_0 = core loss.

Then $P_0 = V_1 I_0 \cos \phi_0$

The ammeter records the no-load current or exciting current I_0. Since I_0 is quite small (2 to 6% of rated current), the primary leakage impedance drop is almost negligible, and for all practical purposes, the applied voltage V_1 is equal to the induced e.m.f. E_1. The I_0 is made up of two components, I_w and I_c (sec. 3.27.1). The vector diagram of the transformer on this no-load is shown in Fig. 3.69, which gives that $I_w = I_0 \sin \phi_0$, and $I_c = I_0 \cos \phi_0$. The value of ϕ_0 is,

$$\phi_0 = \cos^{-1}\left[\frac{W_0}{V_0 I_0}\right]$$

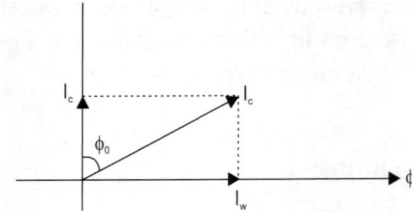

Fig. 3.64. No. load vector diagram.

As there is no current in the transformer because it is open, there is no appreciable copper loss occurring in the machine. The full supply goes to establish the magnetic flux in the core. So, the wattmeter reading goes to supply the entire core loss.

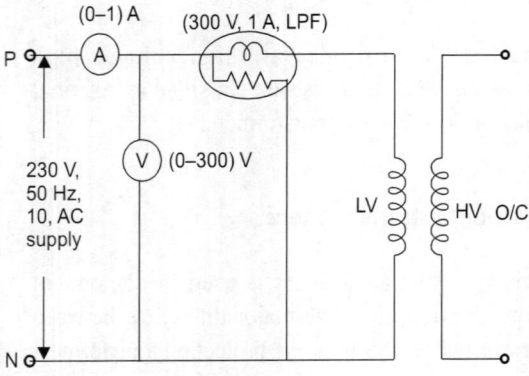

Fig. 3.63

The input power given by the wattmeter reading consists of core loss and ohmic loss. The exciting current being about 2 to 6 per cent of the full load current, the ohmic loss in the primary ($I_0^2 R_0$) varies from 0.04 per cent (2/100 × 2/100 × 100) to 0.36 per cent of the full-load primary ohmic loss. In view of this fact, the ohmic loss during open circuit test is negligible is comparison with the normal core loss (approximately proportional to the square of the applied voltage). Hence the wattmeter reading can be taken as equal to transformer core loss. A negligible amount of dielectric loss may also exist. Error in the instrument readings may be eliminated if required.

$$\therefore \text{No load p.f.} = \cos \phi_0 = \frac{P_0}{V_1 I_0}$$

Short Circuit Test

In this test, the LV winding is short circuited, as shown in Fig. 3.70, and low voltage is applied to the HV (primary) winding. This low voltage is gradually increased so as to supply full-load current at the secondary.

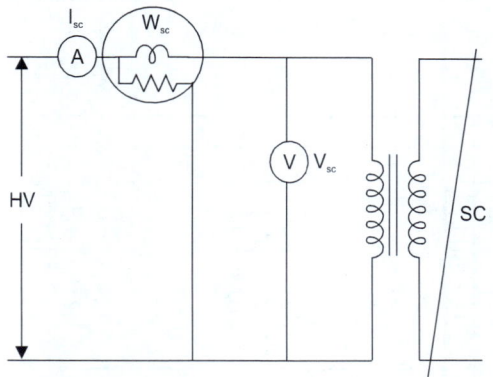

Fig. 3.65. Circuit diagram for S.C. test.

The wattmeter reading represents the full-load copper loss, as the full-load current passes through the winding. The I^2R losses of both the primary and secondary winding are indicated by W_{sc}.

So,

$$W_{sc} = I_{sc}^2 R_{01}$$

$$\therefore R_{01} = \frac{W_{sc}}{I_{sc}^2}, \Omega$$

And, $Z_{01} = \dfrac{V_{sc}}{I_{sc}}, \Omega$

$$\therefore X_{01} = \sqrt{Z_{01}^2 - R_{01}^2}, \Omega$$

So, the equivalent resistance of the transformer and the equal reactance can be found out by sc test.

Illustration 3.78

The ohmic values of the circuit parameters of a transformer, having a turns ratio of 5, are $R_1 = 0.5$ Ω; $R_2 = 0.021$ Ω; $X_1 = 3.2$ Ω; $X_2 = 0.12$ Ω; $R_c = 350$ Ω, referred to the primary; and $X_m = 98$ Ω, referred to the primary. Draw the approximate equivalent circuits of the transformer, referred to (a) the primary and (b) the secondary. Show the numerical values of the circuit parameters.

The circuits are respectively shown in Fig. 13-10 (a) and Fig. 13-10 (b). The calculations are as follows:

(a) $R' = R_1 + a^2 R_2 = 0.5 + (5)^2(0.021) = 1.025$ Ω

$X' = X_1 a^2 X_2 = 3.2 + (5)^2(0.12) = 6.2$ Ω

$R'_c = 350$ Ω

$X'_m = 98$ Ω

(b) $\quad R'' = \dfrac{R_1}{a^2} + R_2 = \dfrac{0.5}{25} + 0.021 = 0.041 \; \Omega$

$\quad\quad X'' = \dfrac{X_1}{a^2} + X_2 = \dfrac{3.2}{25} + 0.12 = 0.248 \; \Omega$

$\quad\quad R''_c = \dfrac{350}{25} = 14 \; \Omega$

$\quad\quad X''_m = \dfrac{98}{25} = 3.92 \; \Omega$

3.34 THREE-PHASE TRANSFORMER CONSTRUCTION

Three-phase transformers are used throughout industry to change values of three phase voltage and current. Since three-phase power is the most common way in which power is produced, transmitted, and used, an understanding of how three phase transformer connections are made is essential. A three-phase transformer is constructed by winding three-single phase transformers on a single core. These transformers are put into an enclosure which is then filled with dielectric oil.

Most power is distributed in the form of three-phase AC. In a DC machine discussions we found that ONE armature coil rotates to induce voltage in it and thus produce electricity. It can now be realized that the produced voltage is by a single winding in the armature and hence it is a SINGLE phase voltage. Of course it is made a DC in the load circuit using the commutator-brush technique. If three such individual windings are connected in the armature either in star or delta then a THREE-phase voltage is induced. Thus, basically all the power generators produce electricity by rotating three coils or windings through a magnetic field within the generator. These coils or windings are spaced 120 degrees apart, as per the three-phase technology. As they rotate through the magnetic field they generate power which is then transmitted out on three lines as a three-phase power.

As shown in Fig. 3.71, the generated 3-phase power in a generator will be transmitted through numerous amount of transmission lines, through a step-up three-phase transformer. Three-phase transformers must have three coils or windings connected in the proper sequence in order to match the incoming power and therefore transform the power voltage to the level of voltage we need and maintain the proper phasing or polarity.

In a three-phase transformer, there is a three-legged iron core as shown in Fig. 3.71. Each leg has a respective primary and secondary winding. The three primary windings (P1, P2, P3) will be connected to provide the proper sequence (or correct polarity) required and will be in a configuration star or delta. The three secondary windings (S1, S2, S3) will also be connected to provide the proper sequence (or correct polarity) required. However, the secondary windings, depending on our voltage requirements, will be in either delta or a star configuration.

There are only 4 possible transformer combinations:

1. Delta to Delta . use: industrial applications
2. Delta to Wye . use : most common; commercial and industrial
3. Wye to Delta . use : high voltage transmissions
4. Wye to Wye . use : rare, causes harmonics and balancing problems.

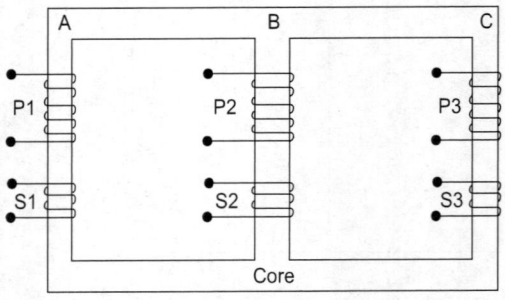

Fig. 3.66. Diagrammatic representation of a 3-phase transformer.

Delta Connections

A delta system is a good short-distance distribution system. It is used for neighbourhood and small commercial loads close to the supplying substation. Only one voltage is available between any two wires in a delta system. A wire from each point of the triangle would represent a three-phase, three-wire delta system. The voltage would be the same between any two wires (see Fig. 3.72).

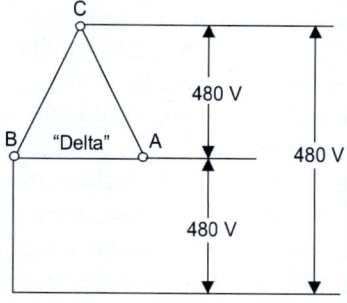

Fig. 3.67. Delta connection.

Wye Connections

In a wye system the voltage between any two wires will always give the same amount of voltage on a three-phase system. However, the voltage between any one of the phase conductors (X1, X2, X3) and the neutral (X0) will be less than the power conductors. For example, if the voltage between the power conductors of any two phases of a three-wire system is 208V, then the voltage from any phase conductor to ground will be 120V. This is due to the square root of three-phase power. In a wye system, the voltage between any two power conductors will always be 1.732 (which is the square root of 3) times the voltage between the neutral and any one of the power phase conductors. The phase-to-ground voltage can be found by dividing the phase-to-phase voltage by 1.732 (See Fig. 3.71).

Three-phase transformers are connected in **delta** or **wye** configurations. A **wye-delta** transformer has its primary winding connected in a **wye** and its secondary winding connected in a **delta** (see Fig. 3.72). A **delta-wye** transformer has its primary, winding connected in **delta** and its secondary winding connected in a **wye** (see Fig. 3.75).

Connecting Single-Phase Transformers into a Three-Phase Bank

If three-phase transformation is needed and a three-phase transformer of the proper size and turns ratio

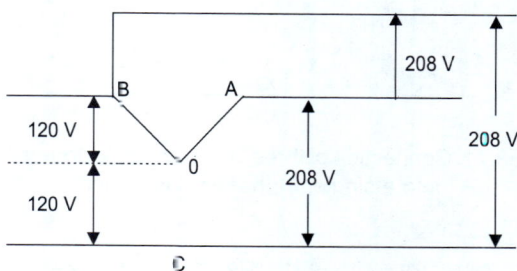

Fig. 3.68. Wye connection.

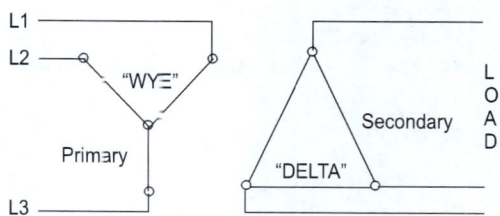

Fig. 3.69. Wye-Delta connection.

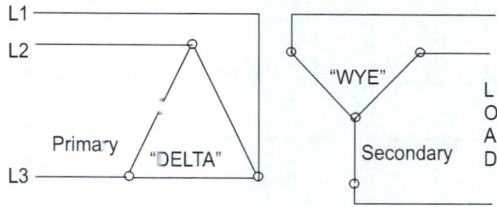

Fig. 3.70. Delta-Wye connection.

is not available, three-single phase transformers can be connected to form a three-phase bank. When three-single phase transformers are used to make a three-phase transformer bank, their primary and secondary windings are conected in a wye or delta conection. The three transformer windings in Fig. 3.76 are labeled H1 and the other end is labeled H2. One end of each secondary lead is labeled X1 and the other end is labeled X2.

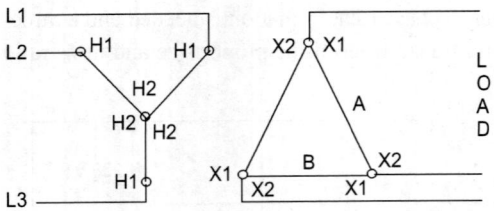

Fig. 3.71. Connection of three, 1-phase transformers, into a single, 3-phase transformer

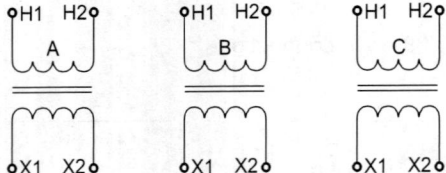

Fig. 3.72

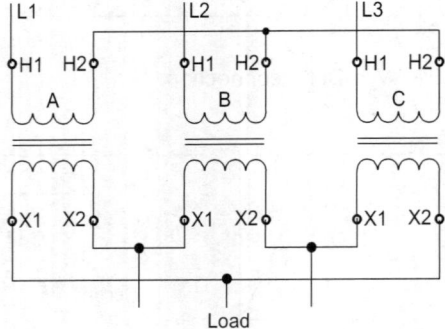

Fig. 3.73

Fig. 3.74 shows three-single phase transformers labeled A, B, and C. The primary leads of each transformer are labeled H1 and H2 and the secondary leads are labeled X1 and X2. The schematic diagram of Fig. 3.71 will be used to connect the three-single phase transformers into a three-phase wye-delta connection as shown in Fig. 3.76.

The primary winding will be tied into a wye coneection first. The schematic in Fig. 3.74 shows, that the H2 leads of the three primary windings are connected together, and the H1 lead of each winding is open for connection to the incoming power line. Notice in Fig. 3.76 that the H2 leads of the primary windings are connected together, and the H1 lead of each winding has been connected to the incoming primary power line.

Fig. 3.76 shows that the X1 lead of the transformer A is connected to the X2 lead of transformer C. Notice that this same conection has been made in Fig. 3.78. The X1 lead of transformer B is conected to X1, lead of transformer A, and the X1 lead of transformer B is connected to X2 lead of transformer A, and the X1 lead of transformer C is connected to X2 lead of transformer B. The load is connected to the points of the delta connection.

Open Delta Connection

The **open delta** transformer connection can be made with only two transformers instead of three (Fig. 3.79). This connection is often used when the amount of three-phase power needed is not excessive, such as a small business. It should be noted that the output power of an open delta connection is only 87% of the rated power of the two transformers. For example, assume two transformers, each having a capacity of 25 kVA, are connected in an open delta connection. The total output power of this connection is 43.5 kVA (50 kVA × 0.87 = 43.5 kVA).

Another figure given for this calculation is 58%. This percentage assumes a closed delta bank containing 3 transformers. If three 25 kVA transformers were connected to form a closed delta connection, the total output would be 75 kVA (3 × 25 = 75 kVA). If one of these transformers were removed and the transformer bank operated as an open delta connection, the output power would be reduced to 58% of its original capacity of 75 kVA. The output capacity of the open delta bank is 43.5 kVA (75 kVA x .58% = 43.5 kVA).

The voltage and current values of an open delta connection are computed in the same manner as a standard delta-delta connection when three transformers are employed. The voltage and current rules for a delta connection must be used when determining line and phase values of voltage current.

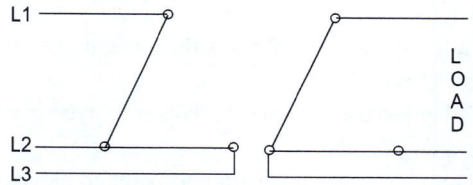

Fig. 3.74. Open Delta connection.

Closing a Delta

When closing a delta system, connections should be checked for proper polarity before making the final connection and applying power. If the phase winding of one transformer is reversed, an extremely high current will flow when power is applied. Proper phasing can be checked with a voltmeter at delta opening. If power is applied to the transformer bank before the delta connection is closed, the voltmeter should indicate 0 volt. If one phase winding has been reversed, however, the voltmeter will indicate double the amount of voltage.

It should be noted that a voltmeter is a high impedance device. It is not unusual for a voltmeter to indicate some amount of voltage before the delta is closed, especially if the primary has been conected as a wye and the secondary as a delta. When this is the case, the voltmeter will generally indicate close to the normal output voltage if the connection is correct and double the output voltage if the connection is incorrect.

Three phase electricity powers large industrial loads more efficiently than single-phase electricity. When single-phase electricity is needed, it is available between any two phases of a three-phase system, or in some systems, between one of the phases and ground. By the use of three conductors a three-phase system can provide 173% more power than the two conductors of a single-phase system. Three-phase power allows heavy duty industrial equipment to operate more smoothly and efficiently. Three-phase power can be transmitted over long distances with smaller conductor size.

CONCLUSION

This chapter discussed various salient features of the fundamentals of DC machines and transformers. The needed derivations, applications of the same, a systematic discussions and associated illustrated examples are the approach of this chapter.

QUESTIONS

DC Machines

1. What is a DC machine?
2. What are the major parts of a DC machine?
3. Draw the cross-sectional view of a 4 pole DC machine and draw the flux path in it.
4. What is the function of field system in a DC machine?
5. What is the function of armature in a DC machine?
6. What are the two major DC machines?
7. How the AC induced voltage become DC in a DC generator? Explain with sketches.
8. Develop an expression for the induced voltage in a DC generator.

9. Write down the voltage equation of a DC generator.
10. What are the types of DC generators? Sketch their electrical equivalent circuits and mark the current directions in them. Write the respective current equations also.
11. What is a separately excited DC generator? Where does it mainly employed?
12. What is the major difference between a flat compounded machine and a over compounded machine?
13. What are the major losses that occur in a DC machine?
14. Write down the power stages in a DC generator.
15. What is O.C.C? What are the other names for the same?
16. Explain how to obtain O.C.C. of a DC generator.
17. What is residual magnetism? Explain.
18. Explain how will you introduce residual voltage if it is absent.
19. What are the causes for the absence of residual voltage? What are the respective remedies?
20. What is critical field resistance?
21. Briefly explain the effect of varying the excitation in the DC generator.
22. Explain the effects that are produced in a DC generator when it is gradually loaded.
23. Describe the load characteristics of a DC generator.
24. What is critical armature resistance?
25. Narrate the applications of DC generators, individually.
26. Explain the working principles of a DC motor.
27. Derive an expression for torque developed by a DC motor.
28. On what factors does the developed torque of a DC motor depend upon?
29. State FRH and FLH.
30. What is the significance of back emf in a DC motor? Discuss.
31. Write down the voltage equation of a DC motor.
32. What are the types of DC motors?
33. Write notes on the speed and speed equation of a DC motor.
34. What are the mechanical characteristics and electrical characteristics of a DC motor?
35. Sketch the mechanical characteristics of (a) shunt motor, (b) series motor and (c) compound motor and justify the shape of the graphs.
36. Sketch the electrical characteristics of (a) shunt motor, (b) series motor and (c) compound motor and justify the shape of the graphs.
37. Write down the speed equation of a DC motor. With reference to that list out the possible methods to control the speed of a DC motor of a (a) shunt motor (b) series motor.
38. Explain the detail the methods you listed in qn. 41.
39. What are starters? What is their essentiality in a DC motor?
40. What are the two major relays employed in a starter? Explain.
41. Draw and explain the 3-point starter employed for DC motors.

Transformations

42. What is a transformer?
43. What are the essential parts of a transformer? What are their functions?
44. On what principles does a transformer work? Explain.
45. What are the types of transformers? Explain.
46. Derive an expression for the emf induced in a transformer.
47. What is the importance of voltage transformation ratio? Write short notes on it.
48. Draw and explain the vector diagram of a transformer on no-load.
49. Explain the behaviour of a transformer when it is gradually loaded.
50. Draw the vector diagram of a transformer on a (a) lagging p.f. load, (b) leading p.f. load.
51. What is the importance for drawing the equivalent circuit of a transformer?

52. Develop equivalent circuit of a transformer in stages.
53. Derive and write down the expression for the equivalent primary reactance referred to secondary.
54. Derive and write down the expression for the equivalent primary reactance referred to secondary.
55. Derive and write down the expression for the total primary resistance referred to secondary.
56. Derive and write down the expression for the total primary reactance referred to secondary.
57. Derive and write down the expression for the equivalent secondary resistance referred to primary.
58. Derive and write down the expression for the equivalent secondary reactance referred to secondary.
59. Derive and write down the expression for the total secondary resistance referred to primary.
60. Derive and write down the expression for the total secondary reactance referred to primary.
61. What is voltage regulation? What is the importance of knowing the status of voltage regulation of a transformer?
62. Derive an expression for the voltage regulation of a transformer.

INTRODUCTION

This chapter is made up of three parts.

In the first part, fundamental aspects of 3-phase alternators is presented. Construction and principle of operation are given at first. This is followed by the expression for the induced emf and introduction to synchronous impedance. The behaviour of a loaded alternator, for varying load power factors is discussed. The importance of voltage regulation in alternators, expression for the same and the associated examples are then presented. A brief discussion on the parallel operation of alternators is also given.

In the second part, introduction to 3-phase induction motors is presented. At first, the types, construction and principle of operation are given, which are followed by the torque producing mechanism and the concept of slip in the machine. Worked illustrations are given. This is then followed by the slip-torque characteristics of the machine. Finally, the starters used for the machine and the speed control theory are presented.

In the third part, an introductory treatment on certain 1-phase machines is given.

CHAPTER OBJECTIVE

At the completion of this chapter, the reader will be able to:
* understand the fundamental theory of alternators;
* understand the fundamental theory of 3-phase induction motors and their speed control; and
* understand 1-phase AC machines.

KEYWORDS

Alternators, Construction and principle of operation, EMF equation, Loaded alternator, Phasor diagram of alternator, Equivalent circuit of alternator, Parallel operation, **3-phase induction motor** (Quantitative), Slip, Types and respective characteristics, starters and **1-phase induction motors** (Quantitative).

4

AC Machines

PART I: SYNCHRONOUS MACHINES

4.1 THE ALTERNATOR CONSTRUCTION AND PRINCIPLE OF OPERATION

The AC generator is a machine which converts the mechanical energy into AC electrical energy. It is sometimes called the synchronous generator. The construction being somewhat similar to DC generator having magnetic field and rotating conductors etc., but main difference being the rotating magnetic field which actually allows more conductors to house in the slots, and more space. In direct-current generators the armature rotates under the poles of an external field system. In modern alternators the construction is the reverse of this, the armature being stationary and the field rotating. Although both types of construction may be used in an alternator, the latter type, with a *stationary armature* and a *rotating field*, is almost universally used for alternating electric power generation. The former type, with the rotating armature, finds its greatest application as a synchronous or rotary converter.

The advantage of this construction is that no difficulty is experienced in inserting the stationary armature winding for very high voltages, as high as 30,000 V in some cases. Also, the current from the armature is collected from fixed terminals. The field winding, of course, requires direct-current excitation.

The alternators are separately excited by an excitor mounted on the same shaft. These excitors supply the magnetising current to the alternator usually, at 110 volts or 250 volts.

Since the magnet rotates, this excitation current has to be fed to the winding by means of two slip rings. This is not a serious matter, as the excitation voltage is low and the power required for excitation is also small.

The stator core is laminated like a DC armature core, ventilating ducts being provided in order to assist cooling. It is not necessary to laminate the field, since this carries a continuous flux, but sometimes the poles of the pole tips are laminated. A typical alternator construction is illustrated in Fig. 4.1.

The rotor of these machines are of two types, viz., the salient pole or projecting pole type, and the smooth cylindrical or non-slient pole type.

Salient Pole Construction

The alternators with salient pole rotor will have large diameter than the length. These constructions are

AC Machines

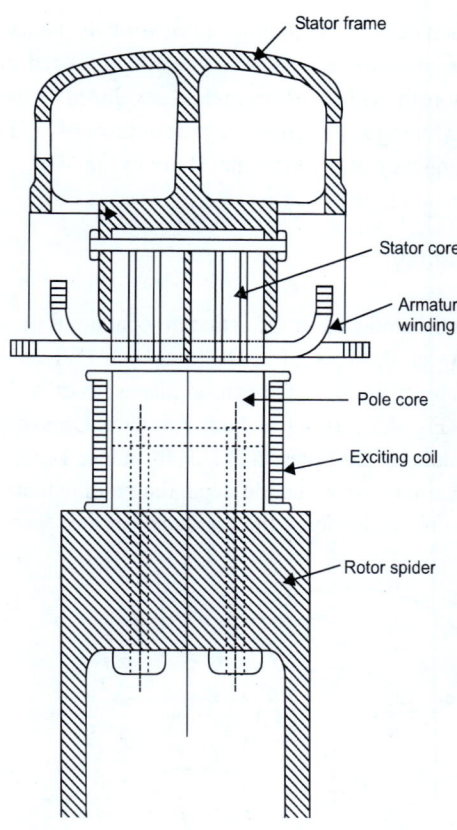

Fig. 4.1. Alternator cross-section.

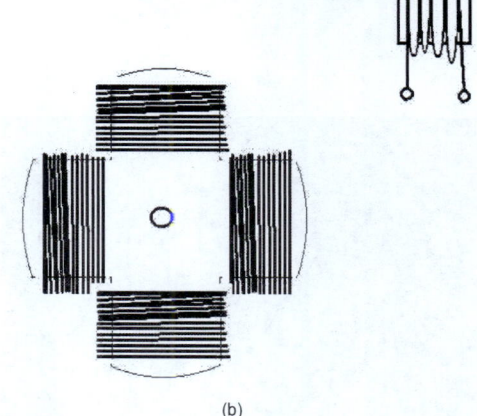

Fig. 4.2. (a) Image of a salient pole alternator with the shaft projection. (b) Schematic representation

suitable for low speed of alternator. They are also having more poles generally from 6 to 40 poles. The pole shoes are fixed or bolted by means of dovetailed joints. The poles are generally made of laminations of silicon steels.

Figure 4.2 shows a salient rotor.

As in Fig. 4.2 the field system will consist of projecting poles. This is called the salient pole type. These are used in the turbo-alternators where the number of poles are very less say 2 or 4. These constructions are specially for high speed. Generally their diameter is less than the axial length.

Smooth Cylindrical Construction

The field system will not have any projecting poles, but is a cylindrical slotted structure. The slots will accommodate field winding to produce the effects of N-S poles alternatively when excited. This is the non-salient type.

If the rotor either *salient* or *non salient* as is rotated at a synchronous speed by a prime mover, the machine will function as an *alternator*, either single-phase or polyphase, depending on the armature connections. The armature stator connections shown in Fig. 4.3 are those required for obtaining a three-phase output using a four-pole rotor.

On the other hand if the stator armature is connected to a single-phase or polyphase AC supply, the machine will function as a *synchronous motor*, and the rotor will rotate at a synchronous speed in synchronism with the rotating field developed by the stator winding as determined by the number of poles and frequency of the machine. Refer to Fig. 4.4.

Frequency

The emf is induced in the armature conductor on a similar method as of a DC machine. When a conductor is opposite the neutral planes, as at A, C and E (Fig. 4.5), its induced e.m.f. is zero. Opposite the middles of the poles, at B,D,F, its induced e.m.f. is a maximum, its direction depending on the name of the pole influencing the conductor at any given

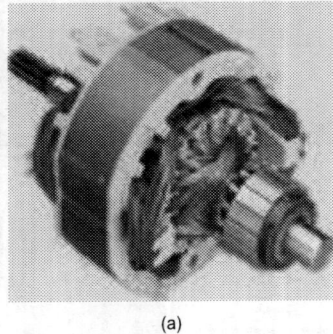

(a)

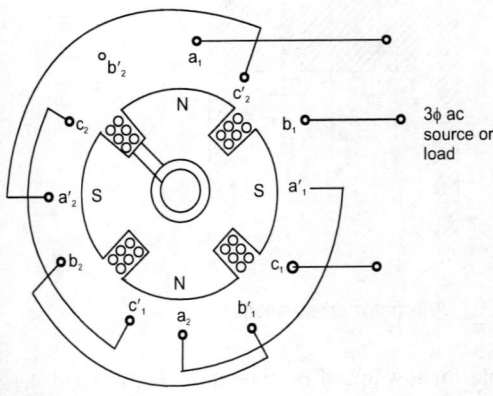

Fig. 4.4.

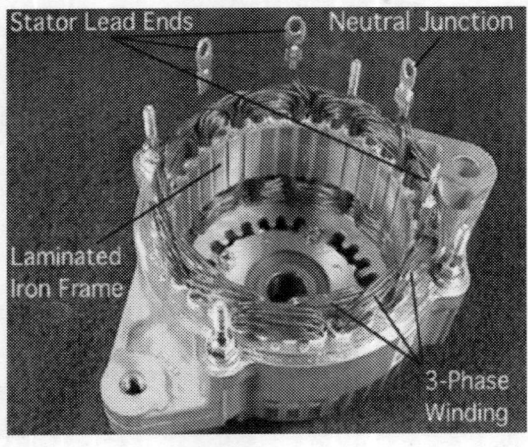

(b)

Fig. 4.3. Image showing the armature winding.

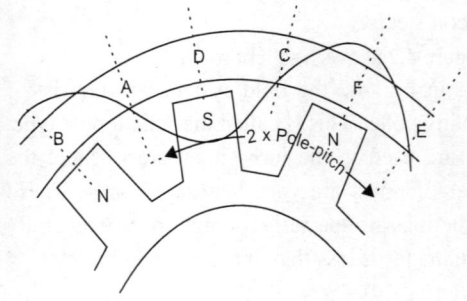

Fig. 4.5

instant. As was discussed in the chapter 2, for an emf to be induced is an armature winding, its conductors must undergo a rate of change of flux-linkage, i.e. $d\phi/dt$. In case of a DC generator it was established by a stationary field and a rotating armature, whereas in an alternator, it is established by a stationary armature and a rotating field. Referring to Fig. 4.5, take for an example the stationary armature conductor A. In the shown position (called inter-polar position), it experiences no flux. So, $e = d\phi/dt = 0$. But soon, when the field poles resolve (say in C.W. direction), this same conductor will be against the south pole and will fully be covered by its pole face. Thus the conductor A cuts maximum of north pole flux inducing a maximum emf in it, that is + E. As the rotor field rotates further, the conductor A will face south pole fully, inducing again a maximum voltage, but in reverse direction, that is –E. In this way a sinusoidally varying emf is induced in each of the armature conductor. The e.m.f. induced in a conductor therefore goes through one complete cycle in an angular distance equal to twice the pole pitch (centre to centre distance of adjacent poles). Obviously, one cycle of e.m.f. is induced in a conductor when one pair of poles passes over it. In other words the e.m.f. in an armature conductor goes through one cycle in an angular distance equal to twice the pole pitch.

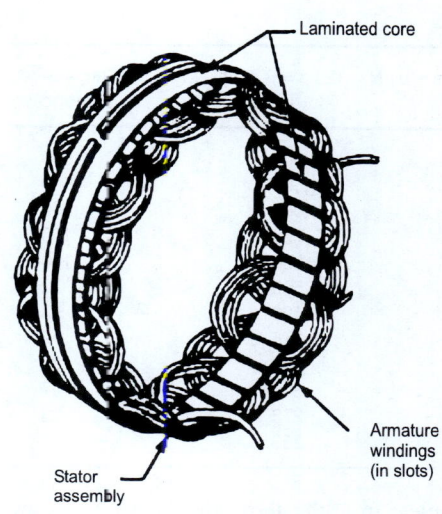

Fig. 4.6 Armature alone shown. The rotating field comes into it.

∴ No. of cycles/revolution = P/2

where P = No. of poles.
Let N = r.p.m.

$$\therefore \frac{N}{60} = \text{r.p.s}$$

∴ Frequency in cycles /second

$$f = \frac{P}{2} \times \frac{N}{60} = \frac{PN}{120} \qquad (4.1)$$

Hence, if the speed and frequency are specified, the number of poles required is fixed. In a DC generator designed to a given specification the number of poles is not fixed, and it would be possible to design, say, 10 and 12 pole machines of equally good performance to the same specification. This definite relationship between f, P, and N has an important bearing on the choice of speed of an alternator. Thus is f = 50, then for P = 2, N = 3000 r.p.m., while for P = 4, N = 1500 r.p.m., and speeds intermediate between 3000 and 1500 are inadmissible. Thus, an alternator otherwise is also said as a machine which will run at a constant speed decided by the constant f and P and that speed is the synchronous speed.

The possible speeds of alternators working on a frequency of 50 c/s are, therefore, as given in the following Table 4.1. These are also the speeds at which synchronous motors run, and are slightly greater than the speeds of induction motors running without any speed-regulating devices in circuit.

It will be seen that when P is large the successive differences are so small that they are of little consequence, but when P is so small as to bring the speed within the range of turbo speeds, the successive differences are very large.

274 Basic Electrical Engineering

Table 4.1

No.of poles = p	N = 120/p = 6000/p	Successive differences
2	3,000	1,500
4	1,500	500
6	1,000	250
8	750	150
10	600	100
12	500	72
14	428	53
16	375	42
18	333	33
20	300	27
22	273	23
24	250	

It is clear from the above that because of slow rotative speeds of engine-driven alternators, their number of poles is much greater as compared to that of the turbo-generators which run at very high speeds.

4.2. ADVANTAGES OF STATIONARY ARMATURE AND ROTATING FIELD CONSTRUCTION

The reader might at first react adversely to the idea of making the armature stationary. After all, it should be a simple matter to bring out the alternating current generated by a moving armature to slip rings on one side of the shaft, and the direct current generated by the same moving armature to commutator bars on the other side of the shaft. In this way, we would have universal machines that could supply either of the shaft. Indeed, this is done in a synchronous converter; but there are several compelling reasons for abandoning idea of a universal machine having a rotating AC armature. Once the armature is stationary, we neither achieve nor require automatic switching from AC to DC by commutation, and only AC is generated. The more significant advantages of the stationary armature and rotating field construction.

Increase armature tooth strength
Reduced armature reactance

Improved insulation
More rigid construction
Reduced number of insulated slip rings
Reduced rotor weight and inertia
Ventilation and improved heat dissipation.

4.2.1 Emf Equations of Alternator

Let Z = No. of conductors or coil sides in series / phase

$\quad\quad$ = 2T where T is the No. of coils or turns per phase
(remember one turn or coil has two sides)

P = No of rotor poles

f = frequency of induced emf in Hz

ϕ = flux / pole in webers

N = rotative speed of the rotor in rpm

In one revolution of the rotor (i.e., in 60 / N second), each sector conductor is cut by a flux of ϕP, webers.

$\therefore d\phi = \phi P$ and $dt = 60 / N$

$\therefore$ average emf induced per conductor

$$= \frac{d\phi}{dt} = \frac{\phi P}{60/N} = \frac{PN\phi}{60} \text{ volt}$$

Now, we know that $f = PN / 120$ or $N = 120 f / P$

Eliminating N in the above equation, we have

Average emf per conductor

$$= \frac{\phi P}{60} \times \frac{120f}{P} = 2f\phi \text{ V}$$

If there are Z conductors in series / phase, then average emf/phase = $2 f Z \phi = 4 f \phi T$ V.

R.M.S. value of e.m.f / phase

$= 1.11 \times 2 \phi f Z = 2.22 \phi Z f$ V

$= 4.44 f \phi T$ V (4.2)

If the alternator is star-connected (as is usually the case), then the line voltage is $\sqrt{3}$ times the phase voltage (as found from the above formula).

Distribution Factor

This is otherwise called as breadth factor. The total emf taken out from the coil is the sum of the emf induced in both coil sides. If the coil is full pitched, the emf induced in both the coil sides will be equal and will act in the same direction around the coil. The emf generated E will be twice that generated in one coil side. But if the coils are not full pitched i.e. short pitch, the e.m.f. in turn will be less than twice that of the generated in one coil side.

(Let there be two coils or say two slots per pole then there are two coils which are inducing the emf Let it be E volts. The coil BB' is 'a' deg. apart, so the maximum emf induced will certainly by a° or say the emf induced will differ in phase from 'A' and resultant will not be the $2E$ but will have some less value, say E_r.)

In actual generators the winding is spread over in several slots per pole pitch and is not concentrated in one slot. So the distribution factor is of great importance, and is defined as the ration of the resultant emf Er to the emf which would be generated if the windings were concentrated in one slot.

So

$$K_d = \frac{\text{Resultant emf with conductor in n slots}}{\text{Resultant emf with conductors in 1 slot}}$$

$$= \frac{Er}{nE}$$

The emf equation (4.2) this will get modified as

$$E = 4.44 f \phi T K_w \text{ V}$$

where k_w is the winding factor

4.3 WORK ILLUSTRATION XIV

Illustration 4.1

Calculate the synchronous speed of a four-pole 50 Hz alternator.

Solution

$$f = \frac{PN}{120} \quad \text{or} \quad N = \frac{120f}{P}$$

$N = 120 \times 50 / 4 = $ **1500 rpm**

Illustration 4.2

What is the frequency of voltage generated by an alternator having 10- poles and rotating at 720 rpm?

Solution

$f = PN / 120 = 10 \times 720 / 120 = $ **60 Hz.**

Illustration 4.3

A 3-phase, 16 - pole alternator has a star - connected winding with 144 slots and 10 conductors per slot. the flux per pole is 30 mWb sinusoidally distributed and the speed is 375 rpm. Find the frequency, the phase and the line emf.

Solution

Formula used

$E = 2.22 f Z$ V - per phase

$f = PN/120 = 16 \times 375/120 = $ **50 Hz**

No. of slots per phase = 144 / 3 = 48

No. of conductors / slot = 10

No. of conductors / phase $Z = 48 \times 10 = 480$

emf per phase = $2.22 \times 50 \times 30 \times 10^{-3} \times 480 = 1{,}600$ V

For a star-connection, $V_L = V_{ph}$

Line voltage = $\times 1600 = $ **2770 V**

Illustration 4.4

A 3-phase water-wheel generator is rated at 100 MVA. unity p.f., 11 kV, star-connected, 50 Hz, 120 r.p.m. Determine:

(i) the number of poles
(ii) the kW ratings
(iii) the current rating
(iv) the input at rated kW load if the efficiency is 97 per cent (excluding the field loss)
(v) prime-mover torque applied to the generator shaft.

Solution

(i) $f = PN/120$ or $P = 120 \times 50 / 120 = $ **50 Hz**

(ii) Since the power factor is unity (cos φ = 1) the kW rating is

= 100 × 1.0 = 100 MW = 100,000 kW
(**kVA × power factor = kW**)

(iii) Current rating is given by the relation,

$W = V_L I_L \cos \phi$

= $100 \times 10^6 = 11{,}000 \times I_L$, $I_L = $ **5,250 A**

(iv) Input = 100 / 0.97 = **103.1 MW**
This is input to the alternator which means that this is the output of its prime moover. Thus it also represents the output of the prime mover.

(v) $T_{sh} = 9.55 \times 103.1 \times 10^6 / 120 = $ **10⁶ N**

Illustration 4.5

Calculate the speed and the open-circuit line and phase voltages of a 4 pole, 3 phase, 50 Hz star connected alternator with 36 slots and 30 conductors per slot. The flux per pole is 0.05 wb.

Solution

$f = \dfrac{PN}{120}$ and thus $N = \dfrac{120 f}{p} = \dfrac{120 \times 50}{4} = 1500$ rpm

Turns per phase = (conductor/ph)/2

= $[(36 \times 30)/3]/2 = 180$

$E_{ph} = 4.44 f\phi T = 4.44 \times 50 \times 0.05 \times 180$

= 1920 V and is $\sqrt{3}(1920)$, V-line.

Illustration 4.6

A synchronous generator runs at 250 rpm and generates at 50 Hz. There are 216 slots, each arranged with 5 conductors for a 3-phase star connection. All the conductors of each phase are in series and flux for pole at no-load is 30 mwb. The distribution factor is 0.96. Deduce the induced voltage in each phase winding and the terminal voltage.

Solution

No of conductors $Z = 216 \times 5 = 1080$

No. of conductor, per phase, $Z_{ph} = 1080/3 = 360$.

$\therefore T_{ph} = 360/2 = 180$. $k_w = 0.96$, given

Thus, $E_{ph} = 4.44 \times 50 \times 30 \times 10^{-3} \times 180 \times 0.96$
$= 1150$ V.

Illustration 4.7

A 4 pole, 50 Hz star connected alternator has 15 slots per pole and each slot has 10 conductors. The winding factor is 0.95. When running on no load for a certain flux per pole the terminal emf was 1825 V. If now the windings are lap connected as a DC machine, what would be the emf between the brushes for the same speed and the same flux per pole.

Solution

From the alternator conditions, we can get the flux for pole, as

$$\phi = \frac{E_{ph}}{4.44 \, f \, T_{ph} k_w}$$

Slots: $15 \times 4 = 60$. Total conductors, $Z = 60 \times 10 = 600$.

$Z_{ph} = 600/3 = 200$. $T_{pH} = Z_{ph}/2 = 200/2 = 100$.

$\therefore \phi = \frac{(1825)}{\sqrt{3}} / (4.44 \times 50 \times 0.95 \times 100) = 50$ mwb.

As a DC generator

$$E_g = \frac{P\phi ZN}{60A}, \text{ where } N = \frac{120f}{P} = 1500 \text{ rpm}$$

$$= \frac{4 \times 50 \times 10^{-3} \times 600 \times 1500}{600 \times 4} = 750 \text{ V}.$$

4.4 ALTERNATOR ON LOAD

As load on an alternator is varied, its terminal voltage V is also found to vary as in DC generators. Infact, it drops due to the increased voltage drop in the armature circuit. This variation in V is due to the following reasons:

1. voltage drop due to armature resistance R_a;
2. voltage drop due to armature leakage reactance X_L;
3. voltage drop due to armature reaction

The armature resistance voltage drop, $I_a R_a$, will increase as the load is increased, because an increase in load means more armature current.

The term X_L is the leakage reactance of the alternator. The alternator being an AC device, the total opposition to the flow of armature current is the impedance and not just resistance alone. As the resistance of the armature is comparatively negligible than its reactance, the impedance reduces to the reactance. Thus

$$Z_a = R_a + j X_L = j X_L$$

The leakage flux in alternator is a flux which does not react with the available air gap field flux for useful energy transformation, but just links in the armature winding itself. The concept is the same as was

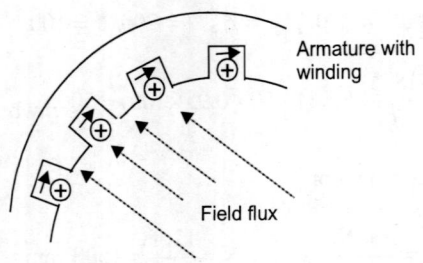

Fig. 4.7

brought out in the Sec. 3.25, hence the leakage phenomena is replaced with an equal reactance of X_L in the armature winding. It causes a voltage drop of $I_a X_L$ volts.

Note that any electric machine will undergo weakening of the field flux available for the energy conversion, as it is gradually loaded. This reaction is called armature reaction, and alternator will also undergo such a reaction when it is loaded.

Consider the Fig. 4.7. The arrow marks shown against each conductor is the flux produced in the current carrying armature conductors; by applying cork-screw rule, this direction is marked. The field flux provided by the pole excitation is also marked.

When the armature is not loaded, there will be no current in the armature winding, and hence the flux around the winding will be zero. The field flux is thus unaffected by any external flux.

But when the armature is gradually loaded, its current also will proportionally increase, causing its flux to increase. Now, the field flux (which is the essential flux cut by the armature conductors for energy transformation) will be distorted by the newly introduced armature flux.

The undesired distortion in the main field flux is called 'armature reaction'.

The drop due to armature reaction is theoretically accounted for by assuming a fictitious reactance X_a in the armature winding. The vector sum of X_L and X_a gives synchronous reactance X_s of the alternator.

Hence, it can now be said that an alternator possesses (i) resistance R_a and (ii) synchronous reactance, X_s. Their vector sum gives synchronous impedance Z_s.

$$Z_s = \sqrt{R_a^2 + X_s^2}, \Omega \qquad (4.3)$$

Thus when an alternator is on load, there is voltage drop due to R_a and X_s or due to Z_s. If V is the terminal voltage / phase on load and E the generated emf/phase on no-load, then, by the voltage equation of the generator

$$E = V + I_a (R_a + jX_S) = V + I_a Z_S \qquad (4.4)$$

Z_s is found out using the open-circuit and short-circuit characteristics of the alternator. The open-circuit characteristic is the curve of induced voltage against exciting current, with the alternator running at normal speed. The short-circuit characteristics is the curve of armature current under short-circuit conditions against exciting current. To determine this, the armature terminals are short-circuited through an ammeter, and a very reduced excitation applied. The excitation is then carefully increased until full armature current is flowing, readings of excitation and armature current having been taken. The curve is a straight line and it can therefore be projected, as shown dotted in Fig. 4.8 into the normal working portion of the diagram.

Erect a perpendicular *PQN* at normal excitation *ON*; then the emf induced in the armature at this excitation is *PN*. On short circuit the whole of this

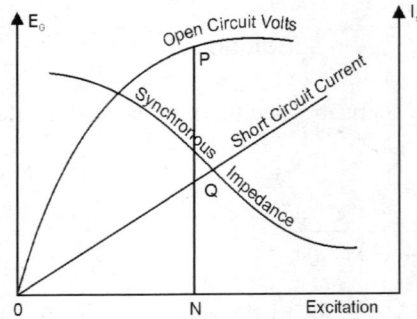

Fig. 4.8. OCC and SCC of alternator.

emf is used in driving the short-circuit current through the armature winding against the impedance, because the terminal voltage on short circuit is zero.

$$\therefore \text{Impedance } Z = \frac{PN(\text{volts})}{QN(\text{amp})} = \frac{E_0}{I_s} \quad (4.5)$$

This is called the "synchronous" impedance to indicate that it refers to working conditions.

Now

$$Z = \sqrt{R_s^2 + (L_a \omega)^2}$$

$$\therefore L_a \omega = \sqrt{\left(\frac{E_0}{I_s}\right) - R_a^2}$$

The quantity $L_a \omega$ is called the "synchronous reactance". The synchronous impedance and reactance as determined by this method are rather higher than the values under absolutely normal conditions, due to the fact that on short circuit such a very low excitation has to be applied that the field is unsaturated.

If the synchronous impedance for different excitations is determined and plotted against excitation the curve has the form shown in Fig. 4.8.

4.5 PHASOR DIAGRAM OF A LOADED ALTERNATOR

Let E = no-load emf/phase

V = terminal voltage/phase

I_a = armature current/phase

f = load p.f angle

Z_s = armature synchronous impedance/phase

An alternator, which is an AC generator has the total circuit voltage induced in the armature, called E,

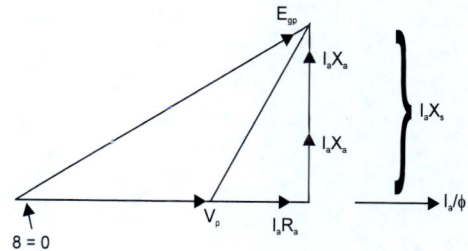

Fig. 4.9. Vector diagram on a resistive load.

volts, which after all kinds of voltage drops discussed above, is made available as the terminal voltage V, volts at the load. So, $E = V + I_a Z_s$. With this knowledge, the following discussion is done.

(a) Unity Load p.f

In a resistive load, the load voltage V and load current I are in phase thus causing the phase angle to be zero and power factor to be unity. In Fig 4.9, V is taken as the reference vector. Current vector I_a is in phase with V. The voltage drop $I_a R_a$ is added in phase with I_a whereas drop $I_a X_S$ is added at right angles to it, their vector sum giving $I_a Z_S$. Summation of $I_a Z_s$ and V_2 yields E.

(b) Lagging Load p.f

An inductive load is assumed to be connected to the alternator. In this case, I_a lags behind V by angle a phase angle of ϕ. As usual I_a, R_a is in phase with I_a whereas $I_a X_S$ is at right angles to it. When $I_a Z_S$ is combined with V, we get E as shown in Fig. 4.10.

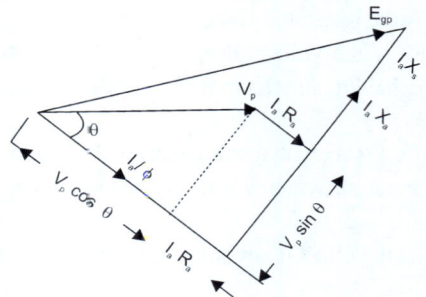

Fig. 4.10 Vector diagram on a lagging load.

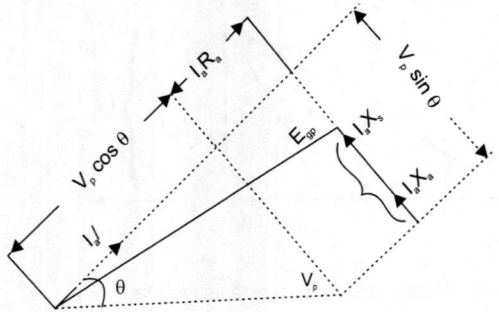

Fig. 4.11. Vector diagram on a leading load.

(c) **Leading Load p.f**
The connected load to the alternator is a capacity load. Such a case is shown in Fig. 4.11. Here, I_a leads V by ϕ. As usual, vector sum of V and $I_a Z_S$ gives E.

4.6. EQUIVALENT CIRCUIT FOR A SINGLE-PHASE AND/OR POLYPHASE SYNCHRONOUS GENERATOR

The relation between the terminal and generated voltage of an alternator was given in by Eq. (4.4) and the circuit is represented in Fig. 4.12. For a single-phase or polyphase alternator, Eq. (4.4) may be expanded and rewritten as the phasor sum:

$$V = E_g - I_a R_a - I_a (jX_a) \pm E_{ar} \qquad (4.6)$$

where V is the terminal voltage per phase

E_g is the generated voltage per phase

$I_a R_a$ is the voltage drop across the armature winding, having an effective (AC) resistance of R_a per phase

$I_a (jX_a)$ is the voltage drop across the reactance of the armature winding due to leakage reactance per phase.

E_{ar} is the effect of armature reaction per phase.

It should be noted that essentially little difference exists between the equivalent circuit of a single-phase

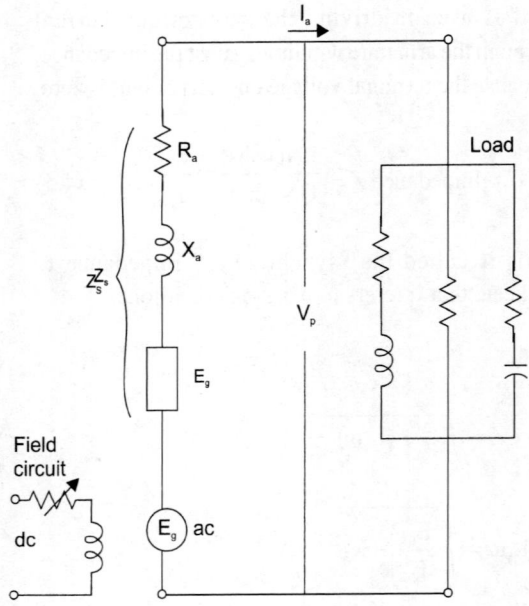

Fig. 4.12. Equivalent circuit representation of an alternator, per phase.

alternator and the three-phase alternator. Each phase winding of a three-phase alternator is assumed to have an effective armature resistance per phase of R_a an armature reactance per phase of X_a and a generated voltage per phase of E_g. Furthermore, it may be assumed that the voltage drop due to the effect of armature reaction is the same in each phase, or volt drop per phase. The components of Eq. (4.6) apply equally well both to polyphase and to single-phase alternators, as the quantities are in per phase.

Before we do so, however, it might be well to consider the factors that may account for difference between the no-load generated voltage per phase (E_g) and the terminal voltage per phase (V_p). The equivalent circuit shown in Fig. 4.13 employs separate DC excitation for the rotating field windings of both and single-phase and polyphase synchronous alternators. consequently, any change in terminal voltage as a result of loading does not effect the excitation of the field emf. In this respect, the alternator is similar to the separately excited DC

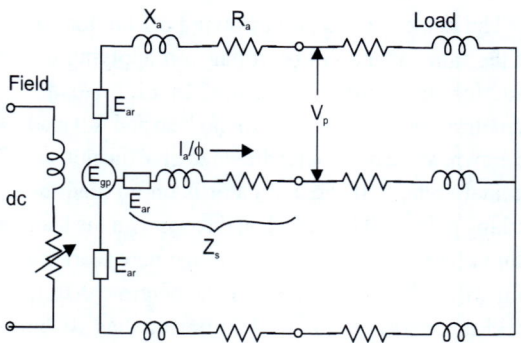

Fig. 4.13.

generator, and a comparison between the two will reveal similarities as well as differences.

For unity power factor load, the relationship among the various voltage drops producing a difference between the generated and the terminal voltage are shown in Fig. 4.9. At unity PF, the phase current in the armature I_a is in phase with the terminal phase voltage V_p by definition. The voltage drop per phase across the effective resistance of the armature $I_a R_a$ is also always in phase with the armature current $I_a X_a$ is always 90° leading with respect to the current through it (since the current lags the voltage by 90° in a circuit possessing inductive reactance only). At unity PF, the armature reaction voltage drop, E_{ar}. The basis generator equation, Eq. (4.3), may now be written for unity power factor loads in complex from as the phasor sum:

$$E_g = (V_p + I_a R_a) + j(I_a X_a + E_{ar}) \text{ Volts (V)} \quad (4.7)$$

When inductive loads, which work at a lagging p.f. is considered, the armature phase current I_a (by definition) lags the terminal phase voltage phase voltage V_p by some angle θ. The voltages may be represented by the diagram shown in Fig. 4.10. The $I_a R_a$ drop is still in phase with the armature phase current, and the quadrature reactance and armature reaction voltage drops lead the armature current by 90°.

$$E_g = (V_p \cos θ + I_a R_a) + j(V_p \sin θ + I_a X_s)$$
$$\text{Volts (V)} \quad (4.8)$$

AC Machines 281

From the diagram of Fig. 4.10 and the above two equations it would appear that to obtain the same rated terminal voltage per phase (V_p), a higher induced voltage per phase (E_g) is required at lagging p.f.s. than at unity p.f.

If the armature phase current I_a (by definition) leads the terminal phase voltage V_p (by some angle θ as a result of an external load (containing a capacitive component) across the AC synchronous alternator, the voltage may be represented as shown in Fig. 4.11. The $I_a R_a$ drop is always in phase with the phase current in the armature, and quadrature synchronous reactance drop $I_a X_s$ leads the armature current by 90°. By indicating E_g in terms of the horizontal and vertical components, we find

$$E_g = (V_p \cos θ + I_a R_a) + j(V_p \sin θ - I_a X_s) \text{ volts (V)} \quad (4.9)$$

It would appear that for the same rated terminal phase voltage, less generated voltage is required for a leading PF than for a lagging PF.

4.7 VOLTAGE REGULATION

The terminal voltage V of as alternator is found to vary with the applied load and its power factor. As was discussed in Sec. 3.28, it is the difference of induced voltage in the armature of the alternator (E) to the terminal voltage as the load terminals (V, which has to be kept at a reasonably minimum value. Thus, (E – V) expressed as the % of V is the voltage regulation.

Thus difference between the generated voltage E_g and the terminal voltage V_p per phase of an alternator, is the synchronous impedance voltage drop Ia Za. This same difference in point of fact exists between V_p and E_g for any power factor and any load, as shown by the various diagrams of Fig. 4.9 through 4.11.

It can be noted that, the synchronous impedance voltage drop is, at all times, the phasor sum of the effective armature resistance voltage drop per phase and the quadrature equivalent voltage drops due to armature reactance and armature reaction per phase for the same load.

The concept of an internal equivalent synchronous impedance possessed by an AC alternator is similar to that of an internal equivalent armature circuit resistance possessed by a DC machine. If the armature circuit resistance of a DC generator is known, it is possible to compute the terminal voltage a DC generator or the counter emf of a DC motor for any value of load. In a similar manner, if the effective armature resistance per phase and synchronous reactance per phase are known, it is also possible to compute the generated emf of a synchronous alternator or motor.

The advantage of the synchronous impedance concept is that it is possible to treat the quadrature voltage drop necessary to overcome the voltage due to armature reaction as a reactive impedance component. This is permissible since this voltage is always in quadrature with the armature current.

The rise of voltage on throwing off the load is not the same as the fall of voltage on applying the load, for the following reason. In Fig. 4.6 the magnetization characteristics for no load and full load are drawn, and a horizontal drawn to give the normal terminal voltage. If the alternator is on no load the working point will be A, and on throwing on the load without increasing the excitation, the new working point will be B, the voltage drop is therefore AB. If working on full load, the excitation has to be increased in order that the terminal voltage may now be normal, the working point being C. On throwing off the load the rise of voltage is CD. Since the two curves are not exactly parallel, AB and CD are not equal.

It is now usual to express the regulation of an alternator as the percentage rise of voltage when full load is thrown off, and not by the percentage fall when full load is thrown on. The former value is called "regulation up" and the latter "regulation down".

Average values of the percentage regulation are 12 or 14 per cent on full non-inductive load, and 30 per cent on an inductive load of power factor 0.8.

Illustration 4.8

An alternator has an armature resistance of 0.3 Ω. When a small excitation is applied the open-circuit voltage is 50 V, the short-circuit current for the same excitation being 40 A. Find the synchronous impedance and reactance.

$$Z = \frac{\text{open-circuit volts}}{\text{corresponding short-circuit current}} = \frac{50}{40}$$

$$= 1.25 \, \Omega$$

$$\therefore L_a \omega = \sqrt{Z^2 + R_a^2} = \sqrt{(1.25)^2 - (0.3)^2}$$

$$= 1.22 \, \Omega$$

It is usual to express the voltage regulation of an alternator in terms of the rise in voltage when full load is thrown off.

Illustration 4.9

A 500 VA, 1100 V, 50 Hz, Y-connected, 3-φ alternator has armature resistance / phase of 1.0 Ω and synchronous reactance / phase of 1.5 Ω. Find its voltage regulation for (a) unity p.f. (b) 0.9 lagging p.f and (c).8 leading p.f. Also, calculate voltage regulation in each case.

Solution

$$I_L = 500{,}000 / \sqrt{3} \times 1100 = 262 \, A$$

armature current,

$$I_a = 262 \text{ A}$$

$$I_a R_a = 262 \times 0.1 = 26.2 \text{ V}$$

$$I_a X_S = 262 \times 1.5 = \mathbf{393 \text{ V}}$$

$$V = V_L / \sqrt{3} = 1100 / \sqrt{3} = \mathbf{635 \text{ V}}$$

(a) Unitiy p.f
As shown in Fig 4.9

$$E = \sqrt{(V+I_a R_a)^2 + (I_a X_S)^2}$$

$$= \sqrt{(635+26.2)^2 + (393)^2} = 269.2 \text{ V}$$

$$\% \text{ regn} = \frac{769.2 - 635}{635} \times 100 = 21.1\%$$

(b) Lagging p.f
This case is shown in Fig. 4.10
It can be easily found from the right angle triangle OAB that

$$OA = V \cos \phi + I_a R_a \text{ and}$$

$$AB = V \sin \phi + I_a X_s$$

Therefore

$$E = \sqrt{V(\cos\phi+I_a R_a)^2 + (V \sin\phi \pm I_a X_s)^2}$$

where (+) is for lagging power factor and (–) is for leading power factor. Given that $\cos \phi = 0.9$ $\phi = 25.83°$; $\sin \phi = \sin 25.83° = 0.436$

$$E = \sqrt{635 \times 0.9 + 26.2)^2 + (635 \times 0.436 + 392)^2} = 898 \text{ V}$$

$$\% \text{ regn.} = \frac{898 - 635}{635} \times 100 = 41.4\%$$

(c) Similarly for the leading power factor E = 534, V and % VR = – 15.9%

Illustration 4.10

3-phase star-connected alternator is rated at 1600 kVA, 13.5 kV. The armature effective resistance are synchronous reactance and 1.5 Ω and 30 Ω respectively per phase. Calculate the percentage regulations for a load of 128- kW at p.f of 0.8 leading.

Solution

Let us first find the phase current of the star-connected alternator for a load of 1280 kW at p.f. of 0.8 leading.

$$1280 \times 10^3 = \sqrt{3} \times 13{,}500 \times I_L \times 0.8, I_L = 68.4 \text{ A}$$

Since the alternator is star-connected, it also represents the phase current. Hence, armature phase current $I_a = 68.4 \text{ A}$

$$I_a R_a = 68.4 \times 1.5 = 103 \text{ V}, I_a X_S = 68.4 \times 30 = 2050 \text{ V},$$

$$V = 13500 \sqrt{3} = 7795 \text{ V}$$

$$E = \sqrt{(7795 \times 0.8 + 103)^2 + (7795 \times 0.6 - 2050)^2}$$

$$= 6910 \text{V}$$

$$\% \text{regn} = \frac{6910 - 7795}{7795} = -0.1135 \text{ or} -11.35\%$$

Illustration 4.11

A 1000 kVA, 4600 V, 3 fY-connected alternator has an armature resistance of 2 Ω per phase and synchronous reactance of 20 Ω/phase. Find the full-load generated voltage per phase at a. Unity p.f. (b) 0.75 PF lagging.

Solution

$$V_{ph} = \frac{V_l}{\sqrt{3}} = \frac{4600\,V}{1.73} = 2660\,V$$

$$I_{ph} = \frac{kVA \times 1000}{3V_{ph}} = \frac{1000 \times 1000}{3 \times 2660} = 125\,A$$

$I_a R_a$ drop/phase = 125 A × 2 Ω = 250 V

$I_a X_s$ drop/phase = 125 A × 2 Ω = 2500 V

(a) At unity p.f.,

$$E_g = (V_p + I_a R_a) + j I_a X_a$$

$$(2660 + 250) + j\,2500 = 2910 + j\,2500$$

$$= 3836\,V/phase$$

(b) At 0.75 p.f. lagging,

$$Eg = (V_p \cos\theta + I_a R_a)$$

$$+ j\,(V_p \sin\theta + I_a X_s) \quad (6\text{-}4)$$

$$= [(2660 \times 0.75) + 250]$$

$$+ j\,[(2660 \times 0.661) + 2500]$$

$$= 2245 + j\,4259$$

$$= 4814\,V/phase$$

Illustration 4.12

Repeat illustration 4.11 to determine the generated voltage per phase at full load with

(a) A load of 0.75 p.f. leading
(b) A load of 0.40 p.f. leading

Solution

From Illustration 4.11

V_p = 2660 V

$I_a R_a$ phase = 250 V

$I_a X_a$ phase = 2500 V

(a) At 0.75 PF leading

$$E_g = (V_p \cos\theta + I_a R_a) + j(V_p \sin\theta - I_a X_s)$$

$$= [(2600 \times 0.75) + 250] + j\,(2660 \times 0.661)\,2500]$$

$$= 2245 - j\,742 = 2364\,/\,V\,phase$$

(b) At 0.40 PF leading

$$E_g = (2660 \times 0.4) + 250) + j[(2660 \times 0.9165) - 2500]$$

$$= 1314 - j\,62 = 1315\,V\,/\,phase$$

Illustrations 4.10 and 4.11 serve to illustrate two facets of the effects of leading or lagging loads on alternator generated voltage and, in turn, voltage regulations, namely.

1. The lower the leading power factor, the greater the voltage rise from no load (E_{gp}) to full load (V_p)
2. The lower the lagging power factor, the greater the voltage decrease from no load (E_{gp}) to full load (V_p)

This may also be seen in the graphical representation taken from the data of these examples as shown in Fig. 4.14.

Illustration 4.13

Calculate the voltage regulations for Fig. 4.14

Solutions

(a) At 0.75 PF lagging

$$VR = \frac{4815 - 2660}{2660} \times 100 = 81.0 \text{ percent}$$

(b) At unit PF

$$VR = \frac{3836 - 2660}{2660} \times 100 = 44.2 \text{ percent}$$

(c) At 0.75 PF leading

$$VR = \frac{2364 - 2660}{2660} \times 100 = 11.13 \text{ percent}$$

(d) at 0.4 power factor lead, $(1315 - 2660)/2660 = -50.6\%$

Illustration 4.14

A 3-phase, 3300 V, 50 Hz, 500 kVA, alternator has a star-connected stator. The effective resistance per phase of the stator winding is 0.4 Ω. The synchronous reactance per phase is 3.8 Ω. Calculate the regulation at full load (i) at unity power factor and (ii) 0.8 p.f. lagging.

Solution

(i) Terminal voltage (line value) = 3300 V

$$\text{Voltage per phase} = \frac{3300}{\sqrt{3}}$$

$$= 1905 \text{ V}$$

$$\text{Full load current } I_a = \frac{500 \times 10^3}{\sqrt{3} \times 3300}$$

$$= 87.5 \text{ A}$$

Open circuit voltage per phase is given by,

$$E = [(V \cos \phi + I_a R_a)^2 + (V \sin \phi + I_a X_s)^2]^{1/2}$$

$$= [(1905 \times 1.0 + 87.5 \times 0.4)^2 + (1905 \times 0 + 87.5 \times 3.8)^2]^{1/2}$$

$$= 1971.6 \text{ V}$$

$$\text{Regulation} = \frac{1971.6 - 1905}{1905} \times 100$$

$$= 3.5\%$$

(ii) Open circuit voltage per phase at 0.8 pf lagging is,

$$E = [(1905 \times 0.8 + 87.5 \times 0.4)^2 + (1905 \times 0.6 + 87.5 \times 3.8)^2]^{1/2}$$

$$= 2149 \text{ V}$$

$$\text{Regulation} = \frac{2149 - 1905}{1905} \times 100$$

$$= 12.8\%$$

Illustration 4.15

A 400 V, 10 kVA, 3 phase alternator with star-connected stator winding has an effective armature resistance per phase of 1.0 Ω. The alternator generates an open circuit voltage per phase of 90 V with a field current of 1.0 A. During the short circuit test, with 1.0 A of field current, the short circuit current flowing in the armature is 15 A. Calculate (i) the synchronous impedance, (ii) synchronous reactance, (iii) if the alternator is supplying a load current of 15 A at 0.8 power factor lagging, to what value would the terminal voltage rise if the load is thrown-off, (iv) Calculate the regulation at (a) 0.8 p.f. lagging (b) unity p.f.

Solution

(i) Open crcuit voltage per phase with 1.0 A of field current = 90 V

Short circuit current with 1.0 of field current = 15 A

Thus synchronous impedance $Z_s = \dfrac{90}{15} = 6\,\Omega$

(ii) Effective armature resistance per phase

$R_a = 1.0\,\Omega$

Synchronous reactance $X_s = \sqrt{(6.0)^2 - (1.0)^2}$

$= 5.9\,\Omega$

(iii) Phase voltage at the terminals of the load,

$V = \dfrac{400}{\sqrt{3}}$

$= 231\,V$

Load current $I_a = 15\,A$

The open circuit voltage per phase is given by

$E = \sqrt{(V\cos\phi + I_a R_a)^2 + (V\sin\phi + I_a X_s)^2}$

$= \sqrt{(231 \times 0.8 + 15 \times 1.0)^2 + (231 \times 0.6 + 15 \times 5.9)^2}$

$= 302\,V$

Thus the voltage rises from 231 to 302 V, when the load is thrown off.

(vi) (a) At 0.8 pf lagging

$E = 302\,V$

$\text{Regulation} = \dfrac{302 - 231}{231} \times 100$

(b) At unity pf

$E = [231 \times 1.0 + 15 \times 1.0)^2 + (231 \times 0 + 15 \times 5.9)^2]^{1/2}$

$= 261.4$

$\text{Regulation} = \dfrac{261.4 - 231}{231} \times 100$

$= 13.2\%$

4.8 PARALLEL OPERATION OF ALTERNATORS

Before an incoming machine can be switched on to the bus-bars the following conditions have to be fulfilled:

(a) The voltage of the incoming machine must be the same as the bus-bar voltage.
(b) The phase of the machine voltage must be identical with the phase of the bus-bar voltage relative to the feeders, i.e. opposite in phase relative to the local circuit through the armatures and bus-bars. This circuit is shown dotted in Fig. 4.14.
(c) The frequency of the incoming machine must be the same as the bus-bar frequency.

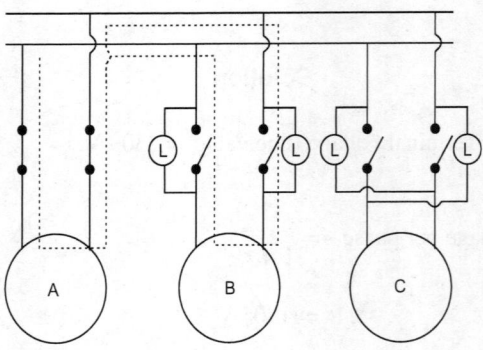

Fig. 4.14. Lamp connections for parallel operation of two 1-phase alternators.

Condition (a) is indicated by a voltmeter, and conditions (b) and (c) are both indicated by synchronizing gear.

Consider first of all the case of single-phase alternators. The simplest form of synchronizer consists of two lamps, LL, connected across the main switch as indicated in Fig. 4.14. If the frequencies of the alternators A and B are not equal, the phase angle between the voltages of A and B will be continually changing, and therefore, the current through the lamps and through the local circuit shown dotted will be changing. The resultant voltage will undergo changes similar in character to the beats produced when two sources of sound of slightly different frequencies are sounding together. This is indicated in Fig. 4.15. In consequence, the lamps will flicker, the alternations in brightness being rapid when there is a large difference in the frequencies, and slow when the frequencies are nearly equal. In the middle of a dark period the two voltages will be in opposition with respect to the local circuit. Hence, the speed of the incoming machine is adjusted until the lamps go in and out very slowly, the incoming voltage is adjusted equal to the bus-bar voltage, and then the switch is closed in the middle of a dark period. It is somewhat easier to judge the middle of the bright than the middle of the dark, period, and some engineers prefer to synchronize "lamps bright". This necessitates the crossing over of the lamp connections, as in the case of machine C.

Conditions Necessary for Successful Parallel Operation

Since, with respect to the local circuit, the emf of an alternator is in phase opposition to the emf of another alternator with which it is working in parallel, the machines run as synchronous motors relative to one another. Hence, if one machine gets into difficulties, say, through a failure in steam supply, it must receive wattful motoring current from the other.

(a) Consider two machines having resistance but no reactance. Their emf-s E_1 and E_2 (Fig. 4.16) will be practically in phase opposition, so that their resultant E_r will be almost in quadrature with both E_1 and E_2. The synchronizing current I will be in phase with E_r and therefore, practically in quadrature with E_1 and E_2, so that it will be an idle current and will therefore convey no real power to the machine needing help.

(b) Suppose that the armatures have reactance only. Then the synchronizing current I will be in quadrature with E_r, and therefore, practically in phase with one of the machine voltages; E_2 in Fig. 4.16. Thus machine II will supply real power to machine I, so that the latter will keep running. This shows that for successful parallel operation, reactance in the armatures is absolutely necessary.

Now consider actual machines with both resistance and reactance. Let the angular phase difference of the two induced emf-s be o and let the circulating current I lag an angle ϕ behind E_2 behind E_r as shown in Fig. 4.16. then so long as 0 is small and the emf-s are equal, we can write

$$E_1 = E_2 = E, \text{ and } E_r = E$$

Circulating current

$$I = \frac{E_r}{Z} = \frac{E}{Z}$$

where Z is the combined impedance per phase of the two armatures.

Synchronizing power $W_s = E_2 I \cos(90 - \phi)$

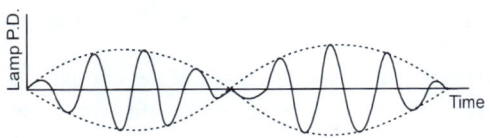

Fig. 4.15.

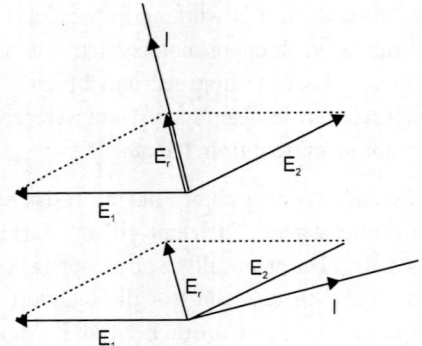

Fig. 4.16. a and b.

$$I = \frac{E^2}{Z} 0 \sin\phi$$

$$\therefore \frac{dW_e}{d0} = \frac{E^2}{Z} \sin\phi$$

$$= \frac{E^2}{\sqrt{R^2 + X^2}} \times \frac{X}{\sqrt{R^2 + X^2}}$$

and this is a maximum when X = R, showing that the maximum synchronizing power would be given with an armature reactance equal to the armature resistance. This condition is, of course, never fulfilled in actual practice, the resistance always being small in comparison with the reactance. Nor is it desirable that this condition should be fulfilled, since it would give rise to an excessively high restoring torque whenever a machine deviated from the steady angular position.

Synchronizing to Mains

For synchronizing to mains the machine is run as a generator with terminals arranged to have the same phase sequence as the mains. Its speed and field current are adjusted such that

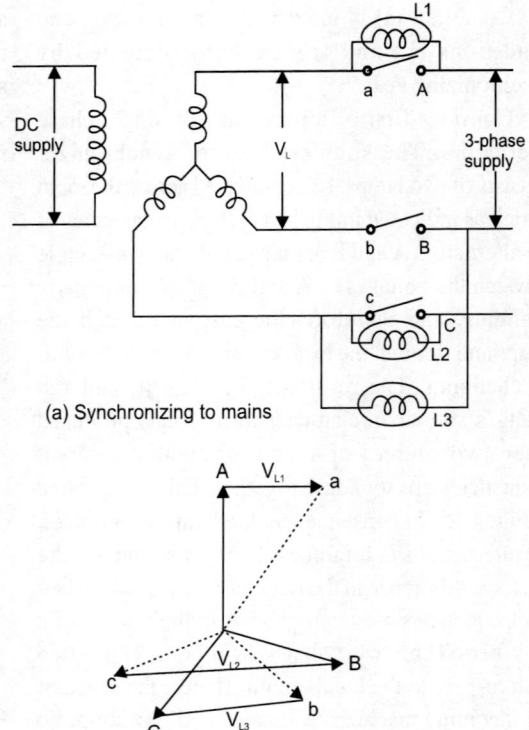

(a) Synchronizing to mains

(b) Determining instant of synchronization

Fig. 4.17.

- the machine terminal voltage is nearly equal to that of the mains, and
- the machine frequency is nearly equal to that of the mains, i.e. its speed is close to synchronous.

The connection diagram is shown in Fig. 4.17. The two sets of 3-phase phasors rotate with respect to each other at the difference in their frequencies synchronization, when the two sets of phasors are co-phasal, $V_{L1} = 0$, $V_{L2} = V_{L3}$, i.e. lamp L_1 is dark and L_2, L_3 are equally bright. the machine is switched on the mains.

Acceptable phase difference in the two phasor sets is about 5°. For larger angular difference the machine would get a current and torque jolt and may not synchronize (falls out of step).

The machine after synchronization would act as a generator or motor depending upon the mechanical conditions at its shaft.

It is immediately obvious from the above that to start a synchronous motor, a small pilot motor (induction type) must be coupled to it to bring it to the speed for synchronization.

Damper Winding

Spring-like synchronous link along with rotor inertia results in oscillations, called *hunting*, initiated by disturbances of the electrical or mechanical sides of the machine. these oscillations are very undesirable electrically and would also fatigue the shaft. These are damped out by providing short-circuit copper bars, known as damper or *armortessur winding*, placed in the rotor pole faces. Damper winding because of induce currents when rotor oscillates wrt the rotating field produces the desired damping effect (damper torque always opposes the oscillatory movement).

Induced currents in the damper winding when the stator is switched on to the supply provide the starting torque (induction principle) for a synchronous motor. The field is switched on after the rotor reaches close to synchronous speed.

Reflection Section

1. The pitch factor for a full-pitch winding is _____.

2. Conceptually explain zero power factor lagging and zero power factor leading.

3. The phasor diagram of synchronous machine connected to an infinite busbars is shown. The machine acts as _____.

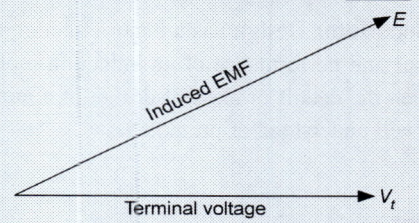

4. Discuss the relation between the connected load and the terminal voltage of the alternator working in leading power factor.

5. The AC armature winding of an alternator operates at much higher voltage than its field. Why and justify.

6. Why the field rotates in an alternator?

PART II: THREE-PHASE INDUCTION MOTOR

4.9 INTRODUCTION

Induction motors use shorted wire loops on a rotating armature and obtain their *torque* from currents *induced* in these loops by the changing *magnetic field* produced in the stator (stationary) coils. Thus an induction motor action involves induced currents in coils on the rotating armature.

At the moment illustrated, the current in the stator coil is in the direction shown (Fig. 4.18) and

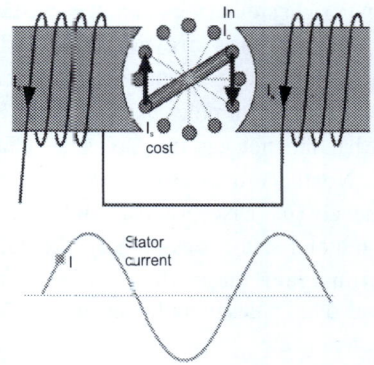

Fig 4.18. Stator field.

increasing. The *induced voltage* in the coil shown drives current and results in a clockwise torque.

Note that this simplified motor will turn once it is started in motion, but has no starting torque. Various techniques are used to produce some asymmetry in the fields to give the motor a starting torque.

4.10 CONSTRUCTION OF A THREE-PHASE AC INDUCTION MOTOR

Induction Motor Design has a major effect on the behaviour and performance of an induction motor. Very often the details or class of design of a motor are not well understood or promoted.

(i) Stator

The stator is the outer body of the motor (Fig. 4.19) which houses the driven windings on an iron core. In a single speed three-phase motor design, the standard stator has three windings, while a single phase motor typically has two windings.

The stator core is made up of a stack of round pre-punched laminations pressed into a frame which may be made of aluminium or cast iron. The laminations are basically round with a round in side through which the rotor is positioned. The inner surface of the stator is made up of a number of deep slots or grooves right around the stator. It is into these slots that the windings are positioned. The arrangement of the windings or coils within the stator determines the number of poles that the motor has. A standard bar magnet has two poles, generally known as North and South. Likewise, an electromagnet also has a North and a South pole. As the induction motor stator is essentially like one or more electromagnets depending on the stator windings, it also has poles in multiples of two. i.e. 2 pole, 4 pole, 6 pole etc.

Fig. 4.19. Image of stator of the induction motor with winding.

The winding configuration, slot configuration and lamination steel all have an effect on the performance of the motor. The voltage rating of the motor is determined by the number of turns on the stator and the power rating of the motor is determined by the losses which comprise copper loss and iron loss, and the ability of the motor to dissipate the heat generated by these losses. The stator design determines the rated speed of the motor and most of the full load, full speed characteristics.

(ii) Rotor

The rotor comprises a cylinder made up of round laminations pressed onto the motor shaft, and a number of short-circuited windings, Fig. 4.20. The rotor windings are made up of rotor bars passed through the rotor, from one end to the other, around the surface of the rotor. The bars protrude beyond the rotor and are connected together by a shorting ring at each end. The bars are usually made of aluminium or copper, but sometimes made of brass. The position relative to the surface of the rotor, shape,

AC Machines 291

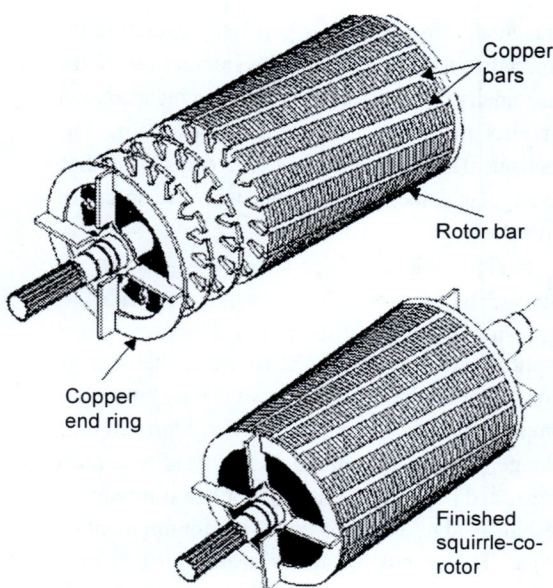

Fig. 4.20 Cage rotor construction.

impedance of the outer portion of the bar is lower than the inner impedance at high frequencies lifting the effective impedance of the bar relative to the impedance of the bar at low frequencies where the impedance of both edges of the bar will be lower and almost equal. The rotor design determines the starting characteristics.

Figure 4.21 is an image showing both the stator and the rotor.

Another type of rotor of a 3-phase induction motor is called the 3-phase slip ring rotor or wound-rotor motors. Unlike the cage rotor, the slip ring rotor will not contain a copper bar in a slot but will consist of a 3-phase winding itself arranged in the rotor slots (Fig. 4.22). Even though the principle of operation of both is the same, there is a good deal of advantage

Fig. 4.21. Image of stator and rotor.

cross-sectional area and material of the bars determine the rotor characteristics. Essentially, the rotor windings exhibit inductance and resistance, and these characteristics can effectively be dependent on the frequency of the current flowing in the rotor.

A bar with a large cross-sectional area will exhibit a low resistance, while a bar of a small cross-sectional area will exhibit a high resistance. Likewise a copper bar will have a low resistance compared to a brass bar of equal proportions.

Positioning the bar deeper into the rotor, increases the amount of iron around the bar, and consequently increases the inductance exhibited by the rotor. The impedance of the bar is made up of both resistance and inductance, and so two bars of equal dimensions will exhibit a different AC impedance depending on their position relative to the surface of the rotor. A thin bar which is inserted radialy into the rotor, with one edge near the surface of the rotor and the other edge towards the shaft, will effectively change in resistance as the frequency of the current changes. This is because the AC

Fig. 4.22. 3-phase windings terminating outside the rotor of an induction motor.

with the slip ring induction motor. External resistances can be added (or reduced) to the 3-phase winding of the rotor. This is not possible with the cage rotor as the copper bars are short circuited at both the ends. Addition of external resistance to the rotor helps a lot in an induction motor to tailor-make the developed torque and also for speed control.

Although the squirrel-cage induction motor is relatively inflexible with regard to speed and torque characteristics, a special wound-rotor version has controllable speed and torque. Application of wound-rotor motors is markedly different from squirrel-cage motors because of the accessibility of the rotor circuit. Various performance characteristics can be obtained by inserting different values of resistance in the rotor circuit.

Wound rotor motors are generally started with secondary resistance in the rotor circuit. This resistance is sequentially reduced to permit the motor to come up to speed. Thus the motor can develop substantial torque while limiting locked rotor current. The secondary resistance can be designed for continuous service to dissipate heat produced by continuous operation at reduced speed, frequent acceleration, or acceleration with a large inertia load. External resistance gives the motor a characteristic that results in a large drop in rpm for a fairly small change in load. Reduced speed is provided down to about 50%, rated speed, but efficiency is low.

4.11 PRINCIPLE OF OPERATION

The AC induction motor is a rotating electric machine designed to operate from a three-phase source of alternating voltage. The stator is a classic three phase stator with the winding displaced by 120°. The most common type of induction motor has a squirrel cage rotor in which aluminium conductors or bars are shorted together at both ends of the rotor by cast aluminium end rings. When three currents flow through the three symmetrically placed windings, a sinusoidally distributed air gap flux generating the rotor current is produced. The interaction of the sinusoidally distributed air gap flux and induced rotor currents produces a torque on the rotor. The mechanical angular velocity of the rotor is lower than the angular velocity of the flux wave by so called slip velocity.

In adjustable speed applications, AC motors are powered by inverters. The inverter converts DC power to AC power at the required frequency and amplitude. The inverter consists of three half-bridge units where the upper and lower switch are controlled complimentarily. As the power device's turn-off time is longer than its turn-on time, some dead-time must be inserted between the turn-off of one transistor of the half-bridge and turn-on of its complementary device. The output voltage is mostly created by a pulse width modulation (PWM) technique. The 3-phase voltage waves are shifted 120° to each other and thus a 3-phase motor can be supplied.

4.12 CONCEPT OF SLIP

Induction motors are probably the simplest and most rugged of all electric motors. They consist of two basic electrical assemblies: the wound stator and the rotor assembly.

The rotor consists of laminated, cylindrical iron cores with slots for receiving the conductors. On early motors, the conductors were copper bars with ends welded to copper rings known as end rings. Viewed from the end, the rotor assembly resembles a squirrel cage, hence the name squirrel-cage motor is used to refer to induction motors. In modern induction motors, the most common type of rotor has cast-aluminium conductors and short-circuiting end rings. The rotor turns when the moving magnetic field induces a current in the shorted conductors. The speed at which the magnetic field rotates is the synchronous speed of the motor and is determined by the number of poles in the stator and the frequency of the power supply.

$$N_s = \frac{120f}{P}$$

Where:
   N_s = synchronous speed
   f = frequency
   P = number of poles

Synchronous speed is the absolute upper limit of motor speed. At synchronous speed, there is no difference between rotor speed and rotating field speed, so no voltage is induced in the rotor bars, hence no torque is developed. Therefore, when running, the rotor must rotate slower than the magnetic field. The rotor speed is just slow enough to cause the proper amount of rotor current to flow, so that the resulting torque is sufficient to overcome windage and friction losses, and drive the load. This speed difference between the rotor and magnetic field, called slip, is normally referred to as a percentage of synchronous speed:

$$s = \frac{100(N_s - N)}{N_s}$$

Where:
   s = slip
   N_s = synchronous speed
   N = actual speed

Illustration 4.16

A 4 pole 3 phase induction motor operates from a supply whose frequency is 50 Hz. Calculate a) synchronous speed with which the magnetic field of the stator is rotating and (b) speed of the rotor when the slip is 0.04.

Solution

N_s = 120f / P = 120 × 50/4 = 1,500 rpm.
From the slip expression,

N = (1−s)N_s = (1−0.04) 1,500 = 1440 rpm.

Illustration 4.17

A 3-phase induction motor is wound for 4 poles and is supplied from 50 Hz system. Calculate a) N_s and b) slip when the rotor speed is 600 rpm.

Solution

N_s = 120f / P = 120 × 50/4 = 1,500 rpm.
Using the slip expression,

s = (N_s−N)/N_s = (1500−600)/1500 = 0.6 (or 60%)

Illustration 4.18

A 12 pole 3 phase alternator driven at a speed of 500 rpm supplies power to an 8 pole 3 phase induction motor. If the full load slip of the motor is 3%, calculate the full load speed of the motor.

Solution

Frequency of the alternator

f = NP/120 = 500 × 712/120 = 50 Hz

Synchronous speed of the induction motor,

N_s = 120 × 50/8 = 750 rpm

Speed of the motor at 3% slip is

(1-s) N_s = (1−0.03) 750 = 727.5 rpm

Illustration 4.19

A 4 pole 3 phase induction motor is supplied from 50 cps supply. Determine its synchronous speed. On full load, its speed is observed to be 1410 rpm. Calculate its full load slip.

Solution

$N_s = 120 \times 50/4 = 1500$ rpm.
Slip, $s = (N_s - N) / N_s = (1500 - 1410) / 1500 = 0.06$ (or 6%)

Illustration 4.20

A 4 pole, 50 Hz, 3-phase induction motor has a star connected rotor. If the full load speed is 1460 rpm, Calculate the slip.

Solution

We can get Ns = 1500 rpm.
So slip = 0.0267.

4.13 TORQUE IN A 3-PHASE INDUCTION MOTOR

As the torque developed in any machine has a direct proportions with the equivalent electric power, let the power develop in the rotor be written. It is

$P_2 = E_2 I_2 \cos \phi_2$

As the voltage induced in the rotor at slip s is sE_2, the current induced is E_2/Z_2 and the power factor is R_2/X_2, we can write the above expression as

$$P_2 = sE_2 \cdot (E_2 / Z_2) (R_2/Z_2)$$
$$= sE_2^2 R_2 / (R_2^2 + X_2^2)$$

As X which is $L\omega$, is also a dependent on slip s, the output power equation finally become,

$P_2 = sE_2^2 R_2 / (R_2^2 + s^2 X_2^2)$

Note importantly that the variable in the above expression is the slip and hence the output power has a direct as well as indirect relation with the slip.

The machine will have a very negligible slip at start. So $s^2 X_2^2$ will be very small leaving the torque T proportional to slip S alone. The machine's torque at start will vary linearly with slip. But when the machine speed falls as it is gradually loaded, the slip will be notable and the term $s^2 X_2^2$ will not be negligible. This will make the torque T to be inversely proportional to the slip, S. So during the running time T will be in inverse proportion with that of S. During the running condition, the machine's torque will vary parabolically with slip.

This is further discussed below.

Starting Characteristics

In order to perform useful work, the induction motor must be started from rest and both the motor and load accelerated up to full speed. Typically, this is done by relying on the high slip characteristics of the motor and enabling it to provide the acceleration torque. Induction motors at rest, appear just like a short circuited transformer, and if connected to the full supply voltage, draw a very high current known as the "Locked Rotor Current". They also produce torque which is known as the "Locked Rotor Torque". The **L**ocked **R**otor **T**orque (**LRT**) and the **L**ocked **R**otor **C**urrent (**LRC**) are a function of the terminal voltage to the motor, and the motor design. As the motor accelerates, both the torque and the current will tend to alter with rotor speed if the voltage is maintained constant.

The starting current of a motor, with a fixed voltage, will drop very slowly as the motor accelerates and will only begin to fall significantly when the motor has reached at least 80% full speed. The actual curves for induction motors can vary considerably between designs, but the general trend is for a high current until the motor has almost reached full speed. The LRC of a motor can range from 500% Full Load Current (FLC) to as high as 1400% FLC. Typically, good motors fall in the range of 550% to 750% FLC.

The starting torque of an induction motor starting with a fixed voltage, will drop a little to the minimum torque known as the *pull up* torque as the motor accelerates, and then rise to a maximum torque

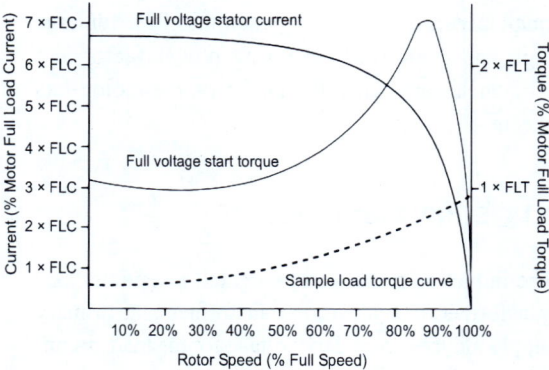

Fig. 4.23

known as the *breakdown* or *pull out* torque at almost full speed and then drop to zero at synchronous speed. The curve of start torque against rotor speed is dependent on the terminal voltage and the motor/rotor design.

The LRT of an induction motor can vary from as low as 60% **F**ull **L**oad **T**orque (**FLT**) to as high as 350% FLT. The pull-up torque can be as low as 40% FLT and the breakdown torque can be as high as 350% FLT. Typical LRTs for medium to large motors are in the order of 120% FLT to 280% FLT.

The power factor of the motor at start is typically 0.1–0.25, rising to a maximum as the motor accelerates, and then falling again as the motor approaches full speed.

A motor which exhibits a high starting current, i.e. 850% will generally produce a low starting torque, whereas a motor which exhibits a low starting current, will usually produce a high starting torque. This is the reverse of what is generally expected.

The induction motor operates due to the torque developed by the interaction of the stator field and the rotor field. Both of these fields are due to currents which have resistive or in phase components and reactive or out of phase components. The torque developed is dependent on the interaction of the in phase components and consequently is related to the I^2R of the rotor. A low rotor resistance will result in the current being controlled by the inductive component of the circuit, yielding a high out of phase current and a low torque.

Figures for the locked rotor current and locked rotor torque are almost always quoted in motor data, and certainly are readily available for induction motors. Some manufacturers have been known to include this information on the motor name plate. One additional parameter which would be of tremendous use in data sheets for those who are engineering motor starting applications, is the starting efficiency of the motor. By the starting efficiency of the motor, it is referred to the ability of the motor to convert amps into newton meters. This is a concept not generally recognised within the trade, but one which is extremely useful when comparing induction motors. The easiest means of developing a meaningful figure of merit, is to take the locked rotor torque of the motor (as a percentage of the full load torque) and divide it by the locked rotor current of the motor (as a percentage of the full load current).

If the terminal voltage to the motor is reduced while it is starting, the current drawn by the motor will be reduced proportionally. The torque developed by the motor is proportional to the current squared, and so a reduction in starting voltage will result in a reduction in starting current and a greater reduction in starting torque. If the start voltage applied to a motor is halved, the start torque will be a quarter, likewise a start voltage of one-third will result in a start torque of one-ninth.

Running Characteristics

Once the motor is up to speed, it operates at low slip, at a speed determined by the number of stator poles. The frequency of the current flowing in the rotor is very low. Typically, the full load slip for a standard cage induction motor is less than 5%. The actual full load slip of a particular motor is dependent on the motor design with typical full load speeds of four pole induction motor varying between 1420 and 1480 RPM at 50 Hz. The synchronous speed of a four pole machine at 50 Hz is 1500 RPM and at 60 Hz a four pole machine has a synchronous speed of 1800 RPM.

The induction motor draws a magnetising current while it is operating. The magnetising current is independent of the load on the machine, but is dependent on the design of the stator and the stator voltage. The actual magnetising current of an induction motor can vary from as low as 20% FLC for large two pole machines to as high as 60% for small eight pole machines. The tendency is for large machines and high speed machines to exhibit a low magnetising current, while low speed machines and small machines exhibit a high magnetising current. A typical medium sized four pole machine has a magnetising current of about 33% FLC.

A low magnetising current indicates a low iron loss, while a high magnetising current indicates an increase in iron loss and a resultant reduction in operating efficiency.

The resistive component of the current drawn by the motor while operating, changes with load, being primarily load current with a small current for losses. If the motor is operated at minimum load, i.e. open shaft, the current drawn by the motor is primarily magnetising current and is almost purely inductive. Being an inductive current, the power factor is very low, typically as low as 0.1. As the shaft load on the motor is increased, the resistive component of the current begins to rise. The average current will noticeably begin to rise when the load current approaches the magnetising current in magnitude. As the load current increases, the magnetising current remains the same and so the power factor of the motor will improve. The full load power factor of an induction motor can vary from 0.5 for a small low speed motor up to 0.9 for a large high speed machine.

The losses of an induction motor comprise: iron loss, copper loss, windage loss and frictional loss. The iron loss, windage loss and frictional losses are all essentially load independent, but the copper loss is proportional to the square of the stator current. Typically the efficiency of an induction motor is highest at 3/4 load and varies from less than 60% for small low speed motors to greater than 92% for large high speed motors. Operating power factor and efficiencies are generally quoted on the motor data sheets.

4.14 EQUIVALENT CIRCUIT

The induction motor can be treated essentially as a transformer for analysis. In transformers, the primary supply induces secondary voltage for the load circuit; similarly, in an induction motor too, the stator supply induced rotor emf and hence torque to be used in the load circuit. Hence it is otherwise called a rotating transformer.

The equivalent circuit of the induction motor hence can be drawn in a similar way as that of a transformer, replacing all primary quantities with stator and all the secondary quantities with rotor.

The induction motor has stator leakage reactance, stator copper loss elements as series components, and iron loss and magnetising inductance as shunt elements. The rotor circuit likewise has rotor leakage reactance, rotor copper (aluminium) loss and shaft power as series elements.

The transformer in the centre of the equivalent circuit can be eliminated by adjusting the values of the rotor components in accordance with the effective turns ratio of the transformer.

From the equivalent circuit and a basic knowledge of the operation of the induction motor, it can be seen that the magnetising current component and the iron loss of the motor are voltage dependent, and not load dependent. Additionally, the full voltage

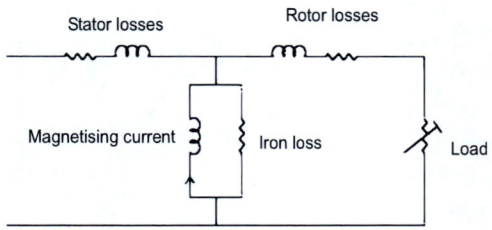

Fig. 4.24. Equivalent circuit of a 3-phase induction motor.

starting current of a particular motor is voltage and speed dependent, but not load dependent.

The magnetising current varies depending on the design of the motor. For small motors, the magnetising current may be as high as 60%, but for large two pole motors, the magnetising current is more typically 20–25%. At the design voltage, the iron is typically near saturation, so the iron loss and magnetising current do not vary linearly with voltage with small increases in voltage resulting in a high increase in magnetising current and iron loss.

4.15 STARTING ARRANGEMENTS FOR MOTORS

(i) Direct On Line

The simplest form of motor starter for the induction motor is the **Direct on Line** starter. The DOL starter comprises a switch and an overload protection relay.

When the stator windings of an induction motor are connected directly to its 3-phase supply, a very large current (5–8 times full load current) flows initially. This surge current reduces as the motor accelerates up to its running speed.

Induction motors can be Direct-on Line (DOL) started in this way. The starting current will not cause damage to the motor unless the motor is repeatedly started and stopped in a short space of time. This is called 'fast cycling'. When very large motors are started direct-on-line they cause a disturbance of voltage (voltage dip) on the supply lines due to the large starting current surge. This voltage disturbance may result in the malfunction of other electrical equipment connected to the supply.

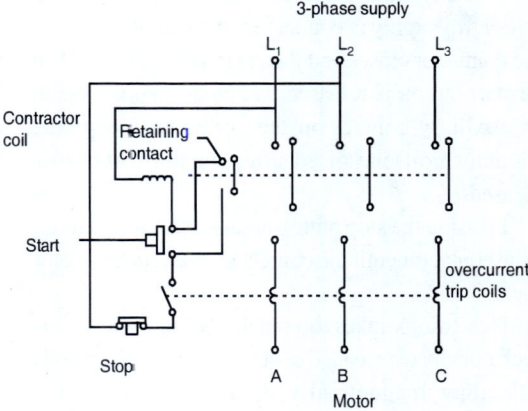

Fig. 4.25. DOL Starter

To limit the starting current some large induction motors are started at reduced voltage and then have the full supply voltage reconnected when they have run up to near rated speed.

Direct-on-line starting is the method most commonly used, the most usual consideration being whether the generator and the distribution system can withstand the starting current without excessive voltage dips. In the case of loads involving considerable inertia, such as centrifugal oil separators, the starting time may also be a factor. In case of doubt, the motor manufacturer should be consulted.

The starting current as we have already seen may be five to eight times the full load current, and the heating of the windings is proportional to the square of the current. At starting it will therefore be 25–64 times normal. Furthermore, at the instant of start there is no windage and no radiation. Therefore a very long starting period may result in overheating.

For these reasons it is also undesirable to make repeated successive starts without intervening periods for cooling.

The contactor coil is connected in series with a start button, stop button and overload trip contacts. This is called the control circuit and is energised from two lines of the 3-phase supply—usually via a step-down transformer. When the start button is pressed the control supply is connected to the contactor coil. The contactor closes and then starts the motor. When the start button is released its contacts spring open. An auxiliary contact on the contractor keeps the contactor coil energised after the start button is released.

Pressing the stop button breaks the control circuit to the contactor coil; the contactor trips and the motor stops.

If the motor takes too much current because it is mechanically overloaded or stalled, the overload coils will either magnetically or thermally open the overload trip contacts which will stop the motor and prevent overheating. Note, the correct term is 'overcurrent' rather than the commonly used 'overload'.

Provided the torque developed by the motor exceeds the load torque at all speeds during the start cycle, the motor will reach full speed. If the torque delivered by the motor is less than the torque of the load at any speed during the start cycle, the motor will cease accelerating. If the starting torque with a DOL starter is insufficient for the load, the motor must be replaced with a motor which can develop a higher starting torque. The acceleration torque is the torque developed by the motor minus the load torque, and will change as the motor accelerates due to the motor speed torque curve and the load speed torque curve. The start time is dependent on the acceleration torque and the load inertia.

DOL starting results in maximum start current and maximum start torque. This may cause an electrical problem with the supply, or it may cause a mechanical problem with the driven load.

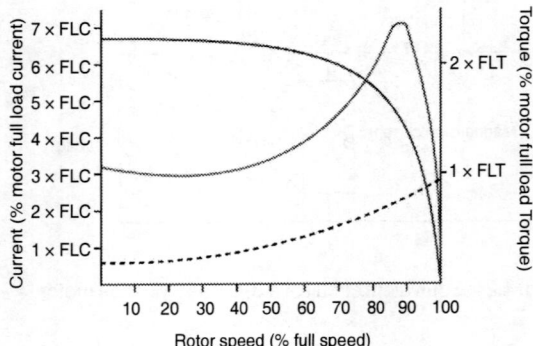

Fig. 4.26

(ii) Primary Resistance

The primary resistance starter will have one or more sets of resistors which, during start, are connected in series with the supply to the motor. The series resistors limit the starting current drawn by the motor, and thus reduce the starting torque of the motor.

Once the motor is up to full speed (or after a period of time) the resistors are bridged by a contactor to apply full voltage to the motor. If the full details of the motor starting characteristics are known, and the starting characteristics of the load are also known, it is practical to determine the correct value of the resistors to provide enough start torque for the load while minimising the starting current. A primary resistance starter correctly designed and constructed, will cause the motor to accelerate the load to almost full speed with the resistors in circuit before they are bridged out. In this case, the transition to full voltage only occurs once the impedance of the motor has risen, and the resulting current is much less than the LRC of the motor. In a poorly designed system, the transition to full voltage will occur at less than 80% full speed, and the current will then step up to almost DOL current, resulting in little gain from the use of

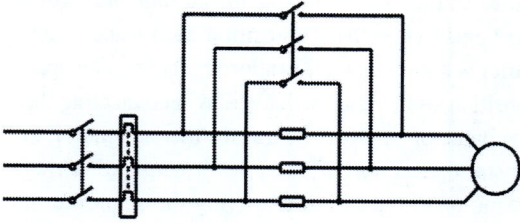

Fig. 4.27. Primary resistance starter

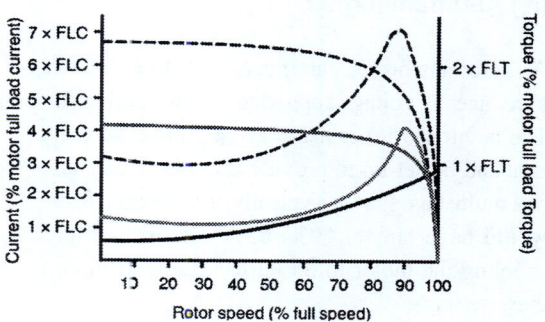

Fig. 4.28. Slip-torque characteristics with primary resitance variation.

the primary resistance starter other than the increased cost of the starter. (advantageous to the starter supplier, not to the end user.) Improved starting characteristics with some loads can be achieved by the use of several stages of resistance and bridging out increasing amounts of resistance as the motor accelerates.

With the primary resistance starter, it is not easy to alter the resistance and hence the starting characteristics once the starter is built. Therefore, it is important that the correct resistors are selected in the first place.

The primary resistance starter reduces the voltage applied to the motor terminals while passing the full starting current to the motor. Consequently, there is a very high power dissipation in the resistors, resulting in the requirement for very high power rated resistors. Typically, the resistors will dissipate as much as 150% – 200% the power rating of the motor for the duration of the start.

The resistors may be either metallic resistors, or liquid resistors. Metallic resistors have a positive temperature coefficient and as a result, as they heat up, their resistance increases. Liquid resistors, such as saline solution, have a negative temperature coefficient and so consequently, as they heat up, their resistance reduces. The heat build up in the resistors during start, and their temperature dependent resistance characteristics, make it essential, the resistors are allowed to fully cool between starts. This restricts the starting frequency and the minimum time between the starts.

(iii) Primary Reactance

A primary reactance starter is similar to a primary resistance starter except that the resistors are replaced by a three-phase reactor to limit the starting current. The operation of the primary reactance starter is essentially the same as that of the primary resistance starter, but the use of a three-phase reactor in place of the resistors offers the advantage of reduced heat loss and greater ease of start current setting due to the ability to change taps on the reactor.

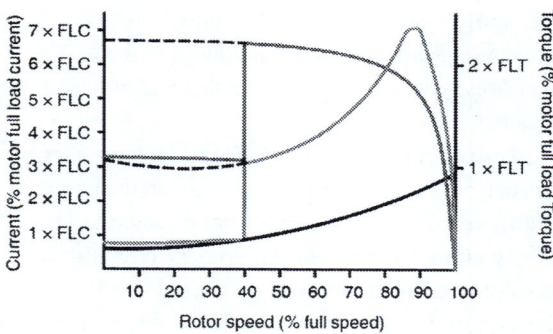

(iv) Autotransformer

An autotransformer starter uses an auto transformer to reduce the voltage applied to a motor during start. The autotransformer may have a number of output taps and be set-up to provide a single stage starter, or a multistage starter. Typically, the autotransformer would have taps at 50%, 65% and 80% voltage, enabling the motor to be started at one or more of these settings.

There are two ways of connecting an autotransformer starter, the most obvious way is to apply full voltage to the transformer via a contactor, and connect the motor to the tap by means of a contactor. When the motor has accelerated to full speed, or has run out of acceleration torque, the tap contactor opens, disconnecting the motor from the transformer and another contactor closes connecting the motor to the supply. The transformer can now be disconnected from the supply. This format is known as an open transition starter and is less than ideal due to the fact that the motor is disconnected for a short period of time during the start period. While the motor is connected and accelerating, there is a rotating magnetic field in the stator which causes flux in the rotor and thus a rotor current to flow. At the instant the motor is disconnected, there is a magnetic field in the rotor which is spinning within the stator winding. The motor acts as a generator until the rotor field decays. The voltage generated by the motor is not synchronised to the supply, and so on reconnection to the supply, the voltage across the contactor at closure can be as much as twice the supply voltage resulting in a very high current and torque transient. This open transition switching is often known as the auto-reclose effect as it yields similar characteristics to opening and closing a breaker on a supply to one or more motors. The consequences of open transition switching can be as bad as broken shafts and stripped gears.

By a rearrangement of the power circuit, it is possible, at no extra cost, to build a closed transition starter and thereby eliminate the current and torque transients. The closed transition auto transformer starter is known as the Korndorffer starter. The open transition switching is achieved by reconnecting the tap contactor between the transformer and motor, to the star connection of the transformer, hard wiring the motor to the tap, and altering the sequence of contactor control. To start the machine, the main contactor and the star contactors are closed applying reduced voltage to the motor. When the motor has reached full speed, (or run out of acceleration torque) the star contactor is opened effectively converting the autotransformer starter into a primary reactance starter. Next the primary reactance is bridged by a contactor applying full voltage to the motor. At no time does the motor become disconnected from the supply.

The transformer is generally only intermittent rated for the starting duty, and so the frequency and duration of the starts is limited. With a transformer starter, it is relatively easy to change taps and thereby increase the starting voltage if a higher torque is required. The auto transformer starter is a constant voltage starter, so the torque is reduced by the voltage reduction squared over the entire speed range, unlike the primary resistance or primary reactance starters which are constant impedance starters and where the start voltage is dependent on the ratio of the motor impedance to the motor plus starter impedance. As the motor accelerates, its impedance rises and consequently, the terminal voltage of the motor also rises, giving a small torque increase at higher speeds.

Unlike the primary resistance and primary reactance starter, the current flowing into the motor is different from that flowing from the supply. The supply current flows into the primary circuit of a transformer, and the secondary current is applied to the motor. The transformer reduces the primary current by the same ratio as the voltage reduction. If the motor is connected to the 50% tap of the transformer, the voltage across the motor terminals will be 50%. Assuming an LRC of 600%, there will be 300% current flowing into the motor. If 300% current flows into the motor, then the current into the transformer will be 150%. This would suggest that the lowest starting current will be achieved by the use of an autotransformer starter. In most instances, the load will require an increasing torque as it accelerates, and so often a higher tap must be selected in order to accelerate the load to full speed before the step to full voltage occurs. If a multistage transformer starter is employed, then the primary current will certainly be lower than other forms of induction motor starter.

Starting large motors with long-run up periods demands a very high current surge from the supply generator. This causes a severe voltage dip which affects every load on the system. Reduced voltage starting will limit the starting surge current.

One way to reduce the initial voltage supplied to the motor is to step it down using a transformer. Then, when the motor has accelerated up to almost full speed, the reduced voltage is replaced by the full mains voltage. The transformer used in this starter is not the usual type with separate primary and secondary windings. It is an autotransformer which uses only one winding for both input and output. This arrangement is cheaper, smaller and lighter than an equivalent double-wound transformer. For induction motor starting, the autotransformer is a 3-phase unit, and because of expense, this method is only used with large motor drives, e.g. electric cargo pumps.

The autotransformer with its range of tapping points gives a set range of starting voltages to limit the motor starting surge current to a reasonable value.

As with the star-delta starter, the autotransformer may use what is called an open transition switching sequence or a closed transition switching sequence between the start and run conditions. In the former, the reduced voltage is rapidly reconnected to the motor.

The circuit diagram below shows a manually operated open transition, autotransformer starter.

The problem with open transition is that a very large surge current can flow after the transition from reduced to full voltage.

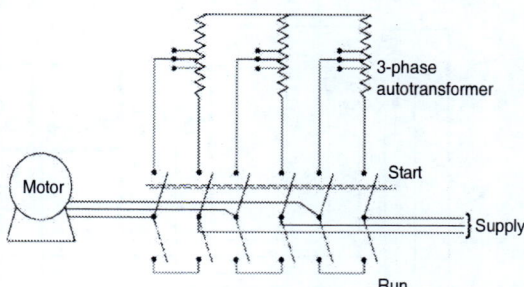

Fig. 4.29. An auto-transformer starter

(v) Star Delta

The Star Delta starter can only be used with a motor which is rated for connection in delta operation at the required line voltage, and has both ends each of the three windings available individually. At Start, the line voltage is applied to one end of each of the three windings, with the other end bridged together, effectively connecting the windings in a star connection. Under this connection, the voltage across each winding is $1/\sqrt{3}$ of line voltage and so the current flowing in each winding is also reduced by this amount. The resultant current flowing from the supply is reduced by a factor of 1/3 as is the torque. i.e. A motor which exhibits a LRC of 600% and an LRT of 180% will exhibit characteristics of: LRC_{star} of 200% and LRT_{star} of 60%. In some cases, this may be enough to get the motor up to full speed, but most, as this is a constant voltage starter, the transition to full voltage will occur at part speed resulting in a virtual DOL type start. To step to full voltage, the star connection is opened, effectively open circuiting the motor, and the ends of the windings are then connected to the three phase supply in a fashion to create a delta connection. This type of starter is an open transition starter and so the switch to delta is

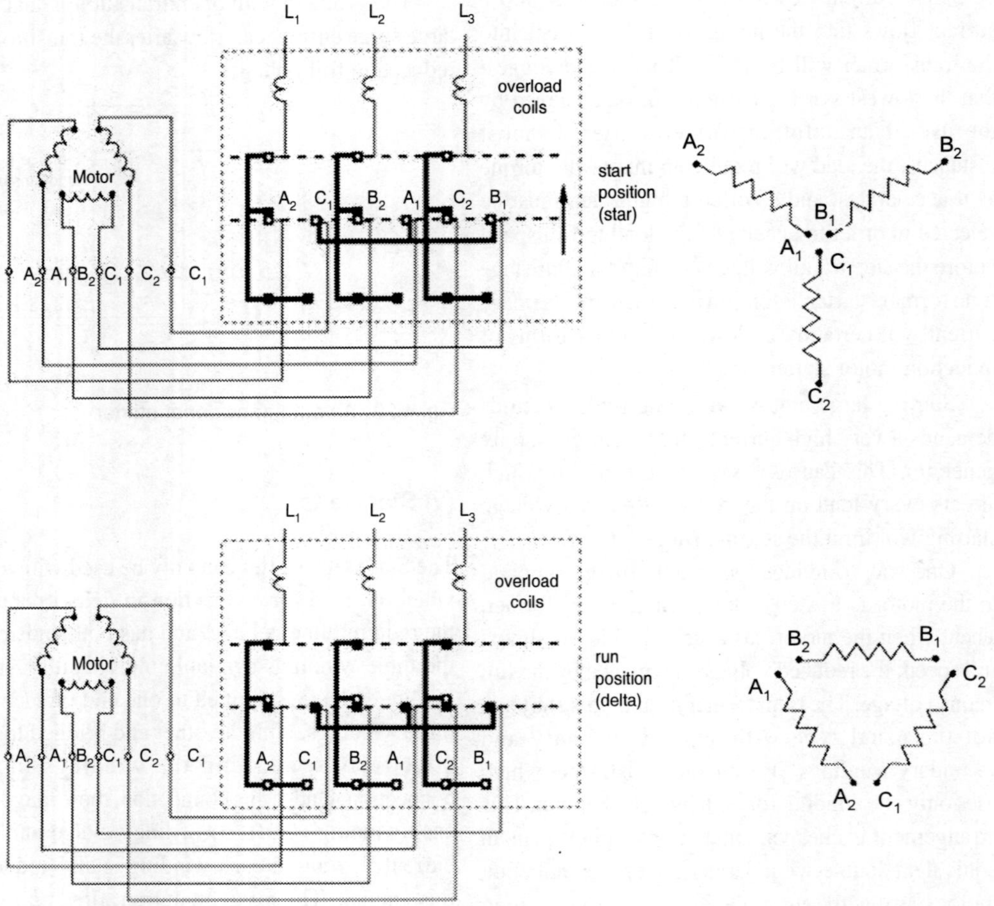

Fig. 4.30. A star-delta starter

accompanied by a very high torque and current transient. In most situations, there would be less damage to the equipment and less interference to the supply if a DOL starter was employed.

The star delta is not easily converted to a closed transition starter, and even the closed transition (Wanchop) star delta starter still has the problem that the start voltage cannot be altered. If there is insufficient torque available in star, then it will go DOL. The star delta starter does get around the regulations in some countries where there is a requirement for a reduced voltage starter, but in reality, in many situations results in more severe transients than DOL. The main benefits of the star delta starter are that it puts more money in the pockets of the switchgear supplier, and it is politically correct.

Reduced voltage starting is used for large motors driving loads like cargo pumps and bow thrusters. Two methods of reduced voltage starting are star-delta starting and autotransformer starting.

After DOL starting, the next most common method is the star-delta method.

Both ends of each phase of the motor starter windings must be brought out and connected to starter. In the start position the windings are connected in star; in the running position they are reconnected in delta. The voltage across each phase winding in the start position is 58% ($1/\sqrt{3}$) of line voltage, with consequent reduction of starting current. The starting torque is also reduced to one-third of that which would obtain with DOL starting. With a single-cage or double-cage rotor of average performance, this represent about 80% of full-load torque, assuming normal line voltage, but if there is appreciable line drop the torque will be proportionately lower. These factors must be taken into account when deciding whether star-delta starting is acceptable for the driven machine. It will be acceptable for centrifugal fans and pumps if, in the latter case, the friction at starting is not excessive.

When the operating handle is placed in the 'start' position the motor stator windings are connected in star across the supply. As the motor approaches normal running speed the operator must quickly change the handle to run position which changes the motor connection from star to delta. If the operator does not move the handle quickly from start to run the motor may be disconnected from the supply long enough for the motor speed to fall considerably. When the handle is eventually put into the run position the motor will take a large current may be large current and accelerate up to speed again. This surge current may be large enough to cause appreciable voltage dip. The prevent this, a mechanical interlock is fitted to the operating handle.

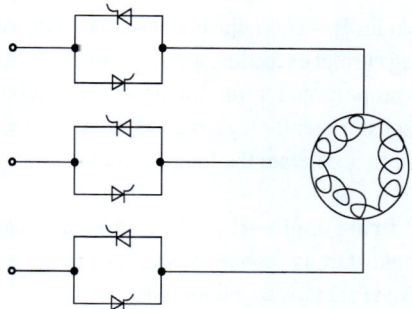

Fig. 4.31. Electronic starters

The handle must be moved quickly from start to run otherwise the interlock jams the handle in the start position.

Electronic starters often referred to as 'soft start', are finding acceptance in the marine industry. Solid-state technology is employed to provide a method of starting without the current and torque surges mentioned previously. Thyristors or a combination of thyristors and diodes are used to control the current flow during motor starting. The basic circuit diagrams for these two alternatives are shown below.

The electronics for controlling the firing of the thyristors is normally accommodated on a small printed circuit board within the motor controller. Although the thyristor/diode configuration is cheaper it has the disadvantage that it generates third and even harmonic currents in the motor windings, whereas the all-thyristor arrangement restricts the even harmonics.

With this type of starter there are normally three adjustments that have to be set to suit the drive machinery:

1. Voltage ramp—This sets the time for the starter to achieve full voltage output. It should be noted that the ramp time is the time taken for the output voltage to reach its maximum and not for the motor to reach its full speed.

If a motor is lightly loaded it may well achieve full speed before full voltage is applied.

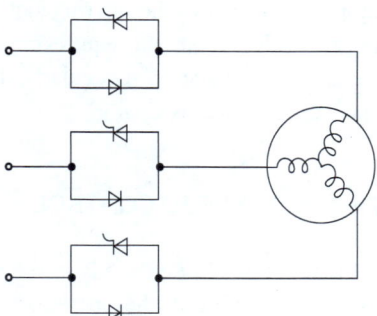

2. Current limit—This adjustment is used to prevent the starting current exceeding a preset value. Because torque is proportional to the square of the current it must be set sufficiently high that adequate torque is developed to accelerate the load from rest.

3. Initial firing angle—It is often important that a drive should start as soon as voltage is applied, e.g. if the drive is standby to a duty unit.

If the initial firing angel is set too small there will be a delay in starting the drive until the voltage has been ramped to a value permitting sufficient torque to be developed to accelerate it from the rest. If the initial firing angle is set too large the load may be suddenly grabbed rather than accelerated smoothly.

Triacs can also be used for electronic starters. However, since they have relatively low current ratings and breakdown voltages they are generally suitable only for low-current low-voltage applications.

(vi) Slip Ring Motors

The slip ring motor is essentially similar to the standard cage induction motor except that the winding on the rotor has far more turns and instead of being short circuited, is brought out to a set of slip rings for external connection.

The operation of the slip ring motor is the same as that of the standard cage induction motor in that torque is generated by the interaction of the stator field and the rotor field. The rotor field being generated by current flowing in the rotor which is caused by the slip between the rotor and the stator field. The torque speed curve and the current speed curve can be altered by the rotor winding termination. A very high value of resistance on the rotor termination will give a very low locked rotor current, and a low locked rotor torque. Reducing the termination resistance, will increase both the locked rotor current and the locked rotor torque up to the point where maximum torque is available under locked rotor conditions. Further reduction of termination resistance will reduce the locked rotor torque by shifting the maximum (Pull Up) from zero speed towards synchronous speed. A short circuit across the rotor will result in the maximum torque occurring at a very low slip, and a locked rotor current as high as 1400% for a locked rotor torque of as low as 50%. It is imperative that there must be resistance in the rotor circuit of a slip ring motor during start if any starting torque is to be developed at a realistic starting current.

Typically, the slip ring motor is started by a multistage starter, developing as high as 300% torque at 250% current. By stepping through lower resistor values as the motor accelerates, the maximum torque is kept in step with the actual rotor speed and thus the maximum acceleration of a very difficult load can be achieved in minimum time and with maximum efficiency.

The slip ring starter comprises an isolation contactor for the stator circuit, the stator being effectively DOL controlled, and a series of rotor resistors and contactors controlled by a sequencer. The resistors must be sized to suit the driven load. The total power dissipated in the resistors will be at least equal to the kinetic energy of the driven load at full speed.

Slip ring motors typically have an open circuit rotor voltage (or frame voltage) of between 400 volts and 500 volts to keep the current to a manageable level.

The major disadvantage of the slip ring motor is that the ring gear suffers wear and requires regular maintenance, as does the starter, particularly if an electromechanical sequencer is used.

4.16 METHODS OF SPEED CONTROL

The speed of a driven load often needs to run at a speed that varies according to the operation it is performing. The speed in some cases such as pumping may need to change dynamically to suit the conditions, and in other cases may only change with a change in process. Electric motors and coupling combinations used for altering the speed will behave as either a "*Speed Source*" or a "*Torque Source*". The "Speed Source" is one where the driven load is driven at a constant speed independent of load torque. A "Torque Source" is one where the driven load is driven by a constant torque, and the speed alters to the point where the torque of the driven load equals the torque delivered by the motor. Closed loop controllers employ a feedback loop to convert a "Torque Source" into a "Speed Source" controller.

(i) Mechanical

There are a number of methods of mechanically varying the speed of the driven load when the driving motor is operating at a constant speed. These are typically:

- Belt Drive
- Chain Drive
- Gearbox
- Idler wheel drive

All of these methods exhibit similar characteristics whereby the motor operates at a constant speed and the coupling ratio alters the speed of the driven load. Increasing the torque load on the output of the coupling device, will increase the torque load on the motor. As the motor is operating at full voltage and rated frequency, it is capable of delivering rated output power.

There is some power loss in the coupling device resulting in a reduction of overall efficiency. The maximum achievable efficiency is dependent on the design of the coupling device and sometimes the way it is set up. (e.g. belt tension, no of belts, type of belts etc.)

Most mechanical coupling devices are constant ratio devices and consequently the load can only be run at one or more predetermined speeds. There are some mechanical methods that do allow for a dynamic speed variation but these are less common and more expensive.

Mechanical speed change methods obey the 'Constant Power Law' where the total power input is equal to the total power output. As the motor is capable of delivering rated power output, the output power capacity of the combination of motor and coupling device (provided the coupling device is appropriately rated) is the rated motor output power minus the loss power of the coupling device. Torque 'T' is a Constant 'K' times the Power 'P' divided by the speed 'N'.

$$T = K \times P / N$$

Therefore for an ideal lossless system, the torque at the output of the coupling device is increased by the coupling ratio for a reduced speed, or reduced by the coupling ratio for an increased speed.

(ii) Magnetic

There are two main methods of magnetically varying the speed of the driven load when the driving motor is operating at a constant speed. These are:

- Eddy Current Drive
- Magnetic Coupling

These methods use a coupling method between the motor and the driven load which operates on induced magnetic forces. The eddy current coupling is quite commonly employed, and is easily controlled by varying the bias on one of the windings. In operation, it is not unlike an induction motor, with one set of poles driven by the driving motor, hence operating at the speed of the driving motor. The second set of poles are coupled to the driven load, and rotate at the same speed as the driven load. One set of poles comprises a shorted winding in the same manner as the rotor of an induction motor, while the other set of poles is connected to a controlled DC current source. When the machine is in operation, there is a difference in speed between the two sets of poles, and consequently there is a current induced in the shorted winding. This current establishes a rotating field and torque is developed in the same way as an induction motor. The coupling torque is controlled by the DC excitation current. This method of coupling is essentially a torque coupling with slip power losses in the coupling.

(iii) Hydraulic

There are two main methods of hydraulically varying the speed of the driven load when the driving motor is operating at a constant speed. These are:

Hydraulic pump and motor
Fluid coupling

The **fluid coupling** is a torque coupling whereby the input torque is equal to the output torque. This type of coupling suffers from very high slip losses, and is used primarily as a torque limited coupling during start with a typical slip during run of 5%. The constant power law still applies, but the power in the driven load reduces with speed. The difference between the input power and the output power is power dissipated in the coupling.

In an extreme case, if the load is locked (stationary) and the motor is delivering full torque to the load via a fluid coupling, the load will be doing no work and hence absorbing no power, with the motor operating at full speed and full torque, the full output power of the motor is dissipated in the coupling. In most applications, the torque requirement of the load at reduced speed is much reduced, so the power dissipation is much less than the motor rating.

In the case of a **hydraulic pump** and motor, the induction motor operates at a fixed speed, and drives a hydraulic pump which in turn drives a hydraulic motor. In many respects, this behaves in a manner similar to a gear box in that the hydraulic system transfers power to the load. The torque will be higher at the load than at the motor for a load running slower than the motor.

Reflection Section

1. The no-load slip will always be less than 1%. Why?
2. The shaft of an induction motor is usually made hollow. Justify.
3. Slip ring of an induction motor is usually made up of the material _____.
4. With reference to space, the rotor speed and stator speed are same. Justify.
5. Why the starting torque of a cage induction motor is low?
6. Three-phase induction motor is analogous to a secondary short-circuited two winding transformer. How?
7. Compare the types of starters used for 3-phase induction motor.

Part III: Single-Phase Induction Motor

4.17 INTRODUCTION

In Part II, 3-phase, AC motors, which are used for high power rating applications, have been discussed. For reasons of economy, most homes, offices and also rural areas are supplied with single-phase AC, as the power requirements of individual load items are rather small. Even though input point power to homes or offices may be 3 phase, the inside wiring is single-phase, 220–230 V for reasons of safety. This has led to the availability of a wide variety of small-size motors of fractional kilowatt ratings. These motors are employed in fans, refrigerators, mixers, vacuum cleaners, washing machines, other kitchen equipment, tools, small farming appliances, etc. individual room air conditioners (single-phase) for home and office (where there is no central air conditioning) are being increasingly used.

Obviously the total number of such small motors in use exceeds the number of integral kW motors in industrial use. Though these motors are simpler in construction as compared to their 3-phase counterparts, their analysis happens to be more complex and requires certain concepts which have not been developed so far. Also the design of such motors is carried out by trial and error till the desired prototype is achieved. Because of the vast numbers in which these motors are produced, even a fractional efficiency increase or a marginal cost saving is extremely important. Nowadays, as in other fields, computers are employed for more accurate and optimum paper designs.

The treatment of the fractional-kW motors as presented in this part is concerned mainly with their method of operation, classification and characteristics of various types, and their typical applications. The analysis of the performance of such motors is beyond the scope here.

By far the vast majority of single-phase induction motors are built in the fractional-horsepower range. Single-phase motors are found in countless applications doing all sorts of jobs in homes, shops, offices, and on the farm. An inventory of the appliances in the average home in which single-phase motors are used would probably number a dozen and more. An indication of the volume of such motors can be had from the fact that the sum total of all fractional-horsepower motors in use today far exceeds the total of integral horsepower motors of all types.

4.18 SINGLE-PHASE INDUCTION MOTORS

A single-phase induction motor comprises a single-phase winding on the stator and a squirrel-cage rotor as shown in Fig. 4.32. The stator winding is connected to a single-phase source. The winding mmf is

$$F = F_m \cos \omega t \cos \theta; \quad (4.10)$$

where $F_m = NI_m$

This pulsating space-distributed field can be split into time rotating fields as

$$F = \frac{1}{2}F_m \cos(\omega t + \theta) + \frac{1}{2}F_m \cos(\omega t - \theta) \quad (4.11)$$

$$= F_f + F_b$$

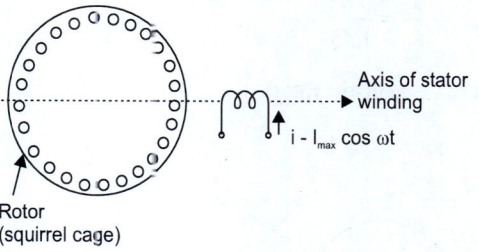

Fig. 4.32. Single-phase induction motor

or vectorially

$$\mathbf{F} = \mathbf{F}_f \mathbf{F}_r \quad (4.12)$$

where F_f and F_b are respectively forward and backward rotating fields rotating at synchronous speed ($\omega = 2\pi f$ rad (elect/s)). Each field has the same peak mmf equal to $1/2\ F_m$.

Rotor Slip with Respect to Two Rotating Fields

Let the rotor be assumed to run at a speed n in the direction of the forward field as shown in Fig. 4.33. It easily follows that

Rotor slip wrt forward field,

$$s_f = \frac{n_s - n}{n_s} = s \quad (4.13)$$

Rotor slip wrt backward field,

$$s_b = \frac{n_s - (-n)}{n_s} = \frac{2n_s 0(n-n)}{N_s}$$

$$= (2 - s) \quad (4.14)$$

At $s = 0.05$ say

$$\frac{s_f}{s_b} = \frac{0.05}{1.45} = \frac{1}{39}$$

At $s = 1$ (standstill rotor)

$$s_f = s_b = 1$$

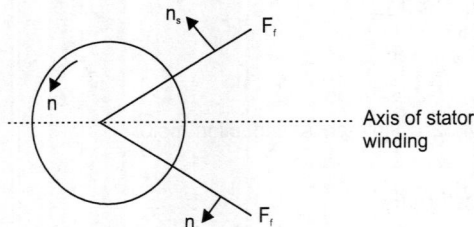

Fig. 4.33.

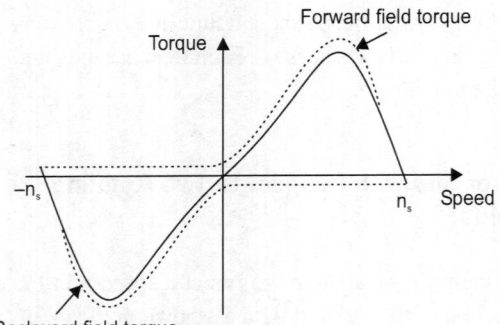

Fig. 4.34. T-s characteristic of a single-winding, single-phase induction motor.

Torque-Speed Characteristic

At standstill the rotor slip is $s = 1$ wrt both the rotating fields. The two fields are therefore equal in strength inducing equal currents in the rotor. As a result, these produce equal but opposite torques with net zero torque. The single-winding, single-phase motor is therefore *non-self starting*. The two rotating fields induce stator emfs which together balance the applied voltage (if low impedance stator is assumed).

If now the rotor is made to run at speed n in the direction of the forward field, the rotor slips wrt the two fields are now vastly different, i.e. $(2 - s) \gg s$. The forward field (low rotor slip) induces low, high p.f. currents in the rotor while the backward field (high rotor slip $(2 - s)$) induces high, low p.f. currents in the rotor. As a consequence, the backward field gets highly attenuated in strength while the strength of the forward field enhances in comparison. The forward torque therefore becomes several times the backward torque (torque being nearly proportional to square of field strength). The single-phase induction motor in this region of lip has T–s characteristic similar to that of a 3-phase motor but has a low efficiency because of the rotor loss caused by the backward field.

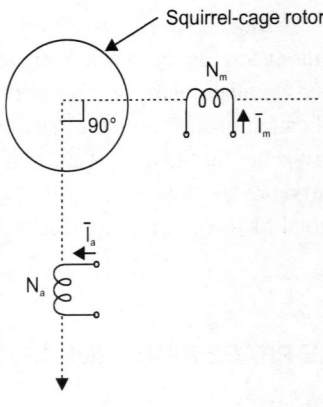

Fig. 4.35.

The T–s characteristic of a single-winding single-phase induction motor as sum of forward and backward field T–s characteristics is shown in Fig. 4.34 from which it is obvious that the motor has no starting torque.

The problem posed now is how to create a starting torque. This will be tackled by strengthening the forward field and weakening the backward field.

Two-phase Motor

Figure 4.35 shows a 2-phase motor. Let

$$\sqrt{2}N_m I_m = \sqrt{2}N_a I_a = F_m$$

The windings are displaced 90° (elect) in space phase and carry currents with 90° time phase difference. The resultant mmf distribution is

$$F = F_m \cos\omega t \cos\theta + F_m \cos(\omega t - 90°)\cos(\theta - 90°)$$

$$= F_m \cos(\omega t - \theta) \qquad (4.15)$$

$$= F_f$$

A single rotating field is thus established rotating at synchronous speed. But the problem now is that to create these currents on the 2-phase windings a 2-phase supply would be necessary, which is not practicable.

Split-Phase Motor

It is a 2-winding, single-phase motor in which the two windings are placed at 90° (elect) but are fed from single phase. The time phase difference in winding currents is obtained by placing suitable impedance in series with one of the windings called the *auxiliary* winding a while the other winding is called the *main* winding m. The current I_a in the higher impedance auxiliary winding is less than the current I_m in the main winding. The auxiliary winding has fewer turns of thinner wire. Unbalanced 2-phase field conditions are thus created at the start and as a result the forward rotating field becomes sufficiently stronger than the backward field resulting in production of starting torque. The auxiliary winding may or may not be left in circuit after the motor starts. For opening the auxiliary winding after motor starts, a centrifugal switch is employed. After starting the motor runs only on the main winding.

Depending on the method of *phase-splitting* (causing time phase difference in the currents of the two windings) there are two types of single-phase motors.

Resistance split-phase motor: The schematic diagram of the resistance split-phase motor is shown in Fig. 4.36 Here high R/X ratio is used for the auxiliary windings. A phase difference of about 30° is achievable as shown in Fig. 4.36(a)

It is a low efficiency, low pf motor and is available in sizes of 1/20 – 1/2 kW.

Capacitor split-phase motor: For phase splitting a capacitor is placed in series with the auxiliary

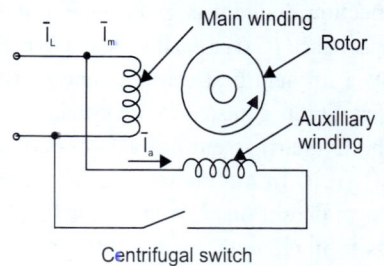

Fig. 4.36. Resistance split-phase motor.

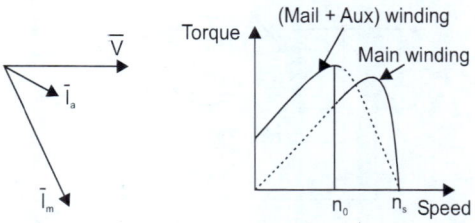

Fig. 4.37. (a) Phasor diagram at start (b) T–s characteristic.

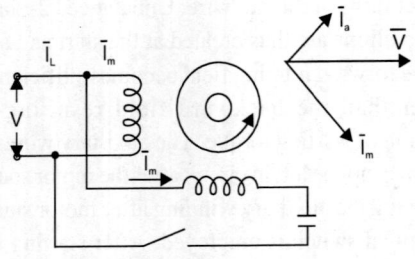

Fig. 4.38. Capacitor-start motor.

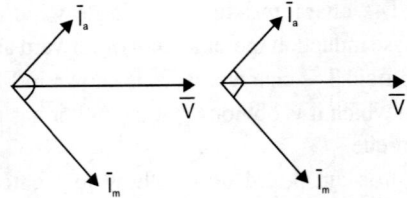

Fig. 4.40. Phasor diagram two-value capacitor motor (a) At start (b) During running

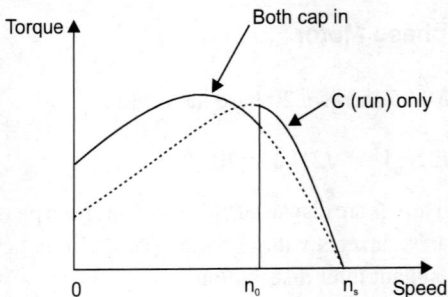

Fig. 4.41

winding as shown in Fig. 4.38 alongwith phasor diagram auxiliary winding is leading and it is possible to make the phase difference between them as 90° at start. During running the auxiliary winding is cut out so that capacitor is only short-time rated. Such a motor is known as capacitor-start motor. It has a far larger starting torque compared to a resistance-start motor. It has wide applications in machine tools, refrigeration, air-conditioning, etc. and is available in up to 5 kW size.

Two-Value Capacitor Motor

The connection diagram is given in Fig. 4.39. A larger capacitance (C (run) and C (start) in parallel) is employed to yield best starting conditions. The phase separation is adjusted to more than 90° (Fig. 4.40). The C (start) is cut out at a certain speed leaving C (run) in circuit to give best running performance; phasor diagram of Fig. 4.40. C (run) also helps to improve the overall pf of the motor. While C (run) is continuous rated, C (start) need only the short-time rated.

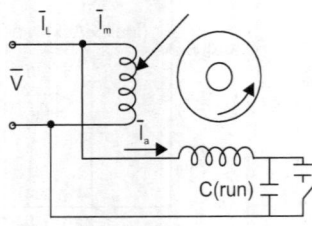

Fig. 4.39. Two-value capacitor motor.

4.19 HOW THE ROTATING FIELD IS OBTAINED

In its pure and simple form the single-phase motor usually consists of a distributed stator winding (not unlike one phase of a three-phase motor) and a squirrel-cage rotor. The AC supply voltage is applied to the stator winding, which in turn creates a field distribution. Since there is a single coil carrying an alternating current, a little thought reveals that the air-gap flux is characterized by being fixed in space and pulsating in magnitude. If hysteresis is neglected, the flux is maximum when the current is instantaneously and it is zero when the current is zero. Such an arrangement gives the single-phase motor no starting torque. To understand this refer to Fig. 4.41. As a matter of convenience the distributed single-phase winding is represented by a coil wrapped around protruding pole pieces; some motors do in fact use this configuration. Assume instantaneously that the flux-density wave is increasing in the upward direction as shown. Then by transformer action a voltage is induced in the rotor having that distribution which enables the corresponding rotor mmf to oppose the changing flux. The accomplish this, current flows out of the right-side conductors and into the left-side conductors as illustrated in Fig. 4.42 Note that this resulting ampere-conductor distribution corresponds to a space phase angle of $\psi = 90°$. By Eq. (17–39) the net torque is therefore zero. Of course what this means is that beneath each pole piece there are as many conductors producing clockwise torque as there are producing counter-clockwise torque. This condition, however, prevails only at standstill. If by some means the rotor is started in either direction, the motor will develop a nonzero net torque in the direction and thereby cause the motor to achieve normal speed. The problem therefore is to modify the configuration of Fig. 4.42 in such a way that it imparts to the rotor a nonzero starting torque.

The answer to this problem lies in so modifying the motor that it closely approaches the conditions prevailing in the *two-phase* induction motor. We know that to obtain a revolving field of constant linear velocity in a two-phase induction motor two conditions must be satisfied. One, there must exist two coils (or windings) whose axes are space-displaced by 90 electrical degrees. Two, the currents flowing through these coils must be time-displaced by 90 electrical degrees and they must have magnitudes such that the mmf's are equal. If the currents are less than 90° apart in time but greater than 0°, a rotating field will still be developed but the locus of the resultant flux vector will be an ellipse rather than a circle. Hence in such a case the linear velocity of the field varies from one point in time to another. Also, if the currents are 90° apart but the mmf's of the two coils are unequal, an elliptical locus for the rotating field again results. Finally, if the currents are neither 90° time-displaced nor of a magnitude to furnish equal mmf's, a rotating magnetic field will continue to be developed but now the locus will be more elliptical than in the previous cases. However, the important aspect of all this is that a revolving field can be so obtained even if its amplitude is not constant during its time history, and satisfactory performance can be achieved with such a revolving field. Of course such performance items as power factor and efficiency will be poorer than for the ideal case, but this is not too serious because the motors are of relatively small power.

Appearing in Fig. 4.42 is the schematic diagram which shows the modifications 4.43 needed to give the single-phase motor a starting torque. A second winding called the *auxiliary* winding is placed in the stator with its axis in quadrature with that of the main winding. Usually the main winding is made to occupy two-thirds of the stator slots and the auxiliary winding is placed in the remaining one-third. In this way the space-displacement condition is met exactly. The time displacement of the currents through the two

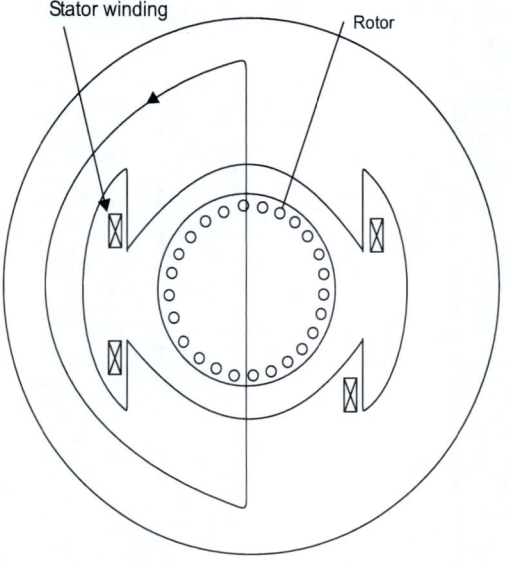

Fig. 4.42 Simplified diagram of the single-phase motor.

windings is obtained at least partially by designing the auxiliary winding for high resistance and low leakage reactance. This is in contrast to the main winding, which has low resistance and higher leakage reactance. Figure 4.44 depicts the time displacement exciting between the auxiliary winding current $\bar{I}_A$ and the main winding current $\bar{I}_M$ at standstill. Frequently in motors of this design the $\bar{I}_A$ and $\bar{I}_M$ phasors are displaced by about 45° in time. Thus with the arrangement of Fig. 4.43 a revolving field results

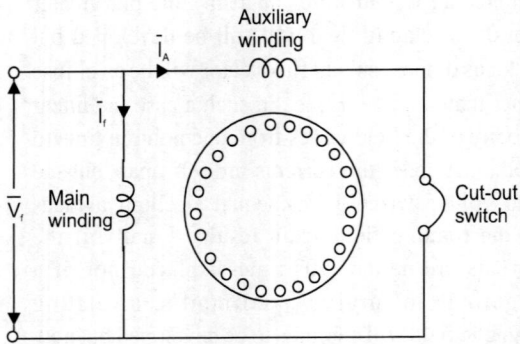

Fig. 4.43. Schematic diagram of the resistance split-phase motor.

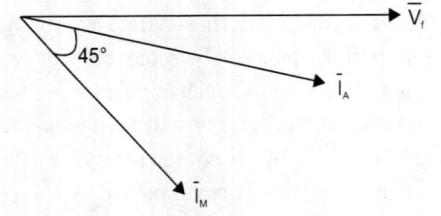

Fig. 4.44. Phasor diagram showing the time-phase displacement between the auxiliary and main winding currents at standstill.

and so the motor achieves normal speed. Because of the high-resistance character of the auxiliary winding, this motor is called the *resistance-start split-phase induction motor*. Also, the auxiliary winding used in these motors has a short time power rating and therefore must be removed from the line once the operating speed is reached. To do this a cut-out switch is placed in the auxiliary winding circuit which, by centrifugal action, removes the auxiliary winding from the line when the motor speed exceeds 75% of synchronous speed.

4.20 THE DIFFERENT TYPES OF SINGLE-PHASE MOTORS

Many different types of single-phase motors have been developed primarily for two reasons. One, the torque requirements of the appliances and applications in which they are used vary widely. Two, it is desirable to use the lowest-priced motor that will drive a given load satisfactorily. For example, a high-torque version of the split-phase induction motor just discussed is designed almost exclusively for washing-machine applications. This motor is available in a single speed rating and in just two horsepower ratings. This, coupled with the large volume of sales, enables it to be the least expensive motor in its category. See the second line of Table 4.2 for more details.

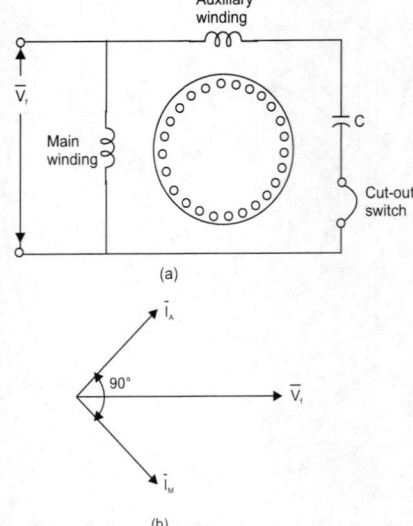

Fig. 4.45. Capacitor-start induction-run motor: (a) schematic diagram; (b) phasor diagram at starting.

The Capacitor-Start Induction-Run Motor

The chief difference between the various types of single-phase motors lies in the method used to start them. In the case of this motor a capacitor is placed in the auxiliary winding circuit so selected that it brings about a 90° time displacement between $\bar{I}_A$ and $\bar{I}_M$. See Fig. 4.45. The result is a much larger starting torque than is achievable with resistance split-phase starting. This motor is widely used for general-purpose applications.

The Capacitor-Start Capacitor-Run Motor

As Fig. 4.46 shows, two capacitors are used in the auxiliary circuit of this motor. By keeping one

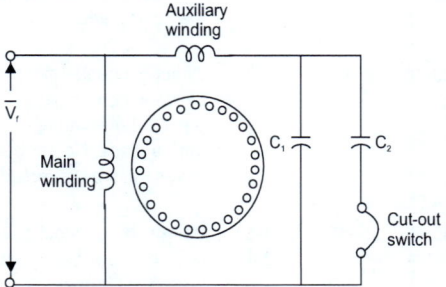

Fig. 4.46. Capacitor-start capacitor-run motor.

capacitor in during normal operation, improved performance is obtained because the motor then behaves more like the balanced two-phase motor. The improved performance is manifested in terms of less noise and higher efficiency and power factor. A second capacitor is needed at starting because the reactive component of the input impedance of the auxiliary circuit is considerably different at standstill than at full speed.

The Permanent-Split Capacitor Motor

In this motor a single capacitor is used both for starting and for running. To take advantage of improved running performance, the value of the capacitor used is that needed at full speed. Consequently, starting torque must be sacrificed. The schematic wiring diagram appears in Fig. 4.47.

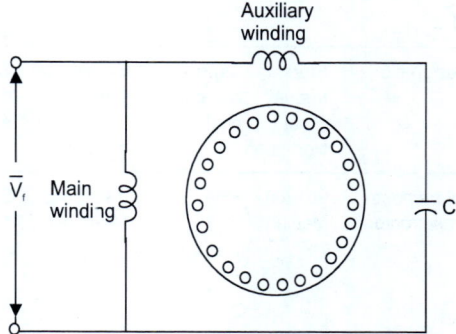

Fig. 4.47. Permanent-split capacitor motor.

The Shaded-Pole Motor

The shaded-pole motor is very extensively used in applications that require 1/20 hp or less. The construction is extremely rugged, as Fig. 4.48 reveals. There is little that can go wrong with this motor aside from overheating. Note that it contains no cut-out switch, which could be a source of trouble, nor does it have an auxiliary winding, which could burn up—especially if the cut-out switch became faulty. The shaded-pole motor consists essentially of copper wound on iron and not much more.

The manner of obtaining a moving-flux field, however, is different in this motor than it is in those considered previously. It is made of a copper ring embedded in each salient-pole piece. As the gap flux is changing in response to the alternating coil current, transformer action causes currents to be induced in

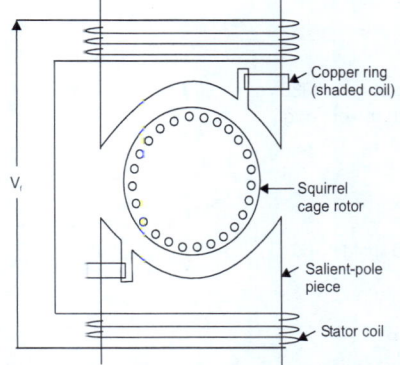

Fig. 4.48. Construction features of the shaded-pole motor.

Table 4.2

Type designation	Starting torque (% of Normal)	Approx. comparative price	Breakdown torque (% of normal)	Starting current at 115v	Power factor (%)	Efficiency (%)	Hp range	Application and general remarks
General-purpose split phase motor	90–200 Medium	85%	185–250 Medium	23 1/4 hp	56–65	62–67	1/20 to 3/4	Fans, blowers, office appliances, food-preparation machines, low- or medium-starting torque, low-inertia loads. Continuous-operation loads. May be reversed.
High-torque split-phase motor	200–275 High	65%	Up to 350	12 High 1/4 hp	50–62	46–61	1/6 to 1/3	Washing machines, pumps, home workshops, Oil burners. Medium- to high-starting torque loads. May be reversed.
Permanent-split capacitor motor	60–75 Low	155%	Up to 225	Medium	80–95	55–65	1/20 to 3/4	Direct-connected fans, blowers, centrifugal pumps. Low-starting-torque loads. Not for belt drives. May be reversed.
Permanent-split capacitor motor	Up to 200 Normal	155%	260		80–95	55–65	1/6 to 3/4	Belt-driven or direct-drive fans, blowers, centrifugal pumps, oil burners. Moderate-starting-torque loads. May be reversed.
Capacitor-start general-purpose motor	Up to 435 Very high	100%	Up 400		80–95	55–65	1/8 to 1/4	Dual voltages. compressers, stokers, conveyors, pumps. Belt-driven loads with high static friction. May be reversed.
Capacitor-start capacitor-run motor	380 High	190%	Up to 260		80–95	55–65	1/8 to 3/4	Compressors, stokers, coneyors, pumps. High torque loads. High power factor. Speed may be regulated.
Shaded-pole motor	50	—	150		30–40	30–40	1/300 to 1/20	Fans, toys, hair dryers, unit heaters. Desk fans. Low-starting-torque loads.

the copper rings, always so directed as to oppose the changing flux. The net effect of this action is that for one portion of the cycle the ring currents cause the flux to concentrate in that part of the pole piece which is free of the ring. At a subsequent portion of the cycle the ring currents act to crowd the gap flux through that part of the pole piece around which the ring is wrapped. In this fashion the air-gap flux undergoes a sweeping motion across the pole face. It appears to be moving from the unshaded to the shaded (or ring) portion of the pole. The sweeping action of the flux occurs periodically, thereby producing a starting as well as a running torque. The starting torque is normally 50% of rated value. The breakdown torque is also relatively low.

CONCLUSION

This chapter, in three parts presented respectively the construction, principle of operation and characteristics of 3-phase alternators, 3-phase induction motors and 1-phase machines. Worked illustrations on the fundamental aspects were given wherever needed.

QUESTIONS

1. Describe the constructional features of a 3-phase alternator.
2. What are the two types of alternator?
3. Which type of alternator is used in hydro stations? Justify.
4. What is the importance of frequency in alternators? Discuss.
5. Describe how an alternator produces electricity.
6. Derive an expression for the emf induced in a alternator.
7. Discuss the role of winding factors in the emf equation.
8. Derive an expression for the pitch factor and distribution factor for alternators.
9. What is the significance of synchronous reactance?
10. Mention the possible modes of voltage drops in an alternator.
11. What is voltage regulation? How much it is important in deciding the performance of a machine?
12. Write short notes on armature reaction in alternators.
13. Discuss, for varying load power factors, the behaviour of alternators, with reference to armature reaction.
14. What is OCC and what is SCC?
15. What are the various methods of finding out the voltage regulation of an alternator?
16. What is the need for putting the alternators in parallel?
17. Discuss the major conditions to be satisfied when placing alternators in parallel.
18. Discuss the dark lamp or bright lamp methods of synchronization.
19. What are the major parts in a 3-phase induction motor?
20. Why the machine is called a 3-phase induction motor?
21. Explain the types of 3-phase induction motor.
22. Explain the principle of operation of a 3-phase induction motor in detail.
23. What is the concept of slip? Develop an expression for the same.
24. Develop an expression for the torque developed in a 3-phase induction motor.
25. On what factors does the torque developed in a 3-phase induction motor depend upon?
26. What is torque slip characteristics? Develop it with due explanations.
27. What is the shift in the torque slip characteristics for (1) varying rotor resistance (2) varying frequency.

28. What are the major starters used for 3-phase induction motors? List them and discuss them individually with near sketches.
29. What are the possible methods of speed control in 3-phase induction motors? Discuss them.
30. Why a 1-phase induction motor is not a self-starting machine? Explain.
31. How to make a 1-phase induction machine a self starting one? Explain.

INTRODUCTION

A detailed introduction to the concepts and fundamentals of control systems is given in this chapter. Modeling of simple physical systems is presented. The derivation of transfer function for such a system is given. A treatment on various control components, such as servomotors, potentiometers, encoders etc., are discussed. First order and second order systems are introduced and their time response characteristics are also presented.

CHAPTER OBJECTIVE

At the completion of this chapter, the reader will be able to:

* model physical systems on the basis of control systems approach, with the implementation of both the open loop and the closed loop approaches;
* derive transfer function for any physical system; and
* understand higher order systems and their theory.

KEYWORDS

Control system, Open and closed loop, Mathematical modeling of electrical, mechanical and thermal systems, Analogous systems, Concepts of transfer function, Encoders, Potentiometer, Servo and stepper motors and Time response analysis.

5

Control Systems

5.1. INTRODUCTION

Control plays a vital role in both engineering and science. From the traffic light control to the space vehicle systems, control has become part and parcel of day-to-day life. The control system is that means by which any quantity of interest in a machine, mechanism or other equipment is maintained or altered in accordance in a desired manner.

5.1.1. System

Before introducing the concept of control system, it is desirable to discuss the features of a system. A system is a co-ordinated unit of individual elements performing a specific function. It produces an output corresponding to a given input according to some rule. Thus, electrical, mechanical, electronic, hydraulic, pneumatic, chemical, and/or interacting combinations may be regarded as systems.

Examples

(i) A Thermometer:
On analyzing the action of a thermometer, the length of the mercury column is directly dependent upon the temperature around the mercury bulb. Here the thermometer is the system with temperature, θ as the input and the length of the mercury thread in the capillary, l as the output.

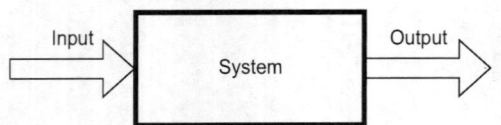

Fig. 5.1.1. Block diagram of a system

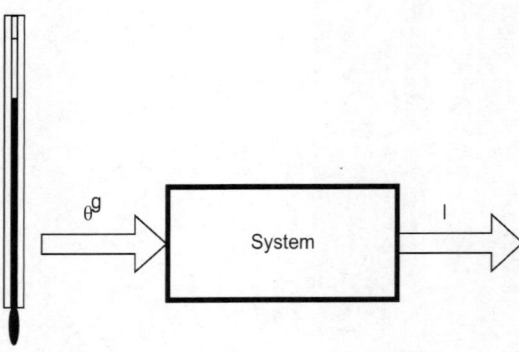

Fig. 5.1.2. Block diagram of a temperature system

---

*This chapter is contributed by Mrs. U. Sabura Banu.

(ii) Mechanical system

Consider a spring mass damper system subjected to an input force. Here force, F(t) is the input and the displacement, x(t) is the output.

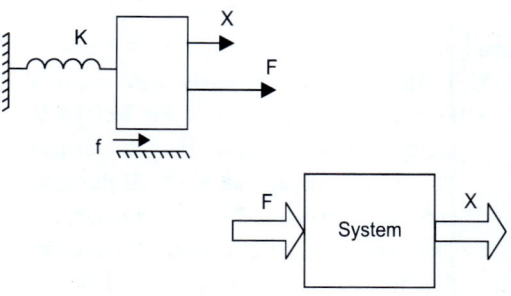

Fig. 5.1.3. Block diagram of a mechanical system

5.1.2. Control System

A control system is in general a combination of elements or subsystems which tends to maintain a quantity or a set of quantities termed output, suitably related to another quantity or a set of quantities termed input.

The reference input is a signal or set of signals in some form, which acts upon the system in such a way that the response at the output takes place in a desired manner. Output quantity is controlled by varying the input quantity.

Control systems may be classified into two types depending upon whether the controlled variable changes the actual input to the control system or not, i.e. whether a feedback is used or not. When feedback is not used, the system is called an open loop system and when a feedback is used it is called a closed loop system.

5.1.3. Open Loop System

The open loop systems are those in which the output has no effect on the control action. In other words, the output is neither measured nor fed back for comparison with the input. The output remains constant for a constant input signal provided the external conditions remain unaltered. The output may be changed to any desired value by appropriately changing the input signal but variations in external conditions or internal parameters of the system may cause the output to vary from the desired value in an uncontrolled fashion. Thus, any physical system, which does not automatically correct for variation in its output, is an open loop system. It is the simplest and most economical type of control system. But, they are generally inaccurate and unreliable and as such are not preferred. Often a system operating in the open loop may be structurally unstable.

Example:
1. One practical example is a washing machine. Soaking, washing and rinsing in the washer operate on a time basis. The machine does not measure the output signal that is the cleanliness of the clothes.
2. Traffic control operated on a time basis. The density in a particular roadside is not taken care.
3. Tank level control by allowing the flow of the liquid for a fixed time interval.

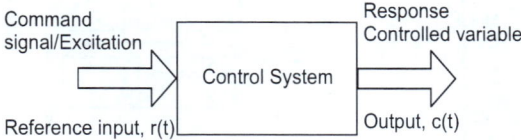

Fig. 5.1.4. Block diagram of a control system

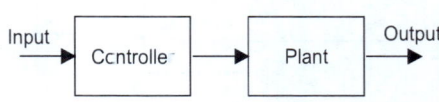

Fig. 5.1.5. Block diagram of an open loop system

5.1.4. Closed Loop Control Systems

Feedback control systems are referred to as closed loop control systems. In a closed loop control system, the actuating error signal which is the difference between the input signal and the feedback signal is fed to the controller so as to reduce the error and bring the output of the system to a desired value.

Often control systems are designed for the specific purpose of maintaining a correspondence between an output and an input under varying conditions while the input remains at a constant value. Such a control system is known as a regulator.

When a closed loop control is applied to industrial processes involving factors such as rate of flow, variation of reaction rate, variation of temperature, pressure, viscosity etc, it is termed process control. The difference between the input signal and the feedback signal is called Error. This error is fed to the controller so as to take control action to reduce it and bring the output of the system to the desired value.

Elements of the closed loop control

(i) To control any physical variable, one must know the present value of the variable. The system used for the measurement of the variable is called a **sensor**.
(ii) **Plant** is the system to be controlled
(iii) To make a control signal, the present level of the physical variable is compared with the desired level and this function is done by **the error detector** and this will produce the error signal.
(iv) The error signal is sent to **the controller** and it will produce control signal so that the error signal can be minimized to zero.

Examples:
1. Traffic control system can be made as closed loop system if the signals are decided on the density of the traffic. Here, the density of the traffic is measured on all the sides and the information is fed to a controller. The timing of the traffic control signals are decided by the controller based on the density of traffic, which is better than open loop system.
2. The liquid level of the tank is to be maintained at a desired value. The liquid level is sensed by a float, which positions the slider arm B on a potentiometer. The desired level is positioned on slider arm A on another potentiometer. The potential difference occurs when the liquid level rises or falls. This potential difference is given as error voltage, which is proportional to the level change. The error voltage actuates the

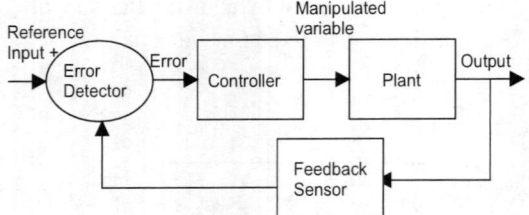

Fig. 5.1.6. Block diagram of a closed loop control system

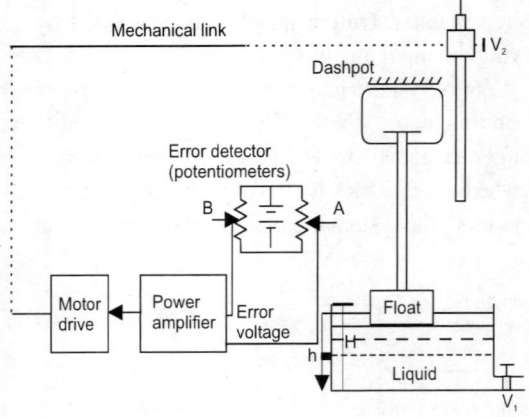

Fig. 5.1.7. Block diagram of a closed loop level control system

motor through a power amplifier, which in turn controls the input flow so that output can be maintained at the desired value.

5.1.5. Comparison of Open Loop and Closed Loop Systems

1. Since the open loop has no feedback, its construction and maintenance are very easy whereas it is not so in the case of closed loop.
2. In closed loop control system, the use of feedback makes the system response relatively insensitive to external disturbances and internal variations in system parameters. But in the open loop control system, the output will be affected if there is any disturbance in the input or any variations in system parameters.
3. To obtain accurate control of a given plant, it is possible to use relatively inaccurate and inexpensive components in closed loop control since it has feedback whereas doing so is impossible in open loop case.
4. System stability is not a major problem in open loop control system. On the other hand, stability is a major problem in the closed loop control system, which may tend to overcorrect errors that can cause oscillations of constant or changing amplitude.
5. Open loop control system is convenient when output is hard to measure or economically not feasible.

5.2. MATHEMATICAL MODELING

A mathematical model of a system is defined as a set of equations that represents the dynamics of the system accurately. The dynamics of any systems, whether they are mechanical, electrical, thermal, electronic, biological, and so on, may be described in terms of differential equations. Such differential equations may be obtained by using physical laws governing a particular system, for example, Newton's laws for mechanical systems and Kirchhoff's laws for electrical systems.

When the mathematical model of a physical system is solved for various input conditions, the result represents the dynamic response of the system. The mathematical model of a system is linear, if it obeys the principle of superposition and homogeneity. This superposition principle says that if a system model has responses $y_1(t)$ and $y_2(t)$ to any two inputs $x_1(t)$ and $x_2(t)$ respectively, the system response to the linear combination of these inputs, $\alpha_1 x_1(t) + \alpha_2 x_2(t)$ is given by the linear combination of the individual outputs, $\alpha_1 y_1(t) + \alpha_2 y_2(t)$ where α_1 and α_2 are constants.

Mathematical models of most physical systems are characterized by differential equations. A mathematical model is linear, if the differential equation describing it has coefficients which are either functions only of the independent variable or are constants. It the coefficients of the describing differential equations are function of time, the mathematical model is linear time-varying. On the other hand, if the co-efficients of the describing differential equations are constants, the model is linear time-invariant.

5.2.1. Mathematical Modeling of Physical Systems

The physical systems are classified as Mechanical systems, Electrical systems etc. based upon the physical law like Newton's laws, Kirchhoff's laws, etc., that govern the particular system.

5.2.2. Mechanical Systems

Mass, Spring and Damper are the three idealized elements used for analyzing mechanical system.

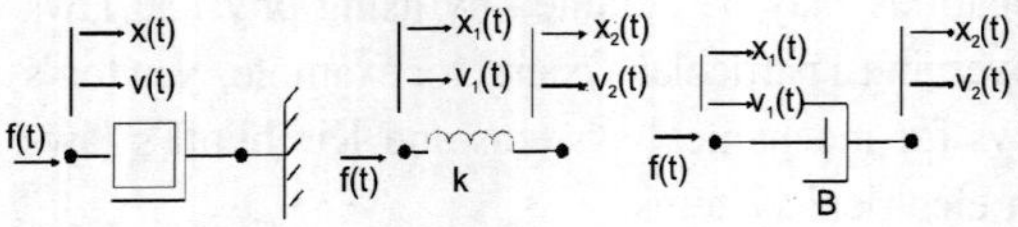

Fig. 5.2.1. Mechanical system

(a) Mass

The ideal mass element represents a particle of mass, which is the lumped approximation of the mass of a body concentrated at the centre of mass. The mass has two terminals, one is free, attached to motional variable v and the other represents the reference.

In Fig. 5.2.1 f(t) represents the applied force, x(t) represents the displacement, and M represents the mass.

Then, in accordance with Newton's second law,

$$f(t) = Ma(t) = \frac{Mdv(t)}{dt} = \frac{Md^2x(t)}{dt^2}$$

Where v(t) is velocity and a(t) is acceleration. It is assumed that the mass is rigid at the top connection point and that cannot move relative to the bottom connection point.

$$f(t) = \frac{Md^2x(t)}{dt^2}$$

(b) Damper

Damper is the damping element and damping is the friction existing in physical systems whenever mechanical system moves on sliding surface.

The friction encountered is of many types:

Friction: The force required at startup.
Coulomb Friction: The force of sliding friction between dry surfaces. This force is substantially constant.

Viscous friction force : The force of friction between moving surfaces separated by viscous fluid or the force between a solid body and a fluid medium. It is linearly proportional to velocity over a certain limited velocity range.

In friction elements, the top connection point can move relative to the bottom connection point. Hence two displacement variables are required to describe the motion of these elements.

A physical realization of this phenomenon is the viscous friction associated with oil and so forth. A physical device that is modeled as friction is a shock absorber. The mathematical model of friction is given by

$$f(t) = B\left(\frac{dx_1(t)}{dt} - \frac{dx_2(t)}{dt}\right)$$

Where B is the damping coefficient.

(c) Spring

The final translational mechanical element is a spring. The ideal spring gives the elastic deformation of a body.

The defining equation from Hooke's law, is given by

$$f(t) = K(x_1(t) - x_2(t))$$

Note here that the force developed is directly proportional to the difference in the displacement of one end of the spring relative to the other.

These equations apply for the forces and the displacements in the directions shown by the arrowheads in Fig. 5.2.1. If any of the directions are reversed, the sign on that term in the equations must be changed.

For these mechanical elements, friction dissipates energy but cannot store it. Both mass and a spring can store energy but cannot dissipate it.

5.2.2.1. Mechanical Rotational Systems

$$T = \frac{J d^2\theta}{dt^2} = \frac{J d\omega}{dt^2}$$

$$T = K(\theta_1 - \theta_2)$$

$$T = B\left(\frac{d\theta_1}{dt} - \frac{d\theta_2}{dt}\right) = B(\omega_1 - \omega_2)$$

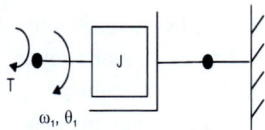

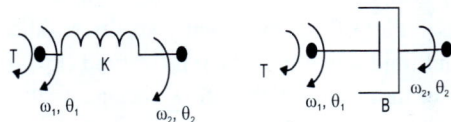

Fig. 5.2.2 Mechanical rotational system

5.2.3. Mathematical Model of Electrical Systems

The basic components that are used in electrical systems are resistance, inductance and capacitance. These systems are analyzed by the application of Kirchhoff's voltage and current laws. In this section, models are developed for simple electrical circuits.

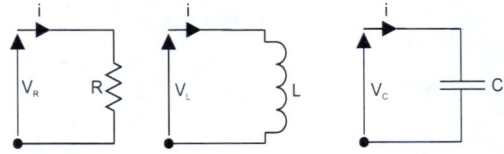

Fig. 5.2.3 Electrical system

The governing equations are

$$V_R(t) = R\, i(t)$$

$$V_L(t) = L \frac{di(t)}{dt}$$

$$V_C(t) = \frac{1}{C}\int i\, dt + V(0)$$

5.2.4. Thermal Systems

Thermal systems are those that involve the transfer of heat from one substance to another. Thermal system may be analyzed in terms of thermal resistance and thermal capacitance although they may not be represented as lumped parameters. But by making some assumptions, they can be represented as distributed parameters, which makes the analysis simple.

Consider the simple thermal system as shown in Fig. 5.2.4.

Assumptions are made as follows.

1. Fluid in the tank is perfectly mixed so that it is at uniform temperature

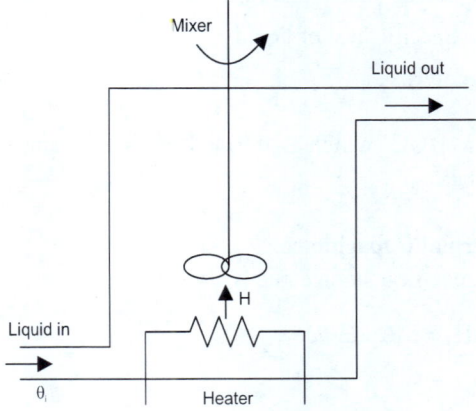

Fig. 5.2.4 Thermal system

2. The tank is insulated to eliminate heat loss to the surrounding air.
3. There is no heat storage in the insulation.

Now a single temperature may be used to describe the thermal state of the entire liquid.

In this system,
θ_i = Steady state temperature of inflowing liquid,
θ = Steady state temperature of out-flowing liquid,
H = Steady state heat input rate from heater.

Let ΔH be a small change in the heat input rate from its steady state value. This change in H will result in the following changes.
(i) Change in heat output rate by an amount ΔH_1.
(ii) Change in heat storage rate of liquid in the tank by an amount ΔH_2.
(iii) Change in temperature of out-flowing liquid by an amount $\Delta\theta$.

• **Thermal Resistance**
Change in outflow heat rate is given by

$$\Delta H_1 = Q C_s \Delta\theta$$

Where

Q = Steady state liquid flow rate
C_s = Specific heat of liquid
$\Delta H_1 = \Delta\theta/R$

If $R = 1/QC_s$ which is defined as the Thermal Resistance.

• **Thermal Capacitance**
Change in heat storage rate is given by

$$\Delta H_2 = MC_s \, d\Delta\theta/dt$$

Where
M = mass of the liquid in the tank
$\Delta d\theta/dt$ = rate of rise of temperature in the tank

$$\Delta H_2 = C \, d\Delta\theta/dt$$

where $C = MC_s$ which is defined as Thermal Capacitance.

$$\Delta H = \Delta H_1 + \Delta H_2$$

$$\Delta H = \frac{\Delta\theta}{R} + C\frac{d\Delta\theta}{dt}$$

5.2.5. Hydraulic System

The dynamics of the fluid systems is represented by ordinary differential linear equations only, if the fluid is incompressible and flow is laminar.

Velocity of sound is a key parameter in fluid flow to determine the compressibility property. If the fluid velocity is much less than the velocity of sound, compressibility effects are usually small. Another important fluid property is the type of fluid flow—laminar or turbulent. Laminar flow is characterized by smooth motion of fluid, while turbulent flow is characterized by an irregular and random motion of fluid.

Reynold found that pipe flow will be laminar for Re, the Reynolds number less than 2,000 and turbulent for Re greater than 3,000. When the Reynolds number is between 2,000 and 3,000, the type of flow is unpredictable.

Pressure drop across a pipe section is given by

$$P = \frac{128 l \mu}{\pi D^4} Q; \text{ for laminar flow}$$

$$= RQ$$

$$P = \frac{8 K_t \rho l}{\pi^2 D^5} Q^2; \text{ for turbulent flow}$$

$$= K_T Q^2$$

where
l = length of pipe section(m)

D = Diameter of pipe (m)
μ = Viscosity (Newton-sec/m²)
Q = Volumetric flow rate (m³/sec)
K_t = a constant to be determined experimentally
ρ = mass density (kg/m³)

In terms of liquid head in a tank, the fluid pressure is given by

$$P = \rho g H$$

Where g = gravitational constant

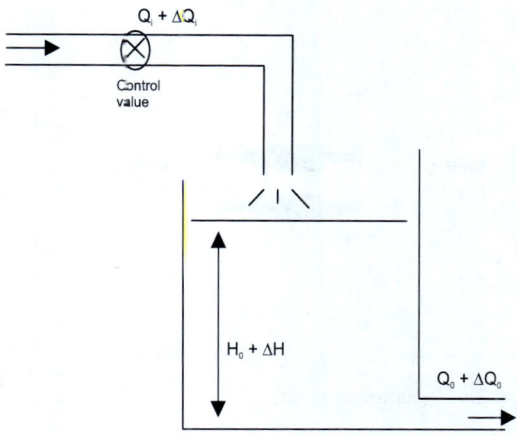

Fig. 5.2.5 Liquid level system

Combining equations 1 & 2,

$$H = RQ \text{ for laminar flow}$$

Where

$$R = \frac{128 l \mu}{\pi D^4 \rho g}$$

$$\Delta H = R \, \Delta Q$$

Where

$$R = \frac{2 K_T Q_0}{\rho g}$$

$$K_T = \frac{8 K_t \rho l}{\pi^2 D^2} \text{ and}$$

Q_0 = value of Q at operating point

The parameter R in the above two equations is referred to as Hydraulic Resistance.

The rate of fluid storage in a tank = A dH/dt
= C dH/dt

Where C = A(m²) = hydraulic capacitance of the tank.

flow rate into the tank and Q_0 be the outflow rate, while H_0 is the steady liquid head in the tank.
At steady state $Q_i = Q_0$

Let $Q_0 = \Delta H'R$

The system dynamics is described by the liquid flow rate balance equation:
Rate of liquid storage in the tank = rate of liquid inflow – rate of liquid outflow.
Rate of liquid storage in the tank = A dH/dt.
Where A = Cross sectional area of the tank.

$$A \frac{dH}{dt} = \frac{A}{\rho g} \frac{dp}{dt} = C \frac{dp}{dt}$$

Where C = A/ρg = Capacitance of tank
Therefore,

$$C \frac{d \Delta H}{dt} = \Delta Q_i - \frac{\Delta H}{R}$$

5.2.5.1. Liquid level system

Consider a simple liquid level system as shown in Fig. 5.2.5, where a tank is supplying liquid through an outlet. Under steady conditions, let Q_i be the liquid

5.2.6. Pneumatic Systems

In number of engineering applications, velocities of gases are small fraction of the velocity of sound and

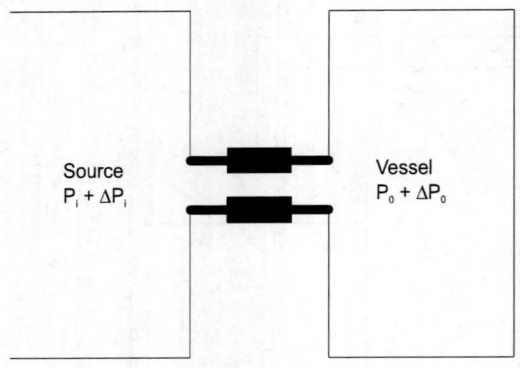

Fig. 5.2.6 Pneumatic system

hence one can treat pneumatic flow as nearly incompressible. Therefore, the results presented for hydraulic systems are directly applicable to this class of pneumatic systems.

Consider a pneumatic system shown in Fig. 5.2.6. A pneumatic source is supplying air to the pressure vessel through a pipeline.

In this system,

P_i = air pressure of the source at steady state (newton/m^2)

P_0 = air pressure in the vessel at steady state (newton/m^2)

ΔP_i = small change in air pressure of the source from its steady state

ΔP_0 = small change in air pressure of the vessel from its steady state

System dynamics is described by the equation:
Rate of gas storage in vessel = rate of gas inflow

$$C \frac{d\Delta P_0}{dt} = \frac{\Delta P}{R} = \frac{\Delta P_i - \Delta P_0}{R}$$

5.2.7. Analogous Systems

Analysing the mechanical translational systems, mechanical rotational systems and electrical systems, it is seen that they are identical. Such systems with identical differential equations are called analogous systems. The force (torque) and voltage e are the analogous variables.

Table 5.2.1. Analogous quantities in force (torque)-voltage analogy

Mechanical Translational Systems	Mechanical rotational Systems	Electrical System
Force F	Torque, T	Voltage, e
Mass, M	Moment of inertia J	Inductance L
Viscous Friction Coefficient B	Viscous Friction Coefficient B	Resistance R
Spring stiffness K	Torsional spring stiffness K	Reciprocal of capacitance 1/C
Displacement x	Angular Displacement, θ	Charge q
Velocity, v	Angular Velocity, ω	Charge i

Table 5.2.2. Analogous quantities in force (torque)-current analogy

Mechanical Translational Systems	Mechanical Rotational Systems	Electrical System
Force F	Torque, T	Current, i
Mass, M	Moment of inertia J	Capacitance C
Viscous Friction Coefficient B	Viscous Friction Coefficient B	Reciprocal of Resistance 1/R
Spring stiffness K	Torsional spring stiffness K	Reciprocal of inductance 1/L
Displacement x	Angular Displacement, θ	Magnetic flux linkage ϕ
Velocity, v	Angular Velocity, ω	Voltage e

Table 5.2.3. Analogous quantities in electrical-thermal-liquid-pneumatic

Electrical systems	Thermal Systems	Liquid level systems	Pneumatic systems
Charge, coulombs	Heat flow, joules	Liquid flow, cub-m	Air flow, cub-m
Current, amps	Heat flow rate, joules/min	Liquid flow rate, cub-m/min	Air flow rate, cub-m/min
Voltage, volts	Temperature, °C	Head, m	Pressure, Newton/m²
Resistance, ohms	Resistance, °C/(joules/min)	Resistance, m/(cub-m/min)	Resistance, (Newton/m²)(cub-m/min)
Capacitance, farads	Capacitance, joules/ °C	Capacitance, cub-m/m	Capacitance (cub-m/min)(Newton/m²)

5.3. TRANSFER FUNCTION

In control system analysis and design, the Laplace transform is used to transform linear differential equations with constant coefficient into algebraic equations. The algebraic equations are much easier to manipulate and analyse, simplifying the analysis of the differential equations. Thus Laplace transform simplifies the analysis and design of linear systems.

The transfer function of a linear time-invariant system is defined to be the ratio of the Laplace transform of the output variable to the Laplace transform of the input variable under the assumption that all initial conditions are zero.

$$G(s) = \frac{L[c(t)]}{L[r(t)]} / \text{System relaxed at } t_0 = 0$$

$G(s) = C(s)/R(s)$ / System relaxed at $t_0 = 0$

where

$C(s)$ = Laplace transform of the output variable, $c(t)$
$R(s)$ = Laplace transform of the input variable, $r(t)$
$G(s)$ = Transfer function of the system

5.3.1. Transfer Function of A Mechanical Translational System

Consider the Mass-Spring-Damper System, the mathematical model shown in figure is given as:

$$f(t) = M\frac{d^2x(t)}{dt^2} + B\frac{dx(t)}{dt} + kx(t)$$

Applying Laplace transform to this equation:

$$F(s) = Ms^2X(s) + BsX(s) + kX(s)$$

i.e. $F(s) = (Ms^2 + Bs + k)X(s)$

Transfer Function

$$G(s) = X(s)/F(s)$$

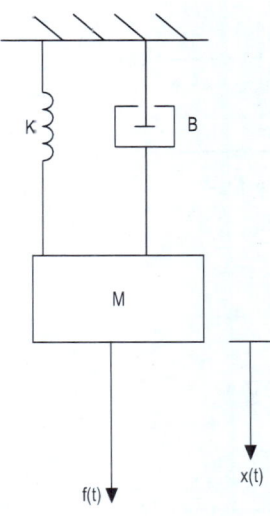

Fig. 5.3.1 Mass spring damper mechanical system

$$G(s) = \frac{X(s)}{F(s)} = \frac{1}{Ms^2 + Bs + k}$$

5.3.2. Transfer Function of A Electrical System

Consider the electrical system with Resistor, Inductor and Capacitor in Series shown in Fig. 5.3.2.

Applying Kirchhoff's voltage law, the mathematical model is

$$v_i(t) = Ri(t) + L\frac{di(t)}{dt} + \frac{1}{C}\int i(t)dt$$

$$v_o(t) = \frac{1}{C}\int i(t)dt$$

Applying Laplace transform to the above equations

$$V_i(s) = RI(s) + LsI(s) + \frac{1}{Cs}I(s)$$

$$V_o(s) = \frac{1}{Cs}I(s)$$

$$V_i(s) = \left[R + Ls + \frac{1}{Cs}\right]I(s)$$

$$\frac{V_o(s)}{V_i(s)} = \frac{1}{\left[R + Ls + \frac{1}{Cs}\right]\frac{1}{Cs}}$$

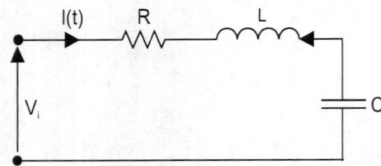

Fig. 5.3.2 Electrical system with R,L,C in series

The transfer function of the electrical system discussed above is

$$\frac{V_o(s)}{V_i(s)} = \frac{1}{\left[LCs^2 + RCs + 1\right]}$$

5.3.3. Transfer Function of A Thermal System

The mathematical model of a thermal system shown in Fig. 5.2.4 is

$$\Delta H = \frac{\Delta\theta}{R} + C\frac{d\Delta\theta}{dt}$$

Applying Laplace transform,

$$\Delta H(s) = \frac{\Delta\theta(s)}{R} + Cs\Delta\theta(s)$$

$$= \left[\frac{1}{R} + Cs\right]\Delta\theta(s)$$

$$= \left[\frac{1 + RCs}{R}\right]\Delta\theta(s)$$

The transfer function is given by

$$\frac{\Delta\theta(s)}{\Delta H(s)} = \left[\frac{R}{1 + RCs}\right]$$

5.3.4. Transfer Function of A Liquid Level System

The mathematical model of a simple liquid level system shown in figure 5.2.5 is given by

$$C\frac{d\Delta H}{dt} = \Delta Q_i - \frac{\Delta H}{R}$$

Applying Laptace transform,

$$Cs\Delta H(s) = \Delta Q_i(s) - \frac{\Delta H(s)}{R}$$

$$\Delta Q_i(s) = Cs\Delta H(s) + \frac{\Delta H(s)}{R}$$

$$= \left[Cs + \frac{1}{R}\right]\Delta H(s)$$

$$= \left[\frac{1 + RCs}{R}\right]\Delta H(s)$$

Therefore the transfer function is

$$\frac{\Delta H(s)}{\Delta Q_i(s)} = \left[\frac{R}{1 + RCs}\right]$$

5.3.5. Transfer Function of a Pneumatic System

The mathematical model of a simple pneumatic system shown in Fig. 5.2.6. is given by

$$C\frac{d\Delta P_0}{dt} = \frac{\Delta P}{R} = \frac{\Delta P_i - \Delta P_0}{R}$$

Applying Laplace Transform, we get

$$Cs\Delta P_0(s) = \frac{\Delta P_i(s) - \Delta P_0(s)}{R}$$

$$(RCs + 1)\Delta P_0(s) = \Delta P_i(s)$$

The transfer function is given by

$$\frac{\Delta P_0(s)}{\Delta P_i(s)} = \frac{1}{(RCs + 1)}$$

Problems

1. Determine the transfer function $V_0(s) / V_i(s)$ of the circuit shown below.

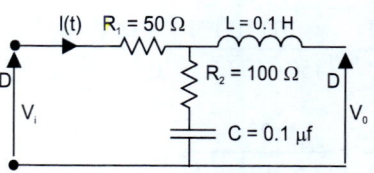

Applying Kirchhoff's voltage law,

$$V_i(t) = R_1 i(t) + R_2 i(t) + \frac{1}{C}\int i(t)dt$$

$$V_0(t) = R_2 i(t) + \frac{1}{C}\int i(t)dt$$

Note: Since the circuit is not closed through L, L term is not included

Applying Laplace transform

$$V_i(s) = \left[R_1 + R_2 + \frac{1}{Cs}\right]I(s)$$

$$V_0(s) = \left[R_2 + \frac{1}{Cs}\right]I(s)$$

$$I(s) = \frac{V_0(s)}{\left[R_2 + \frac{1}{Cs}\right]}$$

Put $I(s)$ in $V_i(s)$

$$V_i(s) = \left[R_1 + Ls + \frac{1}{Cs}\right]\frac{V_0(s)}{\left[R_2 + \frac{1}{Cs}\right]}$$

The Transfer function is

$$\frac{V_0(s)}{V_i(s)} = \frac{R_2 Cs + 1}{[(R_1 + R_2)Cs + 1]}$$

Substituting the value

$$\frac{V_0(s)}{V_i(s)} = \frac{100*0.1*10^{-6}s+1}{\left[(50+100)0.1*10^{-6}s+1\right]}$$

$$\frac{V_0(s)}{V_i(s)} = \frac{1*10^{-5}s+1}{\left[1.5*10^{-5}s+1\right]}$$

2. Find the transfer function of a mechanical system shown in figure

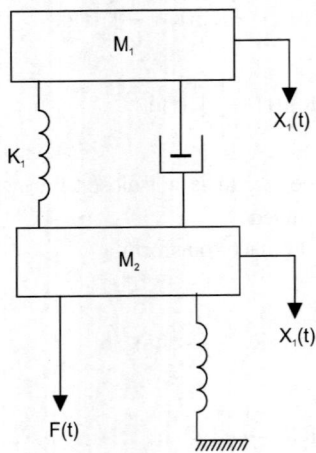

Consider Mass M_1:

Elements attached to mass M_1:

(i) Friction element, B (It is attached to M_2 also)
(ii) Spring element, K_1 (It is attached to M_2 also)
(iii) Displacement in Mass $M_1 = x_1(t)$
(iv) Displacement in Mass $M_2 = x_2(t)$

Forces acting on mass M_1:

Force on Mass $= M_1 \dfrac{d^2 x_1(t)}{dt^2}$

Frictional Force $= B\left[\dfrac{dx_1(t)}{dt} - \dfrac{dx_2(t)}{dt}\right]$

Spring Force $= K_1\left[x_1(t) - x_2(t)\right]$

Force applied = 0

$$0 = M_1 \frac{d^2 x_1(t)}{dt^2} + B\left[\frac{dx_1(t)}{dt} - \frac{dx_2(t)}{dt}\right]$$
$$+ K_1\left[x_1(t) - x_2(t)\right]$$

Consider Mass M_2:

Elements attached to mass M_2:

(i) Friction element, B (It is attached to M_1 also)
(ii) Spring element, K_1 (It is attached to M_1 also)
(iii) Spring element, K_2 (One end is fixed)
(iv) Displacement in Mass $M_2 = x_2(t)$

Forces acting on mass M_2:

Force on Mass; $M_2 \dfrac{d^2 x_2(t)}{dt^2}$

Frictional Force $= B\left[\dfrac{dx_2(t)}{dt} - \dfrac{dx_1(t)}{dt}\right]$

Spring Force (due to K_1) $= K_1\left[x_2(t) - x_1(t)\right]$

Spring Force (due to K_2) $= K_2 x_2(t)$

Force applied $= F(t)$

$$f(t) = M_2 \frac{d^2 x_2(t)}{dt^2} + B\left[\frac{dx_2(t)}{dt} - \frac{dx_1(t)}{dt}\right]$$
$$+ K_1\left[x_2(t) - x_1(t)\right] + K_2 x_2(t)$$

Solution

(i) For mass M_1.

$$0 = M_1 \frac{d^2 x_1(t)}{dt^2} + B\left[\frac{dx_1(t)}{dt} - \frac{dx_2(t)}{dt}\right]$$
$$+ K_1\left[x_1(t) - x_2(t)\right]$$

(ii) At mass M_2.

$$f_{(f)} = M_2 \frac{d^2x_2(t)}{dt^2} + B\left[\frac{dx_2(t)}{dt} - \frac{dx_1(t)}{dt}\right]$$

$$+ K_1[x_2(t) - x_1(t)] + K_2 x_2(t)$$

Applying Laplace transform

$$F(s) = M_2 s^2 X_2(s) + B[sX_2(s) - sX_1(s)]$$

$$+ K_1[X_2(s) - X_1(s)] + K_2 X_2(s) \quad (1)$$

$$0 = M_1 s^2 X_1(s) + B[sX_1(s) - sX_2(s)]$$

$$+ K_1[X_1(s) - X_2(s)] \quad (2)$$

$$M_1 s^2 X_1(s) + Bs X_1(s) + K_1 X_1(s)$$
$$= K_1 X_2(s) + Bs X_2(s)$$

$$(M_1 s^2 + Bs + K_1) X_1(s) = (K_1 + Bs) X_2(s)$$

Writing X_2 in terms of X_1

$$X_2(s) = \frac{(M_1 s^2 + Bs + K_1) X_1(s)}{(K_1 + Bs)} \quad (3)$$

Substitute 3 in 1

$$F(s) = \frac{(M_2 s^2 + Bs + K_1 + K_2)(M_1 s^2 + Bs + K_1) X_1(s)}{(K_1 + Bs)}$$

$$- (Bs + K_1) X_1(s)$$

The Transfer function

$$\frac{X_1(s)}{F(s)} = \frac{(K_1 + Bs)}{M_1 M_2 s^4 + B(M_1 + M_2) s^3 +}$$

$$\overline{+ (K_1 M_2 + K_2 M_1 + K_1 M_1) s^2 + K_2 Bs + K_1 K_2}$$

5.4. CONTROL SYSTEM COMPONENTS

A closed loop control system is represented by the general block diagram. Such a system is composed of three basic elements: the feedback element, controller and controlled system.

In a closed loop control system, the reference input is an input signal proportional to desired output. The feedback signal is a signal proportional to current output of the system. The error detector compares the reference input and feedback signal and if there is a difference it produces an error signal. The controller modifies the error signal for better control action. Depending on the input to the plant, the output will change. This process continues as long as there is a difference between reference input and feedback signal. If the difference is zero, then there is no error signal and the output settles at the desired value.

The feedback element is a device which converts the output (controlled variable) c electrical or digital signal into suitable, such as displacement, pressure or voltage which is then compared with the input (command) signal. Usually the feedback system consists of sensor, tachogenerators, transducer, etc.

The controller consists of an error detector and control elements. The error detector compares the feedback signal obtained from the plant output with the input (command) signal and determines there from the deviation known as the actuating signal. Examples of error detector are potentiometer, LVDT

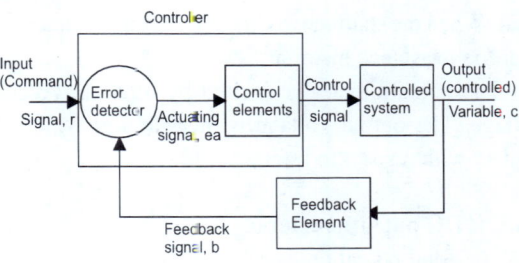

Fig. 5.4.1. Block diagram of control system components

(Linear Variable Differential Transformer), synchros, etc.

It is suitably manipulated by the control elements to produce a control signal. The manipulation may involve amplification, generation of a suitable function of the actuating signal and a power stage. The actuation of the signal to produce a signal acceptable by the plant is done by pneumatic motor/valve, hydraulic motor or electric motor. Examples of electric motors employed are DC servomotor, AC servomotor and stepper motor.

The controllers employed may be electrical, electronic, hydraulic or pneumatic depending on the nature of error signal. If the error signal is electrical then the controller may be electrical or electronic. If error signal is mechanical then the controller may be hydraulic or pneumatic and they are designed using hydraulic servomotors or pneumatic flapper valves.

5.4.1. Potentiometers

A potentiometer is a device that can be used to convert a linear or angular displacement into a voltage. A potentiometer is a variable resistance whose value varies according to the angular/linear displacement of the wiper contact (movable contact).

The resistance element can be constructed by winding resistance wire on a former or by depositing a conducting material on a plastic base. The potentiometer has an input shaft to which a wiper is attached. The displacement is applied to the input shaft. When the shaft moves, the wiper contact slides over the resistance material.

The potentiometer is excited by a DC or AC voltage. The output voltage is measured at wiper contact with respect to reference.

5.4.1.1. DC potentiometers
R_p = Total potentiometer resistance
X_t = Total displacement of the potential over which R_p is spread out

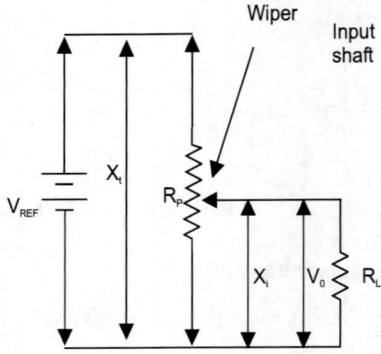

Fig. 5.4.2 A potentiometer with load

X_i = Displacement of the jockey from the ground terminal G (input)
R_L = Load resistance
V_{REF} = Reference voltage
V_0 = Output voltage

As the resistance is spread out uniformly

$$R_i = (X_i / X_t) R_p$$

$$R_{eq} = R_i \,//\, R_p = \cfrac{1}{\left[\cfrac{1}{R_L} + \cfrac{x_t}{x_i}\cfrac{1}{R_p}\right]}$$

Then by voltage dividing action

$$V_0 = V_{REF} \cfrac{R_{eq}}{R_{eq} + (R_p - R_i)}$$

Then,

$$V_0 = \cfrac{V_{REF}}{\cfrac{x_t}{x_i} + (R_p / R_L)[1 - (x_i / x_t)]}$$

If R_L = (Jockey open i.e., no load)

$$V_0 = V_{REF} \frac{x_i}{x_t} \text{ ; a linear relationship}$$

5.4.1.2. Charcteristics of potentiometer

1. The ideal characteristics of a potentiometer is linear variation of resistance with displacement. This is best realized by having very large radius, more number of turns and high resistance elements.
2. The device which measures the output voltage of potentiometer should have high impedance to avoid loading error. If necessary an isolation amplifier with high impedance may be used.
3. When the wiper slides over resistance it makes simultaneous contact with adjacent turns to avoid discontinuity in output. Consequently the output is in the form of staircase steps. Hence, we define a term called resolution which specifies the output voltage per step. The resolution of a potentiometer is defined as the ratio of number of steps to total number of turns. The resolution of potentiometer is an important factor in the determination of minimum value of output voltage.

5.4.1.3. Specification of potentiometer

The specifications of potentiometers used in control system are the following:

1. Turns per unit length is in the range of 6 to 30 turns per mm.
2. Torque required for wiper movement is in the range of 1×10^{-3} kg-m to 1×10^{-2} kg-m.
3. The total resistance of the potentiometer is in the range of 25 ohms to 1 mega-ohms.
4. Power rating is 1 to 10 W.
5. Heat dissipation is ½ W per cm^2.
6. Excitation voltage is 4 to 20 V.
7. Voltage gradient is 0.01 to 0.05 V per degree.

5.4.1.4. AC potentiometer

In potentiometers excited by AC supply the output will be a modulated voltage. The carrier is the excitation voltage. The envelope of the carrier is modulated by the movement of the wiper arm. Hence the information is available in the envelope of the carrier. The AC potentiometer will have inductive effect in addition to resistance which leads to difficulty in balancing the potentiometers used as error detectors.

5.4.1.5. Application of potentiometers

Potentiometers can be used either to convert a mechanical motion to proportional voltage or as an error signal. A single potentiometer excited by DC or AC voltage is used to produce an output voltage proportional to displacement of the input shaft.

When potentiometers are used as error detector, two identical potentiometers are required as shown in Fig. 5.4.3. Both the potentiometer's are excited

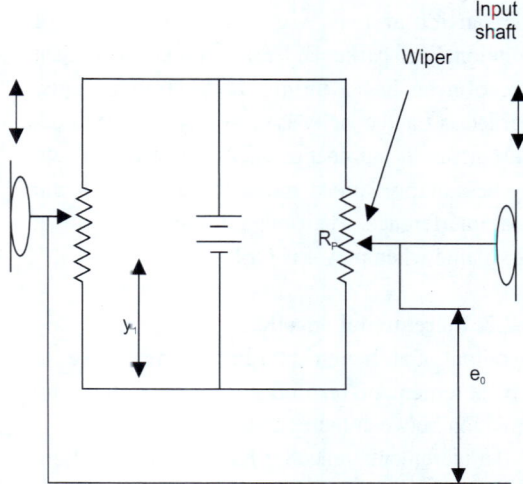

Fig. 5.4.3 A potentiometer as error detector.
Refer the diagram given already

by the same source and at the same potential. Hence, if the wiper arm of both the potentiometers are in the same position then the voltage between two wiper arms is zero.

The position of one wiper arm is kept as reference input. The displacement to be compared is applied to the wiper arm of another potentiometer. Hence, the output voltage which is measured between two wiper arms is proportional to the difference between the displacements of both the wiper arms.

5.4.2. Encoders

Encoders are used to convert linear or rotary displacement into digital code or pulse signals. The two main types of encoders are Absolute and Incremental Encoders.

5.4.2.1 Absolute encoders

Absolute encoder is a digital transducer which gives the position directly. It has binary code etched on to the rotating disc which has as many tracks as the number of bits in the code. Their output is a digitally coded signal with distinct digital code indicative of each particular least significant increment of resolution. Each particular least significant increment of resolution has a unique code; transparency regarded as 1 and opacity as 0. As many photodiode-LED pairs as the number of tracks are needed. These may be suitably spread round the tracks to avoid signal interference. Their main disadvantage is that it is volatile when power is OFF.

5.4.2.2. Incremental encoders

The output of an Incremental Encoder is a pulse for each increment of resolution but these make no distinction between increments.

An incremental encoder has four parts: a light source (LED), a rotary (or translatory) disc, a stationary mask and a sensor (photodiode). The disc has alternate opaque and transparent sectors (of equal width), which are etched by means of photographic process on to a plastic disc. As the disc rotates during half of the increment cycle the transparent sectors of rotating and stationary discs align permitting light from LED to reach the sensor thereby generating an electrical pulse. For fine resolution encoders, multi-slit mask is often used to maximize the reception of shutter light.

The waveform of the sensor output of an encoder is generally triangular or sinusoidal depending upon the resolution required. Square wave signal compatible with digital logic are obtained from it by means of linear OPAM and comparator. Alternate transparent/opaque sectors of the disc and the square

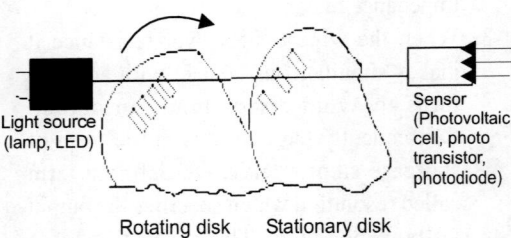

Fig. 5.4.4. Schematic of incremental encoder

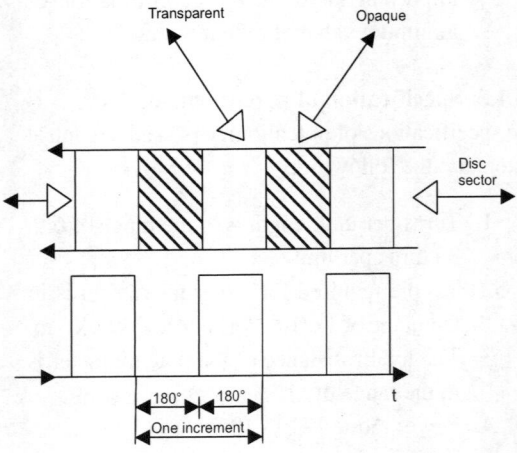

Fig. 5.4.5. Pulse train of incremental encoder

wave pulse train in synchronous with the disc is shown in Fig. 5.4.4. The resolution of such an incremental encoder is given as

$$\text{Basic resolution} = (360°/N)$$

where N = number of sectors of disc; each sector is half transparent and half opaque.

5.4.3. Servomotors

The servo system is one in which the output is some mechanical variable like position, velocity like position, velocity or acceleration. Such systems are generally automatic control systems which work on the error signals. The error signals are amplified to drive the motors used in such systems. These motors used in servo systems are called servomotors.

5.4.3.1. Requirements of a good servomotor
1. Linear relationship between electrical control signal and the rotor speed over a wide range.
2. Linear torque speed characteristics
3. It should be easily reversible
4. Fast response
5. Steady state stability without oscillations or overshoots
6. Low mechanical and electrical inertia.

5.4.3.2. Types of servomotors
The servomotors are basically classified depending upon the nature of the electric supply for its operation. The types of servomotors are as shown in the following chart:

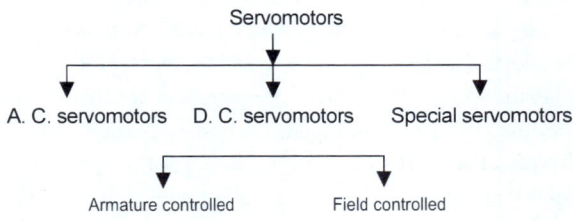

5.4.3.3. AC servomotor
Most of the servomotors used in low power servomechanisms are AC servomotors. The AC servomotor is basically two-phase induction motor. The output power of AC servomotors varies from fraction of watt to few hundred watts. The operating frequency is 50 Hz to 400 Hz.

5.4.3.3.1. Construction: It consists of two parts, namely, stator and rotor. The stator carries two windings, uniformly distributed and displaced by 90°, in space. One winding is called main winding or fixed winding or reference winding. This is excited by a constant voltage AC supply. The other winding is called control winding. It is excited by variable control voltage, which is obtained from a servo amplifier. This voltage is 90° out of phase with respect to the voltage applied to the reference winding. This is necessary to obtain rotating magnetic field. The supply used to drive the motor is single phase and so a phase advancing capacitor is connected to one of the phases to produce a phase difference of 90°. The schematic stator is shown in Fig. 5.4.6.

The rotor construction is usually squirrel cage or drag cup type. Squirrel cage rotor is made of laminations. It has small diameter and large length. Aluminium conductors are used to provide less weight with very high resistance. The rotor bars are

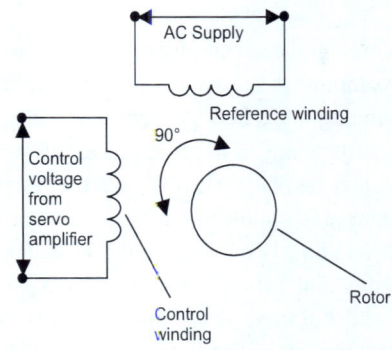

Fig. 5.4.6. Schematic of AC servomotor stator

Fig. 5.4.7a. Squirrel cage rotor

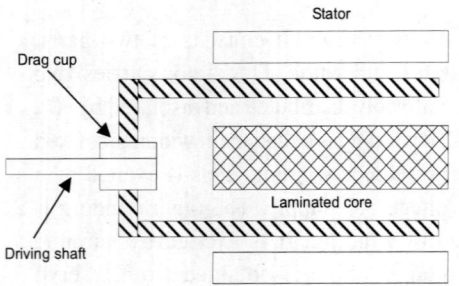

Fig. 5.4.7b. Drag cup type rotor

placed on the slots and short-circuited at both ends by end rings. The squirrel cage rotor and Drop Cup Type Rotor as shown in Figs. 5.4.7 and 5.4.8. The rotor of the servomotor is built with high resistance, so that its X/R (Inductive reactance/Resistance) ratio is small which results in linear speed torque characteristics as shown in Fig. 5.4.9. Drag cup type rotor have two air gaps. Such a construction reduces inertia considerably and hence such type of rotor is used in very low power applications. Aluminium is used for the cup construction.

5.4.3.3.2. Working: The symbolic representation of an AC servomotor is shown in Fig. 5.4.8. The reference winding is excited by a constant voltage source with a frequency in the range 50 to 1000 Hz. By using frequencies of 400 Hz or higher, the system can be made less susceptible to low frequency noise. The control winding is excited by the modulated control signal and this voltage is of variable magnitude and polarity. The control signals are usually of low frequency in the range of 0 to 20 Hz. To provide rotating magnetic field, the control phase

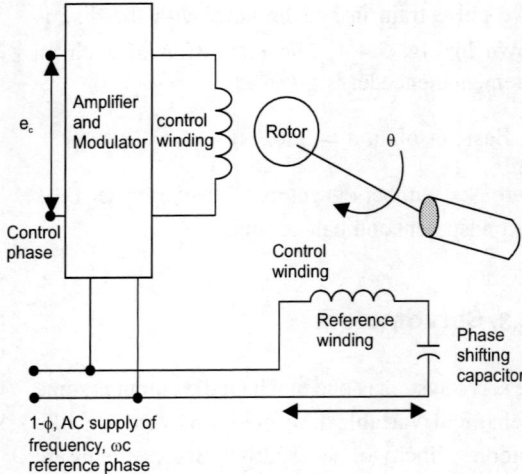

Fig. 5.4.8. Symbolic representation of a.c. servomotor

voltage must be of the same frequency as the reference phase voltage and in addition the two voltages must be in time quadrature. Hence the control signal is modulated by a carrier whose frequency is same as that of reference voltage and then applied to control winding. The AC supply itself is used as a carrier signal for modulation process. The 90° phase difference between the control phase and reference phase voltages is obtained by the insertion of a capacitor in reference winding.

Let e_c = Control signal

$E_{carrier} = E \cos \omega_c t$ = carrier signal

E_{cm} = Modulated control signal.

The speed torque curves of a typical AC servomotor is plotted for fixed reference phase voltage $E \cos \omega_c t$ and different values of constant input voltages $e_c \leq E$. On investigation it is found that all curves have negative slope and the curve for $e_c = 0$ passes through origin, which means the motor develop a decelerating torque and so the motor stops. Also the speed torque curves are nonlinear except in the low speed region. Figure 5.4.10 shows the speed Torque characteristics of servomotor for varying control signal.

Control Systems 337

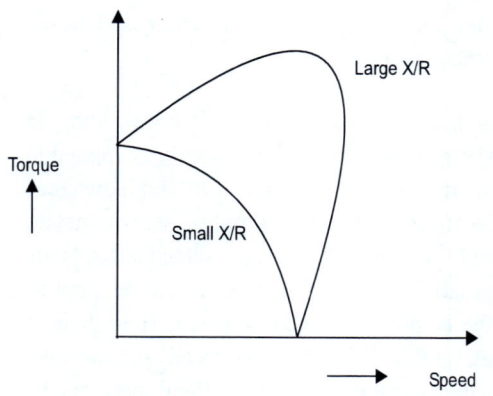

Fig. 5.4.9. Speed torque characteristics for varying X/R

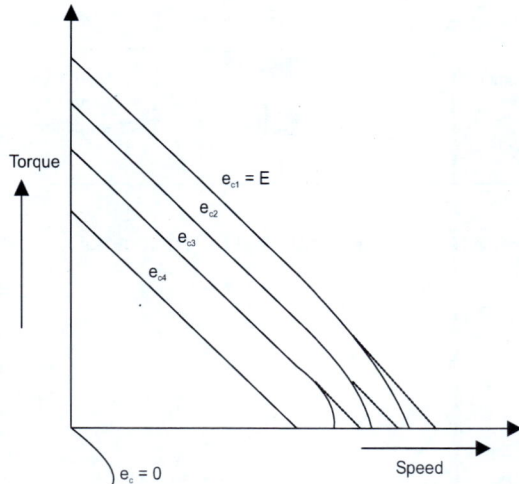

Fig. 5.4.10. Speed torque characteristics of servomotor for varying control signal

5.4.3.3.3. Transfer function of AC servomotor:
Let
T_m = Torque developed by servomotor
q = Angular displacement of rotor
w = $d\theta/dt$ = Angular speed
T_l = Torque required by the load

J = Moment of inertia of load and the rotor
B = Viscous frictional coefficient of load and the rotor
K_1 = Slope of control phase voltage Vs Torque characteristic
K_2 = Slope of speed torque characteristics.

On investigation of the speed torque characteristics, we can find that for speeds near zero, the curves are straight lines parallel to the characteristics at rated input voltage (e_c =E) and are equally spaced for equal increments of the input voltage. Under this assumption, the torque developed by the motor is

Torque developed by motor,

$$T_m = K_1 e_c - K_2 \frac{d\theta}{dt}$$

The rotating part of the motor and the load can be modeled by the equation given below.

Load torque,

$$T_l = J\frac{d^2\theta}{dt^2} + B\frac{d\theta}{dt}$$

At equilibrium the motor torque is equal to load torque.

$$J\frac{d^2\theta}{dt^2} + B\frac{d\theta}{dt} = K_1 e_c - K_2 \frac{d\theta}{dt}$$

On taking laplace transform of equation

$$Js^2\theta(s) - Bs\theta(s) = K_1 E_c(s) - K_2 s\theta(s)$$

$$[Js^2 + Bs + K_2 s]\theta(s) = K_1 E_c(s)$$

$$\frac{\theta(s)}{E_c(s)} = \frac{K_1}{s(Js + B + K_2)} = \frac{K_m}{s(\tau_m s + 1)}$$

where

$$K_m = \frac{K_1}{B+K_2} = \text{Motor gain constant}$$

$$\tau_m = \frac{J}{B+K_2} = \text{Motor time constant}$$

5.4.3.3.4. Applications: It is widely used in instrument servomechanisms, remote positioning devices, process control systems, self balancing recorders, computers, tracking and guidance systems, robotics, machine tools etc.

5.4.3.4. DC servomotor

All DC servomotors are essentially separately excited type. This ensures linear torque speed characteristics. The control of DC servomotor can be from field side or from armature side. Depending upon this, these are classified as field controlled DC servomotor and armature controlled DC servomotor.

5.4.3.4.1. Field controlled DC servomotor: In this motor, the controlled signal obtained from the servo amplifier is applied to the field winding. With the help of constant current source, the armature current is maintained constant. When the armature voltage is constant the torque is directly proportional to field flux. Since the field current is proportional to flux, the torque of the motor is controlled by controlling the field current. This type of motor has large L_f/R_f ratio where L_f is reactance and R_f is resistance of field winding. Due to this the time constant of the motor is high. Reversible operation is possible by reversing the field current. Field inductance slows down the response.

5.4.3.4.2. Armature controlled DC servomotor: In this type of motor the input voltage V_a is applied to the armature with a resistance of R_a and inductance L_a. The field winding is supplied with constant current I_f. Thus armature input voltage controls the motor shaft output. The constant field can be supplied with the help of permanent magnets. Here there is no need for field coils. Speed is directly proportional to armature current. Hence, torque and speed can be controlled by armature voltage. Reversible operation is obtained by reversing the armature current. Armature voltage is controlled by a variable

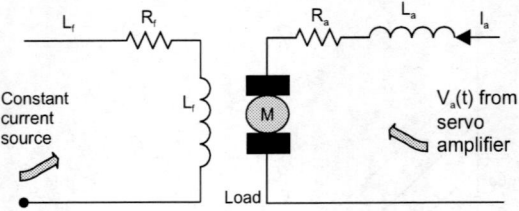

Fig. 5.4.12. Armature controlled DC servomotor

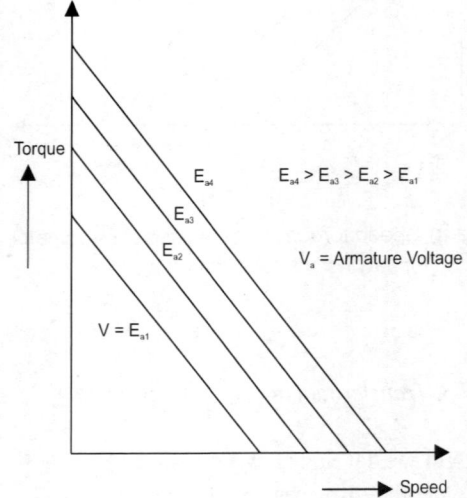

Fig. 5.4.13. Speed torque characteristics of armature controlled DC servomotor

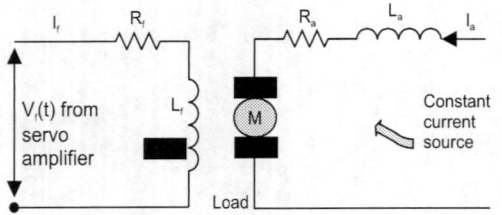

Fig. 5.4.11. Field controlled DC servomotor

resistance in small motors and by thyristors in large motors.

5.4.3.4.3. Applications of DC servomotor: They are used in aircraft systems, electromechanical systems, robot, machine tools, etc.

5.4.4. Stepper Motor

A stepper motor is an electromechanical device which actuates a train of step angular (or linear) movements in response to a train of input pulses on one to one basis—one step actuation for each pulse input. A stepper motor is the actuator element of incremental motion control systems—computer peripherals like printers, tape drives, capstan drives etc., machine tool and process control systems. Stepper motors are used in quartz crystal watches.

The two most widely used types of stepper motors are:

1. Permanent magnet motor
2. Variable reluctance motor

5.4.4.1. Permanent magnet stepper motor

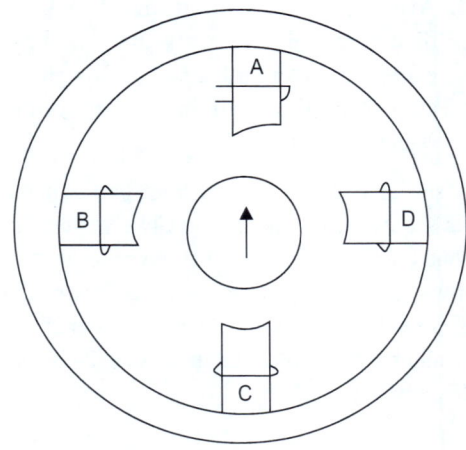

Fig. 5.4.14a. Permanent magnet stepper motor

The stator of this type of stepper motor has salient poles with varying control windings. Each pole carrying a control windings are connected in series and called a phase. Stepper motors may be wound for any number of phases, most popularly being two, three and four phase stepper motors. The rotor is made in the form of a permanent magnet spider cast integral or assembled of a number of permanent magnets.

Consider a stepper motor having 4-pole stator with two-phase windings. Let the rotor be made of permanent magnet with 2 poles. The stator poles are marked A,B,C,D and they are excited with pulses supplied by power transistors. The power transistors are switched by digital controllers or computers. Each control pulse applied by the switching device causes a stepped variation of the magnitude and polarity of voltage fed to the control windings.

If the first excitation is applied to A and C, they develop the magnetic polarities indicated for step 1, as shown in Fig. 5.4.15, and the rotor sets itself vertically. (The magnetic polarity developed in the stator pole can be determined from right hand rule). If now A and C are switched OFF and B and D excited as in step 2, an alignment torque is developed on the rotor to turn its axis to the horizon by a 90° step. With B and D OFF, A & C re-energised with reverse polarity, the rotor turns a further 90° and so on. The direction of rotation is anticlockwise. The direction can be reversed by changing the current directions suitably. The stator currents with uniform pulse frequency and equal ON/OFF periods are shown to a time base.

Further reduction in step angle can be achieved by increasing number of rotor poles.

The step angle is given by

$$A_s = 360°/2PM$$

where 2P = number of poles

M = number of phases in the exciting winding.

Permanent magnet stepper motor with large number of poles cannot be made in small size. To

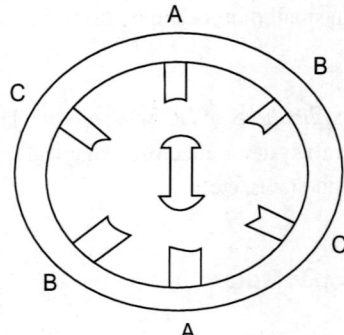

Fig. 5.4.15. Schematic representation of variable reluctance stepper motor

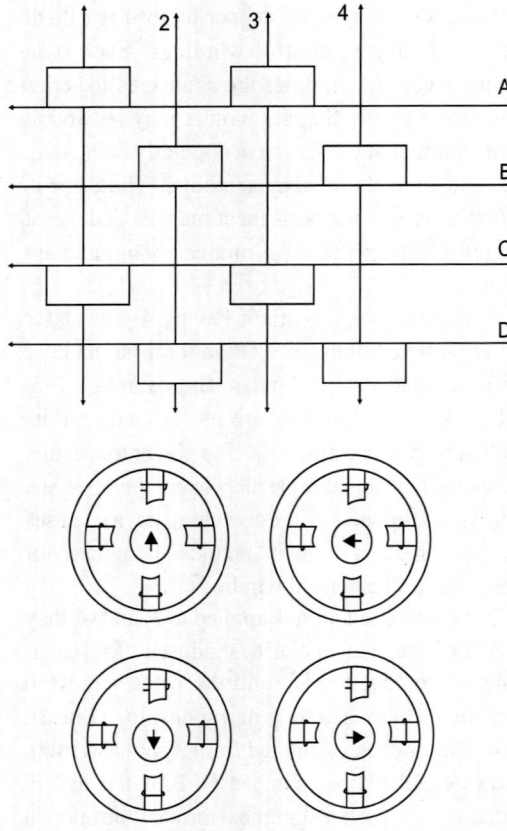

Fig. 5.4.14b. Permanent magnet stepper motor steps

overcome this a variable reluctance type stepper motor is used.

5.4.4.2. Variable reluctance stepper motor

The stator in this type is wound for three phases. The stator has six salient poles (teeth) with concentrated exciting windings around each one of them. It has soft iron rotor with a number of teeth, giving it the appearance of a gear without any exciting winding. The stator also has teeth in addition to a number of wound poles.

Working

The coils are wound around diametrically opposite poles are connected in series and the three phases are energized from a DC source with the help of switches.

1. When the phase A-A′ is excited with switch SW_1 closed, the rotor tries to adjust itself in a minimum reluctance postion between stator and rotor.
2. Next when the phase B-B′ is also excited with switch SW_2 closed, keeping A-A′ energized the magnetic axis of stator moves 30° in clockwise direction and hence rotor also rotates through 30° step in clockwise direction to attain new minimum reluctance position.
3. After that the excitation of AA′ is disconnected and only BB′ is kept energized. Rotor further moves through 30° step to adjust itself in new minimum reluctance position.

By successively exciting the three phases in the specific sequence, the motor takes twelve steps to make one complete revolution. If the rotor were to have four salient projections instead of two, the successive voltage pulses would cause a step of 15° and the number of steps would increase to 24 in one complete revolution.

5.4.4.3. Stepper motor characteristics

The characteristics are classified as:

1. Static characteristics
2. Dynamic characteristics

5.4.4.3.1. Static characteristics

1. Torque displacement characteristics

This gives the relationship between an electromagnetic torque developed and displacement angle θ. The torque increases from zero reaches a maximum value called holding torque and again falls back to zero.

2. Torque current characteristics:

The holding torque of the motor increases with the exciting current. The relation between the holding torque and the current is called torque-current characteristics.

In variable reluctance motors, torque is zero when current is zero. Initially it increases according to parabolic nature and later becomes linear. In permanent magnet motor, the static torque occurs though motor is unexcited and is called detent torque. Further it increases linearly with the current.

5.4.4.3.2. Dynamic characteristics

Response to pulse input: The response of a stepper motor to slow, intermediate and fast stepping rates are shown in Fig. 5.4.17. At low rates the motion is a succession of the step response, while at high rates the motors run at a nearly constant slewing speed. The maximum speed is limited by the load torque and inertia.

Torque vs stepping rate characteristics: This is used to predict the shaft angle at any instant following the applications of one step or a train of steps.

The curve 1 shows the start and synchronization of motor while curve 2 corresponds to loss of synchronism.

f_1 = Definite stepping rate corresponding to given load torque at which motor can start and synchronize without losing step.

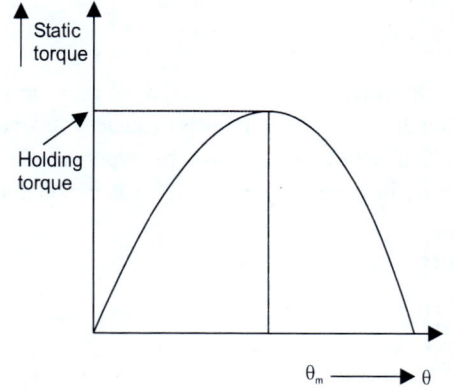

Fig. 5.4.16a. Torque displacement characteristics

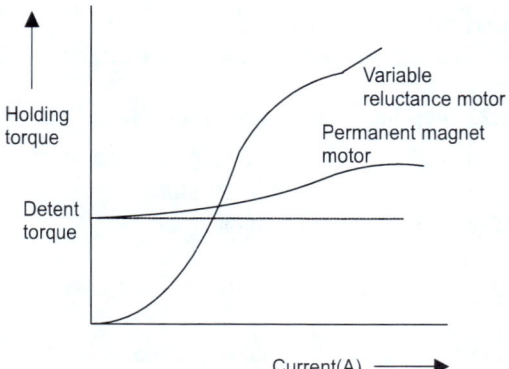

Fig. 5.4.16b. Torque current characteristics

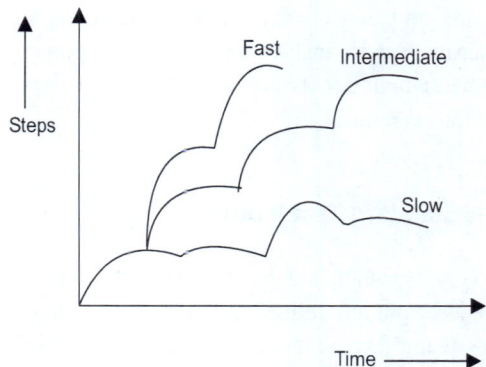

Fig. 5.4.17. Stepper motor response

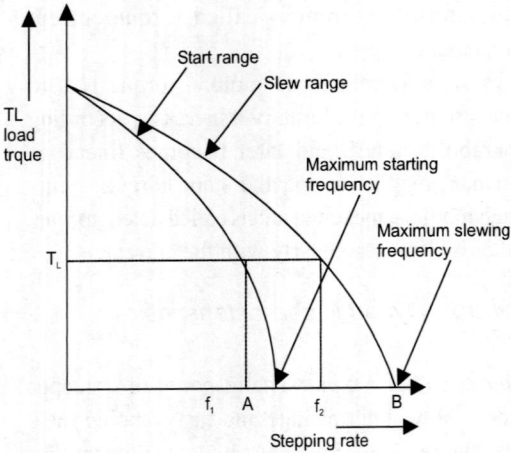

Fig. 5.4.18. Torque stepping rate characteristics

f_2 = Once started stepping rate can be increased upto f_2 after which motor starts losing the steps.

5.5. TIME RESPONSE ANALYSIS OF CONTROL SYSTEM

The first step in analyzing a control system is to derive a mathematical model of the system. One can analyse the system performance after deriving the mathematical model. In practice, the input signal is not known prior and it is random in nature. In some special cases, the input signal is known well in advance and can be expressed analytically or graphically. System analysis is performed by applying certain test input signals and comparing the response of various systems.

5.5.1. Standard Test Inputs

The signals which are most commonly used as reference input are defined as standard test inputs. Some of the test signals are step functions, ramp functions, acceleration functions, impulse functions and sinusoidal functions. With these test signals, mathematical and experimental analyses of control systems can be carried out easily since these signals are very simple functions of time.

Depending on the requirement of a particular situation, the input has to be selected. For example, if in a control system if the input were to change with time, then a ramp function can be used. Similarly, if a system is subjected to sudden disturbances, step function of time can be applied and for a system subjected to inputs for a very small time, an impulse function can be used.

The use of such test signals enables one to compare the performance of any type of system on the same basis.

5.5.1.1. Transient response and steady state response

Transient Response: It is defined as the output variation during the time. It takes to achieve its final value. Transient response may be exponential or oscillatory in nature. Symbolically it is denoted as $c_t(t)$.

For stable operating system:

$$\lim_{t \to \infty} c_t(t) = 0$$

Steady State Response: It is that part of the time response which remains after complete transient response vanishes from the system output. Symbolically it is denoted as $c_{ss}(t)$

Total time response, $c(t) = c_{ss}(t) + c_t(t)$

5.5.1.2. Step input (Position function)

The step signal is one whose value changes from one level to another level A in zero time.

Mathematical representation

$r(t) = Au(t)$ for $t = 0$ 0 for $t<0$

If $u(t) = 1$, then it is called unit step, denoted by $u(t)$

The graphical representation of a step signal is shown in Fig. 5.5.1.

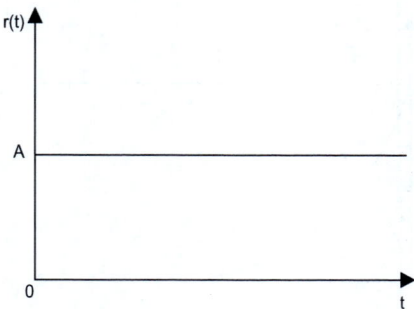

Fig. 5.5.1. Graphical representation of step input

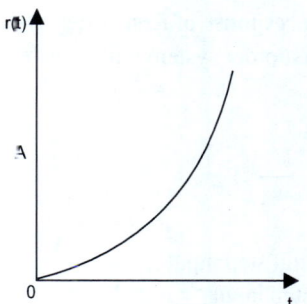

Fig. 5.5.3. Graphical representation of parabolic input

5.5.1.3. Ramp input (Velocity function)

The ramp is a signal which starts at a value of zero and increases linearly with time.

$r(t) = At$ for $t \geq 0$

$= 0$ for $t < 0$

If $A = 1$, it is called unit ramp input

The graphical representation of a ramp signal is shown in Fig. 5.5.2. From equation 1 and 2, it is seen that a ramp signal is integral of a step signal.

The graphical representation of a parabolic signal is shown in Fig. 5.5.3. From equation 2 and 3, it is seen that a parabolic signal is integral of a ramp signal.

5.5.1.5. Impulse Input

A unit impulse is defined as a signal which has zero value everywhere except at $t = 0$, where its magnitude is infinite. It is represented by δ-function and has the following property:

$\delta(t) = 0$ for $t \neq 0$

$$\int_{-\varepsilon}^{+\varepsilon} \delta(t) dt = 1$$

where ε tends to zero

Mathematically, an impulse function is the derivative of a step function

$\delta(t) = du(t)$

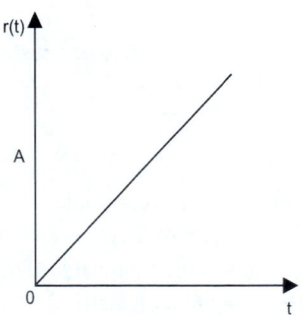

Fig. 5.5.2. Graphical representation of ramp input

5.5.1.4. Parabolic input (Acceleration function)

The mathematical representation of this signal is

$r(t) = \dfrac{At^2}{2}$, $t \geq 0$

$= 0$ for $t < 0$

Table 5.5.1. The Laplace transform of inputs

r(t)	R(s)
Step Input	$1/s$
Ramp Input	$1/s^2$
Parabolic Input	$1/s^3$
Impulse Input	1

5.5.1.6. Time response of first order systems

Consider a first order system with transfer function given by

$$\frac{C(s)}{R(s)} = \frac{1}{Ts+1}$$

Response to unit step input
For unit step input

$r(t) = 1$

$R(s) = 1/s$

The output response is

$$C(s) = \frac{1}{Ts+1} \; / \; C(s)\frac{1}{s} = \frac{A}{s} + \frac{B}{Ts+1}$$

$1 = A(Ts+1) + Bs$

Put $s = 0$; $A = 1$

Put $s = 1$; $1 = 1(T+1)+B = T+1+B$

$B = -T$

$$C(s) = \frac{1}{s} - \frac{T}{Ts+1}$$

Taking inverse Laplace transform, we get

$c(t) = 1 - e^{-t/T}$

Note:

$L^{-1}(1/s) = 1$

$L^{-1}(1/(1(/T)+s)) = e^{-t/T}$

which is plotted in figure.

The output rises exponentially from zero value to the final value of unity.

At

$$t = 0, \; \left.\frac{C(s)}{T(s)}\right|_{t=0} = \frac{1}{T}e^{-t/T}\bigg|_{t=0} = \frac{1}{T}$$

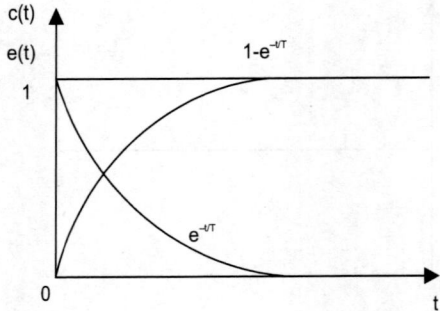

Fig. 5.5.4. Step response of first order system

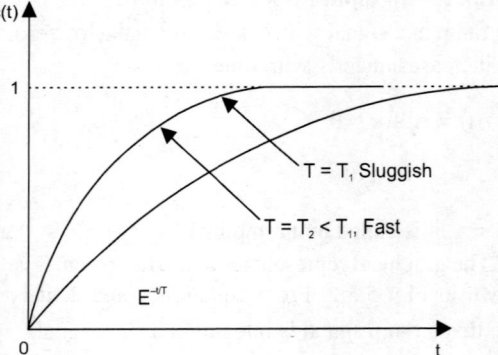

Fig. 5.5.5. Step response of first order system with different time constants

where T is known as the time constant.

The time constant is indicative of how the fast the system tends to reach the final value. The speed of response can be quantitatively defined as the time for the output to become a particular percentage of its final value. A large time constant corresponds to a sluggish system and a small time constant corresponds to a fast response.

5.5.1.7. Response of first order system to unit ramp input

Ramp input : $r(t) = t$

Laplace transform: $R(s) = 1/s^2$.

Output response

$$C(s) = \frac{1}{Ts+1} \cdot \frac{1}{s^2} = \frac{A}{s^2} + \frac{B}{s} + \frac{C}{Ts+1}$$

Taking partial fraction

$$C(s) = \frac{1}{s^2} + \frac{T}{s} + \frac{T^2}{Ts+1}$$

Taking inverse Laplace

$$c(t) = t - T(1 - e^{-t/T})$$

The response is shown in Fig. 5.5.6.

5.5.1.8. Time response of second order system to unit step

The transfer function of a second order system is

$$\frac{C(s)}{R(s)} = \frac{\omega_n^2}{s^2 + 2\zeta\omega_n s + \omega_n^2}$$

where ζ is the damping factor and ω_n is the undamped natural frequency.

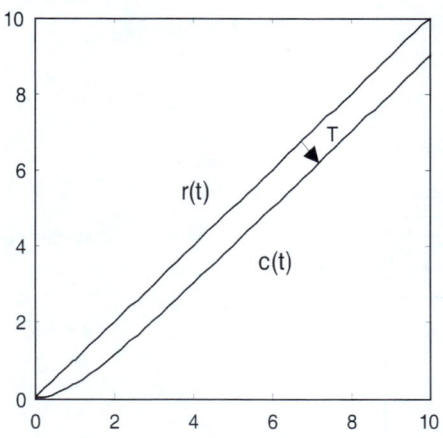

Fig. 5.5.6. Ramp response of first order system

5.5.1.8.1. Analysis of second order system:
Every system has tendency to oppose the oscillatory behaviour of the system called damping. Damping is measured by a factor or a ratio called damping ratio of the system. If it is low, then system will oscillate but slowly, i.e., with damped frequency.

The time response of any system is characterized by the roots of the denominator polynomial $q(s)$, which in fact are the poles of the transfer function. The denominator polynomial $q(s)$ is called the characteristic polynomial and $q(s) = 0$ is called the characteristic equation.

The characteristic equation of the second order system is

$$s^2 + 2\zeta\omega_n s + \omega_n^2 = 0$$

If s_1 and s_2 are the roots of this characteristic equation, then

$$s_1, s_2 = -\zeta\omega_n \pm j\omega_n\sqrt{1-\zeta^2}$$
$$= -\zeta\omega_n \pm j\omega_d$$

where $\omega_d = \omega_n\sqrt{1-\zeta^2}$ is called the damped natural frequency.

5.5.1.8.2. Unit step response:

$$r(t) = 1$$

$$R(s) = 1/s \qquad \frac{C(s)}{R(s)} = \frac{\omega_n^2}{s^2 + 2\zeta\omega_n s + \omega_n^2} \cdot \frac{1}{s}$$

Applying partial fraction, the solution is

$$C(s) = \frac{1}{s} - \frac{s+\zeta\omega_n}{(s+\zeta\omega_n)^2 + \omega_d^2} - \frac{\zeta\omega_n}{(s+\zeta\omega_n)^2 + \omega_d^2}$$

(i) For underdamped case:
Taking Inverse Laplace transform

346 Basic Electrical Engineering

Sl. No.	Range of ζ	Type of closed loop poles	Nature of response	System classification
1	$\zeta = 0$ imaginary	Purely	Oscillation with constant frequency and amplitude	Undamped
2	$0 < \zeta < 1$	Complex conjugates with negative real part	Damped Oscillation	Under-damped
3	$\zeta = 1$	Real, equal and negative	Critical and pure exponential	Critically damped
4	$1 < \zeta < \infty$	Real, unequal and negative	Purely exponential slow and sluggish	Over-damped

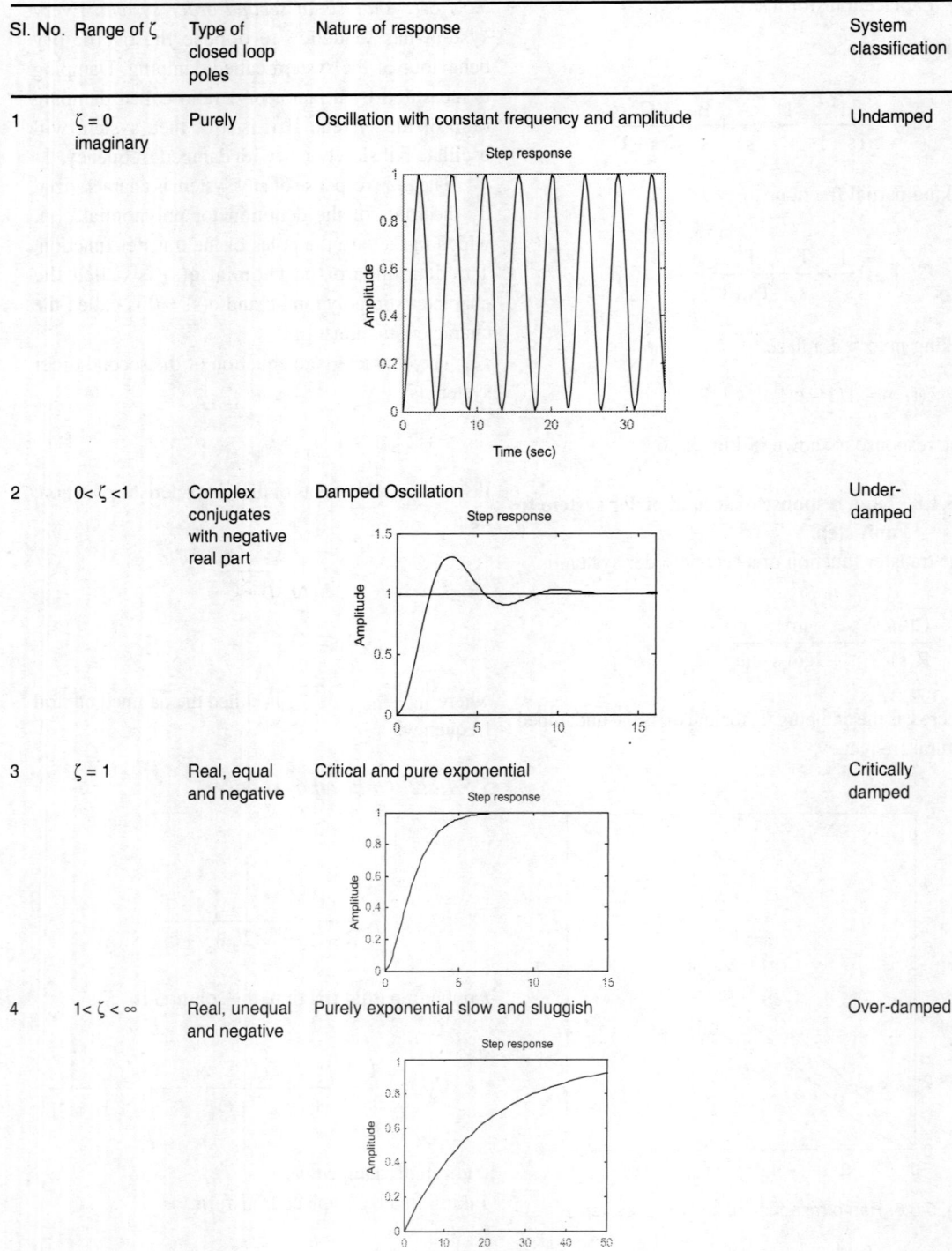

$$c(t) = 1 - \frac{e^{\zeta\omega_n t}}{\sqrt{1-\zeta^2}}[\sqrt{1-\zeta^2}\cos\omega_d t + \zeta\sin\omega_d t]$$

Now consider the right angled triangle

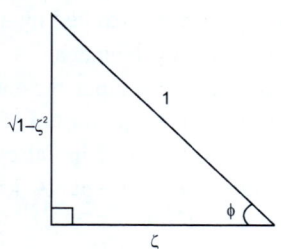

Fig. 5.5.7.

From the above diagram

$\cos\varphi = \zeta$

$\sin\varphi = \sqrt{1-\zeta^2}$

$\tan\varphi = \sqrt{1-\zeta^2}/\zeta$

Applying these equations

$$c(t) = 1 - \frac{e^{\zeta\omega_n t}}{\sqrt{1-\zeta^2}}[\sin\varphi\ \cos\omega_d t + \cos\varphi\ \sin\omega_d t]$$

$$= 1 - \frac{e^{\zeta\omega_n t}}{\sqrt{1-\zeta^2}}[\sin\omega_d t + \varphi]$$

$$c(t) = 1 - \frac{e^{\zeta\omega_n t}}{\sqrt{1-\zeta^2}}[\sin\omega_d t + \tan^{-1}\frac{\sqrt{1-\zeta^2}}{\zeta}]$$

The output response for $\zeta < 1$ which is underdamped.

(ii) For undamped case: $\zeta = 0$

$$c(t) = 1 - \sin(\omega_n t + \tan^{-1}\infty)$$
$$= 1 - \cos\omega_n t$$

(iii) Critically damped case: $\zeta = 1$

$$c(t) = 1 - e^{-w_n t}(1 + w_n t)$$

Overdamped case

$$c(t) = 1 + \frac{\omega_n}{2\sqrt{\zeta^2-1}}\left(\frac{e^{-s_1 t}}{s_1} - \frac{e^{-s_2 t}}{s_2}\right)$$

$$s_1 = (\zeta + \sqrt{\zeta^2-1})\omega_n$$

$$s_2 = (\zeta - \sqrt{\zeta^2-1})\omega_n$$

The response c(t) includes two decaying exponential terms

For underdamped system, expression for c(t):

$$c(t) = 1 - \frac{e^{-\zeta\omega_n t}}{\sqrt{1-\zeta^2}}\sin(\omega_d t + \theta)$$

$$\omega_d = \omega_n\sqrt{1-\zeta^2}$$

$$\theta = \tan^{-1}\left\{\frac{\sqrt{1-\zeta^2}}{\zeta}\right\} \text{ radians}$$

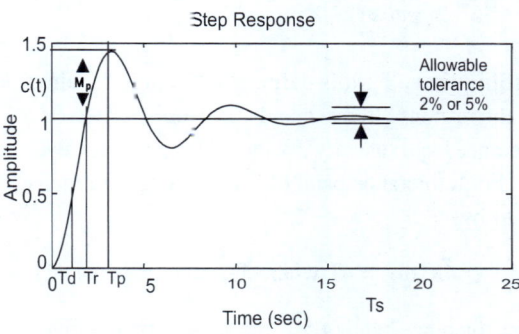

Fig. 5.5.8. Time domain specification of a second order system

5.5.2. Time Response Specifications

Delay time, T_d: It is the time required for the response to reach 50% of the final value in first attempt.

$$T_d = \frac{1+0.7\zeta}{\omega_n}\sec$$

Rise time, T_r: It is the time required for the response to rise from 10% to 90% of the final value.

$$T_r = \frac{\pi - \theta}{\omega_d}\sec$$

Peak time, T_p: It is the time required for the response to reach the peak of time response or the peak overshoot

$$T_r = \frac{\pi}{\omega_d}\sec$$

Peak overshoot M_p: It indicates the normalized difference between the peak response and the steady output and is defined as

Peak percent overshoot

$$M_p = \frac{c(t_p) - c(\infty)}{c(\infty)} \times 100; \quad M_p = e^{-\zeta\pi/\sqrt{1-\zeta^2}}$$

Settling time, T_s: It is defined as the time required for the response to reach and stay within a specified tolerance band (usually 2% or 5%) of its final value.

For a tolerance band of 2%, the settling time is given by

$$T_s \approx 4/(\zeta\omega_n) = 4T \text{ where T is the time constant}$$

For a tolerance band of 5%, the settling time is given by

$$T_s \approx 3/(\zeta\omega_n) = 3T$$

5.6. DESIRABLE POLE LOCATIONS OF TRANSFER FUNCTIONS AND SYSTEM STABILITY

A linear time invariant system is stable if the following two notions of system stability are satisfied.

(i) When the system is excited by a bounded input, the output is bounded.
(ii) In the absence of the input, the output tends towards zero (the equilibrium state of the system) irrespective of initial conditions. This stability concept is known as asymptotic stability).

If a system is subjected to an unbounded input and produces an unbounded response, nothing can be said about its stability. But if it is subjected to a bounded input and produces an unbounded response, it is by definition unstable. The output of an unstable system may increase to a certain extent and then the system may break down or become nonlinear after the output exceeds a certain magnitude.

Let

$$G(s) = \frac{C(s)}{R(s)} = \frac{b_0 s^m + b_1 s^{m-1} + \ldots + b_m}{a_0 s^n + a_1 s^{n-1} + \ldots + a_n}; m < n$$

With initial conditions assumed zero, the output of the system is given by

$$c(t) = L^{-1}[G(s)\,R(s)]$$

Therefore

$$c(t) = \int_0^\infty g(\tau)r(t-\tau)d\tau$$

Where $g(t) = L^{-1}\,G(s)$ is the impulse response of the system.

Taking the absolute value on both sides we get

$$|c(t)| = \left|\int_0^\infty g(\tau)r(t-\tau)d\tau\right|$$

Since the absolute value of integral is not greater than the integral of the absolute value of the integrand,

$$|c(t)| \leq \int_0^\infty |g(\tau)| \, |r(t-\tau)| \, d\tau$$

Bounded input, $|r(t)| \leq M_1 < \infty$

Then

$$|c(t)| \leq M_1 \int_0^\infty |g(\tau)| \, d\tau$$

So, if the impulse response g(t) is absolutely integrable, the the output will be bounded one and hence the system will be stable one (first notation).

The nature of g(t) is dependent on the poles of the transfer function G(s) which are the roots of the characteristic equation. These roots may be both real and complex conjugate and may have multiplicity of various orders. The nature of response terms contributed by all possible types of roots are tabulated and illustrated below.

All the roots which have nonzero real parts [cases (i), (ii), (iii) and (iv)], contribute response terms with a multiplying factor of $e^{\sigma t}$. If $\sigma < 0$ (i.e., the roots have negative real parts), the response terms vanish as $t > \infty$ and if $\sigma > 0$ (i.e., the roots have positive real parts), the response terms increase without bound. Roots on the $j\omega$ axis with multiplicity two or higher [cases (vi) and (viii)] also contribute terms which increase without bound as $t \to \infty$.

(i) Single root at $s = -\sigma$

   Nature of response = $A \, e^{-\sigma t}$

   The nature of roots and the output response is shown

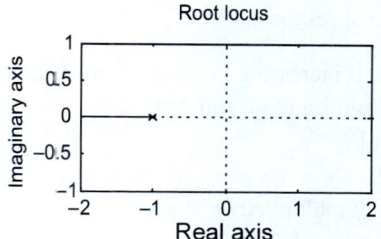

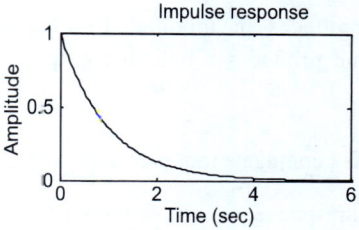

Since the output is decreasing to the value of zero, it is bounded one and hence the system will be STABLE

(ii) Roots having positive real part

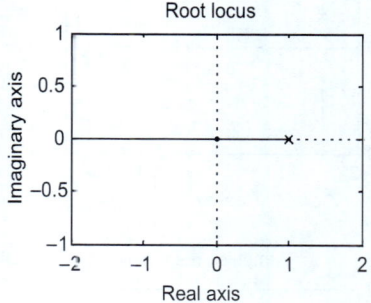

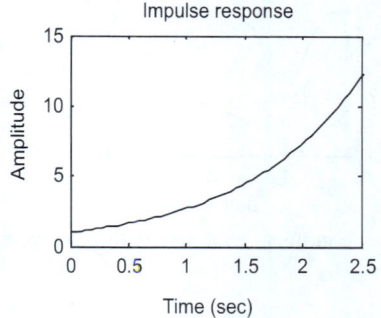

Nature of response = $A e^{-\sigma t}$

Output is increasing towards infinity. Hence the output is unbounded and hence the system is UNSTABLE.

(iii) Roots of multiplicity k at s = σ

Output response = $(A_1 + A_2 t + ... + A_k t^{k-1})e^{\sigma t}$

This produces an unbounded output if σ is positive and produces a bounded output if σ is negative.

(iv) Complex conjugate root pair at $s = -\sigma \pm j\omega$

Output response = $A e^{-\sigma t} \sin(\omega t + \beta)$

Bounded output and hence stable system

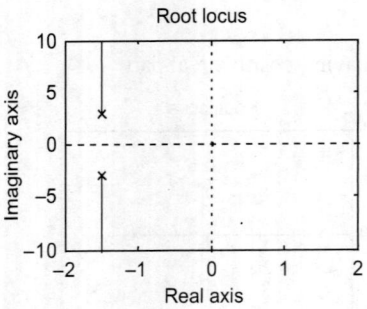

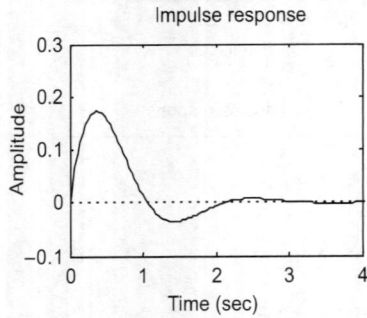

(v) Complex conjugate root pair with positive real part at $s = \sigma \pm j\omega$

Output response = $A e^{\sigma t} \sin(\omega t + \beta)$

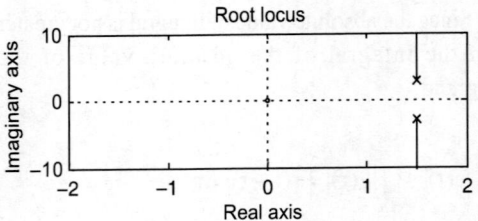

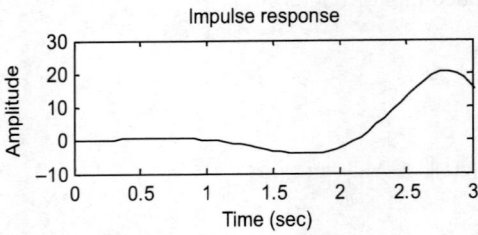

Output : Unbounded
Result : Unstable

(vi) Complex conjugate repeated roots with positive real parts
Unbounded output

(vii) Complex conjugate repeated roots with negative real parts:
Bounded output

(viii) Single complex conjugate root pairs on the jω – axis $s = \pm j\omega$

Reponse = $A \sin(\omega t + \beta)$

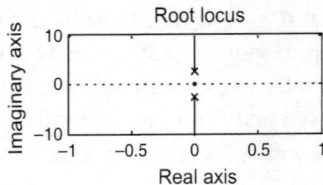

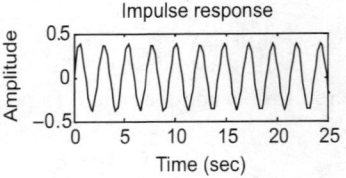

(ix) Complex conjugate root pair of multiplicity k on the jω axis

Response $A_1 \sin(\omega t + \beta_1) + A_2 t \sin(\omega t + \beta_2) + \ldots + A_k t^{k-1} \sin(\omega t + \beta_k)$

Output : Unbounded:
Result : Unstable

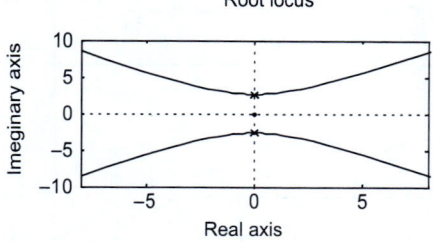

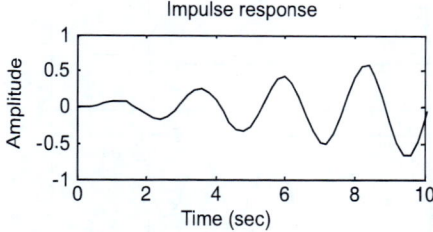

(x) Single root at origin s = 0
Response = A

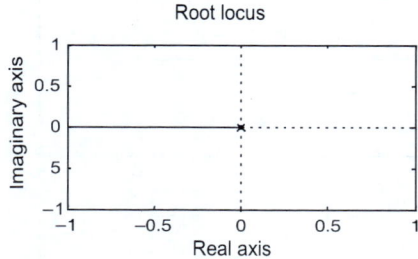

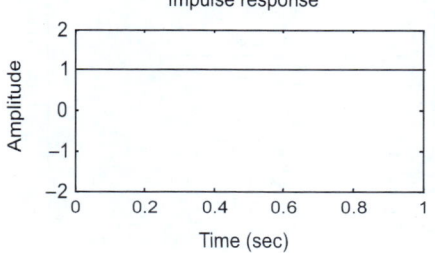

Output: Unbounded
Result : Unstable

(xi) Repeated roots at origin
Output Unbounded
Result : Unstable

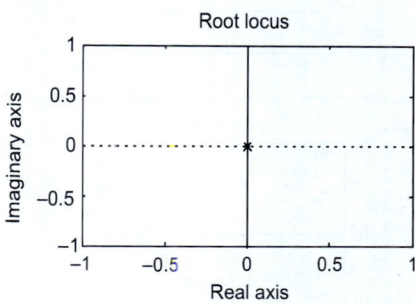

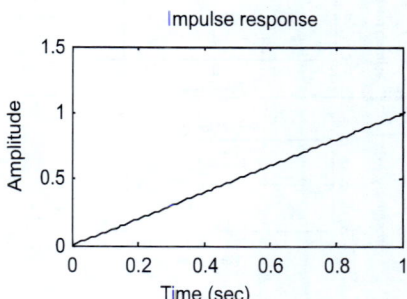

Single root at origin [case (vii)] or multiple root pairs [case (vi)] on the jω axis contribute response terms which are constant amplitude or constant amplitude oscillation. These observations lead us to the following general conclusions regarding system stability.

1. If all the roots of the characteristic equation have negative real parts, then the impulse response is bounded and eventually decreases to zero.
2. If any root of the characteristic equation has a positive real part, g(t) is unbounded and the system is unstable.

352 *Basic Electrical Engineering*

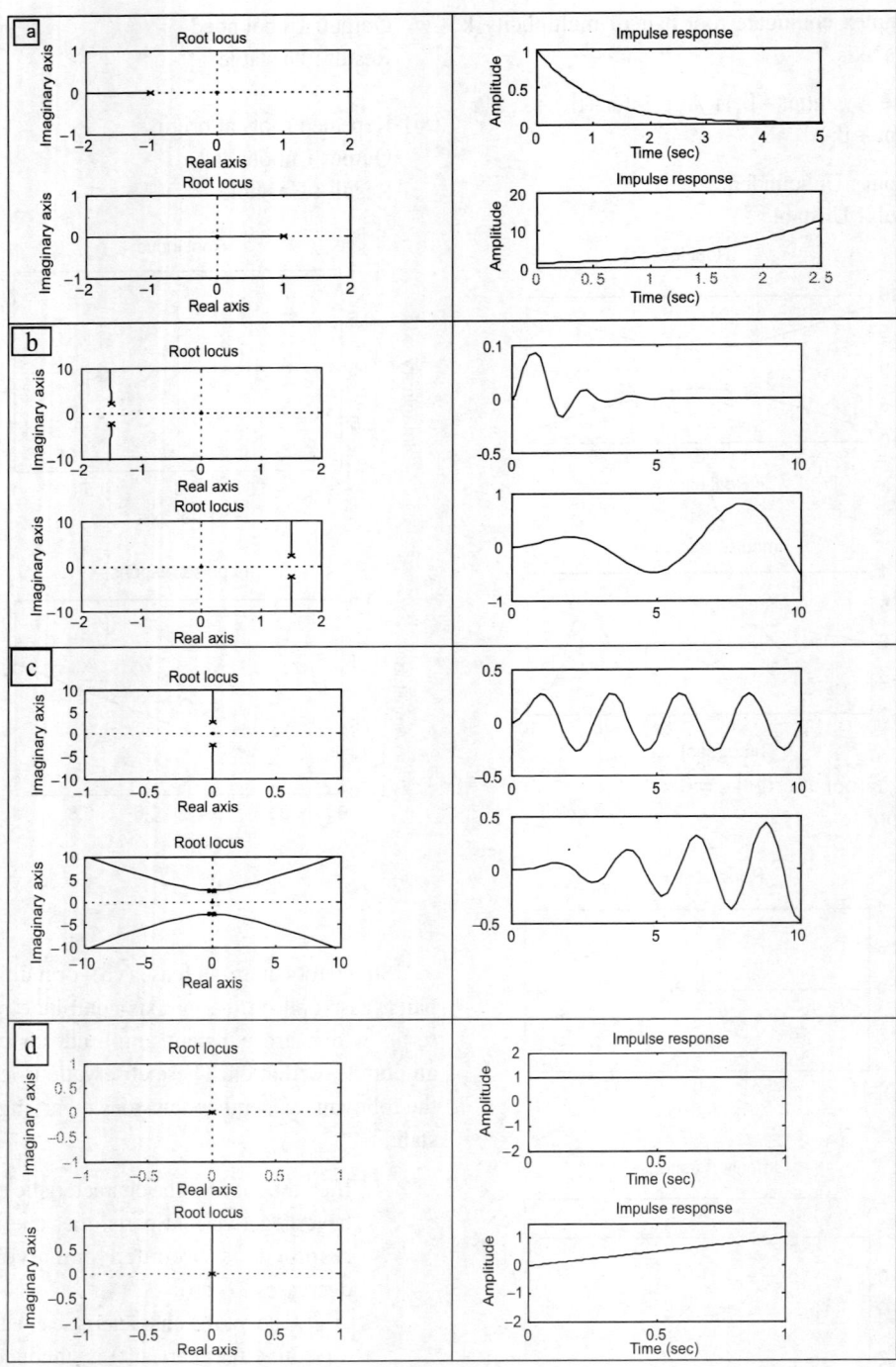

Figure 5.6.1. Response terms obtained by various types of roots

Table. 5.6.1 Response terms contributed by various types of roots

Sl. No.	Type of roots	Nature of response terms contributed
i.	Single root at $s = -\sigma$	$A e^{\sigma t}$
ii.	Roots of multiplicity k at $s = \sigma$	$(A_1 + A_2 t + ... + A_k t^{k-1}) e^{\sigma t}$
iii.	Complex conjugate root pair at $s = \sigma \pm j\omega$	$A e^{\sigma t} \sin(\omega t + \beta)$
iv.	Complex conjugate root pair of multiplicity k at $s = \sigma \pm j\omega$	$(A_1 \sin(\omega t + \beta_1) + A_2 t \sin(\omega t + \beta_2) + ... + A_k t^{k-1} \sin(\omega t + \beta_k)) e^{\sigma t}$
v.	Single complex conjugate root pair on the jù axis (i.e., at $s = \pm j\omega$)	$A \sin(\omega t + \beta)$
vi.	Complex conjugate root pair of multiplicity k on the $j\omega$ axis	$A_1 \sin(\omega t + \beta_1) + A_2 t \sin(\omega t + \beta_2) + ... + A_k t^{k-1} \sin(\omega t + \beta_k)$
vii.	Single root at origin (i.e., at $s = 0$)	A
viii.	Roots of multiplicity k at origin	$A_1 + A_2 t + ... + A_k t^{k-1}$

3. If the characteristic equation has repeated roots on the $j\omega$ axis is unbounded and the system is unstable.
4. If one or more nonrepeated roots of the characteristic equation are on the $j\omega$-axis, then $g(t)$ is bounded and the system is unstable.

System stability with different pole locations:

(i) If all the roots of the characteristic equation have negative real parts, the system is stable.
(ii) If any root of the characteristic equation, has a positive real part or if there is a repeated root on the $j\omega$-axis, the system is unstable.
(iii) If the condition (i) is satisfied except for the presence of one or more nonrepeated roots on the $j\omega$-axis, the system is limitedly stable.

Further subdivision of the concept of stability, a linear system is characterized as:

(i) Absolutely stable with respect to a parameter of the system if it is stable for all values of this parameter.
(ii) Conditionally stable with respect to a parameter, if the system is stable for only certain bounded ranges of values of this parameter.

Problems

1. Find the damping ratio of a system given by

$$\frac{C_{(s)}}{R_{(s)}} = \frac{20}{s^2 + 50.58 + 20}$$

Undamped Natural Frequency; $\omega_n^2 = 20$

$$\omega_n = \sqrt{20} = 4.472 \text{ rad/sec.}$$

Damping ratio

$$2\zeta\omega_n = 50.5$$

$$\zeta = \frac{50.5}{2\omega_n} = \frac{50.0}{2 \times 4.472} = 5.46$$

2. Calculate Undamped Frequency of oscillations, damped frequency of oscillations, damping ratio. Maximum overshoot. Settling time when o/p is is settled within 2% of the final value of the system given below

$$G_{(s)} = \frac{120}{s^2 + 6s + 25}$$

(i) Undamped Natural Frequency

$\omega_n^2 = 25$

$\omega_n = \sqrt{25} = 5$ rad/sec.

(ii) Damping ratio

$2\zeta\omega_n = 6$

$\sigma = 0.6$

(iii) Damped Frequency

$\omega_d = \omega_n\sqrt{1-\zeta^2}$

$= 5\sqrt{1-0.6^2} = 5 \times 0.8 = 4.$

Damping Factor

$\zeta\omega_n = 0.6 \times 4 = 2.4$

(iv) Maximum Overshoot

$M_p = e^{-\pi\zeta/\sqrt{1-\zeta^2}} = e^{-0.6\pi/\sqrt{1-0.6^2}}$

$= e^{-1.885/0.8} = e^{-2.356}$

$= 0.0947$

(v) Settling Time

$T_s = \dfrac{4}{\zeta\omega_n} = \dfrac{4}{5 \times 0.6} = \dfrac{4}{3}$ sec.

$T_s = 1.33$ sec.

3. Calculate the time domain specification of the System given below.

$G_{(s)} = \dfrac{320}{s^2 + 5s + 7.68}$

(i) Undamped Natural Frequency

$\omega_n^2 = 7.68$

$\omega_n = 2.77$ rad/sec.

(ii) Damping ratio

$2\zeta\omega_n = 5$

$\zeta = \dfrac{5}{2 \times 2.77} = 0.902$

(iii) Damped Frequency

$\omega_d = \omega_n\sqrt{1-\zeta^2}$

$= 2.77\sqrt{1-0.902^2}$

$\omega_d = 1.195$ rad/sec.

(iv) Damping Factor

$\zeta\omega_n = 0.902 \times 2.77$

$= 2.498$

(v) Maximum Overshoot

$M_p = e^{-\zeta\pi/\sqrt{1-\zeta^2}}$

$= e^{-0.902\pi/\sqrt{1-0.902^2}}$

$= 0.0014$

(vi) Time reqired to reach peak o/p.

$T_p = \dfrac{\pi}{\omega_d} = \dfrac{\pi}{1.195} = 2.63$ sec.

CONCLUSION

This chapter provided the need for control system and mathematical modeling for the same in electrical engineering. The open loop and closed loop systems and the respective responses were introduced and discussed. The response plots and the way to study them and understand were presented. The importances of static and dynamic characteristics and the damping and oscillation in such characterizations were also discussed.

QUESTIONS

1. What is a control system?
2. What are the basic elements used for modelling mechanical rotational system?
3. Name the two types of electrical analogs for mechanical system.
4. Write the analogous electrical elements in force voltage analogy for the elements of mechanical translational system.
5. Differentiate between open loop and closed loop control systems with real-time examples.
6. Derive expression for transfer function of an electrical and thermal system.
7. Draw the block diagram of control system components.
8. Describe the characteristics of a potentiometer.
9. Write in brief about incremental and absolute encoders.
10. What is servomechanism?
11. Write short notes on the types of servomotors, construction and their working principles.
12. Define stepper motor and describe the static and its dynamic characteristics.
13. Define damping ratio.
14. How a system is classified depending on the value of damping?
15. Define delay time, rise time, peak time, peak overshoot and settling time.
16. What will be the nature of response of a second order system with different types of damping?
17. Sketch the response of a second order underdamped system.
18. Derive the expressions and draw the response of the first order system for unit step input.
19. Write in detail about ramp input and parabolic input.
20. Distinguish between type and order of a system.

INTRODUCTION

This chapter makes an attempt to introduce a few topics of power electronics. The history of power electronics is initially given followed by a discussion on one of the major topics of power electronics called the converters. The concept of inverters is given. Definition on modulation techniques is discussed.

CHAPTER OBJECTIVE

At the completion of this chapter, the reader will be able to:
* know the evolution of power electronics; and
* describe the basics of converters.

KEYWORDS

History of power electronics, Choppers, Inverters and Modulation (Quantitative).

6

Introduction to Power Electronics

6.1 POWER ELECTRONICS: A SMALL IDEA

Power electronics is the art of converting electrical energy from one form to another in an efficient, clean, compact, and robust manner for convenient utilization. A passenger lift in a modern building equipped with a variable-voltage-variable-speed induction machine drive offers a comfortable ride and stops exactly at the floor level. Behind the scene it consumes less power with reduced stresses on the motor and corruption of the utility mains.

Power electronics involves the study of the following:

- Power semiconductor devices, their physics, characteristics, drive requirements and their protection for optimum utilization of their capacities.
- Power converter topologies involving them.
- Control strategies of the converters.
- Digital, analog and microelectronics involved.
- Capacitive and magnetic energy storage elements.
- Rotating and static electrical devices.
- Quality of waveforms generated.
- Electromagnetic and radio frequency interference.
- Thermal management.

6.2 POWER SEMICONDUCTOR DEVICE: HISTORY

Power electronics and converters had a head start when the first device using silicon controlled rectifier (SCR) was proposed by Bell Labs and commercially produced by General Electric in the early fifties. The mercury arc rectifiers were well in use by that time and the robust and compact SCR first started replacing it in the rectifiers and cycloconverters. The necessity arose of extending the application of the SCR beyond the line commutated mode of action, which called for external measures to circumvent its turn-off incapability via its control terminals. Various turn-off schemes were proposed and their classification was suggested but it became increasingly obvious that a device with turn-off capability was desirable, which would permit it a wider application. The turn-off networks and aids were impractical at higher powers.

The bipolar transistor, which had by the sixties been developed to handle a few tens of amperes and block a few hundred volts, arrived as the first competitor to the SCR. It is superior to the SCR in its turn-off capability, which could be exercised via its control terminals. This permitted the replacement of the SCR in all forced commutated inverters and choppers. However, the gain (power) of the SCR is a few decades superior to that of the bipolar transistor and the high base currents required to switch the bipolar spawned the Darlington. Three or more stage Darlingtons are available as a single chip complete with accessories for its convenient drive. Higher operating frequencies were obtainable with a discrete bipolars compared to the 'fast' inverter-grade SCRs permitting reduction of filter components. But the Darlington's operating frequency had to be reduced to permit a sequential turn-off of the drivers and the main transistor. Further, the incapability of the bipolar to block reverse voltages restricted its use.

The power MOSFET burst into the scene commercially in the late seventies. This device also represents the first successful marriage between modern integrated circuit and discrete power semiconductor manufacturing technologies. Its voltage drive capability – giving it again a higher gain, the ease of its paralleling and most importantly the much higher operating frequencies reaching up to a few MHz saw it replacing the bipolar also at the sub-10 kW range mainly for SMPS type of applications. Extension of VLSI manufacturing facilities for the MOSFET reduced its price vis-à-vis the bipolar also. However, being a majority carrier device its on-state voltage is dictated by the RDS(ON) of the device, which in turn, is proportional to about V 2.3 DSS rating of the MOSFET. Consequently, high-voltage MOSFETs are not commercially viable.

Improvements were being tried out on the SCR regarding its turn-off capability mostly by reducing the turn-on gain. Different versions of the gate-turn-off device, the gate turn-off thyristor (GTO), were proposed by various manufacturers, each advocating his own symbol for the device. The requirement for an extremely high turn-off control current via the gate and the comparatively higher cost of the device restricted its application only to inverters rated above a few hundred kVA.

The lookout for a more efficient, cheap, fast and robust turn-off-able device proceeded in different directions with MOS drives for both the basic thyristor and the bipolar. The insulated gate bipolar transistor (IGBT) – basically a MOSFET driven bipolar from its terminal characteristics has been a successful proposition with devices being made available at about 4 kV and 4 kA. Its switching frequency of about 25 kHz and ease of connection and drive saw it totally removing the bipolar from practically all applications. Industrially, only the MOSFET has been able to continue in the sub-10 kVA range primarily because of its high switching frequency. The IGBT has also pushed up the GTO to applications above 2-5 MVA.

Subsequent developments in converter topologies, especially the three-level inverter, permitted the use of IGBT in converters of 5 MVA range. However, at ratings above that the GTO (6 kV/6 kA device of Mitsubishi) based converters had some space. Only SCR based converters are possible at the highest range where line-commutated or load-commutated converters were the only solution. The surge current, the peak repetition voltage are applicable only to the thyristors making them more robust, especially thermally, than the transistors of all varieties.

Presently, there are a few hybrid devices and intelligent power modules (IPMs) The IPMs have already gathered wide acceptance.

The 4500 V, 1200 A IEGT (injection-enhanced gate transistor) of Toshiba or the 6000 V, 3500 A IGCT (integrated gate commutated thyristor) of ABB are promising at the higher power ranges. However, these new devices must prove themselves before they are accepted by the industry at large.

Silicon carbide is a wide bandgap semiconductor with an energy bandgap wider than about 2 eV that possesses extremely high thermal, chemical, and mechanical stability. Silicon carbide is the only wide bandgap semiconductor among gallium nitride (GaN, E_G = 3.4 eV), aluminum nitride (AlN, E_G = 6.2 eV), and silicon carbide that possesses a high-quality native oxide suitable for use as an MOS insulator in electronic devices. The breakdown field in SiC is about 8 times higher than in silicon. This is important for high-voltage power switching transistors. For example, a device of a given size in SiC will have a blocking voltage 8 times higher than the same device in silicon. More importantly, the on-resistance of the SiC device will be about twice lower than the silicon device. Consequently, the efficiency of the power converter is higher. In addition, SiC-based semiconductor switches can operate at high temperatures (~600°C) without much change in their electrical properties. Thus, the converter has a higher reliability. Reduced losses and allowable higher operating temperatures result in smaller heat sink size. Moreover, the high frequency operating capability of SiC converters lowers the filtering requirement and the filter size. As a result, they are compact, light, reliable, and efficient and have a high power density. These qualities satisfy the requirements of power converters for most applications and they are expected to be the devices of the future.

Ratings have been progressively increasing for all devices while the newer devices offer substantially better performance. With the SCR and the pin-diodes, so called because of the sandwiched intrinsic 'i' layer between the 'p' and 'n' layers, having mostly line-commutated converter applications. Emphasis was mostly on their static characteristics: forward and reverse voltage blocking, current carrying and over-current ratings, on-state forward voltage, etc., and also on issues like paralleling and series operation of the devices. As the operating speeds of the devices increased, the dynamic (switching) characteristics of the devices assumed greater importance as most of the dissipation was during these transients. Attention turned to the development of efficient drive networks and protection techniques which were found to enhance the performance of the devices and their peak power handling capacities. Issues related to paralleling were resolved by the system designer within the device itself like in MOSFETs, while the converter topology was required to take care of their series operation as in multi-level converters.

The range of power devices thus developed over the last few decades can be represented as a tree, Fig. 6.1, on the basis of their controllability and other dominant features.

6.3 POWER CONVERTER TOPOLOGIES

A power electronic converter processes the available form to another having a different frequency and/or voltage magnitude. There can be four basic types of converters depending on the function performed:

Introduction to Power Electronics 361

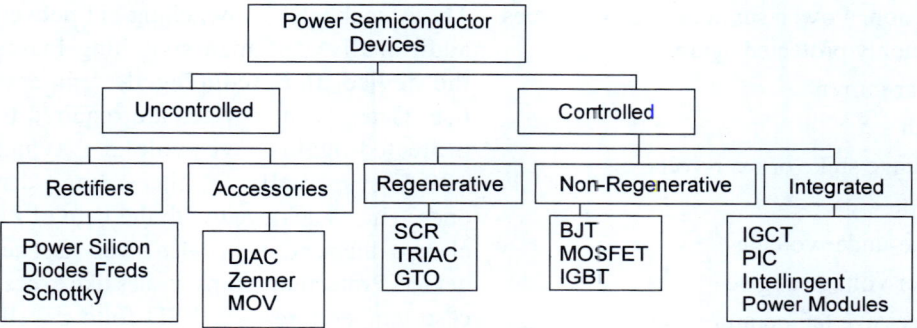

Fig. 6.1. Types of power semiconductor devices

Conversion	Name	Function	Symbol
DC to DC	Chopper	Constant to variable DC or variable to constant DC	
DC to AC	Inverter	DC to AC of desired voltage and frequency	
AC to DC	Rectifier	AC to unipolar (DC) current.	
AC to AC	Cycloconverter, AC-PAC, Matrix converter	AC of desired frequency and/or magnitude from generally line AC	

6.4 PROTECTION OF POWER DEVICES AND CONVERTERS

Power electronic converters often operate from the utility mains and are exposed to the disturbances associated with it. Even otherwise, the transients associated with switching circuits and faults that occur at the load point stress converters and devices. Consequently, several protection schemes must be incorporated in a converter. It is necessary to protect both the main terminals and the control terminals. Some of these techniques are common for all devices and converters. However, differences in essential features of devices call for special protection schemes particular for those devices. The IGBT must be protected against latching, and similarly, the GTO's turn-off drive is to be disabled if the anode current exceeds the maximum permissible turn-off-able current specification. Power semiconductor devices are commonly protected against:

1. Over-current
2. di/dt
3. Voltage spike or over-voltage
4. dv/dt
5. Gate-under voltage
6. Over voltage at gate
7. Excessive temperature rise
8. Electro-static discharge.

Semiconductor devices of all types exhibit similar responses to most of the stresses, however there are marked differences. The SCR is the most robust device on practically all counts. Its internal thermal capacities are excellent. A HRC fuse, suitably selected, and in coordination with fast circuit breakers would mostly protect it. This sometimes becomes a curse when the cost of the fuse becomes exorbitant. All transistors, specially the BJT and the IGBT are actively protected (without any operating cost!) by sensing the main terminal voltage. This voltage is related to the current carried by the device. Further, the transistors permit designed gate current waveforms to minimize voltage spikes as a consequence of sharply rising main terminal currents. Gate resistances have significant effect on turn-on and turn-off times of these devices, permitting optimization of switching times for the reduction of switching losses and voltage spikes.

Protection schemes for over-voltages, the prolonged ones and those of short duration, are guided by the energy content of the surges. Metal oxide varistors (MOVs), capacitive dynamic voltage-clamps and crow-bar circuits are some of the strategies commonly used. For high dv/dt stresses, which again have similar effect on all devices, R-C or R-C-D clamps are used depending on the speed of the device.

These 'snubbers' or 'switching-aid-networks', additionally minimize switching losses of the device, thus reducing its temperature rise. Gates of all devices are required to be protected against over-voltages (typically +20 V) especially for the voltage driven ones. This is achieved with the help of Zener clamps, the zener being also a very fast acting device. Protection against issues like excessive case temperatures and ESD follow well set practices. Forced cooling techniques are very important for the higher rated converters and whole environments are air-cooled to lower the ambient.

6.5 TYPES OF BASIC DC-DC CONVERTERS

Three basic types of DC-DC converter circuits

1. Buck converter
2. Boost converter
3. Buck-Boost converter

In all of these circuits, a power device is used as a switch. This device earlier used was a thyristor, which is turned on by a pulse fed at its gate. In all these circuits, the thyristor is connected in series with load to a DC supply, or a positive (forward) voltage is applied between anode and cathode terminals. The thyristor turns off, when the current decreases below the holding current, or a reverse (negative) voltage is applied between anode and cathode terminals. So, a thyristor is to be force-commutated, for which additional circuit is to be used, where another thyristor is often used. Later, GTOs came into the market, which can also be turned off by a negative current fed at its gate, unlike thyristors, requiring proper control circuit. The turn-on and turn-off times of GTOs are lower than those of thyristors. So, the frequency used in GTO based choppers can be increased, thus reducing the size of filters. Earlier, DC-

DC converters were called choppers, where thyristors or GTOs are used. It may be noted here that buck converter (DC-DC) is called 'step-down chopper', whereas boost converter (DC-DC) is a 'step-up chopper'. In the case of chopper, no buck-boost type was used.

With the advent of bipolar junction transistor (BJT), which is known as self-commutated device, it is used as a switch, instead of thyristor, in DC-DC converters. This device (NPN transistor) is switched on by a positive current through the base and emitter, and then switched off by withdrawing the above signal. The collector is connected to a positive voltage. Nowadays, MOSFETs are used as a switching device in low voltage and high current applications. It may be noted that, as the turn-on and turn-off time of MOSFETs are lower as compared to other switching devices, the frequency used for the DC-DC converters using it (MOSFET) is high, thus, reducing the size of filters as stated earlier. These converters are now used for applications, one of the most important being switched mode power supply (SMPS). Similarly, when application requires high voltage, insulated gate bipolar transistors (IGBT) are preferred over BJTs, as the turn-on and turn-off times of IGBTs are lower than those of power transistors (BJT), thus the frequency can be increased in the converters using them. So, mostly self-commutated devices of transistor family as described are being increasingly used in DC-DC converters.

6.5-A BUCK CONVERTERS (DC-DC)

A buck converter (DC-DC) is shown in Fig. 6.1(a). Only a switch is shown, for which a device as described earlier belonging to transistor family is used. Also a diode (known as freewheeling) is used to allow the load current to flow through it, when the switch (i.e., a device) is turned off. The load is inductive (R-L) one. In some cases, a battery (or back emf) is connected in series with the load (inductive). Due to the load inductance, the load current must be allowed a path, which is provided by the diode; otherwise, i.e., in the absence of the above diode, the high induced emf of the inductance, as the load current tends to decrease, may cause damage to the switching device. If the switching device used is a thyristor, this circuit is called a step-down chopper, as the output voltage is normally lower than the input voltage. The output voltage and current are shown in Fig. 6.1(b).

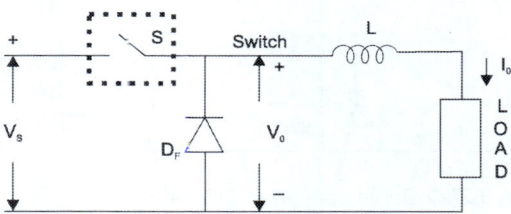

Fig. 6.1(a) Buck converter (DC-DC)

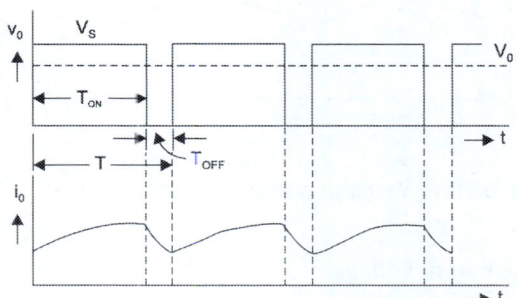

Fig. 6.1(b) Output voltage and current waveforms

The average output voltage,

$$V_0 = \frac{1}{T}\int_0^T v_c dt = \frac{1}{T}\int_0^{T_{ON}} V_s dt = V_s \left(\frac{T_{ON}}{T}\right) = kV_s$$

The duty ratio,

$$k = (T_{ON}/T) = [T_{ON}/(T_{ON} + T_{OFF})]$$

6.5-B. BOOST CONVERTERS (DC-DC)

A boost converter (DC-DC) is shown in Fig. 6.2(a). Only a switch is shown, for which a device belonging to transistor family is generally used. Also, a diode is used in series with the load. The load is of the same type as given earlier. The inductance of the load is small. An inductance (L) is assumed in series with the input supply. Waveforms of source current (i_s) are shown in Fig. 6.2(b).

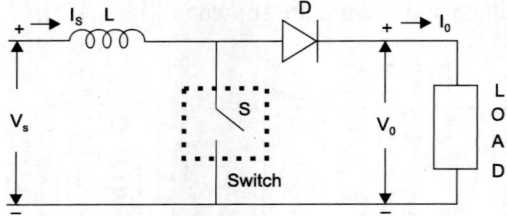

Fig. 6.2(a) Boost converter (DC-DC)

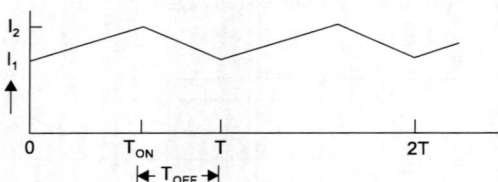

Fig. 6.2(b) Waveforms of source current (i_s)

The output voltage,

$$V_0 = V_S \left(\frac{T}{T_{OFF}} \right)$$

$$= V_S \left(\frac{T}{T - T_{ON}} \right)$$

$$= V_S \left(\frac{1}{1 - (T_{ON}/T)} \right)$$

$$= V_S \left(\frac{1}{1-k} \right)$$

The time period, $T = T_{ON} + T_{OFF}$

The duty ratio,

$$k = (T_{ON}/T) = [T_{ON}/(T_{ON} + T_{OFF})]$$

6.5-C. BUCK-BOOST CONVERTERS (DC-DC)

A buck-boost converter (DC-DC) is shown in Fig. 6.3. Only a switch is shown, for which a device belonging to transistor family is generally used. Also, a diode is used in series with the load. The connection of the diode may be noted, as compared with its connection in a boost converter (Fig. 6.2a). The inductor (L) is connected in parallel after the switch and before the diode. The load is of the same type as given earlier. A capacitor (C) is connected in parallel with the load. The polarity of the output voltage is opposite to that of input voltage here. Inductor current (i_L) waveform are shown in Fig. 6.3(b).

The output voltage is

$$V_0 = V_S \left(\frac{T_{ON}}{T_{OFF}} \right)$$

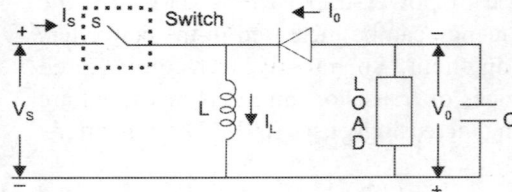

Fig. 6.3(a) Buck-boost converter (DC-DC)

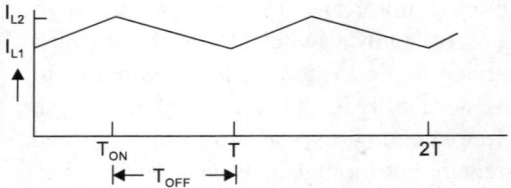

Fig. 6.3(b) Inductor current (i_L) waveform

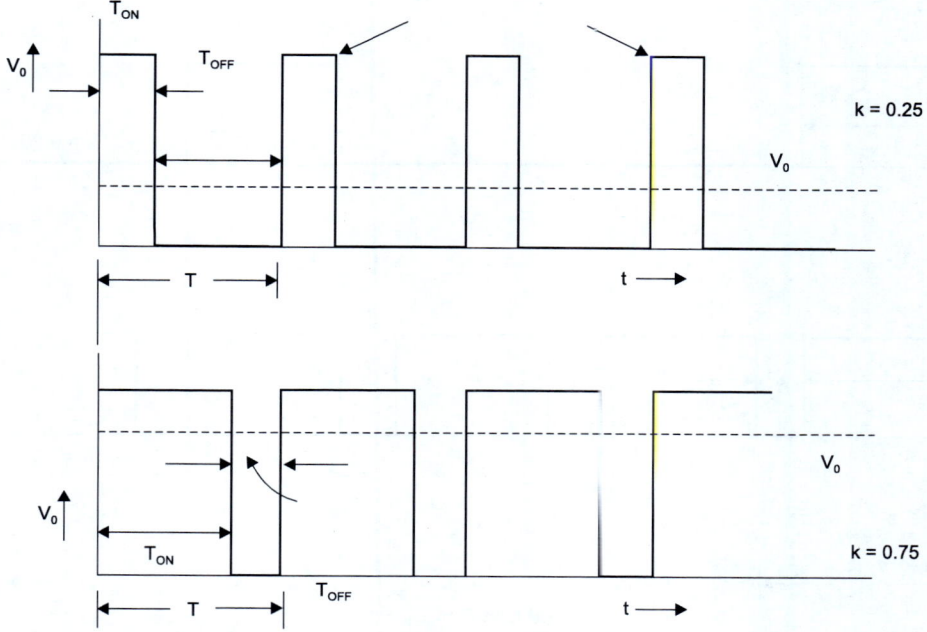

Fig. 6.4 Pulse-width modulation control (constant frequency)

$$= V_s \left(\frac{T_{ON}}{T - T_{ON}} \right)$$

$$= V_s \left(\frac{(T_{ON}/T)}{1 - (T_{ON}/T)} \right)$$

$$= V_s \left(\frac{k}{1-k} \right)$$

The time period, $T = T_{ON} + T_{OFF}$

The duty ratio, $k = (T_{ON}/T) = [T_{ON}/(T_{ON} + T_{OFF})]$

Control Strategies

In all cases, it is shown that the average value of the output voltage can be varied. The two types of control strategies (schemes) are employed in all cases. These are:

(a) time-ratio control, and
(b) current limit control.

Time-ratio control

In the time ratio control the value of the duty ratio, $k = T_{ON}/T$ is varied. There are two ways, which are constant frequency operation, and variable frequency operation.

Constant frequency operation

In this control strategy, the ON time, T_{ON} is varied, keeping the frequency ($f = 1/T$), or time period T is constant. This is also called pulse width modulation control (PWM). Two cases with duty ratios, k as (a) 0.25 (25%), and (b) 0.75 (75%) are shown in Fig. 6.4. Hence, the output voltage can be varied by varying ON time, T_{ON}.

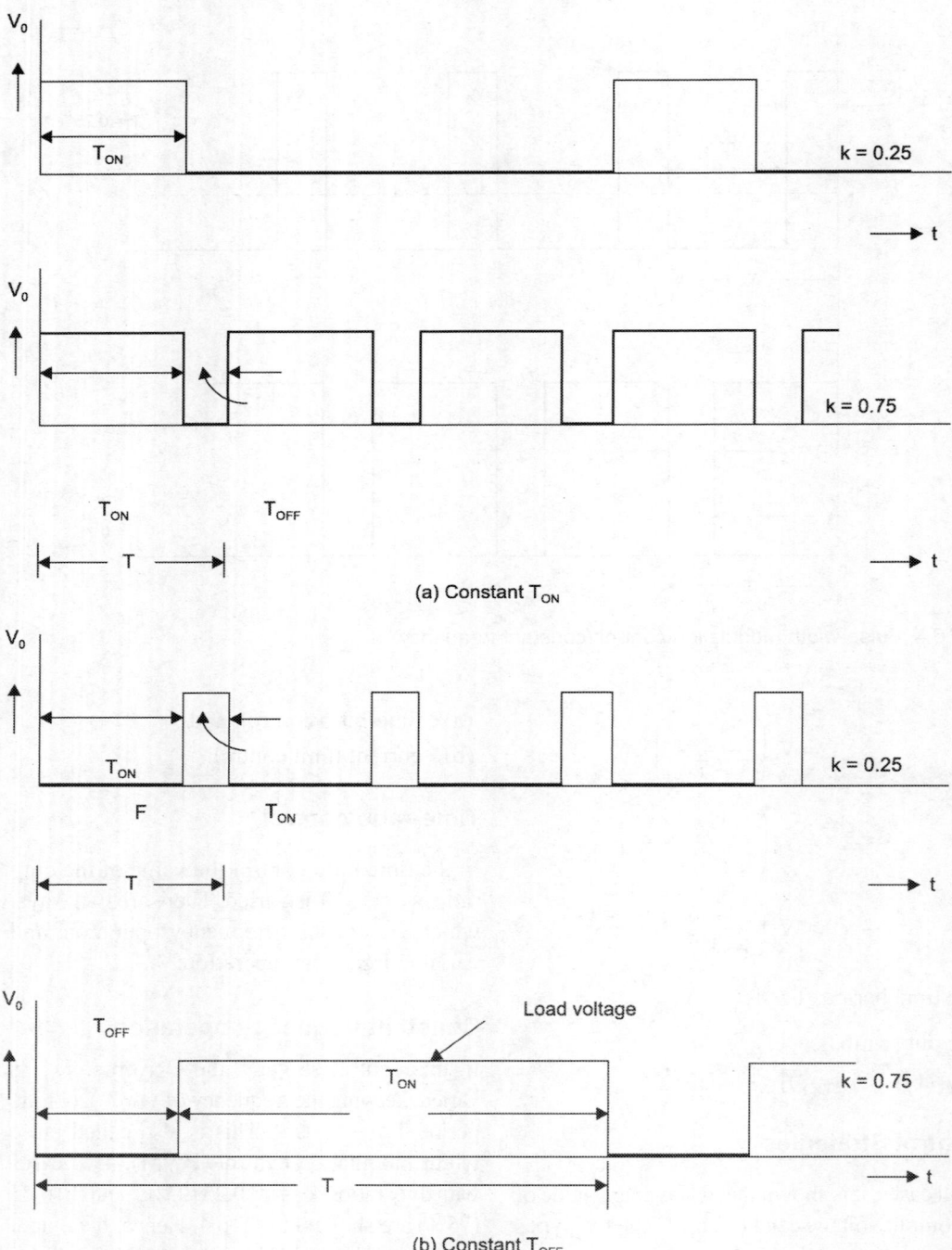

Fig. 6.5 Output voltage waveforms for variable frequency system

Variable frequency operation

In this control, the frequency ($f = 1/T$), or the time period (T) is varied keeping either ON time T_{ON} or OFF time T_{OFF} a constant.

Two cases with (a) the ON time, T_{ON} constant, and (b) the OFF time, T_{OFF} constant, with variable frequency or time period (T), are shown in Fig. 6.5. The output voltage can be varied in both cases, with the change in duty ratio, $k = T_{ON}/T$.

There are major disadvantages in this control strategy. These are:

(a) The frequency has to be varied over a wide range for the control of output voltage in frequency modulation. Filter design for such a wide frequency variation is, therefore, quite difficult.

(b) For the control of duty ratio, frequency variation would be wide. As such, there is a possibility of interference with systems using certain frequencies, such as signalling and telephone line in frequency modulation technique.

(c) The large OFF time in frequency modulation technique, may make the load current discontinuous, which is undesirable.

Thus, the constant frequency system using PWM is the preferred scheme for DC-DC converters (choppers).

Current limit control

As can be observed from the current waveforms for the types of DC-DC converters described earlier, the current changes between the maximum and minimum values, if it (current) is continuous. In the current limit control strategy, the switch in DC-DC converter (chopper) is turned ON and OFF, so that the current is maintained between two (upper and lower) limits. When the current exceeds upper (maximum) limit, the switch is turned OFF. During OFF period, the current freewheels in say, buck converter (DC-DC) through the diode, DF, and decreases exponentially.

When it reaches lower (minimum) limit, the switch is turned ON. This type of control is possible, either with constant frequency, or constant ON time, T_{ON}. This is used only, when the load has energy storage elements, i.e., inductance, L. The reference values are load current or load voltage. This is shown in Fig. 6.6. In this case, the current is continuous, varying between I_{max} and I_{min}, which decides the frequency used for switching. The ripple in the load current can be reduced, if the difference between the upper and lower limits is reduced, thereby making it minimum. This, in turn, increases the frequency, thereby increasing the switching losses.

In this lesson, first one in this module (#3), the three basic circuits— buck, boost and buck-boost, of DC-DC converters (choppers) are presented, along with the operation and the derivation of the expressions for the output voltage in each case, assuming continuous conduction. The different strategies employed for their control are discussed.

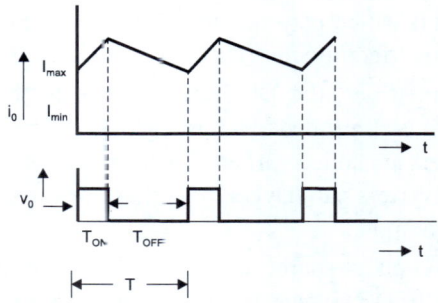

Fig. 6.6 Current limit control

6.6 INVERTER

An inverter is the power electronic circuit, which converts the DC voltage into AC voltage. The DC source is normally a battery or output of the controlled rectifier.

- The output voltage waveform of the inverter can be square wave, quasi-square wave or low distorted sine wave.
- The output voltage can be controlled with the help of drives of the switches.
- The pulse width modulation techniques are most commonly used to control the output voltage of inverters. Such inverters are called PWM inverters.
- The output voltage of the inverter contains harmonics whenever it is not sinusoidal.
- These harmonics can be reduced by using proper control schemes.

Types of inverters

Inverters can be broadly classified into two types. They are:

1. Voltage source inverter (VSI)
2. Current source inverter (CSI)

When the DC voltage remains constant, then it is called voltage source inverter (VSI) or voltage fed inverter (VFI).

When input current is maintained constant, then it is called current source inverter (CSI) or current fed inverter (CFI).

Sometimes, the DC input voltage to the inverter is controlled to adjust the output. Such inverters are called variable DC link inverters. The inverters can have single-phase or three-phase output.

- A voltage source inverter (VSI) is fed by a stiff DC voltage, whereas a current source inverter is fed by a stiff current source.
- A voltage source can be converted to a current source by connecting a series inductance and then varying the voltage to obtain the desired current.
- A VSI can also be operated in current-controlled mode, and similarly a CSI can also be operated in the voltage control mode.
- The inverters are used in variable frequency AC motor drives, uninterrupted power supplies, induction heating, static VAR compensators, etc.

The following table gives us the comparative study between VSI and CSI.

VSI	CSI
VSI is fed from a DC voltage source having small or negligible impedance.	CSI is fed with adjustable current from a DC voltage source of high impedance.
Input voltage is maintained constant.	The input current is constant but adjustable.
Output voltage does not depend on the load.	The amplitude of output current is independent of the load.
The waveform of the load current as well as its magnitude depend on the nature of load impedance.	The magnitude of output voltage and its waveform depends on the nature of the load impedance.
VSI requires feedback diodes.	The CSI does not require any feedback diodes.
The commutation circuit is complicated.	Commutation circuit is simple as it contains only capacitors.
Power BJT, Power MOSFET, IGBT, GTO with self commutation can be used in the circuit.	They cannot be used as these devices have to withstand reverse voltage.

6.7 MODULATION

Communication is the basic asset of mankind through which he exchange the information. It gives the idea of what is going around us. We communicate with many people in our daily life. We also need the entertainment media in our day-to-day life like TV, radio, browsing, newspaper etc., which also acts as a source of communication.

Electronic communication includes TV, radio, Internet, etc. Here we need to transmit the information bearing signal from one place to another. To do this we need to strengthen the signal. So that the signal travels for long distances. This is what is called modulation. Let us study more about the modulation in this section.

Communication is the basic process of exchanging information. The basic components of electronic communication system are:

1. Transmitter
2. Communication channel
3. Receiver.

- A transmitter is a collection of electronic circuits designed to convert the information in to a signal suitable for transmission over a given communication medium.
- A receiver is a collection of electronic circuits designed to convert the signal back to the original information.
- The communication channel is the medium by which the electronic signal is transmitted from one place to another.

Modulation is the process of superimposing the information contents of a modulating signal on a carrier signal (which is of high frequency) by varying the characteristics of carrier signal according to the modulating signal. The types of modulation is shown are Fig. 6.7.

Modulation is a process in which the base band signal modifies another high-frequency signal called, the carrier.

Types of modulation

We can modulate the information bearing signal into two types, namely, called modulation techniques.

1. Analog modulation
2. Digital modulation

- Analog modulation is the process of converting an analog input signal into a signal that is suitable for RF transmission.
- Digital modulation is the process of converting a digital bitstream into an analog signal suitable for RF transmission.

Modulation index

Modulation index indicates the depth of modulation. As the amplitude of the modulating signal increases, modulation index increases.

For amplitude modulation, the modulation index,

$$m = \frac{E_m}{E_c}$$

$$m = \frac{\text{Amplitude modulating signal}}{\text{Amplitude of the carrier}}$$

For frequency modulation,

$$m = \frac{\delta}{f_m}$$

$$m = \frac{\text{Maximum frequency deviation}}{\text{Modulating frequency}}$$

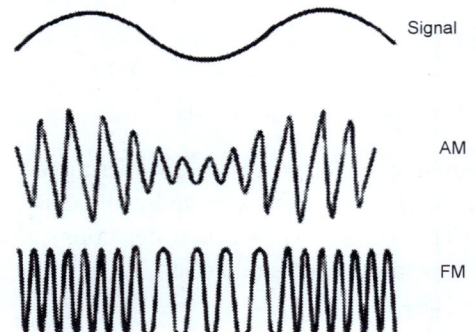

Fig. 6.7 Types of modulation

Analog modulation

The analog carrier signal is modulated by analog information signal so that information bearing analog signal can travel larger distance without the fear of loss due to absorption.

The analog modulation is of two types:

1. Amplitude modulation
2. Angle modulation

The angle modulation is further classified as frequency modulation and phase modulation.

Amplitude modulation

In this type of modulation the strength of the carrier signal is varied with the modulating signal.

Frequency moduation

In this type of modulation the frequency of the carrier signal is varied with the modulating signal.

Phase modulation

In this type of modulation the phase of the carrier signal is varied with the modulating signal. It is the variant of the frequency modulation.

Digital modulation

Digital modulation means an analog signal of carrier is converted by a digital data bit stream. There are two types of bits in binary:

1. Logic 0 (low)
2. Logic 1 (high).

This method is used to convert digital signal to analog and the responding demodulation is applied to convert analog signal to digital signal. Here the analog signal bearing information is transmitted by the digital method.

There are four types of digital modulation which are as below:

1. Pulse code modulation
2. Differential pulse code modulation
3. Delta modulation
4. Adaptive delta modulation.

Pulse modulation

The pulse wave modulation is a process of sampling of the continuous wave at periodic intervals and transmitting a very short pulse of radio frequency carrier for each sample, with the pulse characteristics being varied in some manner proportional to the signal amplitude at the sampling instant. The pulse modulation signals are shown in Fig. 6.8.

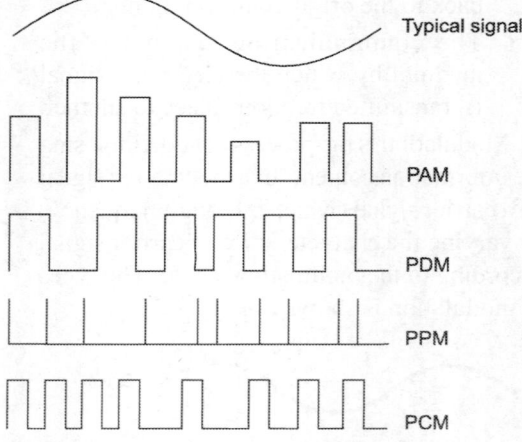

Fig. 6.8. Pulse modulation signals

Demodulation is the process of recovering the signal intelligence from a modulated carrier wave. This process, also called detection, is the reverse process of modulation.

CONCLUSION

Power electronics was introduced in 1970s and since then it has reached summit heights due to sustained research in this area. This chapter is an introduction to this area of electrical engineering to the beginners in a cursory way. The history of power electronics, various blocks of it, major circuits such as converters / inverters and choppers and modulation aspects were introduced in this chapter.

QUESTIONS

1. Can the chopper be termed a transformer? If so, how?
2. Where are the buck converters majorly used? Similarly, where are the boost converters used?
3. Can you specify the need of a converter and the need for an inverter?
4. How does an UPS and an inverter differ?
5. There are various radio stations available. Can you relate frequency modulation and tuned circuit to them and understand how different stations work at different frequencies?
6. Tabulate the major differences between an amplitude modulated signal and a frequency modulated signal.
7. What is pulse modulation?

INTRODUCTION

Fundamental elements of a practical electric power system, namely, the generation, transmission, distribution and utilization, are introduced in this chapter. Major protective elements such as switchgear, fuse and circuit breakers are also brought in the discussion. A preliminary treatment on batteries is given at the end of the chapter.

CHAPTER OBJECTIVE

At the completion of this chapter, the reader will be able to:
* describe the power system and its components; and
* describe the safety components of power system.

KEYWORDS

Stages of a power system, Role of transformers, Substations, Power distribution, Load types, Accessories and Concept of UPS.

7

Power System: An Overview

7.1 POWER SYSTEM

Electric power systems are real-time energy delivery systems. Real time means that power is generated, transported, and supplied the moment you turn on the light switch. Electric power systems are not storage systems like water systems and gas systems. Instead, generators produce the energy as the demand calls for it. Figures. 7.1 and 7.2 show the basic building blocks of an electric power system.

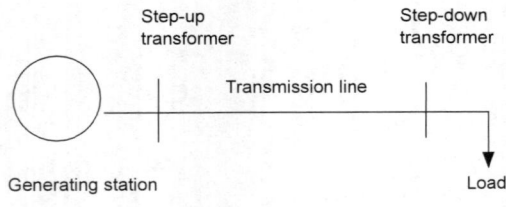

Fig. 7.1

The system starts with generation, by which electrical energy is produced in the power plant and then transformed to the power station to high-voltage electrical energy that is more suitable for efficient long distance transportation. The power plants transform other sources of energy in the process of producing electrical energy. For example, heat, mechanical, hydraulic, chemical, solar, wind, geothermal, nuclear, and other energy sources are used in the production of electrical energy. High-voltage (HV) power lines in the transmission portion of the electric power system efficiently transport electrical energy over long distances to the consumption locations. Finally, substations transform this HV electrical energy into lower-voltage energy that is transmitted over distribution power lines that are more suitable for the distribution of electrical energy to its destination, where it is again transformed for residential, commercial, and industrial consumption.

A full-scale actual interconnected electric power system is much more complex than that shown in Fig. 7.2; however the basic principles, concepts, theories, and terminologies are all the same.

7.2 BASIC STRUCTURE OF POWER SYSTEM

A power system is an interconnected network with components converting nonelectrical energy continuously into the electrical form

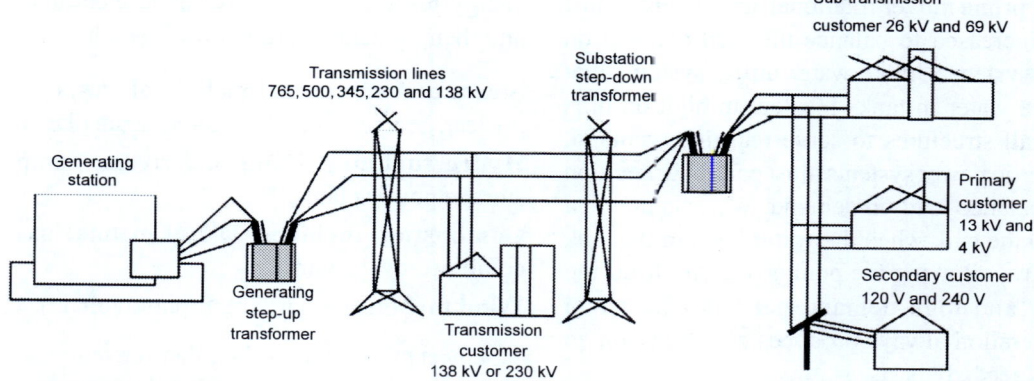

Fig. 7.2 Schematic representation of electric power generation, transmission, distribution and utilization.

and transporting the electrical energy from generating sources to the loads/users. A power system serves one important function, that is, to supply customers with electricity as economically and as reliably as possible. It can be divided into three sub-systems:

Generation: Generating and/or sources of electrical energy.

Transmission: Transporting electrical energy from its sources to load centres with high voltages (115 kV and above) to reduce losses.

Distribution: Distributing electrical energy from substations (44 kV ~ 12 kV) to end users/customers.

Utilisation: The 'electric loads' which consume electricity.

In the generating station, the generator converts nonelectrical energy to electrical energy. The devices connecting generators to transmission system and from transmission system to distribution system are transformers. Their main functions are stepping up the lower generation voltage to the higher transmission voltage and stepping down the higher transmission voltage to the lower distribution voltage. The main advantage of having higher voltage in transmission system is to reduce the losses in the grid. Since transformers operate at constant power, when the voltage is higher, then the current has a lower value. Therefore, the losses, a function of the current square, will be lower at a higher voltage.

Generation

There are basically two physical laws that describe how electric power systems work. (Gravity is an example of a physical law.) One law has to do with generating a voltage from a changing magnetic field and the other has to do with a current flowing through a wire creating a magnetic field. Both the physical laws are used throughout the electric power system from generation through transmission, distribution, and consumption.

Power plants produce electrical energy on a real-time basis. Electric power systems do not store energy such as most gas or water systems do. For example, when a toaster is switched on and drawing electrical energy from the system, the associated generating plants immediately see this as new load and slightly slow down. As more and more loads (i.e., toasters, lights, motors, etc.) are switched on, generation output and prime mover rotational shaft energy must be increased to balance the load demand on the system. Unlike water utility systems that store water in tanks located up high on hills or tall structures to serve real-time demand, electric power systems must control generation to balance load on demand. Water is pumped into the tank when the water level in the tank is low, allowing the pumps to turn off during low and high demand periods. Electrical generation always produces electricity on an "as needed" basis.

Note: Some generation units can be taken off-line during light load conditions, but there must always be enough generation online to maintain frequency during light and heavy load conditions. There are electrical energy storage systems such as batteries, but electricity found in interconnected AC power systems is in a real-time energy supply system, not an energy storage system.

Power generation plants produce the electrical energy that is ultimately delivered to consumers through transmission lines, substations, and distribution lines. Generation plants or power plants consist of three-phase generator(s), the prime mover, energy source, control room, and substation.

The mechanical means of turning the generator's rotor is called the prime mover. Turbines are the usual turbine members in electric power generating stations. The prime mover's energy sources include the conversion process of raw fuel, such as coal, to the end product, steam, that will turn the turbine. The bulk of electrical energy produced in modern interconnected power systems is normally produced through a conversion process from coal, oil, natural gas, nuclear, and hydro. To a lesser degree, electrical power is produced from wind, solar, geothermal, and biomass energy resources. The more common types of energy resources used to generate electricity and their associated prime movers include:

Steam turbines: Fossil fuels (coal, gas, oil), nuclear, geothermal, solar-heated steam plants.
Hydro turbines: Dams and rivers, pump storage power plants.
Combustion turbines: Diesel, natural gas combined cycle plants.
Wind turbines: Solar direct (photovoltaic)

Power production can be either conventional or renewable. Coal based thermal electric power generating stations, water based hydroelectric power stations, nuclear power station and a hybrid of these generating stations form the conventional power production. However, there is a great deal of recommendations for the renewable electric power production, due to the thinning raw materials such as coal and water.

Sun power based solar power production, wind based wind energy stations, biomass power stations and such related stations form the renewable group.

Transformers in power system

Transformers are essential components in electric power systems. They come in all shapes and sizes. Power transformers are used to convert high-voltage power to low-voltage power and vice versa. Power can flow in both directions: from the high-voltage side to the low-voltage side or from the low-voltage side to the high-voltage side.

Generation plants use large step-up transformers to raise the voltage of the generated power for efficient transport of power over long distances. Then step-down transformers convert the power to subtransmission, or distribution voltages, for further transport or consumption.

Distribution transformers are used on distribution lines to further convert distribution voltages down to voltages suitable for residential, commercial, and industrial consumption. There are many types of transformers used in electric power systems. Instrument transformers are used to connect high-power equipment to low-power electronic instruments for monitoring system voltages and currents at convenient levels.

Large power transformers consist of two or more windings for each phase and these windings are usually wound around an iron core. The iron core improves the efficiency of the transformer by concentrating the magnetic field and reduces transformer losses. The high-voltage and low-voltage windings have a unique number of coil turns. The turns ratio between the coils dictates the voltage and current relationships between the high- and low-voltage sides.

Instrument transformers include CTs and PTs (i.e., current transformers and potential transformers). These instrument transformers connect to metering equipment, protective relaying equipment, and telecommunications equipment. Regulating transformers are used to maintain proper distribution voltages so that consumers have stable wall outlet voltage. Phase shifting transformers are used to control power flow between tie lines. Transformers can be single phase, three phase, or banked together to operate as a single unit.

The term instrument transformer refers to current and voltage transformers that are used to scale down actual power system quantities for metering, protective relaying, and/or system monitoring equipment. The application of both current and potential transformers also provides scaled-down quantities for power and energy information.

CTs are used to scale down the high magnitude of current flowing in high-voltage conductors to a level much easier to work with safely. For example, it is much easier to work with 5 amperes of current in the CTs secondary circuit than it is to work with 1000 amperes of current in the CTs primary circuit. Using the CTs turn ratio as a scale factor provides the current level required for the monitoring instrument. Yet, the current located in the high-voltage conductors is actually being measured. Taps (or connection points to the coil) are used to allow options for various turns ratio scale factors to best match the operating current to the instrument's current requirements. Most CTs are located on transformer and circuit breaker bushings.

Potential transformers

Similarly, potential transformers (PTs) are used to scale down very high voltages to levels that are safer to work with. For example, it is much easier to work with 115 V AC than 69 kV AC. The 600:1 scale factor is taken into account in the calculations of actual voltage.

PTs are also used for metering, protective relaying, and system monitoring equipment. The instruments connected to the secondary side of the PT are programmed to account for the turns ratio scale factor. Like most transformers, taps are used to allow options for various turn ratios to best match the operating voltage with the instrument's voltage level requirements.

Transmission lines

After power is generated it must be delivered to the place you need it. This is **done by wires**, and in the most simple model there is a battery with wires attached directly to the device you are using.

The problem is that electric power losses voltage over distance. Early pioneers **Thomas Edison** and **Oskar von Miller** solved this by simply using higher voltages which could make the trip, but this soon proved impractical. Alternating current had much more potential to travel over distance because the voltage could be changed — high voltage for a long distance segment, and lower voltage for smaller local powerlines. DC power has reemerged as the best way to send power over distance in the form of HVDC (high voltage direct current) however AC power remains the most common method.

Now, why use high-voltage transmission lines?

The best answer to that question is that high-voltage transmission lines transport power over long distances much more efficiently than lower-voltage distribution lines for two main reasons.

First, high-voltage transmission lines take advantage of the power equation, that is, power is equal to the voltage times current. Therefore, increasing the voltage allows one to decrease the current for the same amount of power. Second, since transport losses are a function of the square of the current flowing in the conductors, increasing the voltage to lower the current drastically reduces transportation losses. Plus, reducing the current allows one to use smaller conductor sizes.

Bundled conductors are widely used in three-phase HV transmission line. The two-conductors-per-phase option is called bundling. Power companies bundle multiple conductors—double, triple, or more—to increase the power transport capability of a power line. The type of insulation used in this line is referred to as V-string insulation. V-string insulation, compared to I-string insulation, provides stability in wind conditions.

Bundling conductors significantly increases the power transfer capability of the line. The extra relatively small cost when building a transmission line to add bundled conductors is easily justified since bundling the conductors actually doubles, triples, quadruples, and so on the power transfer capability of the line. For example, assume that a right of way for a particular new transmission line has been secured. Designing transmission lines to have multiple conductors per phase significantly increases the power transport capability of that line for a minimal extra overall cost.

Two static wires on the very top of the tower is used to shield itself from lightning. The static wires in this case do not have insulators; instead, they are directly connected to the metal towers so that lightning strikes are immediately grounded to earth. Hopefully, this shielding will keep the main power conductors from experiencing a direct lightning strike.

Raising voltage to reduce current

Raising the voltage to reduce current reduces conductor size and increases insulation requirements. Let us look at the power equation again:

$$Power = Voltage \times Current$$

$$VoltageIn \times CurrentIn = VoltageOut \times CurrentOut$$

From the power equation above, raising the voltage means that the current can be reduced for the same amount of power. The purpose of step-up transformers at power plants, for example, is to increase the voltage to lower the current for power transport over long distances. Then at the receiving end of the transmission line, step-down transformers are used to reduce the voltage for easier distribution. For example, the amount of current needed to transport 100 MW of power at 230 kV is half the amount of current needed to transport 100 MW of power at 115 kV. In other words, doubling the voltage cuts the required current in half. The higher-voltage transmission lines require larger structures with longer insulator strings in order to have greater air gaps and needed insulation. However, it is usually much cheaper to build larger structures and wider right of ways for high-voltage transmission lines than it is to pay the continuous cost of high losses associated with lower-voltage power lines. Also, to transport a given amount of power from point "a" all to point "b", a higher-voltage line can require much less right of way land than multiple lower-voltage lines that are side by side.

Raising Voltage to Reduce Losses

The cost due to losses decreases dramatically when the current is lowered. The power losses in conductors are calculated by the formula I^2R. If the current (I) is doubled, the power losses quadruple for the same amount of conductor resistance (R)!

Again, it is much more cost effective to transport large quantities of electrical power over long distances using high-voltage transmission lines because the current is less and the losses are much less.

Figure 7.3 shows how the electricity is generated at and transmitted at various high voltage levels, till it reaches the domestic loads at 220 V AC.

7.3 SUBSTATIONS

Power is generated at relatively lower AC voltage which is then transmitted at various AC high voltages. As seen, it is economical to transmit power at high voltage level. Distribution of electrical power is done at lower voltage levels as specified by consumers. This is the tail-end of the generation-transmission-distribution path. For maintaining these voltage levels and for providing greater stability, a number of transformation and switching stations have to be created in between generating station and consumer ends. These transformation and switching stations are generally known as electrical substations.

Thus, in simple words, substations can be said as the major link between very HV transmission and LV distribution/utilization,

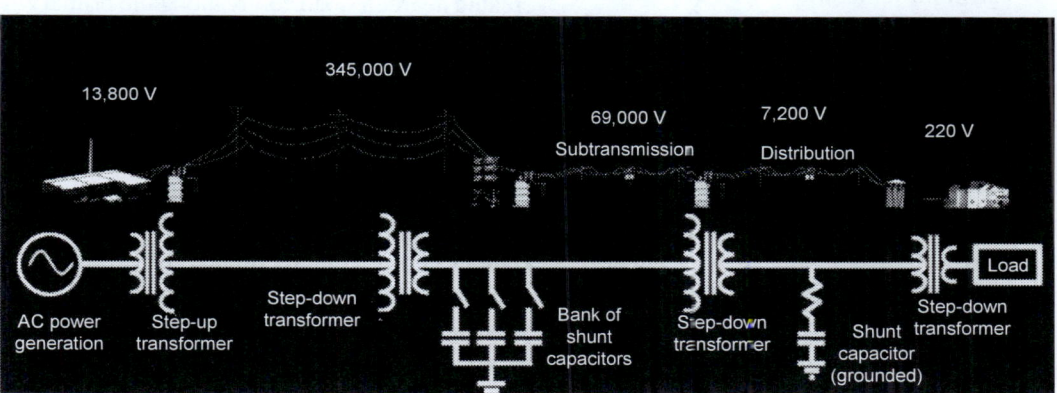

Fig. 7.3 Power system: generation to utilization through transmission and distribution

on the load side. On the generating side, the substations are where the voltages are increased to high values by using step-up transformers, and after the transmission, they are again stepped down for distribution.

In addition to changing the voltages the substations have, a variety of protective devices like circuit breakers and fuses are present to protect the distribution networks. These are designed in such a way that various distribution circuits can be isolated for repairs and load shedding.

Substations are normally outdoors and are enclosed by a wire fence.

While electric substations take part of the distribution of electricity, they have many other functions as follows:

1. Step-up and step-down of the voltage for transmission and distribution. As power is transmitted at a higher voltage over long distances, the current is lower. This results in lower transmission losses but doesn't provide the proper current for homes and businesses to use – thus the need of stepping up and stepping down the voltage.
2. **Switching and isolating the circuits for maintenance:** Switching is also an important function of substations. Closing down a feeder circuit when the load demands are high needs to be done for the safety of the generating plants. Switching high voltages is dangerous work, and special circuit breakers like air circuit breakers and oil circuit breakers for reducing arcs have to be used.
3. **Load shedding:** When the power demand is more than the supply, the substations do load shedding on distribution circuits to maintain balance across the electrical network.
4. **Correction of power factors circuits:** The power factor has to be kept at the correct value when reactive loads are there to protect the generating plant and increase efficiency.

Depending on the purposes, the substations may be classified as given below.

Step-up substation

Step-up substations are associated with generating stations. Generation of power is limited to low voltage levels due to limitations of the rotating alternators. These generating voltages must be stepped up for economical transmission of power over long distance. So there must be a step-up substation associated with generating station.

Step-down substation

The stepped up voltages must be stepped down at load centres, to different voltage levels for different purposes. Depending on these purposes the step-down substation are further categorized in different subcategories.

Primary step-down substation

The primary step-down substations are created nearer to load centre along the primary transmission lines. Here primary transmission voltages are stepped down to different suitable voltages for secondary transmission purpose.

Secondary step-down substation

Along the secondary transmission lines, at load centre, the secondary transmission voltages are further stepped down for primary distribution purpose. The stepping down of secondary transmission voltages to primary distribution levels are done at secondary step-down substation.

Distribution substation

Distribution substations are situated where the primary distribution voltages are stepped down to supply voltages for feeding the actual consumers through a distribution network.

Bulk supply or industrial substation

Bulk supply or industrial substations are generally a distribution substation but they are dedicated for one consumer only. An industrial consumer of large or medium supply group may be designated as bulk supply consumer. Individual step-down substation is dedicated to these consumers.

Fig. 7.4 View of a practical power system

Operation of a substation

Electricity is generated in a thermal power plant, hydroelectric power plant, and nuclear power plant, etc. This electricity is then supplied to a transmission substation near the generating plant. In the transmission substation the voltage is increased substantially using step-up transformers. The voltage is increased to reduce the transmission losses over long distances. This electricity then is supplied to a power substation where it is stepped down using step-down transformers and then supplied to a distribution grid. In the distribution grid there are additional transformers and voltage is further reduced for distributing further down the grid. From here the electricity is supplied to step-down transformers near residential quarters that step down the voltage to 110/220 volts as per each country's requirement.

7.4 ELECTRIC POWER DISTRIBUTION

Power distribution is a process which is used to move electricity from locations where it is generated to people who need it. The process of power distribution starts at the facility where electricity is generated. Distribution takes place through a system known as the electrical grid or simply "grid" which is designed to keep power constantly on call so that it can meet demand.

Managing power distribution is a balancing act, with the goal being to create a steady supply for consumers without overloading the system with too much power. Electricity must be used as it is generated, as most storage techniques are highly inefficient.

A *distribution substation* is located near or inside city/town/village/industrial area. It receives power from a transmission network. The high voltage from the transmission line is then stepped down by a step-down transformer to the primary distribution level voltage. Primary distribution voltage is usually 11 kV, but can range between 2.4 kV and 33 kV depending on the region or consumer.

A typical power distribution system consists of :
- Distribution substation
- Feeders
- Distribution transformers
- Distributor conductors
- Service mains conductors

Along with these, a distribution system also consists of switches, protection equipment, measurement equipment, etc.

Distribution feeders: The stepped-down voltage from the substation is carried to distribution transformers via feeder conductors. Generally, no tappings are taken from the feeders so that the current remains same throughout. The main consideration in designing of a feeder conductor is its current carrying capacity.

Distribution transformer: A distribution transformer, also called service transformer, provides final transformation in the electric power distribution system. It is basically a step-down 3-phase transformer. Distribution transformer steps down the voltage to 400Y/230 volts. Here it means, voltage between any one phase and the neutral is 230 volts and phase-to-phase voltage is 400 volts. However, in USA and some other countries, 120/240 volts split-phase system is used; where voltage between a phase and neutral is 120 volts.

Distributors: Output from a distribution transformer is carried by distributor conductor. Tappings are taken from a distributor conductor for power supply to the end consumers. The current through a distributor is not constant as tappings are taken at various places throughout its length. So, voltage drop along the length is the main consideration while designing a distributor conductor.

Service mains: It is a small cable which connects the distributor conductor at the nearest pole to the consumer's end.

Figure 7.5 shows a **simple radial AC power distribution system**. The figure does not show other equipment like circuit breakers, measuring instruments, etc. for simplicity purpose.

Electrical power distribution is the final stage of an electrical power system, which entails the delivery of electricity to the load. The primary role of this section is to carry the electricity from the transmission lines to the loads in the individual customers to the different strata of society. In the power distribution section of an electrical power system, there are two main subsections: primary distribution and secondary distribution.

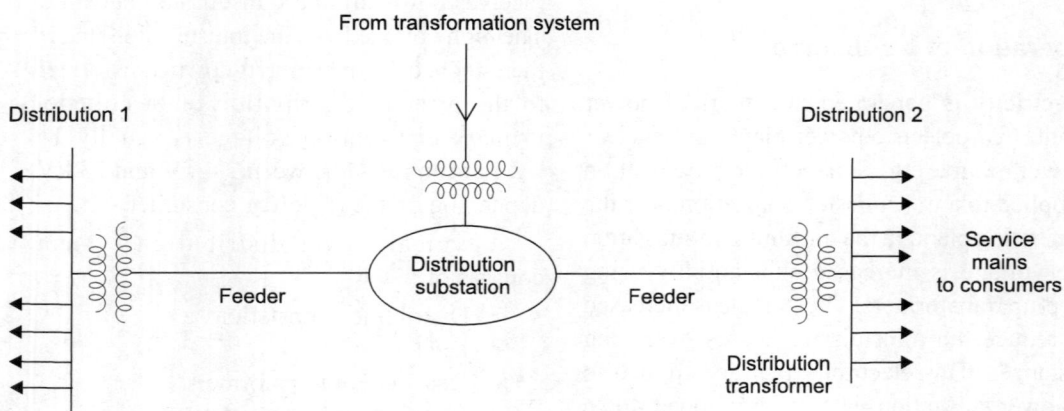

Fig. 7.5 AC power distribution system

Primary distribution

Primary distribution lines contain a distribution transformer present in the locality of the clientele. Primary distribution ranges from 4-kV to 35-kV. Only industries can directly feed the transmission line. Most average consumers are connected to a transformer that brings down the voltage to a useable level.

The distribution network for the primary distribution comes mainly of two types—**radial** or **network**. A radial network is primarily like a tree, where there is only one line of connection for the customer to the source of supply. A network system, on the other hand, has multiple or parallel connections to the source of supply. A radial connection is primarily used in rural areas, while the network connection is primarily used in load-sensitive areas, such as a dense urban area. However, as bad as radial systems sound, based on there only being a single connection to the source. Modern-day radial networks do contain backup options.

When we shift our focus onto to how these radial connections are designed in specific conjuncture to the electrification of rural areas, we notice that the distribution voltages are notably higher in rural areas. This is because the higher the voltages, the fewer the poles that need

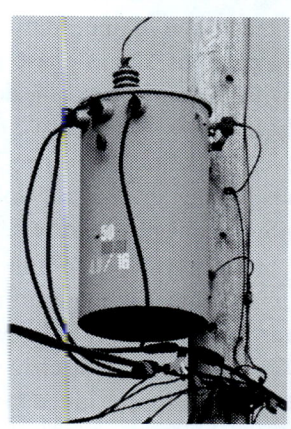

Fig. 7.6

to be erected for the distribution of the electricity. Most of these distribution connections in rural areas are made of galvanized steel. This strong form of steel permits it to be connected over long distances without the need for too many electric distribution poles.

Secondary distribution

The parameters that encompass the properties of electricity are not strictly limited to voltage and current. When it comes to electricity, there is a third important property of electricity—frequency. There are primarily two frequencies in which electricity is produced, either 50 or

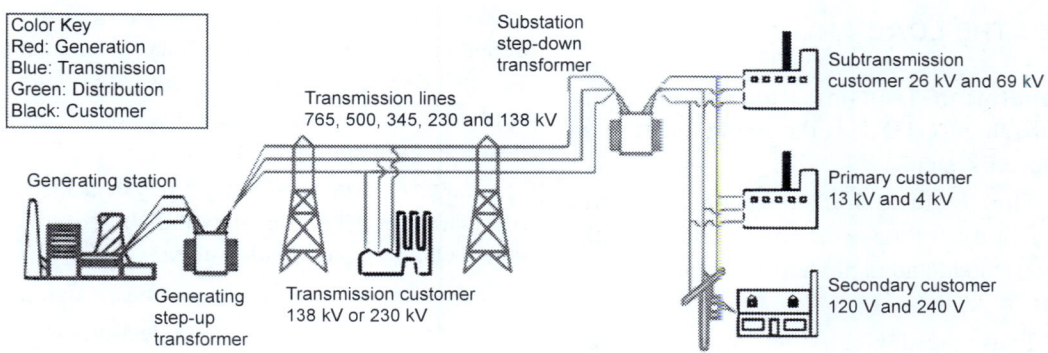

Fig. 7.7 Secondary distribution

60 Hz. This electricity is then delivered to domestic customers as single-phase electric power. Seen with an oscilloscope, the domestic power supply in North America would look like a sine wave, oscillating between −170 volts and +170 volts, giving an effective voltage of 120 volts RMS. In India, it would be 230 volts.

However, in some countries of Europe and India, three-phase power is more efficient in terms of power delivered per cable used and is more suited for running large electric motors. Some large European appliances may be powered by three-phase power, such as electric stoves and cloth dryers. A ground connection is normally provided for the customer's system, as well as for the equipment owned by the utility. The purpose of connecting the customer's system to the ground is to limit the voltage that may develop if high-voltage conductors fall onto lower-voltage conductors, which are usually mounted lower to the ground, or if a failure occurs within a distribution transformer; this process is also famously known as grounding. With this, we can conclude that the distribution of electricity is no small feat and requires a tremendous amount of calculation and engineering with the proper understanding of the geographical area being powered!

7.5 THE LOAD

Generation-Transmission-Distribution-Utilization (GTDU) is the single line flow of a power system.

The 'electric load', which consumes electricity for its workability is called the electric load and is the terminating utilization part.

Thus, the device which takes electrical energy is known as the electric load (Fig. 7.8).

In other words, the electrical load is a device that consumes electrical energy in the form of current and transforms it into other forms like heat, light, work, etc. The electrical load may be resistive, inductive, capacitive or some combination between them. The electric load is schematically indicated in the figure below.

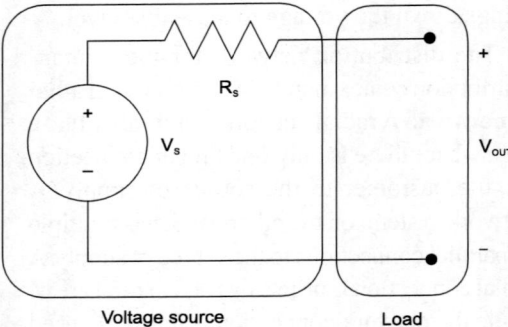

Fig. 7.8 Load representation

In technical terms, electrical load is synonymous with current. That is, loading a generator means *drawing current* from it. In case of DC, load is described by the level of current. The higher the current, the higher is the load. Load is the generic term for something in the circuit that draws power.

The characteristics of a "load" can vary widely. It is normally the largest power draw and most components in the circuit are there to support the load. Anything that uses electricity to do work will draw current. The amount depends on how much resistance to current flow the device has and the amount of voltage applied to it, assuming the source has more power to give than the load will use.

The term is derived from a physical load, like carrying a bundle of wood. It requires work, when you are loaded down.

The term load is used in a number of ways.
- To indicate a device or a collection of the equipment which use electrical energy.

Power System: An Overview 385

- For showing the power requirement from a given supply circuit.
- The electrical load indicates the current or power passing through the line or machine.

The classification of loads are shown in the figure below.

Types of electrical loads

The nature of the load depends on the load factor, demand factor, diversity factor, power factor, and a utilisation factor of the system. The different types of load are explained below in details.

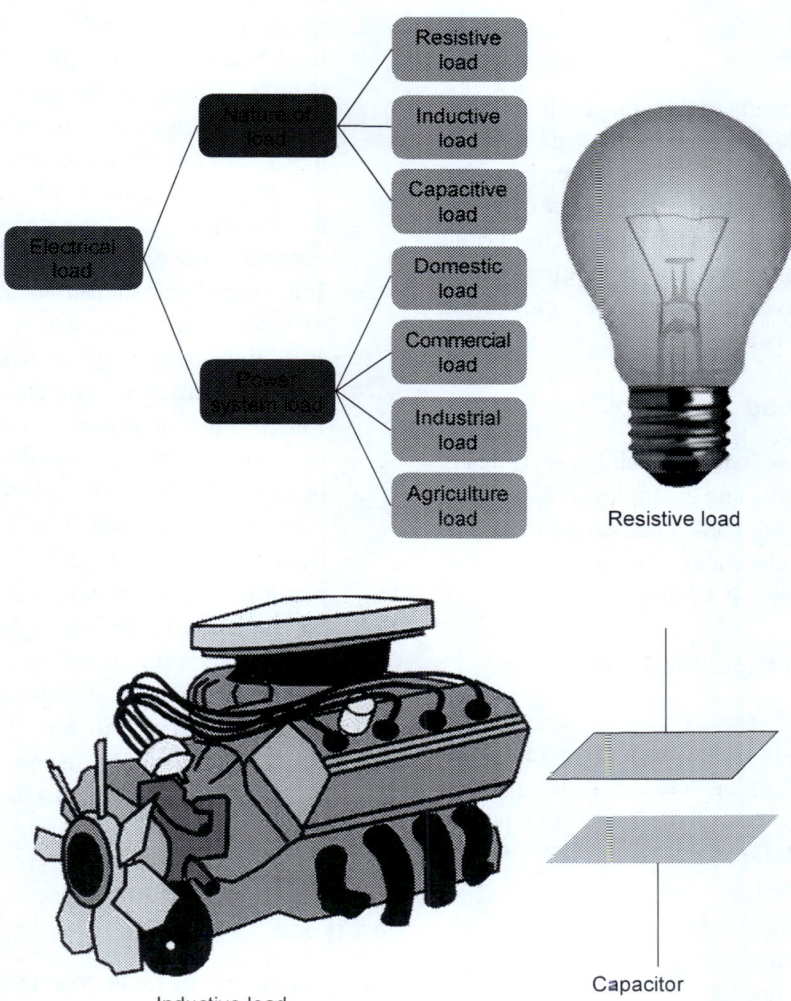

Fig. 7.9 Types of loads

Resistive load

The resistive load obstructs the flow of electrical energy in the circuit and converts it into thermal energy, due to which the energy dropout occurs in the circuit. The lamp and the heater are the examples of the resistive load. The resistive loads take power in such a way so that the current and the voltage wave remain in the same phase. Thus, the power factor of the resistive load remains in unity.

Inductive load

The inductive loads use the magnetic field for doing the work. The transformers, generators, motors are examples of load. The inductive load has a coil which stores magnetic energy when the current passes through it. The current wave of the inductive load is lagging behind the voltage wave, and the power factor of the inductive load is also lagging.

Capacitive load

In the capacitive load, the voltage wave leads the current wave. The examples of capacitive loads are capacitor bank, three-phase induction motor starting circuit, etc. The power factor of such type of loads is leading.

Types of electrical loads in power system

The total loads of an area depend on its population and living standard of the people. The different types of the loads in a power system are as follows.

1. Domestic load
2. Commercial load
3. Industrial load
4. Agriculture load

1. **Domestic load**: The domestic load is defined as the total energy consumed by the electrical appliances in the household work. It depends on the living standard, weather and type of residence. The domestic loads mainly consist of lights, fan, refrigerator, air conditioners, mixer, grinder, heater, ovens, small pumping, motor, etc. The domestic load consumes very little power and also independent of frequency. This load largely consists of lighting, cooling or heating.

2. **Commercial load:** Commercial load mainly consists of lightning of shops, offices, advertisements, etc., Fans, heating, air conditioning and many other electrical appliances used in establishments such as market restaurants, etc., are considered a commercial load.

3. **Industrial load:** Industrial load consists of small-scale industries, medium-scale industries, large-scale industries, heavy industries and cottage industries. The induction motor forms a high proportion of the composite load. The industrial loads are the composite load. The composite load is a function of frequency and voltage and it forms a major part of the system load.

4. **Agriculture load:** This type of load is mainly motor pumps-sets load for irrigation purposes. The load factor of this load is very small, e.g., 0.15 – 0.20.

Also, based up on the usage, the term 'electric load' is given examples as tabulated below:

Electrical load classification and types

1. **According to Load Nature-1**
   - Resistive Electrical Loads.
   - Capacitive Electrical Loads.

- Inductive Electrical Loads.
- Combination Electrical Loads.
2. **According to Load Nature-2**
    - Linear Electrical Load.
    - None -Linear Electrical Load.
3. **According to Load Function**
    - Lighting Load.
    - Receptacles/General/Small Appliances Load.
    - Power Loads.
4. **According to Load Consumer Category**
    - Residential Electrical Loads (Dwelling Loads).
    - Commercial Electrical Loads.
    - Industrial Electrical Loads.
    - Municipal/Governmental Electrical Loads (Street Lighting, Power Required for Water Supply and Drainage Purposes, Irrigation Loads and Traction Loads).
5. **According to Load Grouping**
    - Individual Loads (Single Load).
    - Load Centres (Area Loads).
6. **According to Load Planning**
    - Existing Electrical Loads.
    - Future Electrical Loads (Electrical Loads Growth).
    - New Electrical Loads (Additional Electrical Loads).
7. **According to Load Operation Time**
    - Continuous Electrical Loads.
    - Non-Continuous Electrical Loads.
    - Duty, Intermittent Electrical Loads.
    - Duty, Periodic Electrical Loads.
    - Duty, Short-Time Electrical Loads.
    - Duty, Varying Electrical Loads.
8. **According to Load Importance**
    - Vital Electrical Loads (Life Safety Electrical Loads).
    - Essential Electrical Loads (Emergency Electrical Loads).
    - Non-Essential Electrical Loads (Normal Electrical Loads).
9. **According to Load /Phase Distribution**
    - Balanced Electrical Loads.
    - Non-Balanced Electrical Loads.
    - Neutral Load.
    - Line to Neutral Load.
10. **According to Number of Electrical Loads Phases**
    - Single-phase Electrical Loads.
    - Three-phase Electrical Loads.
11. **According to Actual Electrical Loads Value**
    - Nameplate Load.
    - Full Load.
    - Percent of Full Load.
    - No Load (Open Circuit).
12. **According to Electrical Loads unit**
    - Electrical Loads in kVA.
    - Electrical Loads in kW.
    - Electrical Loads in HP.
13. **According to Electrical Loads Diversity (Simultaneous and Non-simultaneous Operation)**
    - Connected Load.
    - Demand Load.
14. **According to Unity Electrical Load Units**
    - Unity Electrical Load in VA/M2.
    - Unity Electrical Load in VA/Ft2.
    - Unity Electrical Load in VA/Linear Foot.
15. **According to Electrical Loads Operation Coincidence**
    - Non-Coincident Electrical Loads.
    - Coincident Electrical Loads.
16. **According to Electrical Loads Usage Method**
    - Fixed in Place Loads.
    - Portable Loads.
17. **According to Method of Load Reduction/Control**
    - Dimmed Electrical Load.
    - Shed Electrical Load.
    - Shifted Electrical Load.

7.6 SOME IMPORTANT ACCESSORIES IN THE POWER SYSTEM

7.6-A SWICHGEAR

The apparatus used for controlling, regulating and switching on or off the electrical circuit in the electrical power system is known as switchgear. The switches, fuses, circuit breaker, isolator, relays, current and potential transformer, indicating instrument, lightning arresters and control panels are examples of the switchgear devices.

One of the basic functions of switchgear is protection, which is interruption of short-circuit and overload fault currents while maintaining service to unaffected circuits. Switchgear also provides isolation of circuits from power supplies. Switchgear is also used to enhance system availability by allowing more than one source to feed a load.

The fundamental requirements for a protection system areas follows:

Reliability: The ability of the protection to operate correctly. It has two basic elements—dependability, which is the certainty of a correct operation on the occurrence of a fault, and security,which is the ability to avoid incorrect operation during faults.

Speed: Minimum operating time to clear a fault in order to avoid damage to equipment.

Selectivity: Maintaining continuity of supply by disconnecting the minimum section of the network necessary to isolate the fault.

The property of selective tripping is also called 'discrimination' and is achieved by two general methods:

- Time graded systems
- Unit systems

Cost: Maximum protection at the lowest cost possible.

The switchgear system is directly linked to the supply system. It is placed in both the high and low voltage sides of the power transformer. It is used for de-energizing the equipment for testing and maintenance and for clearing the fault.

When the fault occurs in the power system, heavy current flows through equipment due to which the equipment gets damaged, and the service also get interrupted. So to protect the lines, generators, transformers and other electrical equipment from damage automatic protective devices or switchgear devices are required.

The automatic protective switchgear mainly consists of the relay and circuit breaker. When the fault occurs in any section of the system, the relay of that section comes into operation and close the trip circuit of the breaker which disconnects the faulty section. The healthy section continues supplying loads as usual, and thus there is no damage to equipment and no complete interruption of supply.

The switchgear is mainly classified into two types, the outdoors type and the indoor type. For voltage above 66 kV, the output switchgear is used. Because for the high voltage, the building work will unnecessarily increase the installation cost owing to large spacing between the conductor and large size of insulators.

Below 66 kV there is no difficulty in providing the building work for the switchgear at a reasonable cost. The indoor type switchgear is of metal clad type and is compact. Because of the compactness, the safety clearance for operation is also reduced and thus reduced the area required.

Generally, electrical switchgear rated to 1 kV is known as **low voltage switchgear**. The term **LV switchgear** includes all the above-mentioned protective devices such as circuit breaker, etc., of low voltage capacity.

In addition other functions, over-voltage protection and under-voltage protection are provided by specific devices (lightning and various other types of voltage-surge arrester, relays associated with contactors, remotely controlled circuit-breakers, and with combined circuit-breaker/isolators and so on)

Typically, switchgear in substations are located on both the high- and low-voltage sides of large power transformers. The switchgear on the low-voltage side of the transformers may be located in a building, with medium-voltage circuit breakers for distribution circuits, along with metering, control, and protection equipment. For higher voltages (over about 66 kV), switchgear is typically mounted outdoors and insulated by air, although this requires a large amount of space. Gas-insulated switchgear saves space compared with air-insulated equipment, although the equipment cost is higher. Oil insulated switchgear presents an oil spill hazard.

Switches may be manually operated or have motor drives to allow for remote control.

For HV applications, the installation cannot be confined to a building, and it will be more than that, because of the high electrical quantities to be handled. For industrial applications, a transformer and switchgear line-up may be combined in one housing, called a unitized substation (USS).

Fig. 7.10. Switchboard and switchgear

7.6-B. FUSES AND SWITCH FUSE UNIT (SFU)

A **fuse** is an **electric/electronic or mechanical device**, which is used to protect circuits from over current, overload and make sure the protection of the circuit. Electric fuse was invented by Thomas Alva Edison in 1890. There are many types of fuses, but function of all these fuses is same. In this article, we will discuss the different types of fuses, its construction, working and operation and their application in various electronics and electrical system.

A general fuse consists of a low resistance metallic wire enclosed in a noncombustible material. It is used to connect and install in series with a circuit and device which needs to be protected from short circuit and over current, otherwise, electrical appliance may be damaged in case of absence of the fuse and circuit breaker as they are unable to handle the excessive current according to their rating limits.

The **working principle of a fuse** is based on the —'**heating effect of current**', i.e., whenever a short circuit, over-current or mismatched load connection occurs, then the thin wire inside the fuse melts because of the heat generated by the heavy current flowing through it. Therefore, it disconnects the power supply from the connected system. In normal operation of the circuit, fuse wire is just a very low resistance component and does not affect the normal operation of the system connected to the power supply.

Fuses can be classified as —'One Time Only Fuse', 'Resettable Fuse', 'Current limiting and non-current limiting fuses' based on the usage for different appliacations. One time use fuses contain a metallic wire, which burns out, when an over-current, over-load or mismatched load connect event occurs. User has to manually replace these fuses. Switch fuses are cheap and widely used in almost all the electronics and electrical systems.

On the other hand, the resetable fuse automatically reset after the operation when fault occurs at the system.

In the current limiting fuse, they produce high resistance for a very short period while the non-current limiting fuse produce an arc in case of high current flow to interrupts and limit the current in related and connected circuit.

There are **different types of fuses** available in the market and they can be categories on the basis of different aspects, as shown below. It is good to know that fuses are used in AC as well as DC circuits.

7.6-C. SFU

SFU is another kind of fuse + switch unit. It is switched fuse unit. It has one switch unit and

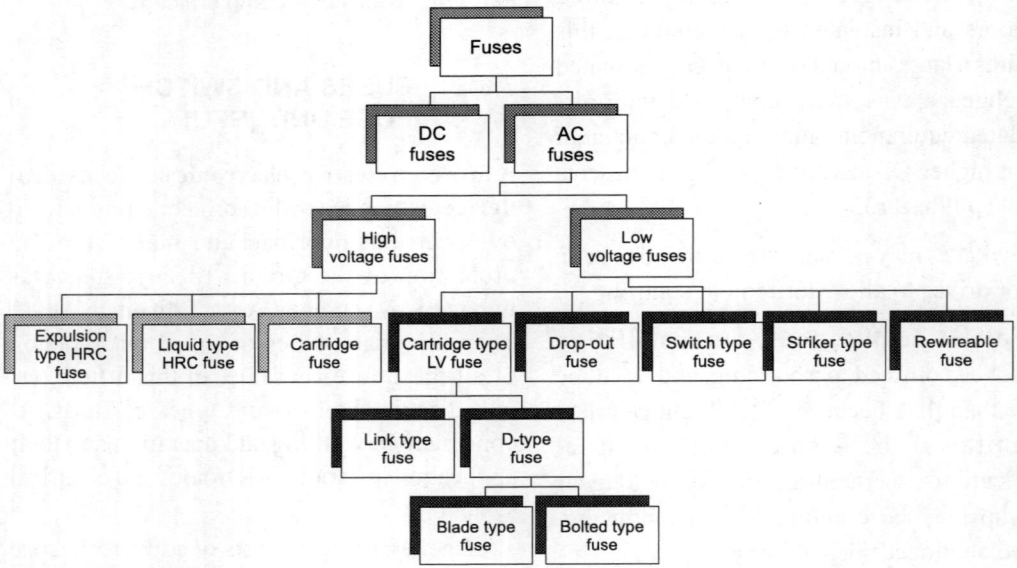

Fig. 7.11 Types of fuses

one fuse unit. When we operate the breaker, the contacts will get closed through switch and then the supply will pass through the fuse unit to the output.

Whereas in a fuse switch unit there is no separate switch and fuse unit. There is only the fuse unit which itself acts as a switch. When we operate it the fuse unit will close the input and output of the breaker. SFU has been used to trip the circuit, particularly for high capacity tripping.

The switch fuse unit (SFU) is used for the following purposes.

1. For isolation of the circuit.
2. For the protection of the equipment in case of the short circuit faults.

Circuit breakers

One of the important parts of switchgear is circuit breakers. A circuit breaker is a switching device that interrupts the abnormal or fault current. It is a mechanical device that disturbs the flow of high magnitude (fault) current and in addition performs the function of a switch and hence it is a switching device which can be operated manually and automatically for controlling and protecting an electrical power system. The circuit breaker is mainly designed for closing or opening of an electrical circuit, thus protects the electrical system from damage under faulty conditions.

The modern power system deals with huge power network and huge numbers of associated electrical equipment. During a short circuit fault or any other type of electrical fault (such as electric cable faults), a high fault current will flow through this equipment as well as the power network itself. This high current may damage the equipment and networks permanently.

For saving these pieces of equipment and the power networks, the fault current should be cleared from the system as quickly as possible. Again after the fault is removed, the system must come to its normal working condition as soon as possible for supplying reliable quality power to the receiving ends. In addition to that for proper controlling of the power system, different switching operations are required to be performed.

So for timely disconnecting and reconnecting different parts of power system network for protection and control, there must be some special type of switching devices which can be operated safely under huge current carrying condition. During the interruption of large current, there would be large arcing in between switching contacts, so care should be taken to quench these arcs in circuit breaker in a safe manner. The **circuit breaker** is the special device which does all the required switching operations during current carrying condition.

Even in houses we use circuit breakers between the input of the electric power and all the loads. In such as building wiring, the hot wire and the neutral wire never touch directly. The charge running through the circuit always passes through an appliance, which acts as a resistor. In this way, the electrical resistance in appliances limits how much charge can flow through a circuit (with a constant voltage and a constant resistance, the current must also be constant). Appliances are designed to keep current at a relatively low level for safety purposes. Too much charge flowing through a circuit at a particular time would heat the appliance's wires and the building's wiring to unsafe levels, possibly causing a fire.

This keeps the electrical system running smoothly most of the time. But occasionally, something will connect the hot wire directly to the neutral wire or something else leading to ground. For example, a fan motor might overheat and melt, fusing the hot and neutral wires together. Or someone might drive a nail into the wall, accidentally puncturing one of the power lines. When the hot wire is connected directly to ground, there is minimal resistance in the circuit, so the voltage pushes a huge amount of charge through the wire. If this continues, the wires can overheat and start a fire.

The circuit breaker's job is to cut off the circuit whenever the current jumps above a safe level. In the following sections, we'll find out how it does this.

Circuit breaker (Fig. 7.12) essentially consists of fixed and moving contacts. These contacts are touching each other and carrying the current under normal conditions when the circuit is closed. When the circuit breaker is closed, the current carrying contacts, called the electrodes, engaged each other under the pressure of a spring.

During the normal operating condition, the arms of the circuit breaker can be opened or closed for a switching and maintenance of the system. To open the circuit breaker, only a pressure is required to be applied to a trigger.

apart from each other by some mechanism, thus opening the circuit.

Once a fault is detected, the circuit breaker contacts must open to interrupt the circuit; this is commonly done using mechanically stored energy contained within the breaker, such as a spring or compressed air to separate the contacts.

7.6-D. MCB (MINIATURE CIRCUIT BREAKER)

MCB (Fig. 7.13) is an electromechanical device which guards an electrical circuit from an over current, that may effect from short circuit, overload or imperfect design. This is a better option to a fuse since it doesn't require alternate once an overload is identified. An MCB can be simply rearranged and thus gives a better operational protection and greater handiness without incurring huge operating cost. The operating principle of MCB is simple.

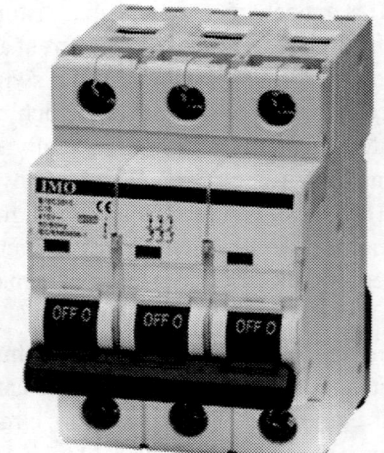

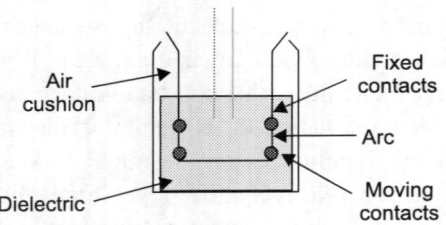

Fig. 7.12 Representation of circuit breaker

Whenever a fault occurs on any part of the system, the trip coil of the breaker gets energized and the moving contacts are getting

Fig. 7.13 Miniature circuit breaker

An MCB function by interrupting the stability of electrical flow through the circuit once an error is detected. In simple conditions this circuit breaker is a switch which routinely turns off when the current flows through it and passes the maximum acceptable limit. Generally, these are designed to guard against over current and overheating.

MCB substitutes the rewirable switch-fuse units for low power domestic and industrial applications in a very quick manner. In wiring system, the MCB is a blend of all three functions such as protection of short circuit, overload and switching. Protection of overload by using a bimetallic strip and short circuit protection by used solenoid.

These are obtainable in different pole versions like single, double, triple and four poles with neutral poles if necessary. The normal current rating ranges from 0.5-63 A with a symmetrical short circuit breaking capacity of 3-10 kA, at a voltage level of 230 or 440 V.

Characteristics of MCB

The characteristics of an MCB mainly include the following:
- Rated current is not more than 100 amperes
- Normally, trip characteristics are not adjustable
- Thermal/thermal magnetic operation

7.6-E. MCCB (MOLDED CASE CIRCUIT BREAKER)

An MCCB is used to control electric energy in distribution n/k and has short circuit and overload protection. This circuit breaker is an electromechanical device which guards a circuit from short circuit and over current. They offer short circuit and over current protection for circuits ranging from 63 to 3000 A. The primary functions of MCCB is to give a means to manually open a circuit, automatically open a circuit under short circuit or overload conditions.

An MCCB is an option to a fuse since it doesn't need an alternate once an overload is noticed. Unlike a fuse, this circuit breaker can be simply reset after a mistake and offers enhanced operator safety and ease without acquiring operating cost. Generally, these circuits have thermal current for over current and the magnetic element for short circuit release to work faster.

Characteristics of MCCB

The characteristics of an MCCB mainly include the following:
- The range of rated current to 1000 A.
- Trip current may be adjusted
- Thermal/thermal magnetic operation

7.6-F. ELCB (EARTH LEAKAGE CIRCUIT BREAKER)

An ELCB (Fig. 7.14) is used to protect the circuit from the electrical leakage. When someone gets an electric shock, then this circuit breaker cuts off the power at the time of 0.1 sec for protecting the personal safety and avoiding the gear from the circuit against short circuit and overload.

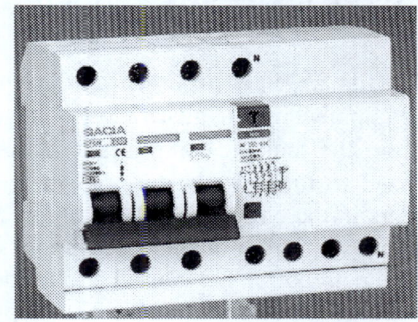

Fig. 7.14

ELCB is a security device used in electrical system with high earth impedance to avoid shock. It notices small stray voltages on the metal fields of electrical gear, and interrupts the circuit if an unsafe voltage is detected. The main principle of the earth leakage protectors is to stop injury to humans and nature due to electric shock.

This circuit breaker is a specialized kind of latching relay that has structures incoming mains power connected through its switching contacts so that this circuit breaker disconnects the power supply in an unsafe condition.

ELCB notices fault currents from live to the ground wire inside the installation it guards. If enough voltage emerges across the sense coil in the circuit breaker, it will turn off the supply, and stay off until reset by hand. A voltage-sensing earth leakage circuit breaker doesn't detect fault currents of any other ground body.

Characteristics of ELCB

The characteristics of an ELCB mainly include the following:
- This circuit breaker connects the phase, earth wire and neutral
- The working of this circuit breaker depends on current leakage

7.6-G RCCB (RESIDUAL CURRENT CIRCUIT BREAKER)

A RCCB (Fig. 7.15) is an essential current sensing equipment used to guard a low voltage circuit from the fault. It comprises a switch device used to turn off the circuit when a fault occurs in the circuit. RCCB is aimed at guarding a person from the electrical shocks. Fires and electrocution are caused due to the wrong wiring or any earth faults. This **type of circuit breaker** is used in situations where there is a sudden shock or fault happening in the circuit.

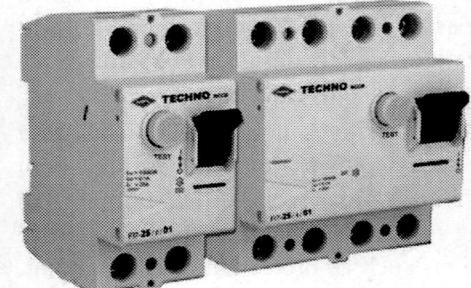

Fig. 7.15 Residual current circuit breaker

For instance, a person suddenly comes in contact with an open live wire in an electrical circuit. In that situation, in the absence of this circuit breaker, a ground fault may occur and the individual is at the hazardous situation of receiving a shock. But, if the similar circuit is defended with the circuit breaker, it will tour the circuit in a second therefore, avoiding a person from the electric shock. Therefore, this circuit breaker is good to install in an electrical circuit.

Characteristics of RCCB

- Whenever there is any ground fault occurs, then it trips the circuit
- The amount of current supplied through the line should go back through neutral
- These are a very effective type of shock protection

7.7 NOTES ON ELECTRIC BATTERIES

Batteries are a collection of one or more cells whose chemical reactions create a flow of electrons in a circuit. All batteries are made up of three basic components: an anode (the '–'

side), a cathode (the '+' side), and some kind of electrolyte (a substance that chemically reacts with the anode and cathode).

While there are many types of batteries, the basic concept by which they function is same. When a device is connected to a battery, a reaction occurs that produces electrical energy. This is known as an electrochemical reaction. Italian physicist Count Alessandro Volta first discovered this process in 1799 when he created a simple battery from metal plates and brine-soaked cardboard or paper. Since then, scientists have greatly improved upon Volta's original design to create batteries made from a variety of materials that come in a multitude of sizes.

When the anode and cathode of a battery are connected to a circuit, a chemical reaction takes place between the anode and the electrolyte. This reaction causes electrons to flow through the circuit and back into the cathode where another chemical reaction takes place. When the material in the cathode or anode is consumed or no longer able to be used in the reaction, the battery is unable to produce electricity. At that point, the battery is "dead."

Batteries that must be thrown away after use are known as primary batteries. Batteries that can be recharged are called **secondary batteries**.

Today, batteries are all around us. They power our wristwatches for months at a time. They keep our alarm clocks and telephones working, even if the electricity goes out. They run our smoke detectors, electric razors, power drills, mp3 players, thermostats – and the list goes on. If you're reading this article on your laptop or smartphone, you may even be using batteries right now!

Take a look at any battery, and you'll notice that it has two **terminals**. One terminal is marked (+), or positive, while the other is marked (–), or negative. In normal flashlight batteries, like AA, C or D cell, the terminals are located on the ends. On a 9-volt or car battery, however, the terminals are situated next to each other on the top of the unit. If you connect a wire between the two terminals, the electrons will flow from the negative end to the positive end as fast as they can. This will quickly wear out the battery and can also be dangerous, particularly on larger batteries.

To properly harness the electric charge produced by a battery, you must connect it to a **load**. The load might be something like a light bulb, a motor or an electronic circuit like a radio.

The internal workings of a battery are typically housed within a metal or plastic case. Inside this case are a **cathode**, which connects to the positive terminal, and an **anode**, which connects to the negative terminal. These components, more generally known as **electrodes**, occupy most of the space in a battery and is the place where the chemical reactions occur. A **separator** creates a barrier between the cathode and anode, preventing the electrodes from touching while allowing electrical charge to flow freely between them. The medium that allows the electric charge to flow between the cathode and anode is known as the **electrolyte**. Finally, the **collector** conducts the charge to the outside of the battery and through the load.

A lot happens inside a battery when you pop it into your flashlight, remote control or other wire-free device. While the processes by which they produce electricity differ slightly from battery to battery, the basic idea is the same (Fig. 7.16).

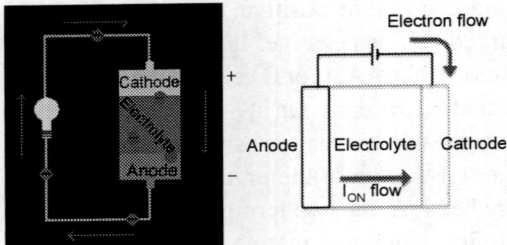

Fig. 7.16 Battery operation

When a load completes the circuit between the two terminals, the battery produces electricity through a series of electromagnetic reactions between the anode, cathode and electrolyte. The anode experiences an oxidation reaction in which two or more ions (electrically charged atoms or molecules) from the electrolyte combine with the anode, producing a compound and releasing one or more electrons. At the same time, the cathode goes through a reduction reaction in which the cathode substance, ions and free electrons also combine to form compounds. While this action may sound complicated, it is actually very simple. The reaction in the anode creates electrons, and the reaction in the cathode absorbs them. The net product is electricity. The battery will continue to produce electricity until one or both of the electrodes run out of the substance necessary for the reactions to occur.

Modern batteries use a variety of chemicals to power their reactions. Common battery chemistries include:

- **Zinc-carbon battery:** The zinc-carbon chemistry is common in many inexpensive AAA, AA, C and D dry cell batteries. The anode is zinc, the cathode is manganese dioxide, and the electrolyte is ammonium chloride or zinc chloride.
- **Alkaline battery:** This chemistry is also common in AA, C and D dry cell batteries. The cathode is composed of a manganese dioxide mixture, while the anode is a zinc powder. It gets its name from the potassium hydroxide electrolyte, which is an alkaline substance.
- **Lithium-ion battery (rechargeable):** Lithium chemistry is often used in high-performance devices, such as cell phones, digital cameras and even electric cars. A variety of substances are used in lithium batteries, but a common combination is a lithium cobalt oxide cathode and a carbon anode.
- **Lead-acid battery (rechargeable):** This is the chemistry used in a typical car battery. The electrodes are usually made of lead dioxide and metallic lead, while the electrolyte is a sulfuric acid solution. The symbol of a battery is shown in Fig. 7.17.

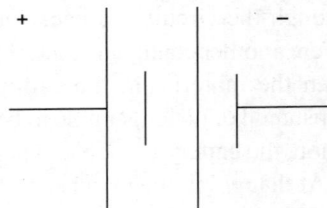

Fig. 7.17 Symbol of battery

Backup batteries play a vital role in providing an undisturbed supply of power to the users.

A **backup battery** provides power to a system when the primary source of power is unavailable. Backup batteries range from small single cells to retain clock time and date in computers, up to large battery room facilities that power uninterruptible power supply systems for large data centres. Small backup batteries may be primary cells; rechargeable backup batteries are kept charged by the prime power supply.

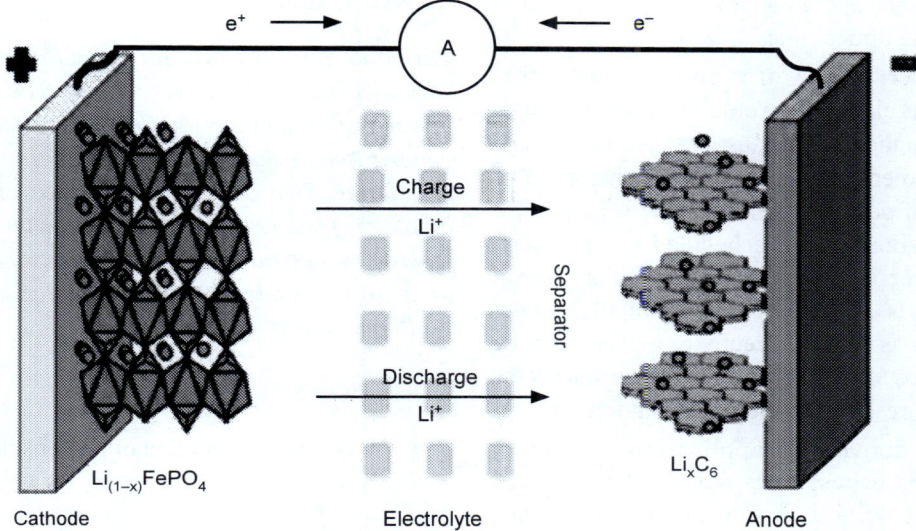

Fig. 7.18 Schematic of lithium ion battery

Backup batteries in aircraft keep essential instruments and devices running in the event of an engine power failure. Backup batteries are almost always used in burglar alarms. The backup battery prevents the burglar from disabling the alarm by turning off power of the building. Additionally, these batteries power the remote cellular phone systems that thwart phone line snipping as well.

Modern personal computer motherboards have a backup battery to run the real-time clock circuit and retain configuration memory while the system is turned off. This is often called the CMOS battery or BIOS battery. The original IBM used a relatively large primary lithium battery, compared to later models, to retain the clock and configuration memory. These early machines required the backup battery to be replaced periodically due to the relatively large power consumption.

Backup batteries are used in uninterruptible power supplies (UPS), and provide power to the computers they supply for a variable period after a power failure, usually long enough to at least allow the computer to be shut down gracefully These batteries are often large valve regulated lead-acid batteries in smaller or portable systems. Data centre UPS backup batteries may be wet cell lead-acid or nickel cadmium batteries, with lithium ion cells available in some ratings.

Server-grade disk array controllers often contain onboard disk buffer, and provide an option for a "backup battery unit" (BBU) to maintain the contents of this cache after power loss. If this battery is present, disk writes can be considered completed when they reach the cache, thus speeding up I/O throughput by not waiting for the hard drive. This operation mode is called "write-back caching".

A local backup battery unit is necessary in some telephony and combined telephony/data applications built with use of digital passive optical networks. In such networks there are active units on telephone exchange side and on the user side, but nodes between them are all passive in the meaning of electrical power usage. So, if a building (such as an apartment/house) loses power, the network continues to function. The user side must have standby power since operating power is not transferred over data optical line.

Power failure in a power station that produces electricity would result in a blackout situation that would cause irreparable damage to equipment such as the turbine-generator. The safety of power station employees is a major concern during an unscheduled power outage at a power plant. A bank of large station backup batteries is used to power uninterruptible power supplies as well as directly power emergency oil pumps for up to 8 hours while normal power is being restored to the power station.

The above said application of backup batteries necessitates several batteries to serve the purpose. A single or a few batteries will not be sufficient to maintain backup for a longer duration of time. So, a battery room is essential to house a group of high power massive batteries.

A battery room is a room in a facility used to house batteries for backup or uninterruptible power systems. Battery rooms are found in telecommunication central offices, and to provide standby power to computing equipment in data centres. Batteries provide direct current (DC) electricity, which may be used directly by some types of equipment, or which may be converted to alternating current (AC) by uninterruptible power supply (UPS) equipment. The batteries may provide power for minutes, hours or days depending on the electrical system design, although most commonly the batteries power the UPS during brief electric utility outages lasting only seconds.

Battery rooms were used to segregate the fumes and corrosive chemicals of wet cell batteries (often lead acid) from the operating equipment; a separate room also allows better control of temperature and ventilation for the batteries. In 1890, the Western Union central telegraph office in New York City had 20,000 wet cells, mostly primary zinc-copper type, in use.

CONCLUSION

This chapter provided an introduction to various aspects of a power system. The huge power system contains four large parts termed generation, transmission, distribution and utilization: Also accessories such as circuit breakers, feeders, batteries and many more forms a successful power system. Such blocks of a power system were introduced and explained in this chapter.

QUESTIONS

1. What is a power plant or power station?
2. What is the difference between the transmission line and distribution line?
3. What are the common sources of energy?
4. Write down the classification of the transmission line?
5. What is a relay? How many types of protection relays are there based on their characteristic?
6. Draw a single line diagram of the power station.
7. What is the difference between a fuse and a breaker?
8. Bring out the difference between the relay and the circuit breaker?
9. What is meant by bus bar protection?
10. What is the purpose of employing instrument transformers in a power system?
11. Write a brief note on the types of electric loads.
12. What is the role of switchgear in a power system?
13. Briefly talk about SF6 circuit breakers and mention why they are better than other CBs?
14. Electric batteries in power system: Write a technical note.

INTRODUCTION

This chapter provides an introductory treatment on electrical measuring instruments. Various types, forces involved in operating a measuring instruments, ammeter, voltmeter, wattmeter, energy meter and resistance measuring instrument are majorly dealt with. Preliminary illustrations help the understanding of concepts.

CHAPTER OBJECTIVE

At the completion of this chapter, the reader will be able to:
* describe fundamental principle of measuring instruments and the operation of their various types; and
* describe the applications of measuring instruments as ammeters, voltmeters, etc.

KEYWORDS

Forces in measuring instruments, Calibration, MC and PMMC instruments. Types of meters, Wattmeters and Megger.

8
Measuring Instruments

8.1 DEFINITION AND CLASSIFICATION

The devices that is commonly used for the measuring of electrical quantities such as voltage, current, power, energy and related quantities are known as measuring instruments in electrical engineering. The instruments indicate the value of these quantities, based on which some technical understanding can be arrived at and also appropriate actions and decisions can be initiated.

Measuring instruments can be classified in various headings. Based on the supply of operation measuring instruments are classified as AC instruments and DC instruments, respectively known as moving iron (MI) instruments and moving coil (MC) instruments.

Based on the method of reading, that is, whether the readings are made through pointer and scale arrangements or through a digital display – the measuring instruments are classified respectively, as analog instruments and digital instruments. The analog instruments indicate the magnitude of the quantity in the form of the pointer movement. They usually indicate the values in the whole numbers, though one can get the readings up to one or two decimal places also. The readings taken in decimals places may not always be entirely correct, since some human error is always involved in reading. Digital display will have a display board on which the quantity being measured is precisely displayed as numerals which can be readily read.

The measuring instruments can also be classified based up on the mode of usage. If the instruments are permanently placed at the place where the reading is to be made, then the type will be called permanently mounted instruments. Most of the industries have such instruments as the machinery is continuously operated and hence the readings are to be noted continuously. In the case of institutions, the instruments are required at the time of conduct of the experiments and otherwise they are to be safely kept at the stores. Such instruments are classified as portable instruments.

There is a classification of instruments which is majorly based on the applications. There are certain instruments which are intended just to show (or indicate) the physical quantity being measured. They are called indicating instruments. These instruments are calibrated against the standard values of the physical quantities. The movement of the pointer directly indicates the magnitude of the quantity, which can be whole numbers or even fractions. Nowadays, the digital instruments are becoming very popular, which indicate the values directly in numerical form and even in decimals thus making them easy to read and more accurate.

A large are number of instruments can be used as the controllers. For instance, when a certain value of the pressure is reached, the measuring instrument breaks the electrical circuit, which stops the running of compressor. Similarly, the thermostat starts or stops the compressor of the refrigeration system depending on the temperature achieved in the evaporator. They fall under the category of measuring instruments used as the controllers.

Measuring instruments are used to record and store the data. In this age of computerization storing the recorded data has become necessary. These data are considered important, so that the recorded items can be retrieved at any time for further processing. A number of instruments are connected to the pen that moves on the paper. As the pointer of the instrument moves as per the changes in the magnitude of the quantity, the pen also moves on the paper making the graph against certain parameter, like time. Attaching small memory to the PCB can also enable recording of the instruments in the chip. These are called recording instruments. Electrocardiogram (ECG) is one such instrument.

The measuring instruments can also be used to transfer the data to some distant places. The instruments kept in unsafe locations like high temperature can be connected by wires and their output can be taken at some distant places which are safe for the human beings. The signal obtained from these instruments can also be used for operating some controls. Take for instance the power system. Electricity, due to various reasons, is generated at remote places and then transmitted to thickly populated areas for utilization. For a successful and safety operation of such a vast power system, data from thousands and thousands of power system junctions are transmitted to a central data centre for processing. A variety of novel and new age instruments are there to perform this task. They are called *transmitting instruments*.

Some measuring instruments carry out a calculations like addition, subtraction, multiplcation, division, etc. Some can also be used to find solutions to highly complex equations. This forms another recent type of instruments.

Measuring instruments operate on a few different modes. Based on the modes of operation they are classified as moving coil (MC) instruments, moving iron (MI) instruments and dynamometer type instruments. This chapter will discuss these modes of instruments in an introductory level in the forthcoming sections.

The principle of operation of a measuring instrument can be purely electromagnetic or it can also be otherwise employing a different technique. The instruments hence are classified as per the principle of operation as (i) electromagnetic, which uses magnetic effects of electric current to move the pointer, (ii) electrostatic, which uses the force between the forces between the electrically charged conductors to establish the deflecting force essential to move the pointer, and (iii) electrothermic which uses the heating effect for the operation.

Thus, the types of measuring instruments can be made in various headings.

8.2 CALIBRATION OF INSTRUMENTS

All the new instruments have to be calibrated against some standard in the very beginning. This will ascertain the accuracy of the design on the new instrument. The scale of the new instrument is thus fixed. Then it is put into regular usage.

Calibration of the measuring instrument is the process in which the readings obtained from the instrument are compared with the sub-standards in the laboratory at several points along the scale of the instrument. As per the readings obtained from the instrument and sub-standards, the curve is plotted. If the instrument is accurate there will be matching of the scales of the instrument and the sub-standard. If there is deviation of the measured value from the instrument against the standard value, the instrument is calibrated to give the correct values.

All the new instruments have to be calibrated against some standard in the very beginning. For the new instrument the scale is marked as per the sub-standards available in the laboratories, which are meant especially for this purpose. After continuous use of the instrument for long, sometimes it loses its calibration or the scale gets distorted, in such cases the instrument can be calibrated again if it is in good reusable condition.

Even if the instruments in the factory are working in the good condition, it is always advisable to calibrate them from time-to-time to avoid wrong readings of highly critical parameters. This is very important especially in the companies where very high precision jobs are manufactured with high accuracy.

8.3 THE MAJOR FORCES (TORQUES) WHICH OPERATE A MEASURING INSTRUMENT

In any instrument that is used to indicate the quantity being measured through a pointer movement, essentially four kinds of forces are required. These forces are expected to operate rapidly and settle the pointer on the scale against the quantity being measured (Fig. 8.1).

The initial driving force is called the deflecting force (or the deflecting torque), which is required to move the needle from its rest (unconnected) position. This is obtained from the simple principle of electromagnetic induction that magnetic field due to current in a coil interacts with the field of magnetic resulting in a deflection. A current carrying coil inside an energized magnet will be able to establish the deflecting torque (Fig. 8.2).

Fig. 8.1 Pointer against a scale

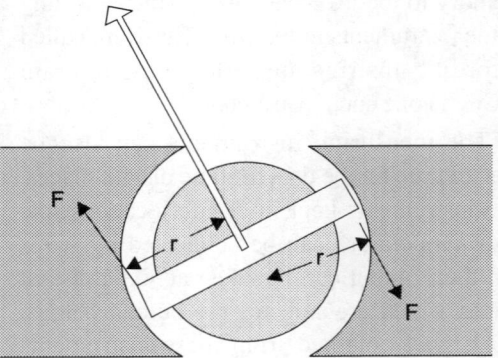

Fig. 8.2 Current carrying coil inside a magnetic field exhibits deflecting torque

Once the deflecting torque is established, the needle connected to the moving coil will move indefinitely. It is expected to move from the earlier equilibrium position to the new position and settle there so as to enable the reader to be sure of the quantity being measured. The force which establishes this is called the *controlling force* (*or controlling torque*). This force controls the movement of the needle against the proper quantity being measured on the scale, and hence the name.

Controlling torque is achievable by a simple spring-balance arrangement at the bottom of the needle. It is a mechanical arrangement to keep something in position against the spring tension. Hair thin spring is employed for this purpose to create a controlling torque (on the needle) just equivalent and opposite to that of the deflecting torque. When these two torques become same, the needle will be positioned on the needle against the quantity being measured.

However, there can be small oscillations on the needle (or pointer) against the final steady position of it. These are to be damped out in order to read accurately. *Damping force* (*or torque*) helps to achieve this. Augmenting the controlling force further will provide a proper damping force.

These are the three major forces or torques to operate a measuring instrument. The following sections discuss a few measuring instruments which are used especially for DC and AC circuits and for the measurements of electric power and energy.

Establishment of the three control torques

Deflecting torque

One important requirement in indicating instruments is the arrangement for producing operating or **deflecing torque** (T_d) when the instrument is connected in the circuit to measure the given electrical quantity.

This is achieved by utilizing the various effects of electric current or voltage. The deflecting torque causes the moving system to move from its zero position. The deflecting torque is produced by utilizing one or more of the following effects of current or voltage:

1. Magnetic effect : Moving-iron instruments.
2. Electrodynamic effect : (i) Moving coil instruments
   (ii) Dynamometer type.
3. Electromagnetic induction effect : Induction type instruments.
4. Thermal effect : Hot-wire instruments.
5. Chemical effect : Electrolytic instruments.
6. Electrostatic effect : Electrostatic voltmeters

The table below gives information about the electrical measuring instruments in which deflecting torque is produced by utilizing the first three effects.

S.No.	Type	Effect	Suitable for	Instrument
1.	Moving-iron	Magnetic effect	DC and AC	Ammeters, Voltmeters
2.	Moving-coil	Electrodynamic effect	DC only	Ammeters, Voltmeters
3.	Dynamometer type	Electrodynamic effect	DC and AC	Ammeters, Voltmeters, Watt-meters. Usually for watt-merers.
4.	Induction type	Electromagnetic induction effect	AC only	Ammeters, Voltmeters, Watt-meters Energy meters. Usually for watt-merers and energy meters

Controlling torque

The controlling torque (T_C) opposes the deflecting torque and increases with the deflection of the moving system. The pointer comes to rest at a position where the two opposing torques are equal, i.e., $T_d = T_c$. The **controlling torque** performs two functions.

- It increases with the deflection of the moving system so that the final position of the pointer on the scale will be according to the magnitude of an electrical quantity (i.e., current or voltage or power) to be measured.
- It brings the pointer back to zero when the deflecting torque is removed. If it were not provided, the pointer once deflected would not return to zero position on removing the deflecting torque. The *controlling torque* in indicating instruments may be provided by one of the following two methods:
1. Spring control.
2. Gravity control.

Spring Control: Refer to Fig. 8.3.

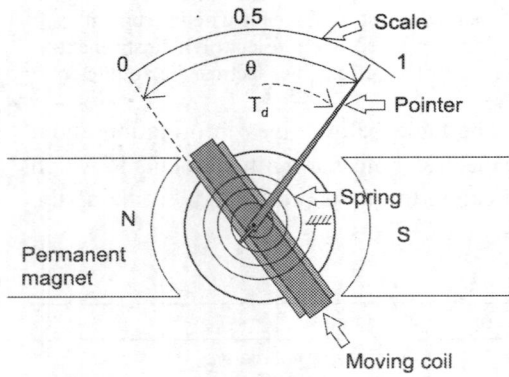

Fig. 8.3 Establishing control torque using spring

This is the most common method of providing controlling torque, in electrical instruments. A spiral hairspring made of some non-magnetic material like phosphor bronze is attached to the moving system of the instrument as shown in the figure.

Springs also serve the additional purpose of leading current to the moving system (i.e., operating coil). With that deflection of the pointer, the spring is twisted in the opposite direction. This twist in the spring provides the controlling torque.

Since the torsion torque of a spiral spring is proportional to the angle of twist, the controlling torque (T_c) is directly proportional to the angle of deflection of pointer (θ), i.e., $T_c \, \alpha \, \theta$.

The pointer will come to rest at a position where controlling torque is equal to the deflecting torque i.e. $T_d = T_c$.

In an instrument where the deflecting torque is uniform, spring control provides a uniform scale over the whole range. The balance weight is attached to counterbalance the weight of the pointer and other moving parts.

Gravity Control Method. Refer to Fig. 8.4.

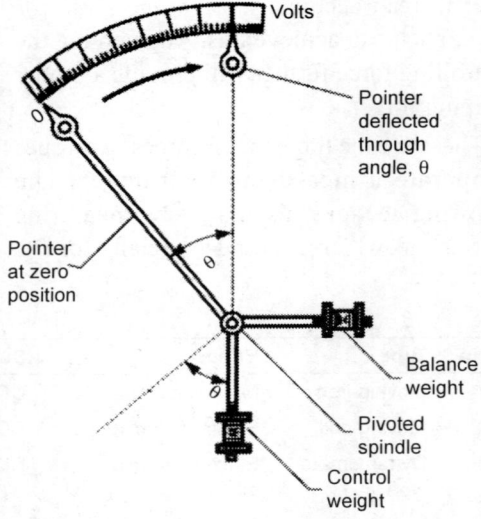

Fig. 8.4 Establishing control torque using gravity weight

In this method, a small weight is attached to the moving system, which provides necessary controlling torque. In the zero position of the pointer, the control weight hangs vertically downward and therefore provides no controlling torque.

However, under the action of deflecting torque, the pointer moves from zero position and control weight moves in opposite direction. Due to gravity, the control weight would tend to come in original position (i.e., vertical) and thus provides an opposing or controlling torque. The pointer comes to rest at a position where controlling torque is equal to the deflecting torque.

In this method, controlling torque (T_c) is proportional to the sine of angle of deflection (θ), i.e., $T_c \propto \sin \theta$.

Because in this method controlling torque (T_c) is not directly proportional to the angle of deflection (θ) but it is proportional to $\sin \theta$ therefore, gravity control instruments have non-uniform scales; being crowded in beginning.

Damping torque

A *damping torque* is produced by a damping or stopping force which acts on the moving system only when it is moving and always opposes its motion. Such a torque is necessary to bring the pointer to rest quickly. If there is no **damping torque**, then the pointer will keep moving to and fro about its final deflected position for some time before coming to rest, due to the inertia of the moving system.

This damping torque acts only when the pointer is in motion and always opposes the motion. The position of the pointer when stationary is, therefore, not affected by the damping torque. The degree of damping decides the behaviour of the moving system.

If the instrument is underdamped, the pointer will oscillate about the final position for some time before coming to rest. On the other hand, if the instrument is overdamped, the pointer will become slow and lethargic.

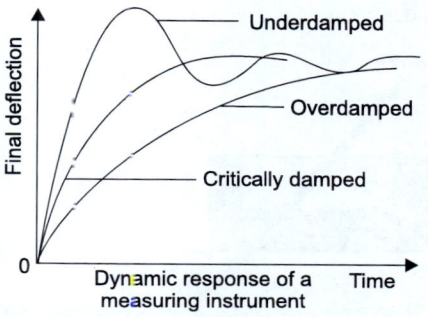

Fig. 8.5 Various damping characteristics

However, if the degree of damping is adjusted to such a value that the pointer comes up to the correct reading quickly without oscillating about it, the instrument is said to be critically damped. The damping torque in indicating instruments can be provided by:

- Air friction damping.
- Fluid friction damping.
- Eddy current damping.

Air Friction Damping: Refer to Fig. 8.6.

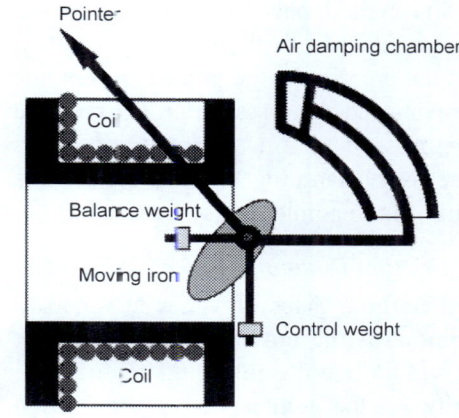

Fig. 8.6 Air friction damping

Arrangements of *air friction damping* are shown in Figs. 8.6 and 8.7. In the arrangement shown in Fig. 8.6, a light aluminum piston is attached to the spindle that carries the pointer and moves with a very little clearance in a rectangular or circular air chamber closed at one end.

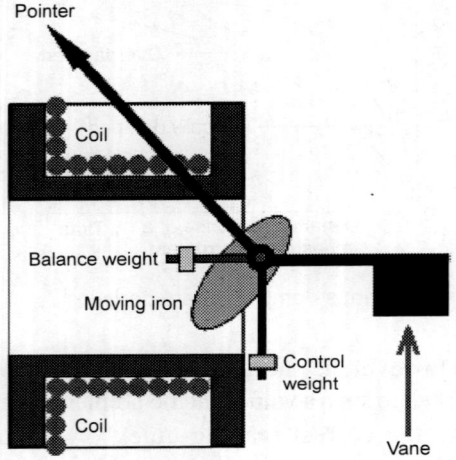

Fig. 8.7 Air friction damping

The cushioning action of the air on the piston damps out any tendency of the pointer to oscillate about the final deflected position. This method is not favoured these days and the one shown in Fig. 8.7 is preferred.

In this method, one or two light aluminum vanes are attached to the same spindle that carries the pointer. As the pointer moves, the vanes swing and compress the air. The pressure of compressed air on the vanes provides the necessary damping force to reduce the tendency of the pointer to oscillate.

Fluid Friction Damping: Refer to Fig. 8.8.

In this method, discs or vanes attached to the spindle of the moving system are kept immersed in a pot containing oil of high viscosity. As the pointer moves, the friction between the oil and vanes opposes the motion of the pointer and thus necessary damping is provided.

The **fluid friction damping** method is not suitable for portable instruments because of the oil contained in the instrument. In general, fluid friction damping is not employed in indicating instrument, although one can find its use in Kelvin electrostatic voltmeter.

Eddy Current Damping: Refer to Fig. 8.9.

Two methods of eddy current damping are generally used.

In the first method, as shown in the figure, a thin aluminum or copper disc is attached to the moving system is allowed to pass between the poles of a permanent magnet. As the pointer moves, the disc cuts across the magnetic field and eddy currents are induced in the disc.

These eddy currents react with the field of the magnet to produce a force which opposes the motion according to **Lenz's law**. In this way, eddy current damping torque reduces the oscillations of the pointer.

In the second method, the coil which produces the deflecting torque is wound on an aluminum former. As coil moves in the field of the instrument, **eddy currents** are induced in the aluminum former to provide the necessary damping torque.

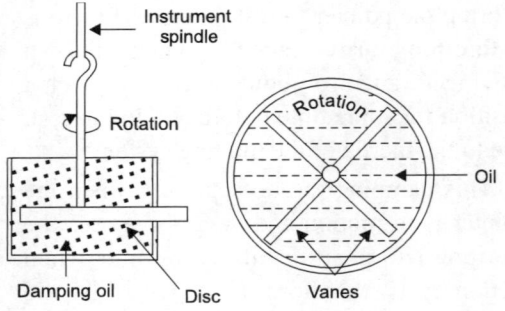

Fig. 8.8 Fluid friction damping

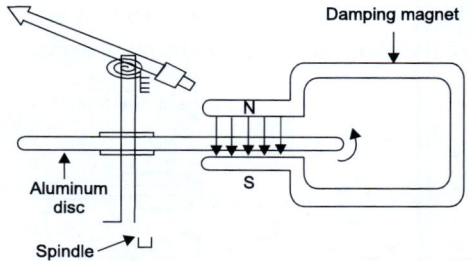

Fig. 8.9 Eddy current damping

Figure 8.10 shown a 3D view of a measuring instrument, with all the three torques implemented.

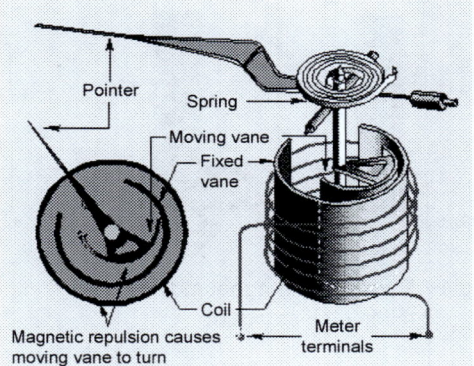

Fig. 8.10 Measuring instrument and the three essential torques.

8.4 PERMANENT MAGNET MOVING COIL (PMMC) INSTRUMENT

The interaction between the induced field and the field produced by the permanent magnet causes a deflecting torque, which results in the rotation.

Production of the three important torques (deflecting torque, controlling torque and damping torque) of PMMC are discussed below.

Deflecting torque

The force (F) which will be perpendicular to both the direction of the current flow and the direction of magnetic field as per Fleming's left hand rule can be written as

$$F = NB\,IL$$

where N: turns of wire on the coil
B: flux density in the air gap
I : current in the movable coil
L : vertical length of the coil

Theoretically, the electromagnetic torque is equal to the multiplication of force with distance to the point of suspension.

Hence, torque on left side of the cylinder, $T_L = NB\,IL \times W/2$ and torque on right side of the cylinder, $T_R = NB\,IL \times W/2$

Therefore, the total torque will be $= T_L + T_R$. That is, $T = NB\,ILW$ or $NBIA$ where A is effective area ($A = L \times W$).

Controlling torque is produced by the spring action and opposes the deflection torque so as the pointer can come to rest at the point where these two torques are equal (electromagnetic torque = control spring torque). The value of control torque depends on the mechanical design of spiral springs and strip suspensions.

The controlling torque is directly proportional to the angle of deflection of the coil. Control torque, $C_t = C\theta$ where, θ = deflection angle in radians and C = spring constant Nm-rad.

Damping torque ensures the pointer comes to an equilibrium position, i.e., at rest in the scale without oscillating to give an accurate reading. In PMMC as the coil moves in the magnetic field, eddy current sets up in a metal former or core on which the coil is wound or in the circuit of the coil itself which opposes the motion of the coil resulting in the slow swing of a pointer and then come to rest quickly with very little oscillation.

Construction: A coil of thin wire is mounted on an aluminum frame (spindle) positioned between the poles of a U-shaped permanent magnet which is made up of magnetic alloys like alnico. The coil is pivoted on the jewelled bearing and thus the coil is free to rotate. The current is fed to the coil through spiral springs which are two in numbers. The coil which carries a current, which is to be measured, moves in a strong magnetic field produced by a permanent magnet and a pointer is attached to the spindle which shows the measured value. Refer to Fig. 8.11.

Principle of operation: When a current flows through the coil, it generates a magnetic field which is proportional to the current in case of an ammeter. The deflecting torque is produced by the electromagnetic action of the current in the coil and the magnetic field.

When the torques are balanced the moving coil will stop and its angular deflection represents the amount of electrical current to be measured against a fixed reference, called scale. If the permanent magnet field is uniform and the spring linear, then the pointer deflection is also linear.

The controlling torque is provided by two phosphorous bronze flat coiled helical springs. These springs serve as a flexible connection to the coil conductors.

Damping is caused by the eddy current set-up in the aluminum coil which prevents the oscillation of the coil, and hence the pointer.

PMMC instrument as an ammeter and a voltmeter

An ammeter is an instrument for measuring current in any branch of an electric circuit. It must be placed in series with the measured branch, and must have very low resistance to avoid significant alteration of the current that is needed to measure. When PMMC is used as an ammeter, except for a very small current range, the moving coil is connected across a suitable low resistance shunt, so that only small part of the main current flows through the coil.

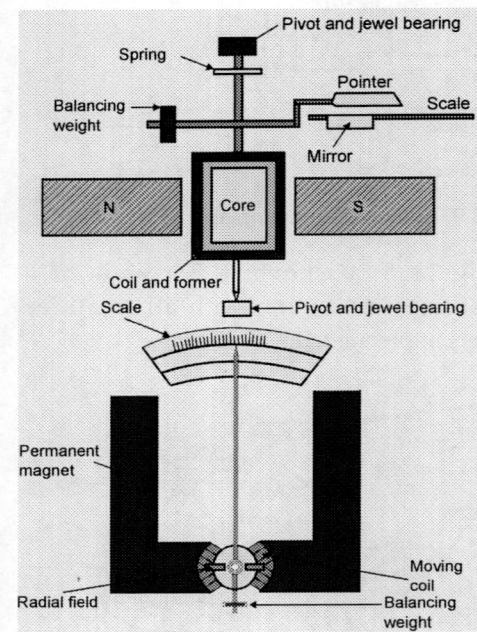

Fig. 8.11 Permanent magnet moving coil (PMMC) instrument

Upon configuring a PMMC as an ammeter with a shunt resistor is connected to it and this is for the purpose of adjusting the range of measurement. The value of the shunt resistor depends on the current range the device is going to measure.

The shunt consists of a number of thin plates made up of alloy metal, which is usually magnetic and has a low-temperature coefficient of resistance, fixed between two massive blocks of copper. A resistor of the same alloy is also placed in series with the coil to reduce errors due to temperature variation.

Consider the below problem for a better understanding of the need of a shunt to configure PMMC as an ammeter.

Given a PMMC instrument with a 100 µA movement and an internal resistance of 100 Ω to be converted into a 0 - 100 mA ammeter, calculate the value of the shunt resistance required.

For the solution of this problem, Fig. 8.12 is drawn.

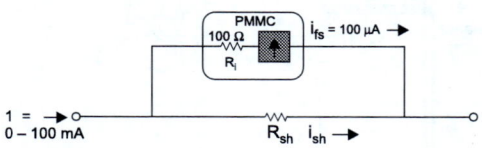

Fig. 8.12

From the given problem, we can create an equivalent circuit as shown on Fig. 8.12. We can see that the current (I) divides into the branches of the PMMC and the shunt resistor. With this set-up, we can apply the current-divider formula.

Using the current-divider formula:

$$I_{FS} = \frac{R_{SH}}{R_{SH} + R_I} I$$

where:

I_{FS} - PMMC's full scale current
I - maximum current to be measured
R_I - PMMC's internal resistance
R_{SH} - shunt resistance

Given from the problem, we can obtain the values:

I_{FS} - 100 μA
I - 100 mA
R_I - 100 Ω
R_{SH} - ?

Substituting these values to the current-divider formula, we get:

$$100 \ \mu A = \frac{R_{SH}}{R_{SH} + 100 \ ohms} 100 \ mA$$

$$R_{SH} = 0.1 \ \Omega$$

Therefore, a 0.1 Ω shunt resistance is required to operate the PMMC as an ammeter measuring current ranging from 0 - 100 mA.

In this way PMMC functions as an ammeter, to measure the current in a circuit.

When PMMC is used as a voltmeter, the coil is connected in series with a high resistance. Rest of the function is same as above. The same moving coil can be used as an ammeter or voltmeter with an interchange of above arrangement.

The galvanometer is used to measure a small value of current along with its direction and strength. It is mainly used onboard to detect and compare different circuits in a system. PMMC can also be used as a galvanometer.

Advantages

- consumes less power and has great accuracy;
- has a uniformly divided scale and can cover an arc of 270 degrees;
- has a high torque to weight ratio;
- can be modified as ammeter or voltmeter with suitable resistance;
- has efficient damping characteristics and is not affected by stray magnetic field; and
- produces no losses due to hysteresis.

Disadvantage

- can only be used on DC supply as the reversal of current produces a reversal of torque on the coil;
- very delicate and sometimes uses AC circuit with a rectifier;
- costly as compared to moving coil iron instruments; and
- may show an error due to loss of magnetism of permanent magnet.

8.5 MOVING IRON (MI) INSTRUMENT

This instrument is one of the most primitive forms of measuring and relay instrument. Moving iron type instruments are of mainly two types. Attraction type and repulsion type instrument.

Whenever a piece of iron is placed near a magnet it is attracted by the magnet. The force of this attraction depends on the strength of the said magnetic field. If the magnet is an electromagnet then the magnetic field strength can easily be increased or decreased by increasing or decreasing current through its coil.

As the piece of iron can be brought into the field of the magnet irrespective of its North or South polarity, the instrument is fit to be used in AC circuits.

Accordingly, the attraction force acting on the piece of iron would also be increased and decreased. Depending on this simple phenomenon attraction type **moving iron instrument** was developed.

Whenever two pieces of iron are kept side by side and a magnet is brought near to them the iron pieces will repel each other. This repulsion force is due to same magnetic poles induced in same sides of the iron pieces due to the external magnetic field. This repulsion force increases if field strength of the magnet is increased. Like case if the magnet is an electromagnet, then magnetic field strength can easily be controlled by controlling input current to the magnet. Hence, if the current increases the repulsion force between the pieces of iron is increased and if the current decreases the repulsion force between them is decreased. Depending on this phenomenon repulsion type moving iron instrument was constructed.

Moving iron attraction type instruments

The basic working principle of these instruments is very simple that a soft iron piece if brought near the magnet gets attracted by the magnet.

The construction of the attraction type instrument is shown in Fig. 8.13.

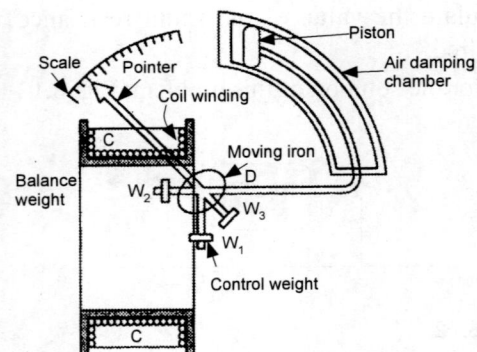

Fig. 8.13 Moving iron attraction type instruments

It consists of a fixed coil (C) and moving iron piece D. The coil is flat and has a narrow slot-like opening. The moving iron is a flat disc which is eccentrically mounted on the spindle. The spindle is supported between the jewelled bearings. The spindle carries a pointer which moves over a graduated scale. The number of turns of the fixed coil depends on the range of the instrument. For passing current through the coil only a few turns are required.

The controlling torque is provided by the springs but gravity control may also be used for vertically mounted panel type instruments.

The damping torque is provided by the air friction. A light aluminum piston is attached to the moving system. It moves in a fixed chamber. The chamber is closed at one end. It can also be provided with the help of van attached to the moving system.

The operating magnetic field in moving iron instruments is very weak. Hence, eddy current damping is not used since it requires a permanent magnet which would affect or distort the operating field.

Moving iron repulsion type instrument

These instruments have two vanes inside the coil, the one is fixed and other is movable.

When the current flows in the coil, both the vanes are magnetised with like polarities induced on the same side. Hence, due to repulsion of like polarities, there is a force of repulsion between the two vanes causing the movement of the moving van. The repulsion type instruments are the most commonly used instruments.

The two different designs of repulsion type instruments are:
(i) Radial vane type
(ii) Co-axial vane type

Radial vane repulsion type instrument

Figure 8.14 shows the radial vane repulsion type instrument. Out of the other moving iron mechanism, this is the most sensitive and has most linear scale.

The two vanes are radial strips of iron. The fixed vane is attached to the coil. The movable vane is attached to the spindle and suspended in the induction field of the coil. The needle of the instrument is attached to this vane.

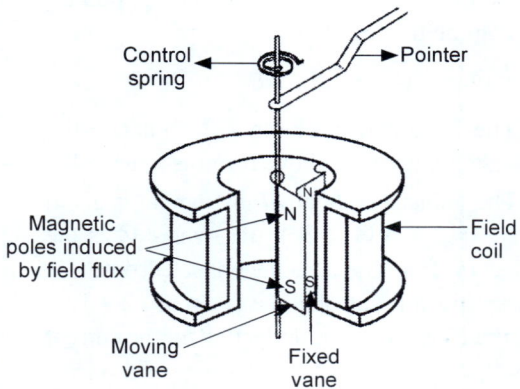

Fig. 8.14 Radial vane repulsion type instrument

Even though the current through the coil is alternating, there is always repulsion between the like poles of the fixed and the movable vane. Hence, the deflection of the pointer is always in the same direction. The deflection is effectively proportional to the actual current and hence the scale is calibrated directly to read amperes or volts. The calibration is accurate only for the frequency for which it is designed because the impedance is different for different frequencies.

Concentric vane repulsion type instrument

Figure 8.15 shows the concentric vane repulsion type instrument. The instrument has two concentric vanes. One is attached to the coil frame rigidly while the other can rotate coaxially inside the stationary vane.

Both the vanes are magnetised to the same polarity due to the current in the coil. Thus, the movable vane rotates under the repulsive force. As the movable vane is attached to the pivoted shaft, the repulsion results in rotation of the shaft. The pointer deflection is proportional to the current in the coil. The concentric vane type instrument is moderately sensitive and the deflection is proportional to the square of the current through the coil. Thus, the instrument is said to have square low response. Thus, the scale of the instrument is non-uniform in nature. Thus, whatever may be the direction of the current in the coil, the deflection in the moving iron instruments is in the same direction. Hence, moving iron instruments can be used for both AC and DC measurements. Due to square low response, the scale of the moving iron instrument is non-uniform.

Advantages

The various advantages of moving iron instruments are:
(1) The instruments can be used for both AC and DC measurements.
(2) As the torque to weight ratio is high, errors due to the friction are less.

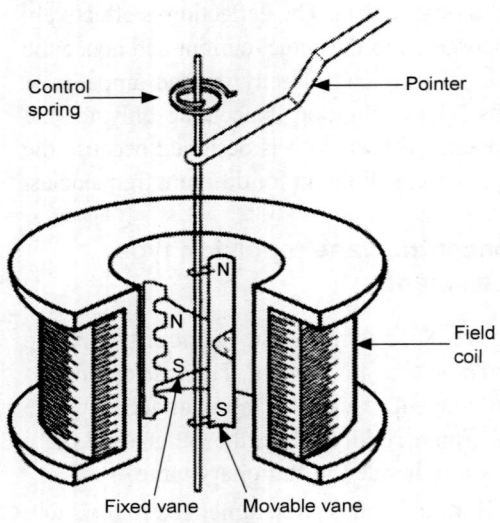

Fig. 8.15 Concentric vane repulsion type instrument

(3) A single type of moving element can cover the wide range hence these instruments are cheaper than other types of instruments.
(4) There are no current carrying parts in the moving system hence these meters are extremely rugged and reliable.
(5) These are capable of giving good accuracy. Modern moving iron instruments have a DC error of 2% or less.
(6) These can withstand large loads and are not damaged even under severe overload conditions.
(7) The range of instruments can be extended.

Disadvantages

The various disadvantages of moving iron instruments are:
(1) The scale of moving iron instruments is not uniform and is cramped at the lower end. Hence, accurate readings are not possible at this end.
(2) There are serious errors due to hysteresis, frequency changes and stray magnetic fields.
(3) The increase in temperature increases the resistance of coil, decreases stiffness of the springs, and the permeability and hence affect the reading adversely.
(4) Due to the nonlinearity of B-H curve, the deflecting torque is not exactly proportional to the square of the current.
(5) There is a difference between AC and DC calibration on account of the effect of inductance of the meter. Hence, these meters must always be calibrated at the frequency at which they are to be used. The usual commercial moving iron instrument may be used within its specified accuracy from 25 to 125 Hz frequency range.
(6) Power consumption is on higher side.

The deflecting torque produced in the MI instrument is given by the formula

$$T_d = \tfrac{1}{2} I^2 \, dL/d\varnothing$$

And the angle of deflection of the pointer is given be the formula

$$\varnothing = \tfrac{1}{2} (I^2/K) \, dL/d\varnothing.$$

The following problem will enhance the understanding of the above formula practically.

The inductance of a MI instrument is given by $L = (12 + 6\varnothing - \varnothing^2)$, µH. where $\varnothing$ is the deflection from the zero unconnected position. The spring constant shall be taken as 12×10^{-6} N·m/rad. Calculate the deflection for the current of 8 A.

Note that the inductance is nonlinear and it is position ($\varnothing$) dependent because of the varying air gap structure of the magnetic circuit.

In this problem $dL/d\varnothing = 6 - 2\varnothing$ µH/rad, which is $(6 - 2\varnothing) \times 10^{-6}$, H/rad.

Substituting all the relevant values in the equation for Ø, the value of Ø is obtained as 2.526 rad or 144°.

The above two sections discussed the major ammeter and voltmeter used in DC as well as AC electric circuits. The next section introduces the concept of electric power measuring instruments.

8.6 DYNAMOMETER TYPE WATTMETER

A dynamometer type wattmeter is most commonly employed for measurement of power in AC as well as DC circuits. The principle of dynamometer type wattmeter is based on the principle that mechanical force exists between two current carrying conductors.

Construction of dynamometer type wattmeter

It essentially consists of two coils, namely, fixed coil and moving coil. The fixed coil is split into two equal parts which are placed close together and parallel to each other. The moving coil is pivoted between the two fixed coils and is placed on the spindle to which the pointer is attached.

The power is a product of the current and voltage in the branch of the circuit where the electric power is actually measured. So the wattmeter will have two coils respectively, to measure the current and the voltage. Current being a series quantity, the current coil will be provided in series in this instrument, and the voltage being an across quantity, the voltage coil will be provided in parallel in this instrument.

The fixed coils are connected in series with the load and carry the circuit current. It is, therefore called current coil. The moving coil is connected across the load and carries current proportional to the voltage. It is therefore called potential coil. Generally, a high resistance is connected in series with potential coil to limit the current through it. Refer to Fig. 8.16.

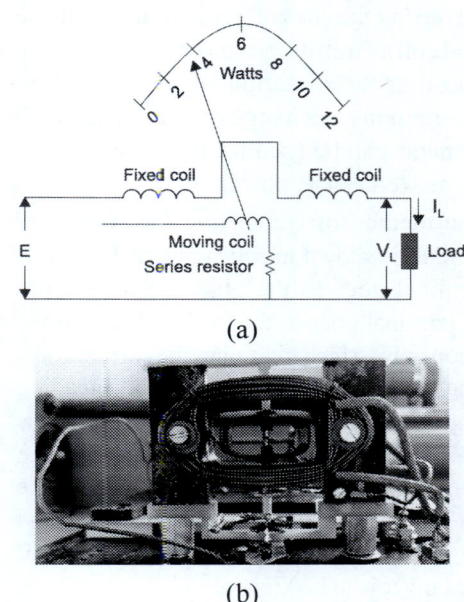

Fig. 8.16 Dynamometer type wattmeter: (a) schematic, (b) assembly

The controlling torque is provided by springs which also serve the additional purpose of leading current into and out of the moving coil. Air friction damping is employed in such instruments.

Working of dynamometer type wattmeter

When power is to be measured in a circuit, the instrument is suitably connected in the circuit. The current coil is connected in series with load so that it carries the circuit current. The potential coil is connected across the load so that it carries current proportional to the voltage.

Due to the current in the coils, mechanical force exists between them. The result is that the moving coil moves the pointer over the scale. The pointer comes to rest at a position where deflecting torque is equal to the controlling torque.

Reversing the current, reverses the field due to fixed coil as well as the current in the moving coil so that the direction of the deflection torque remains unchanged. Therefore, such instruments can be used for the measurement of AC as well as DC power.

The method of connecting the wattmeter in a circuit is shown in schematic of Fig. 8.17. M terminal goes to the phase of the supply and L terminal goes to the load. C is called the common point which multiplies the current coil's reading (current) and the potential coil's reading (voltage) to get the power being measured. V is the voltage terminal, which is chosen from the range of the voltages of the wattmeter as per the circuit's requirement. CV is the voltage coil and hence is kept across the branch of the circuit whose power is to be known.

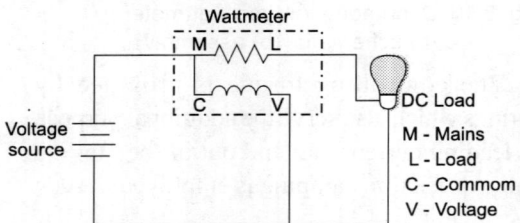

Fig. 8.17 Dynamometer type wattmeter circuit diagram

Dynamometer type wattmeter instruments have uniform scale.

8.7 SINGLE- AND THREE-PHASE WATTMETER AND ENERGY METER

There are four main parts of the operating mechanism. (Refer to Fig. 8.18.) They are:

(i) Driving system
(ii) Moving system
(iii) Braking system
(iv) Registering system

(i) **Driving system:** The driving system of the meter consists of two electromagnets. The core of these electromagnets is made up of silicon steel laminations. The coil of one of the electromagnets is excited by the load current. This coil is called the current coil. The coil of second electromagnet is connected across the supply and, therefore, carries a current proportional to the supply voltage. This coil is called the pressure coil.

Consequently, the two electromagnets are known as series and shunt magnets respectively. Copper shading bands are provided on the central limb. The position of these bands is adjustable. The function of these bands is to bring the flux produced by the shunt magnet exactly in quadrature with the applied voltage.

(ii) **Moving system:** This consists of an aluminum disc mounted on a light alloy shaft. This disc is positioned in the air gap between series and shunt magnets. The upper bearing of the rotor (moving system) is a steel pin located in a hole in the bearing cap fixed to the top of the shaft. The rotor runs on a hardened steel pivot, screwed to the foot of the shaft. The pivot is supported by a jewel bearing. A pinion engages the shaft with the counting or registering mechanism.

(iii) **Braking system:** A permanent magnet positioned near the edge of the aluminum disc forms the braking system. The disc moves in the field of this magnet and thus provides a braking torque. The position of the permanent magnet is adjustable, and therefore braking torque can be adjusted

Measuring Instruments 415

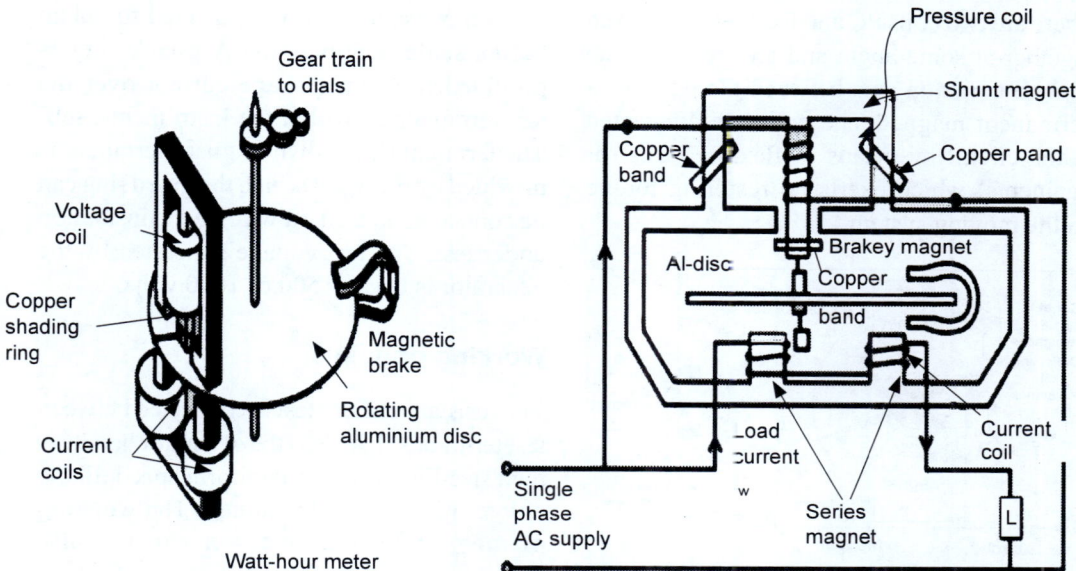

Fig. 8.18 Induction type wattmeter/energy meter

by shifting the permanent magnet to different radial positions as explained earlier.

The construction of induction type instrument, to act as a wattmeter, ends at this stage. For energy meter, the electric energy is obtained as a product of electric power over a period of time. Thus, a continuous record of energy is to be made, over the observation time period. A counting mechanism is employed for this purpose.

(iv) **Registering (counting) mechanism:** The function of a registering or counting mechanism is to record continuously a number which is proportional to the revolutions made by the moving system. By a suitable system, a train of reduction gears, the pinion on the rotor shaft drives a series of five or six pointers. These rotate on round dials which are marked with ten equal divisions.

8.8 INSTRUMENT TO MEASURE RESISTANCE

The megger insulation tester is an instrument used for measuring high resistances of the order of mega ohms and for testing the insulation resistance. Working principle of a megger is based on the working principle of moving coil instruments, which states that when a current carrying conductor is placed in a magnetic field, a mechanical force is experienced by it. The magnitude and direction of this force depend on the strength and direction of the current and magnetic field.

For measurement purpose, ohm meters are connected to a component which is removed from the circuit.

Construction of a megger insulation tester

It consists of (Fig. 8.19) a hand driven DC generator and a direct reading ohm meter.

There are two coils PC and CC which are fixed together at some angle and are free to rotate about a common axis between the poles of a permanent magnet. The coils are connected in the circuit by means of flexible leads (or ligaments) which exerts no restoring torque on the moving system.

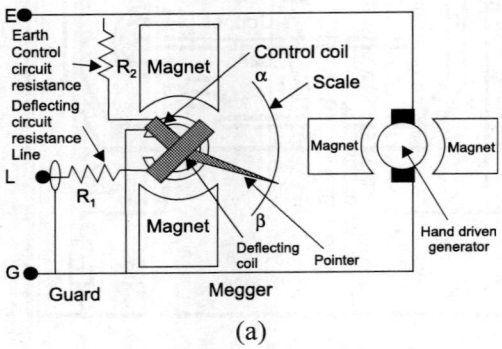

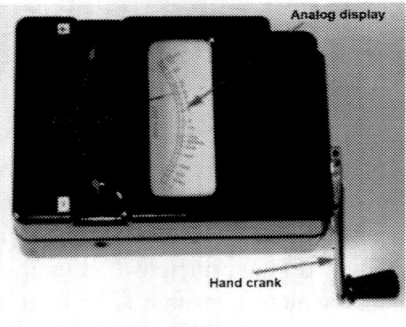

Fig. 8.19 Megger, a high resistance measuring instrument: (a) schematic, and (b) assembly

The current (or deflecting) coil is connected in series with a resistance (R_1) between the generator terminal and the test line terminal. The deflecting circuit resistance (R_1) limits the current and controls the range of the instrument. The pressure (or control) coil is connected across the generator terminals in series with a compensating coil and protection resistance (or control circuit resistance) R_2.

Compensating coil is connected to obtain better scale proportions. A guard ring is provided to shunt leakage current over the test terminals or within the instrument itself. The terminal (G) known as guard terminal, is provided by means of which the guard ring can be connected to a guard wire on the insulation under test. The test voltage generated by the generator is usually 500 or 1000 volts.

Working principle

The resistance under test is connected between test terminals (L and G) the generator handle is then steadily turned at a uniform speed till the pointer gives a steady reading. The **working of megger** insulation tester can be fully understood from the following steps:

- **Step 1.** When the test terminals are open, the resistance to be measured is infinite. In case the generator handle is rotated, the generated voltage sends current through the potential coil and no current flows in the current coil. Therefore, the moving system rotates in such a direction that the pointer rests at infinity end of the scale.
- **Step 2.** If the test terminals are short-circuited and the generator is operated, it sends a large current through the current coil and a very small current flows through the potential coil. Therefore, the resultant torque so produced turns the pointer to zero end of the scale.
- **Step 3.** If the unknown resistance to be measured is connected between test terminals, an appreciable amount of current flows in both the coils. The actual position taken up by the pointer depends on the ratio of currents in the two coils, i.e., upon the unknown resistance.

CONCLUSION

This chapter provided an introductory treatment electrical measuring instruments. Various types, forces involved in operating a measuring instrument, ammeter, voltmeter, wattmeter, energy meter and resistance measuring instrument were majorly dealt with, along with preliminary illustrations for a better understanding.

QUESTIONS

1. Explain the role of three torques that are employed by a measuring instrument.
2. Why a measuring instrument has to be calibrated?
3. Explain the types of controlling torque producing mechanisms employed in a measuring instrument.
4. In the damping torque mechanism, explain what happens to the instrument when it is over-damped, under-damped and critically damped.
5. Explain how a MC meter used as an ammeter and as a voltmeter.
6. Discuss how readings are made in a permanent magnet meter.
7. Discuss why the scale of an MI meter is clumsy near the start position of the needle.
8. Explain how voltage and current of the power being computed by a wattmeter is measured by it.
9. Explain how time is incorporated in the Joules measurement of an energy meter.
10. Discuss the construction and operation of a megger.

INTRODUCTION

This chapter provides information about the wiring subject of electrical engineering and its rudiments to a beginner.

CHAPTER OBJECTIVE

At the completion of this chapter, the reader will be able to:
* understand the wiring, and its various types;
* understand the accessories used for wiring.

KEYWORDS

Types of domestic wiring, Wiring materials, Types of lighting and earthing.

9
Domestic Wiring

9.1 DOMESTIC WIRING

The connection of lamps, fans, domestic appliances like the fridge, mixies, heater, audio/video systems in the customer required fashion and by fulfilling the electrical safety precautions and IS rules is known to be the domestic electrical wiring.

The above said items are generally known as the electrical loads. Thus, an electrical load is a one which will consume electrical power when energised.

These appliances (i.e., loads) are connected to the supply mains by wires. This process is known as the wiring.

The wiring could be done through different methods. It needs wiring materials and accessories to complete the wiring process. Precautionary measures must be undertaken when the wiring is done. Some defined rules are adopted to carry out wiring.

This chapter, as a summary, deals with these points.

9.2 WIRING MATERIALS AND ACCESSORIES

Essential accessories to undertake and complete the domestic wiring are listed below, with a brief explanation about each.

Plier: The pliers are used to hold the wires in position to carry out the job. It is of two types: the **nose plier** which has long tapered slot to hold the wire under work. It is also used to tighten or loosen the small nuts and screws. It usually cannot cut the wires.

The other type is the **cutting plier** which, as its name implies, is used for cutting thin wires or strips, and removing the insulators from the insulated wires.

Note that when the connections are to be made between the wires and a load, the insulator of the wire near joint must be removed; cutting pliers are highly useful at this place.

Flexible wires: These are otherwise simply known as wires. It is the main material in the domestic wiring which links the load to the supply. It basically consists of copper conductors of fewer strands insulated with

Indian rubber and vulcanised. The insulation protects the user even when he/she touches the wire under operation.

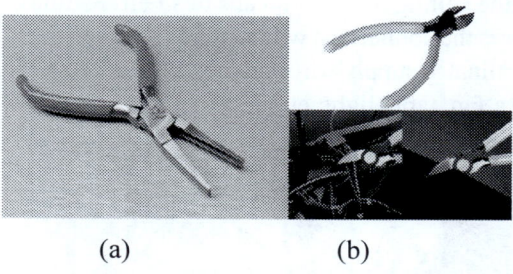

(a)　　　　　　(b)

(a) Nose plier, and (b) cutting plier

Depending on the type of insulation which covers the bare conductor, the wires are classified as PVC wires in which the insulator used is polyvinyl chloride, VIR wires which use the vulcanised Indian rubber, etc.

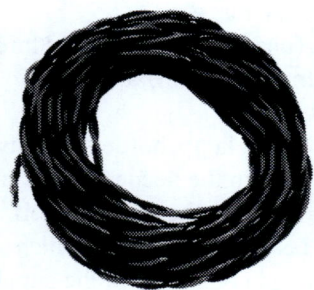

(c) Flexible wires

Screwdriver : It is used to drive-in or drive-out the screw or to keep the screws in position. Its size ranges from 1.5 inches to a maximum 2.5 feet.

The handle of the screwdriver is usually of wood or plastic which serves as the insulator, and is voltage-graded which means that the screwdriver can be used on the live-line up to the mentioned voltage.

Hammer: A most common accessory used in wiring to pass in the nails and for related operations.

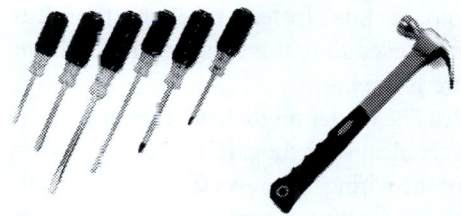

(d) Screwdriver　　(e) Hammer

Hand drill: The hand drill is very often used in domestic wiring. When the wire passes from, say, one room to other, a small hole may be needed to be made in the door beam separating the rooms so that the wire could be taken to the other room; hand drill finds utility here. Also in house wiring it is very often required to drill hole in the wooden boxes to facilitate the passage of wire which terminates into switches or other fittings.

It consists of a tip of hardened steel jaw into which is placed the drill of needed size. Rotation of the handle forward will drill the bit into the wood thus making a hole in the wooden board.

(f) Hand drill

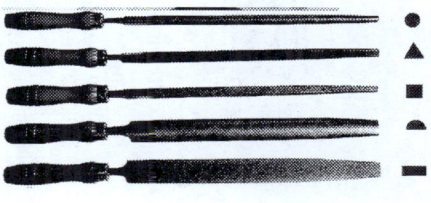

(g) File

Knife: It is a general knife used for removing the insulation in the wire tips to make connection. Instead of pliers, the ideal accessory used to remove the insulation from the wire is the knife.

File: The reader might have used the file in the mechanical workshop. It is also widely used in domestic wiring to remove the waste or rough or extra or unwanted portion of the material under use. It's length can go upto 1 foot.

It has throughout its length, on the surface, rough grooves which when moved on the surface of the job, removes the waste or smoothens its surface. It is mainly used to cut the meter parts such as rods, pipes, stripes, etc., used for wiring. This comes in two fashions, viz., solid frame or non-adjustable frame hack saw and adjustable frame hack saw.

Its main part is the blade with several teeth which cut materials when the saw is pushed forward.

Switches: The switches are used in domestic wiring to connect the load to the supply. They are made up of insulating materials like porcelain, plastic, china clay, etc. They are mainly of two types, viz., single-way switch and two-way switch. The one-way switches are used for wiring circuits in which a load is to be controlled by only one switch on the other hand, two-way switches are used for using circuits in which a load is to be controlled from two points independently.

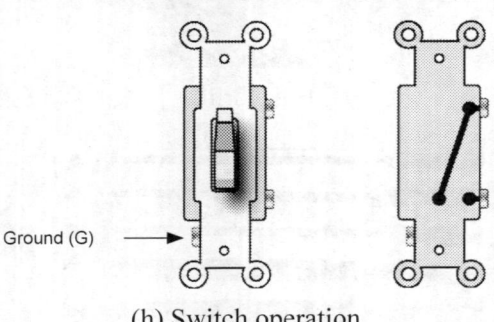

(h) Switch operation

When the contact touches and closes the circuit, the supply will be given for load (ON). If not and if the contactor is left open as shown, then no supply will be given (OFF).

Ceiling rose: Ceiling roses are otherwise can be referred to as tapping points. The wiring made from switch to the load will terminate in the ceiling rose. By means of small flexible wires the connection will be made between the terminated supply wire and the load itself (may be a fan, table light, etc).

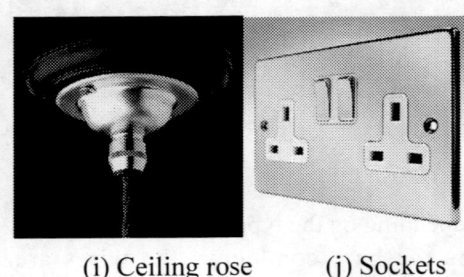

(i) Ceiling rose   (j) Sockets

Sockets: Socket outlets are used where connections are to be made temporarily. They are generally used to connect loads like table lamp, audio systems, etc.

Two-pin (phase and neutral) and three-pin (phase, neutral and earth) are the two main socket types.

Holders: The lamp holders are used to (i) hold the lamp in the needed place, and (ii) connect the same to the supply itself. They come in two fashions. A holder in which a lamp can be fixed and connected to the supply; its position cannot be changed.

Another holder which can be rotated so that the illumination of the light can be concentrated at the needed place. It is of course, an advantageous aspect.

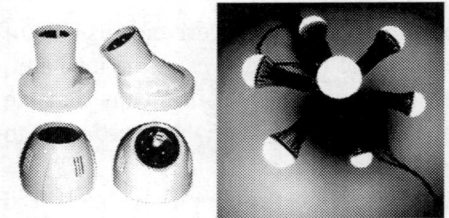

(k) Stationary and rotatable lamp holders

Fig. 9.1.a to 9.1.k - Wiring accessories

Domestic Wiring

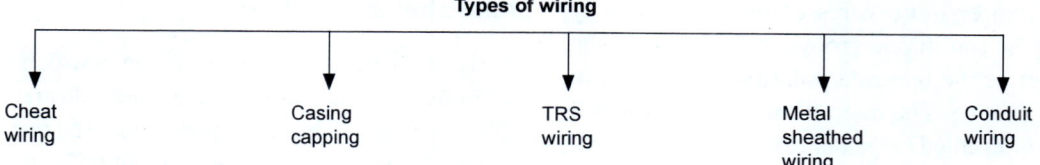

Fig. 9.2 Types of wiring

9.3 TYPES OF WIRING

The wiring system to be chosen depends on various factors like safety, appearance, whether the wiring to be made is permanent or temporary, the users budget, cost, etc. As shown in Fig. 9.2, the wiring system can be broadly classified into five types.

Cleat wiring : In this system the conductor used is usually the VIR (vulcanised Indian rubber) conductor. In cleat wiring the conductors are supported in the porcelain cleats.

The cleats used in this system of wiring are made up of two halves one with grooves to receive the conductor and the other without grooves (Fig. 9.3) The whole assembly is fixed on the walls by means of screws which further tightens the grip of the wire between the two halves of the cleat. The wires must be laid stretched between the cleats, that is, properly held between the cleats to avoid contact with the wall.

The system is most suitable for temporary wiring. The inspection and alterations can be easily made. It is very neat and clean at the time of erection. However, after some time the dust collection will be severe. It also sags at some places over its length.

This kind of wiring is decades old fashion, and is found only in ancestor houses. They are obsolete in the modern trend.

Wooden casing capping: In the cleat wiring cleats were porcelain. In wooden casing type the cleats are wood. This system of wiring is most commonly used for residential buildings.

It consists of rectangular wooden blocks called **casing** made up of teak wood. It has usually two grooves into which the wires are laid. Thin casing is covered at the top by means of a rectangular strip or pad made up of wood of width same as that of casing; this is known as the **capping**. After inserting the wire into the grooves of the casing, the capping will be kept on the casing covering the conductor and is screwed.

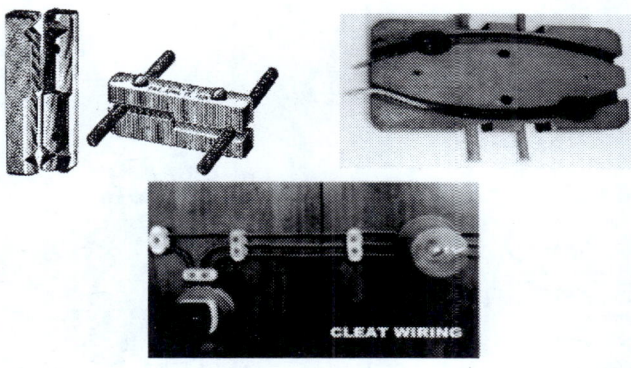

Fig. 9.3 Cleat wiring

Two or more wires of the same polarity may be run in one groove and in no case the wires of the opposite polarity may be run in one groove. The main disadvantage is that it needs a skilled carpenter to carry out this work. Refer to Fig. 9.4.

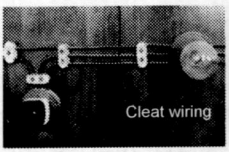 

Fig. 9.4 Wooden wiring

T.R.S.: The T.R.S. (known as the tough rubber sheathed) wiring is a cable system of wiring. (Cable is nothing but a conductor system in which the current carrying conductor will be the innermost part known as the core; it will be covered with several layers of insulators, to protect the cable from tampering and the people from shock.). The T.R.S. cables are available with single, twin (double) or three cores.

The cable is quite flexible so that it can be bent according to the need; the insulation of the cable can withstand rough usage, moisture, acids, etc.

This system does not need any supports as cleats or cappings as in the previous systems. It just can be run on the wall; only point is, it is kept fixed on the wall by means of clamps kept at equal distances. But care must be taken to ensure that the fixing of the clamps does not stress the cable much.

The cable system used in T.R.S. is shown in Fig. 9.5.

Metal sheathed wiring

This is another cable system of wiring in which the conductors used as the core of the cable are rubber insulated. The cable core is covered by a layer of lead alloy containing about 95% of lead which provides a protection to the cable from the mechanical injury.

Care must be taken to earth the lead covering, otherwise the lead covering may become alive causing a severe shock. (The need and the process of earthing is discussed later.) The cable ran on the walls are held by means of metal clips. Sharp bends should be avoided if the cable is to be bent.

Conduit wiring: An excellent system of wiring which provides mechanical protection, safety against shock, firing and provides a clean look, is the conduit wiring.

In this system of wiring the conductors joining the load and the switch are run in metallic tubes or rigid heavy gauge steel tubes or PVC pipes, known as the conduits. The conduits will be held on the wall by means of metallic chips.

Before the brickwork of the wall is plastered by cement, the above-mentioned wiring work will be made. Once the wire is run in the conduit and proper connections are made at the load and switch ends, the wall will be plastered. Thus, the conduits will never be visible, giving a neat look to the building.

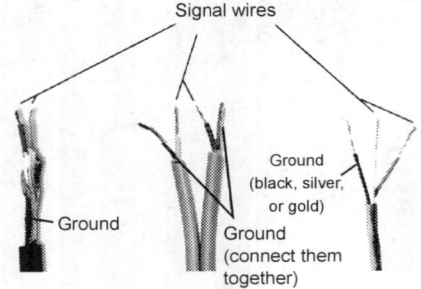

Fig. 9.5 T.R.S. wiring

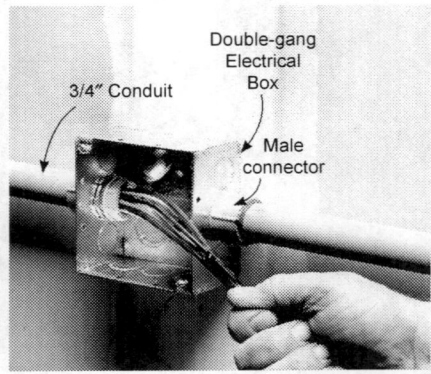

Fig. 9.6 Conduit and wires

Domestic Wiring

The main disadvantage of the system is maintenance. To rectify any major fault in the wiring work, the wall may have to be chiseled; of course, such fault is a rare occurrence.

Fig. 9.7 A complete set of conduit wiring, which will be finally cemented.

9.4 SIMPLE DOMESTIC WIRING LAYOUTS

Some wiring circuits are given below and each is clearly explained. Usually, single-phase supply is given to the domestic loads. A single-phase supply has a phase and a neutral. It is specified here with a P and N.

(a) **Single load wiring:** A lamp controlled by a single-way switch is in Fig. 9.8. Of course, instead of a lamp, a fan or any electrical appliance can be connected.

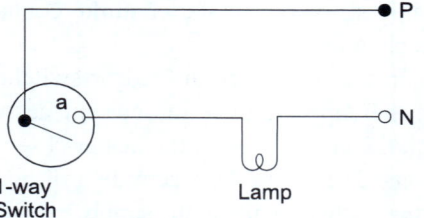

Fig. 9.8 When the contact is at 'a', light is ON. If not, OFF.

(b) **Corridor lighting circuits:** Let the corridor have a number of lamps and let each lamp be controlled by two switches. (One switch may be inside the room and the other outside.) In such case lamps L_1 and L_2 are different lamps having controls at switches S_1–S_2, S_3–S_4, respectively. Refer to Fig. 9.9.

Trace the path. Let lamp 1, L_1, be considered. When the contact of the switch S_1 is at position b and that of the switch S_2 is at position b', the circuit path is $P - q_1 - r_1 - b - b' - N$. The circuit is closed. The bulb glows. If now the contact is moved to a' in switch S2, the above part becomes open; the bulb is made OFF. [Reader may do similar tracing for the lamp 2, L_2, also].

Two-way switches are used.

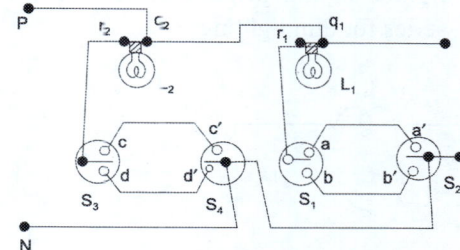

Fig. 9.9 Corridor lighting

(c) **Series lamp control:** In such circuits several lamps will be connected in series to the supply through a switch. When the switch is 'ON' all lights will burn; the brightness will be very poor and the lights will be dim, when the lamps are put in series.

It is depicted in Fig. 9.10. When the contact is in position 'a', the lamps L_1, L_2, L_3 will be ON. (They will give a dim light); otherwise they will be OFF.

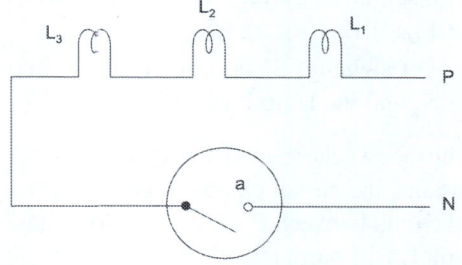

Fig. 9.10 Series connection

(d) Series-parallel connection: Using a two-way switch, such connection can be achieved. In the circuit of Fig. 9.11, the lamp L_1 is lighted to full brightness when the switch contact position is at **a**. When the contact is moved to position **b**, the whole of circuit is 'OFF'. When the contact is at the position **c**, the lamps L_1 and L_2 will be in series so as to give a dim light. Thus, with such an arrangement either lamp L_1 can be lighted to full brightness or both in series for dim lighting.

circuit **P – L – b – d –** is open. (Note still switch S_2 is in position a only; it is undisturbed). Thus, the light L **is made OFF**.

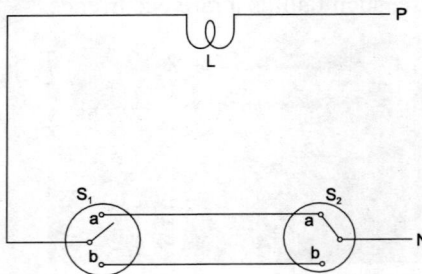

Fig. 9.12 Staircase wiring

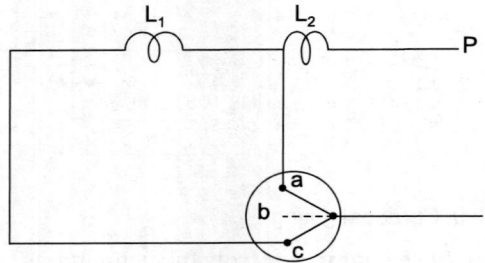

Fig. 9.11 Series-parallel circuit

(e) Stair-case connection: It is another quite interesting type of connection in which a lamp will be controlled by two switches. It gets the name as staircase connection because it is highly useful in the multi-storeyed buildings in which a bulb switched ON at one floor will be switched OFF from the other floor.

Such circuit arrangement is shown in Fig. 9.12.

Let switches S_2 be floor 1 and switch S_1 and the lamp L be in floor 2.

With the switches S_1 and S_2 at their respective **a** positions, the circuit can be found to form a closed circuit between **P – L – a – C – N**. Thus, the light L will burn. ('ON')

After going to the floor 2 if the switch is made 'OFF' (i.e., contact is moved to 'b'), the

9.5 FLUORESCENT TUBE LIGHTING

The fluorescent tube circuit is shown in Fig. 9.13.

The main parts of fluorescent tube are the choke (an inductive circuit, L), the starter, the capacitor C and the closed tube; T which is filled with fluorescent material.

The thermal starter consists of two bimetallic strips and a heater coil, H. When no current passes through the circuit, (that is, when the tube light is in OFF state) the two bimetallic strips of the starter **make** contact with each other.

When the tube in connected or switched ON to the supply mains, the current passes through the choke L → to the heater H → to the electrode **P** → to the electrode q through the strips → then finally to the supply mains N itself, thus completing the circuit. (The reader may trace this path.)

Now, as the current passes through the electrodes of the tube, they are heated and produce electrons into the tube path. At this same instant the heater element will also be heated up and will heat the metallic strips; this heating will cause them to get separated apart. This is the first important stage in the operation.

The separation of the strips causes the current in the circuit to get interrupted. (That is, as the strips are separated there will be no closed path, thus the current gets interrupted.)

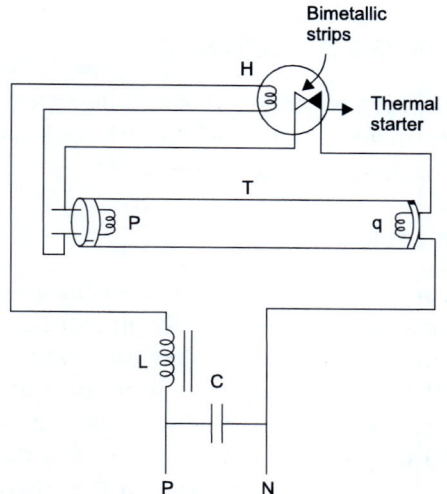

Fig. 9.13 Fluorescent tube-light

9.6 METHODS OF EARTHING

Pipe Earthing

A galvanised iron pipe of approved length and diameter is used as the electrode. The size of the pipe depends on (a) the current to be carried, and (b) the type of soil.

It is seen in an earlier chapter that the change in current induces enough voltage in the inductor of the circuit. Here too it happens. The sudden interruption in the circuit current causes a high voltage surge across the electrodes of the tube, which will be sufficient to put the tube into operation. Thus the choke (L) is just to produce a sudden voltage surge for starting the tube. The tube will glow with full brightness.

The colour of these depends upon the fluorescent powder (illuminating) material used inside the tube. The powder used as a fluorescent material is activated by the ultraviolet rays generated in the tube. The tube is mainly used to convert these ultraviolet rays to visible radiation.

These tubes are very popular as an 80W tube, which are equivalent to a 250W incandescent lamp (the ordinary lamp). Thus the energy consumption (and the bill) get reduced. Technical advantage is that, they produce no glare.

9.7 FUSE

It is an important part of the electrical wiring. It will be kept at the very beginning of the wiring. That is, the electricity board supply will be terminated at the cut-out (mains); from here to all the loads of the house the phase and neutral will be run through a fuse.

Fuse is a protective element. If the current drawn by a wiring or any appliance exceeds from the value which the wiring or the appliance cannot withstand, it is not allowed to flow to the appliance (b) inserting a wire (known as a fuse) which automatically melts and breaks (opens) the circuit. Thus the appliance is protected. Refer to Fig. 9.14.

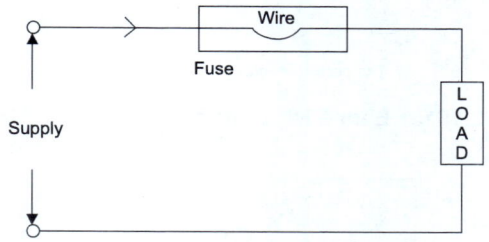

Fig. 9.14 Fuse schematic

If the current (I) drawn from the supply exceeds a limit which cannot be withstood by the load, the fuse wire will melt and disconnect the appliance from the supply.

There are several ampere ranges for the fuse wires. According to the current limit needed, it must be properly chosen.

9.8 EARTHING

It is always a practice to take the earth's potential a zero for all practical purposes. Hence, any electrical appliance, or any part of it or any machine when connected to the earth, that body attains zero potential and is said to be 'earthed'. Thus, the voltage of the earthed body will attain a zero potential.

Always, the frames and bodies of all electrical loads are to be earthed, so that, in case of any leakage of current, the chances of fire or the chances of energising the human body touching is eliminated by allowing the leakage current to flow to the earth immediately (than through the human body in contact). The fuse will then blow off, disconnecting the machine from the supply.

Figure 9.15-a gives the parts of the leakage current to the earth when it is earthed, thus saving the human body in contact.

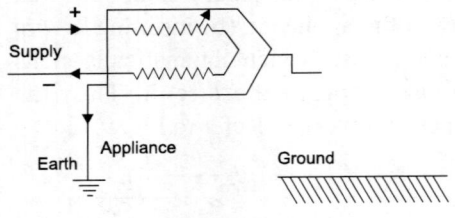

Fig. 9.15 (a) Earthed–the person is safe.

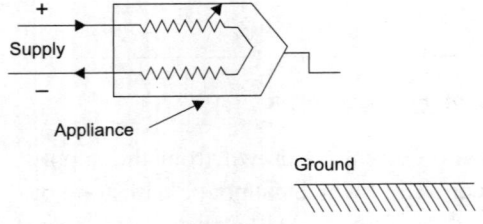

Fig. 9.15 (b) Unearthed–the person received shock.

Figure 9.15-b shows the danger of un-earthing. Here the appliance is not earthed; the leakage current is pasing through the body in contact, which may even kill him.

9.9 TYPES OF EARTHING

The main components used in earthing system mainly include earth cable, earthing joint (earthing lead), and earth plate.

Earth cable

The conductor is used to connect metallic parts of an electrical system like plug sockets, metallic shells, fuses, distribution boxes. Metallic parts of motors, transformers, generators, etc. The range of these conductors depend on the earth cable size used in the wiring circuit. The **earth wire** in the cross-sectional area must be less than the solid wire used in the electrical wiring system.

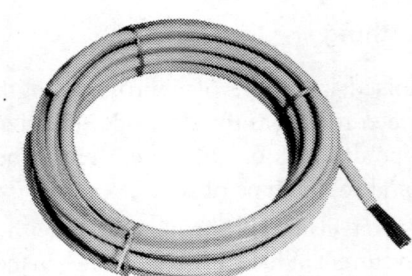

Fig. 9.16 Earthing cable.

In general, the copper wire utilized as an earth continuity conductor size is 3 - standard wire gauge (SWG). Ground wires which are smaller than 14-SWG should not be used. In some situations, copper strips are used instead of a bare copper conductor.

Earthing joint

The 'ground electrode' as well as conductors fixing to the 'ground continuity conductor' is called earthing joint (earthing lead). The tip where the earthing joint connects the ground continuity conductor is known as connecting end. The lead of the ground must be low size, straight, and should include a minimum amount of joints. Although copper wires are usually used as grounding leads; whereas copper strips are selected for high fitting because it carries high fault current values due to its broad region.

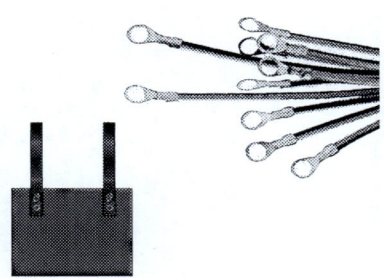

Fig. 9.17 Earthing joint

Earth plate

The last part of the electrical grounding system which is hidden underground and linked to the lead of grounding is known as the earth plate. Earth electrode is a pipe, plate or metallic rod, or plate; which has extremely low resistance for carrying the fault current to the ground safely.

It can be of iron or copper rod and must be placed in wet earth and in case the moisture content of earth is low then put some water in the earth plate. The earth plate is always placed in the vertical, and coat with salt and charcoal lime around the earth plate. This helps in protecting the earth plate as well as maintaining ground moisture around the earth plate. The earth plate must be placed four metres long for the better earthing.

CONCLUSION

This chapter provided rudiments of domestic wiring. Types of wiring and wiring accessories are discussed. The need for earthing and methods to achieve it were also pointed out.

QUESTIONS

1. What is wiring? How does the house wiring and industrial wiring differ?
2. What are the materials used for conduits to do wiring?
3. Complete a domestic wiring with 2 rooms, each having 2 lights, 1 fan and 4 sockets.
4. What are the various types of fluorescent lighting? Narrate them.
5. Explain the process of diverting the current to ground in earthing.
6. What are the advantages and disadvantages of concealed wiring?

Appendix

Table of Laplace Transform

Sl. No.	f(t)	$F_{(s)} = \mathcal{L}[f(t)] = \int_0^\infty f(t)e^{-st}dt$
1.	Unit Impulse $\delta(t)$	1
2.	Unit step	$\dfrac{1}{s}$
3.	t	$\dfrac{1}{s^2}$
4.	t^n	$\dfrac{n!}{s^{n+1}}$
5.	e^{-at}	$\dfrac{1}{s+a}$
6.	te^{-at}	$\dfrac{1}{(s+a)^2}$
7.	$1-e^{-at}$	$\dfrac{a}{s(s-a)}$
8.	$e^{-at}-e^{-bt}$	$\dfrac{b-a}{(s+a)(s+b)}$
9.	$\sin at$	$\dfrac{a}{s^2+a^2}$
10.	$\cos at$	$\dfrac{s}{s^2+a^2}$
11.	$e^{-at}\sin bt$	$\dfrac{b}{(s+a)^2+b^2}$

12.	$e^{-at} \cos bt$	$\dfrac{s+a}{(s+a)^2 + b^2}$
13.	$at - 1 + e^{-at}$	$\dfrac{a^2}{s^2(s+a)}$
14.	$\cosh at$	$\dfrac{S}{S^2 - a^2}$
15.	$\sinh at$	$\dfrac{a}{s^2 - a^2}$
16.	$\dfrac{a-b}{a} + bt - \left(\dfrac{a-b}{a}\right) e^{-at}$	$\dfrac{a(s+b)}{s^2(s+a)}$
17.	$1 + \dfrac{b}{a-b} e^{-at} - \dfrac{a}{a-b} e^{-bt}$	$\dfrac{ab}{s(s+a)(s+b)}$

Index

A

Absolute encoders 334
AC electric machine 188
AC induction motor 292
   running characteristics 295
   starting arrangements for 297
   starting characteristics 294
AC potentiometer 333
AC servomotor 335
Active power 55
   of a pure capacitive circuit 56
   of a pure inductive circuit 55
   of a pure resistive circuit 55
Admittance 97
Agriculture load 386
Air core transformer 259
Air friction damping 405
Alkaline battery 396
Alternating current 35
Alternating quantities, measurement of 39
   representation of 41
Alternator on load 277
Alternator,
   construction of 270
   EMF equation of 274
   principle of operation of 270
Alternators, parallel operation of 286
Ampere 3
Ampere turn formula 152
Amplitude modulation 370
Analog modulation 370
Analogous systems 326
Apparent power 55
Armature control 237
Armature controlled DC servomotor 338
Armature core 190
Armature divertor 238
Autotransformer 259, 300
Average value 39

B

Back EMF 215
Backup battery 396
Basic DC-DC converters 362
Bipolar transistor 359
Blue phase 134
Boost converters (DC-DC) 364
Breakdown 295
Buck converters (DC-DC) 363

Buck-boost converters (DC-DC) 364
Bulk supply 381

C

Cage rotor 291
Capacitive load 386
Capacitor split-phase motor 309
Capacitor-start capacitor-run motor 313
Capacitor-start induction-run motor 313
Capping 423
Ceiling rose 422
Circuit breakers 391
Cleat wiring 423
Closed loop control systems 320
Closed loop systems 321
Coercive force 181
Combustion turbines 376
Commercial load 386
Compensation theorem 115
Compound motors 224
Concentric vane repulsion type instruction 411
Conductance 97
Conductor 5
Conduit wiring 424
Constant frequency operation 365
Control strategies 365
Control system 319
   components of 331
   response analysis of 342
Controlling torque 404, 407
Converters 358
Copper loss 253
Core loss 253
Corridor lighting circuits 425
Critical field resistance 211
Cumulative compound motor 226
Current and voltage, alternative expressions for 7
Current limit 304
   control 367
Current, definition of 2

D

D.C. circuits 2
Damper 322
Damper winding 289
Damping torque 405, 407
DC electric machine 188
DC generator 191
   applications of 196
   characterization of 208
   EMF equations of 196
   no-load voltage characteristics of 208
DC machines 188, 189
   current and voltage equations of 197
DC motor 214
   applications of 223
   characteristics of 223
   electromagnetic torque characteristic of 223
   speed characteristics of 224
   speed control of 236
   speed of 219
   starters for 235
   torque and speed of a motor 220
   torque developed in 216
   types of 218
DC potentiometers 332, 333
DC servomotor 338
Deflecting torque 403, 407
Delay time 348
Delta connections 263
Delta system 134
Diamagnetic materials 177

Differential compound motor 226
Digital modulation 370
Direct magnetization 184
Direct on line starter 297
Distribution factor 275
Distribution feeders 382
Distribution substation 381
Distribution transformer 259, 382
Distributors 382
Domenstic wiring 420
Drag cup type rotor 336
Dynamic devices 188
Dynamic machines 188
Dynamically induced EMF 169
   magnitude of 170
Dynamometer type wattmeter 413

E

Earth cable 428
Earth leakage circuit breaker 393, 394
Earth plate 429
Earthing joint 429
Earthing, methods of 427, 428
Eddy current damping 406
Eddy currents 406
Electric batteries 394
Electric circuit 14
   fundamental laws of 6
Electric current 3
Electric field 149
Electric machines 188
Electric potential 3
Electric potential difference 3
Electric power distribution 381
Electrical loads, types of 385
Electrical system,
   mathematical model of 323
   transfer function of 328

Electrodes 395
Electromagnetic force 168
Electromagnetic induction 166
Electromagnetism 148
   basic terms in 148
Electronic starters 303
Encoders 334
Energy 5
Energy meter 14
Equivalent circuit 296
Equivalent resistance 247

F

Faraday's postulations 166
Ferromagnetic materials 178, 179
   magnetization of 184
Field control 237
Field controlled DC servomotor 338
File 422
First order systems, time response of 344
Flaming's right hand rule 167
Flexible wires 420
Fluid coupling 306
Fluid friction damping 406
Fluorescent tube lighting 426
Flux 149
Flux density 149
Form factor 40
Free electrons, movement of 2
Frequency 272
Frequency modulation 370
Fuse 389, 427
   working principle of 390

G

Generator,
   principle of 192
   types of 195

H

Hammer 421
Hand drill 421
Helmholtz's theorem 111
Holders 422
Hydraulic system 324
Hydro turbines 376
Hysteresis curve of alloys 183
Hysteresis curve of hand alloys 183
Hystresesis loop and magnetic properties 180

I

Impedance 61
Impedance triangle 61
Impulse input 343
Incremental encoders 334
Indirect magnetization 185
Inductive load 386
Inductive reactance 60
Industrial load 386
Industrial substation 381
Initial firing angle 304
Inverter 367
   types of 368, 369
Iron core transformer 259
Iron loss 253

J

Joule 6

K

Kirchoff's current law 11
Kirchoff's voltage law 12
Knife 421

L

Lagging load p.f 279
Lap 190
L-C ac series circuit power in 59
Lead-acid battery 396
Leading load p.f 280
Leakage flux 154
Lenz's law 406
Lines of magnetic flux 149
Liquid level system 325
   transfer function of 328
Lithium-ion battery 396
Load 384
Load shedding 380
Load versus voltage characteristic 213
Loaded alternator 279
Low voltage switchgear 389

M

Magnetic characteristics 176
Magnetic field 148, 149
   energy stored in 173
   strength of 150
Magnetic flux 149
   density 149
Magnetic fringing 155
Magnetic inductions 149
   types of 169
Magnetic leakage 249
Magnetic lines of force 149
Magnetic recording, hysteresis in 182
Magnetization curves 176
Magnetizing force 150
Magnetomotive force 150
Mass element 322
Maximum power transfer theorem 116
MCB (miniature circuit breaker) 392
   characteristics of 393
Measuring instruments,
   calibration of 401
Measurement transformer 259
Measuring instruments 400
Mechanical rotational systems 323

Mechanical systems 321
Mechanical translational system, transfer function of 327
Megger insulation tester 415
Mesh current method 108
Mesh system 134
Metal sheathed wiring 424
Methods of speed control 304
Modulation 368
Modulation index 369
Molded case circuit breaker 393
Moving iron attraction type instruments 410
Moving iron instrument 409
Moving iron repulsion type instrument 410
Multiple voltage control 237
Mutual inductances 171
Mutually induced EMF 171

N

Network analysis and theorems 107
Neutral point 135
Neutral wire 135
Nodal analysis 109
No-load characteristics 250
Norton's equivalent circuit 113
Norton's theorem 113

O

Ohm 4
Ohm's law 6, 62
   in magnetic circuits 151
Open circuit (or no-load) test 260
Open delta connection 264
Open delta transformer 264
Open loop 321
   system 319

P

Parabolic input 343
Parallel ac circuits 63
Parallel admittances 113
Parallel circuit 15
Parallel operation 287
Parallel resonance 93, 98
Parallel resonant circuit, current in 98
Paralleling field winding 238
Paramagnetic substances 177
Peak factor 41
Peak overshoot 348
Peak time 348
Permanent magnet moving coil instrument 407
   advantages 409
   as an ammeter and a voltmeter 408
   disadvantage 409
   principle of operation 408
Permanent magnet stepper motor 339
Permanent magnetic moment 177
Permanent-split capacitor motor 313
Permeability 181
Phase modulation 370
Phase voltage 136
Physical systems, mathematical modeling of 321
Pipe earthing 427
Plant 320
Plier 420
Pneumatic system, transfer function of 329
Pneumatic systems 325, 326
Polar representation 42
Polyphase 134
Polyphase synchronous generator 280
Potential difference 4
Potential transformers 377
Potentiometers 332
Power & energy, expressions for 6
Power 5
Power converter topologies 360

438 Index

Power devices and converters,
 protection of 361
Power electronics 358
Power factor 55
Power semiconductor device 358
Power system 374
 electrical loads in 386
Power transformer 259
Primary distribution lines 383
Primary reactance starter 299
Primary resistance starter 298
Primary step-down substation 380
Protection transformers 259
Pull out torque 295
Pulse modulation 370
Pure capacitor 57
Pure inductor, power in 57
Pure resistor, power in 56

Q
Quality factor 96, 99

R
Radial vane repulsion type instrument 411
Raising voltage to reduce current 378
Ramp input 343
R-C series circuit, power in 59
Reactance 60
Reciprocity theorem 114
Rectangular representations 42
Red phase 134
Registering (counting) mechanism 415
Residual current circuit breaker 394
Residual flux 181
Residual magnetism 181
Resistance 4
 in terms of physical quantities 5
 measuring instrument 415
Resistance split-phase motor 309

Resistive load 386
Resonance 93
Resonant frequency 93, 95
Retentivity 181
Rise time 348
R-l series circuit, power in 58
Root mean square (RMS) value 40
Rotating field 310
Rotor 290
 torque-speed characteristic 308

S
Salient pole, construction of 270
Saturation 179
Screwdriver 421
Second order system, time response of 345
Secondary batteries 395
Secondary distribution 383
Secondary step-down substation 380
Self inductance 171
Self-excited DC compound generator 195
Self-excited DC shunt generators 195
Self-excited series generator 195
Self-excited shunt generator 210
 failure of 212
Self-induced EMF 171
Sensor 320
Series a.c. circuit 63
Series circuit 14
Series lamp control 425
Series magnetic circuits 153
Series motor 223, 225
 starter 236
Series parallel magnetic circuits 153
Series resonance 93, 94
Series resonant circuit, current in 95
Series-parallel circuit 16
Series-parallel connection 426

Index 439

Service mains 382
Servomotors 335
Settling time 348
Shaded-pole motor 313
Shaft torque 217
Short circuit test 261
Shunt generator 213
Shunt motor 223, 225
   starter for 235
Simple domestic wiring layouts 425
Simple harmonic EMF 37
Simple magnetic circuit 153
Sine EMF 37
Single- and three-phase wattmeter 414
Single load wiring 425
Single-phase generator 280
   equivalent circuit for 280
Single-phase induction motor 306, 307
Single-phase motors, types of 312
Sinusoidal EMF 37
Sinusoidal wave 35
   terms associated with 38
Slip ring motors 304
Slip, concept of 292
Smooth cylindrical construction 272
Sockets 422
Speed control, methods of 304
Speed regulations 220
Split-phase motor 309
Spring 322
Squirrel-cage induction motor 292
Stair-case connection 426
Standard test inputs 342
Star 134
Star delta starter 301
Statically induced EMF 170

Stationary armature, advantages of 274
Stator 290
Steady state response 342
Steam turbines 376
Step input (position function) 342
Step-down substation 380
Step-down transformer 259
Stepper motor 339
Step-up substation 380
Step-up transformer 258
Substations 379
Susceptance 97
Switch fuse unit 389, 390
Switches 422
Switchgear 388
Synchronous machines 270
System 318
   stability with different
   pole locations 353

T

Tapped field control 238
Terminals 395
Thermal capacitance 324
Thermal resistance 324
Thermal systems 323
   transfer function of 328
Thevenin's theorem 112
Three control torques 403
Three-phase EMF generation 134
Three-phase induction motor, torque in 294
Three-phase induction motor 289
   construction of 290
Three-phase system 134
Three-phase transformer 262
Time response specifications 348
Time-domain equation representation 44

Time-ratio control 365
Torques 402
Tough rubber sheathed 424
Transfer function 327
  pole locations of 348
Transformer 239
  characteristics of 250
  efficiency of 253
  EMF equation of 242
  losses in 253
  resistance and leakage reactance with 249
  types and construction 239
  voltage regulation of 253
Transient response 342
Transmission lines 377
Two-phase motor 309
Two-value capacitor motor 310

U

Unidirectional torque 215
Unit 6
Unity load p.f 279

V

Variable frequency operation 367
Variable reluctance stepper motor 340
Variable supply voltage 238

Vector representation 42
Velocity function 343
Volt 4
Voltage control 237
Voltage drops, sign convention for 13
Voltage magnification 95, 96
Voltage ramp 303
Voltage regulation 281
Voltage sources, sign convention for 12
Voltage transformation ratio 242

W

Watt 5
Wave form representation 44
Wave windings 190
Wind turbines 376
Wiring materials 420
Wooden casing capping 423
Wye connections 263

Y

Y system 134
Yellow phase 134
Yoke 189

Z

Zinc-carbon battery 396